Soil Magnetism

Soil Magnetism
Applications in Pedology, Environmental Science and Agriculture

Neli Jordanova

Full Professor,
National Institute of Geophysics, Geodesy and Geography,
Bulgarian Academy of Sciences,
Sofia Bulgaria

AMSTERDAM • BOSTON • HEIDELBERG • LONDON
NEW YORK • OXFORD • PARIS • SAN DIEGO
SAN FRANCISCO • SINGAPORE • SYDNEY • TOKYO

Academic Press is an imprint of Elsevier

Academic Press is an imprint of Elsevier
125 London Wall, London EC2Y 5AS, United Kingdom
525 B Street, Suite 1800, San Diego, CA 92101-4495, United States
50 Hampshire Street, 5th Floor, Cambridge, MA 02139, United States
The Boulevard, Langford Lane, Kidlington, Oxford OX5 1GB, United Kingdom

Library of Congress Cataloging-in-Publication Data
A catalog record for this book is available from the Library of Congress

British Library Cataloguing-in-Publication Data
A catalogue record for this book is available from the British Library

ISBN: 978-0-12-809239-2

For information on all Academic Press publications
visit our website at https://www.elsevier.com/

Publisher: Candice Janco
Acquisition Editor: Candice Janco
Editorial Project Manager: Emily Thomson
Production Project Manager: Paul Prasad Chandramohan
Cover Designer: Mark Rogers

Typeset by TNQ Books and Journals

To Diana

Contents

Foreword

I accepted with great honor and pleasure the invitation to write a foreword to the work of Neli Jordanova, titled *Soil Magnetism: Applications in Pedology, Environmental Science and Archeology.* I have to say that I have known Prof. Jordanova for many years, since when she spent some 3 years as a promising young postdoctoral at our institute. During this stay, we had many opportunities to work together and to discuss different issues of soil and environmental magnetism, a field that was emerging in that time. Neli brought to our small team very needed stimulus—a knowledge of soils, which was that necessary complement to our rock-magnetic expertise. During our numerous discussions, we came to clear agreement that soil magnetism is a very promising field of research, with many potential applications, but requires efficient and close interdisciplinary cooperation between soil scientists and (geo)physicists (rock magnetists). Both magnetism and soils have a complex and "heterogeneous" nature, with seemingly no overlap. However, they have much in common—iron-bearing minerals, in particular different forms of iron oxides. Rock magnetists have facilities and expertise to determine (or at least estimate) the type, concentration, and grain-size distribution of iron oxides. Based on this knowledge, one may assess their origin and diagenetic pathways. Then, the partner experienced in soil or environmental sciences may discuss the controlling effects and processes such as, for example, climatic conditions (past or present), erosion, land use, pollution, etc. During the past few decades, soil magnetism research has made evident progress. At the end of the last century, the studies reported on simple observations that, for example, the concentration of iron oxides deposited from the atmosphere in the topsoil decreased exponentially with distance from dominant source of emissions. At present, the authors address and discuss much more complex issues, ranging from determining soil properties with respect to unexploded ordnance detection to arable soil degradation due to agriculture, as well as a number of studies dealing with reconstruction of past climatic changes or human activities. Neli Jordanova has a great advantage that she has always been well aware of this complexity of soil magnetism. With her knowledge and attitude, she has been able to cover a wide range of subjects, and the present work represents the extent of her research through the present. This work is a good compilation of a huge amount of complex information on general soil properties along with specific magnetic properties of a wide range of soil types. The work is well structured into 10 chapters, five of which are devoted to specific groups of soils. Neli Jordanova covered soils from Bulgaria, soils with which she is that familiar, in addition to many useful results and references related to different soils worldwide. Three chapters are devoted to explanation of the advantages and possible applications of soil magnetism. Thus, soil scientists may benefit from learning about magnetic properties, presented in a noncomplex way. On the other hand, rock magnetists may acquire new knowledge (or extend existing knowledge) on processes that take place in different types of soils and affect populations of iron oxides. This interdisciplinarity is, along with the range of reported soils, the main "gunpowder" of this work. However, being a rock magnetist, I would expect somewhat more detailed introduction of various magnetic parameters, which are used in the work. In particular, soil scientists would benefit from explanation of how individual magnetic parameters can contribute to addressing the three main questions: what is the type, concentration and grain-size distribution of iron oxides in soils. Despite this drawback I found this work very useful for soil scientists, environmentalists, and rock magnetists at different levels—from university students to specialized professionals. After few previously published works dealing with environmental magnetism (e.g., Maher and Thompson, 1999; Evans and Heller, 2003; see the Introduction), the present work represents the next logical step, presenting soil magnetism from the point of view of complex soil sciences.

Eduard Petrovsky
Institute of Geophysics, The Czech Academy of Sciences, Prague, Czech Republic

Acknowledgments

Writing this book was a challenging task for me and I would not be able to compile all the data, analyze them, and put into a systematic order without the help of many colleagues and friends. I am grateful to Prof. DSc. Mary Kovacheva for continuous support during the years and many fruitful discussions. I would like to thank my colleagues Petar Petrov and Rositsa Mihajlova from the Palaeomagnetic Laboratory at the National Institute of Geophysics, Geodesy and Geography in Sofia for persistent help over the years in the field sampling campaigns and laboratory magnetic measurements on the soil profiles.

Valuable discussions and common work with Dr. Eduard Petrovský and Dr. Aleš Kapicka from the Institute of Geophysics at the Czech Academy of Sciences on the application of environmental magnetism for evaluation of anthropogenic pollution inspired me to study the natural magnetic signature of soils developed under different environmental conditions.

Dedicated work on magnetism of different soil types benefited greatly from several research projects carried out at the Palaeomagnetic Laboratory in Sofia—the SCOPES Project IB7320-110723 "Environmental Applications of Soil Magnetism for Sustainable Land Use"; 2005–2008 with partners from the Institute of Geophysics, ETH (Zurich), Prof. Dr. Ann Hirt and Institute for Terrestrial Ecology, Soil Chemistry, ETH (Zurich), Prof. Dr. Ruben Kretzschmar, to whom I would like to thank for providing access to their laboratory facilities and for the useful discussions. Our work in the FP7 Collaborative Project No. 211386 "Interactions between soil related sciences–Linking geophysics, soil science and digital soil mapping" (iSOIL) gave us the opportunity to look for the possible applications of soil magnetism from a new point of view and I would like to thank all our partners for helpful cooperation.

I strongly appreciate useful comments and discussions with Assoc. Prof. Dr. Toma Shishkov from the Institute of Soil Science, Agrotechnologies and Plant Protection "N. Pushkarov" (Sofia), who helped in description of the soil profiles, included in this work, and their correlation to the WRB classification. I am also grateful to Assoc. Prof. Dr. Dimo Dimov from the Department of Geology and Geography at the Sofia University "St. Kl. Ohridski" for valuable help during the sampling campaigns and for providing geological information and maps.

I would like to thank Louisa Hutchins, an Associate Acquisitions Editor at Elsevier Ltd., who first invited me to propose this book project. I further highly appreciate the help of Candice Janco (ELS-HBE) for putting the project forward. I am grateful to Emily Thomson (ELS-OXF) for dedicated help and encouragement during preparation of the book.

Last, but not least, I am very grateful and indebted to my family, and especially my sister, for encouragements, constant support, and inspiring discussions.

Neli Jordanova
Sofia, July 2016

Abbreviations

MAGNETIC PARAMETERS

ARM	Anhystereic remanence, units (Am^2/kg)
B_c	Coercive force, units (mT)
B_{cr}	Coercivity of remanence, units (mT)
HIRM	Hard isothermal remanent magnetization, units (Am^2/kg)
IRM	Isothermal remanence, units (Am^2/kg)
M_{rs}	Saturation remanent magnetization, units (Am^2/kg)
M_s	Saturation magnetization, units (Am^2/kg)
SIRM	Saturation isothermal remanent magnetization, units (Am^2/kg)
T_c	Curie temperature, units (°C)
T_{ub}	Unblocking temperature, units (°C)
χ	Mass-specific magnetic susceptibility, units (m^3/kg)
χ_{ARM}	Anhysteretic susceptibility, units (m^3/kg)
χ_{fd}	Frequency-dependent magnetic susceptibility, units (m^3/kg)
$\chi_{fd}\%$	Percent frequency-dependent magnetic susceptibility, units (%)
χ_{hf}	High field magnetic susceptibility, units (m^3/kg)
K	Volume magnetic susceptibility, units SI

OTHER

DCB	Dithionite-citrate-bicarbonate (selective Fe-extraction method)
DRS	Diffuse reflectance spectroscopy
Fe_d	Dithionite-extractible iron
Fe_o	Oxalate-extractible iron
Fe_t (Fe_{tot})	Total iron content
MAP	Mean annual precipitation
MAT	Mean annual temperature
MD	Multidomain
pH	Soil reaction
PSD	Pseudo—single domain
REE	Rear Earth elements
SD	Single domain
SEM	Scanning electron microscopy
SP	Superparamagnetic
XRD	X-ray diffraction
XRF	X-ray fluorescence

Introduction

Soil cover is not simply one of the Earth's compartments; it is the key element in the Critical zone, where complex and dynamic interactions take place between rock, soil, water, air, and living organisms (NRC, 2001). Taking into account that soil formation requires centuries to thousands of years (Schaetzl and Anderson, 2009), soil is considered to be a nonrenewable resource that must be preserved and managed for future generations. Today's challenges of our society, facing climate changes and a growing world population, require detailed knowledge on the role of soils in the dynamics of the ecosystems at both global and local scales. The major threats to ecosystem functioning, such as soil erosion, nutrient cycling, and water quality, call for a wide interdisciplinary approach, including scientific disciplines such as pedology, ecology, atmospheric chemistry and physics, biogeochemistry, hydrology, geology, and geophysics.

Iron in soil is a key element accomplishing the interaction between the major Earth's cycles—carbon cycle (C-cycle) and dust cycle (D-cycle) (Shao et al., 2011). As a building element in the mineral fraction of the soil, iron oxides play an important role in the interactions between different soil compartments. Keeping in mind the relatively low amount of iron in soils (sometimes less than 0.1%), their identification and characterization are not always straightforward (Cornell and Schwertmann, 2003). Due to the high sensitivity of the soil magnetic signature toward even minor amounts of Fe oxides present, the application of magnetic methods in soil studies provides an opportunity to obtain valuable information about soil magnetic mineralogy, geochemistry, redox conditions, and functioning, necessary for the development of soil management strategies and practices in agriculture.

Soil magnetism is one of the branches in environmental magnetism studies, established as a scientific discipline in the mid-20th century and giving valuable contribution for reconstructing processes and factors, that drive the major environmental processes on Earth, as well as on other planets. However, due to the genetic roots of soil magnetism in geophysics, little is known in the soil science community about the applicability of this methodology. Apart from the palaeoclimate reconstructions based on magnetic proxy records, which is a widely used approach, the other aspects of applications of soil magnetic studies in pedology and agriculture are little recognized by the general earth science audience. On the other hand, geophysicists working on soil magnetism face difficulties in understanding and interpreting the mineral magnetic signature within the frame of the real soil system variability and peculiarities of the pedogenic processes. The aim of this book is to bridge this gap, providing an up-to-date synthesis of the magnetism of different soil orders with respect to soil-forming factors and the specific geochemistry of iron compounds in various pedogenic regimes. Revealing the possibilities of the magnetic methodology for resolving specific problems in the studies of soils to wider audience is the ultimate goal, which hopefully is achieved. The author uses mostly examples from her studies on soils from Bulgaria, but she always refers to the international soil classification system (WRB), which allows readers to consider the results in a common perspective. One of the main obstacles in compiling the soil magnetism studies available divided into major soil orders was the correct correlation between the two commonly used classification systems—the WRB (FAO) and the Soil Taxonomy (USA). Some studies use only national classification systems, which imposed the need to look for reliable correlation with the international classifications. In this book, the correlation between the major soil classification systems is made according to *A Handbook of Soil Terminology. Correlation and Classification,* edited by Krasilnikov et al. (2009).

Soil cover in Bulgaria reflects the evolution of the natural processes since the Pliocene, influenced by neotectonic and anthropogenic factors. The wide variability of the modern climate conditions, expressed in five different climate temperature–moisture regimes identified (thermic-xeric; mesic-xeric; mesic-ustic; mesic-udic; cryic-udic), imposed on the wide variations in parent rock lithology, neotectonics, and biodiversity resulted in the development of various soils, classified to 20 of 28 main FAO soil taxonomic units in Bulgaria (Shishkov and Kolev, 2014). This soil diversity contrasts with the relatively small territory of Bulgaria, being about 110,000 km^2.

The structure of the book aims first to introduce the major terminology, facts, and definitions from the magnetism of natural materials in Chapter 1, including the existing theories and hypotheses on the pedogenic magnetic enhancement and pathways of iron transformations in various soil environments, followed by a detailed description of laboratory procedures, measurements, and instrumentation used in the following chapters. Special consideration is given to the differences in sampling strategies in soil science and environmental magnetism. Chapters 2 to 6 present detailed magnetic studies of master soil profiles of the major soil types from Bulgaria in relation to published research on magnetism of soils from the same order in other parts of the world. At the beginning of each chapter, a short overview of the main characteristics [physical, (geo)chemical] and formation mechanism of each soil order and its world spatial distribution is presented. The last section in each of the Chapters 2 to 7 discusses the pedogenesis of the iron oxides in the corresponding soil order, as reflected in soil magnetism. Chapter 7 deals with the magnetic properties of soils from the Antarctic Peninsula (the Livingston Island). Chapter 8 summarizes all major regularities, uniquely characteristic for each particular soil order in an attempt to reveal the possibilities of the magnetic technique for discrimination, classification, and identification of the soil processes and types. Mapping of topsoil magnetic properties in Bulgaria, statistical data analysis revealing the main factors controlling the variability of the data, and the roles of the soil type and lithology are presented in Chapter 9. Chapter 10 introduces the range of possible applications of soil magnetism in pedology, agriculture, environmental pollution studies, paleoclimate reconstructions and evaluation of the effects of fire on soil magnetism, landmine clearance operations and forensic research, and archaeology. Examples from different studies worldwide are presented.

Undergraduate students, postgraduate students, geophysicists, soil scientists, geochemists, and professionals working on interdisciplinary projects in earth science are intended as book's main audience. The presented examples and case studies show how the knowledge of soil properties and genesis, combined with basic physics, chemistry, and magnetic mineralogy of the soil, helps in resolving various environmental and applied problems.

REFERENCES

Cornell, R., Schwertmann, U., 2003. The Iron Oxides. Structure, Properties, Reactions, Occurrence and Uses (Weinheim, New York).

Krasilnikov, P., Martí, J.-J.I., Arnold, R., Shoba, S. (Eds.), 2009. A Handbook of Soil Terminology, Correlation and Classification. Earthscan, UK, ISBN 978-1-84407-683-3 (hardback).

National Research Council (NRC), 2001. Basic Research Opportunities in Earth Science. National Academy Press, Washington, DC.

Schaetzl, R., Anderson, A., 2009. Soils. Genesis and Geomorphology. Cambridge Univ. Press, UK, ISBN 978-0-521-81201-6.

Shao, Y., Wyrwoll, K.-H., Chappel, A., Huang, J., Lin, Z., McTainsh, G.H., Mikami, M., Tanaka, T.Y., Wang, X., Yoon, S., 2011. Dust cycle: an emerging core theme in Earth system science. Aeolian Res. 2, 181–204.

Shishkov, T., Kolev, N., 2014. The Soils of Bulgaria. World Soils Book Series. Springer. http://dx.doi.org/10.1007/978-94-007-7784-2.

CHAPTER

1 MAGNETISM OF MATERIALS OCCURRING IN THE ENVIRONMENT—BASIC OVERVIEW

INTRODUCTION

The magnetic signature of the soil is a complex mixture of contributions from different mineral constituents, including diamagnetic, paramagnetic, and ferromagnetic phases. For detailed information on the magnetism of materials, readers can refer to a number of excellent textbooks such as those by Chikazumi (2010) and Coey (2009). With an emphasis on terrestrial magnetic minerals, Dunlop and Özdemir's (1997) book is also an outstanding reference work. Generally, the magnetic signal of soils is dominated by the presence of minor amounts of strongly magnetic ferrimagnetic iron (Fe) oxides—magnetite (Fe_3O_4), maghemite (γ-Fe_2O_3), titanomagnetites ($Fe_{2-x}Ti_xO_4$), and, rarely, pyrhotite (Fe_3S_4). Although in higher absolute amounts (wt%), iron oxyhydroxide goethite (α-FeOOH) and hematite (α-Fe_2O_3) are antiferromagnetic (e.g., having small saturation magnetization values compared with the ferrimagnets), and their contribution to the total magnetic signal of soils is strongly suppressed. One important feature of the ferromagnetic materials is their different magnetic state depending on the size of the grain—the so-called magnetic domain structure. The very small nano-sized ferrimagnetic grains of magnetite and maghemite (5−15 nm diameter) are called superparamagnetic (SP) and play an important role in the soil's magnetism, as far as the pedogenic magnetic oxides are usually within this size range. Superparamagnetic particles have a distinctive magnetic behavior—they do not retain any remanent magnetization but possess very high magnetic susceptibility compared with the larger grains of the same mineral (Dunlop and Özdemir, 1997). Slightly larger grain sizes lead to a single domain (SD) state. The SD particles of magnetite/maghemite (sizes of about 20 nm diameter) show the highest magnetic stability; that is, their remanent magnetization is most stable against demagnetizing factors (thermal agitation, alternating magnetic fields, time). In larger particles magnetic domains form, and the particles are in a pseudo-SD (PSD) or multidomain (MD) state (Dunlop and Özdemir, 1997).

Soil Magnetism. http://dx.doi.org/10.1016/B978-0-12-809239-2.00001-2

1.1 IRON OXIDES IN SOILS—FORMATION PATHWAYS, PROPERTIES, AND SIGNIFICANCE FOR SOIL FUNCTIONING

Iron is the fourth major element in Earth's crust after O, Si, and Al. It is an important element present in many natural minerals (terrestrial and extraterrestrial) and is vital for living organisms (Ilbert and Bonnefoy, 2013). The excess oxygen at the Earth's surface leads to the persistent dominance of oxide forms of iron in the natural environment. The exact type of Fe oxide that will prevail in certain natural deposits (rocks, sediments, soils, aerosols) is strongly dependent on the origin and environmental conditions, leading to the formation and secondary transformation of the initial mineralogical composition. The Fe oxides in soils are an intimate product of the major soil-forming processes and represent a sensitive "mirror" for the complex biogeochemical interactions between the soil's constituents (mineral, organic, and living) and external factors such as climate (temperature, precipitation), time, and topography. The soil color is primarily determined by the color and concentration of the prevailing Fe oxide. Red, yellow, and brown soils owe their color to the presence of insoluble Fe oxides, while blue-green and pale soils are affected by the presence of reduced forms of Fe oxides and strong leaching. Although constituting a minor amount from the total soil mineralogy (usually less than 1 wt%, sometimes up to 5 wt%), Fe oxides affect soil functioning and properties such as aggregation, phosphorous retention capacity, charge, and exchange capacity (Achat et al., 2016; do Carmo Horta and Torrent, 2007; Duiker et al., 2003; Arias et al., 1995).

1.1.1 LITHOGENIC (PRIMARY) FE OXIDES IN SOILS

The origin of natural magnetic minerals present in soils can be lithogenic (primary) and pedogenic (secondary). Lithogenic magnetic minerals are usually coarse-grained and inherited from the parent rock. They are less prone to chemical weathering because of their small surface/volume ratio and the relatively high stability of Fe oxides in a weathering environment (Schaetzl and Anderson, 2009). Sometimes pedogenic alteration of these coarse grained lithogenic grains is reported to occur, expressed by characteristic cracks on the surface due to reductive dissolution (Grimley and Arruda, 2007; Fisher et al., 2008). The mineralogical composition of the lithogenic magnetic minerals strongly depends on the origin and mineralogy of the parent material. Soils developed on intrusive and volcanic rocks inherit their Fe oxide mineralogy. Usually these are (titano) magnetites and titanohematites (hemoilmenites) (O'Reilly, 1984; Dunlop and Özdemir, 1997).

The titanomagnetites ($Fe_{2-x}Ti_xO_4$; $0 < x < 1$) form a solid solution series with magnetite (Fe_3O_4) and ulvospinel (Fe_2TiO_4) as end members. They have an inverse spinel structure, and their magnetic properties (saturation magnetization, Curie temperature (T_c), coercivity) strongly depend on the titanium (Ti) content. The ulvospinel phase is antiferromagnetic with a zero net moment, while the other end-member—magnetite—is ferrimagnetic with strong magnetization [4 μ_B, where μ_B is Bohr's magneton (e.g., Chikazumi, 2010)]. Curie temperatures decrease linearly with an increasing Ti content and for natural terrestrial titanomagnetites, T_c varies in the range from 150 to 200°C for TM60 up to a T_c of 586°C for stoichiometric magnetite (O'Reilly, 1984). Depending on the cooling history, oxygen availability, and composition of the primary magma, titanomagnetites form inhomogeneous intergrowths of Ti-rich and Ti-poor areas (e.g., exsolution structures) (for a detailed magnetic characterization of the TM series, see Dunlop and Özdemir, 1997). The second solid solution series with the general formula ($Fe_{2-y}Ti_yO_3$)—hemoilmenites—has as end members hematite (α-Fe_2O_3) and ilmenite ($FeTiO_3$). In the range of $0.5 < y < 1$, titanohematite is ferrimagnetic with a T_c between −200°C and 200°C. For $0 < y < 0.5$, titanohematite is antiferromagnetic with weak ferromagnetism and T_c values between 200°C and 680°C.

Hematite (α-Fe_2O_3) is among the most widespread Fe oxides on the Earth's surface environment because of its high thermodynamic stability. It has a rhombohedral crystal structure and antiferromagnetic behavior below the Neel temperature T_N of 675°C. Hematite is characterized by high coercivity and saturation magnetization almost 200 times less than that of magnetite. The physical and magnetic properties of hematite strongly depend on the grain size—coarse-grained minerals are known as specularite, while nano-sized hematite has a typical red/pink color determining the appearance of the soil's color. The smaller the grain size of hematite, the more saturated is the redness (Pailhe et al., 2008).

1.1.2 PEDOGENIC (SECONDARY) FE OXIDES IN SOILS

Pedogenic Fe oxides are of a secondary origin; e.g., formed as a result of soil formation and development. The most characteristic feature of the pedogenic Fe oxides is their small size, low crystallinity, and the widespread presence of substitutions in their lattices (Cornell and Schwertmann, 2003). Hematite and goethite represent the prevailing (by volume or weight) phases of Fe oxides in soils. The two minerals may be simultaneously present in the soil, while their relative abundance is a powerful indicator of the climate during their pedogenic formation. A low-temperature—high-humidity combination leads to the dominant formation of goethite, while a high-temperature—low-humidity combination results in hematite formation (Schwertmann, 1988). Lepidocrocite (γ-FeOOH) is a paramagnetic Fe oxide at room temperature, which is characteristic for reductomorphic soils in temperate and subtropical climates (van Breemen, 1988). It has patchy distribution in soil and is usually found as concretions and crust around roots and voids. Ferrihydrite ($5Fe_2O_3 \cdot 9H_2O$) is another paramagnetic oxide, usually having low crystallinity, and its occurrence is related to cold-to-temperate and humid climate conditions. Ferrihydrite and lepidocrocite are usually formed during the initial stages of pedogenesis in young soils together with an excess of organic matter and dissolved silica, which impede the transformation of ferrihydrite to goethite and lepidocrocite (Cornell and Schwertmann, 2003).

Despite the predominance of goethite and hematite as main Fe oxides in soils, the soil magnetic signature is dictated by the presence of the strongly magnetic pedogenic Fe oxides magnetite and maghemite. The latter are responsible for the soil's magnetic enhancement phenomena (see next section). The formation of pedogenic maghemite in soils can be explained by several pathways. The simplest possibility is through the oxidation of magnetite present in the parent material (Marques et al., 2014; Rümenapp et al., 2015). However, it cannot elucidate the appearance of maghemite in soils developed on parent rocks of low magnetite content. Another pathway of formation is through "green rust" oxidation (Schwertmann, 1988), which is realized in reductomorphic conditions. The latter are typical of deeper (illuvial) soil horizons but not of humic and organic layers. The third pathway for maghemite formation is through the thermal transformation of Fe oxyhydroxides during heating up to 300–500°C in the presence of organic matter (Mullins, 1977). This assumes the importance of wildfires in the ancient past. The formation of pedogenic magnetite can be attributed to biogenic as well as to inorganic processes. Biogenic magnetite in soil has been identified (Fassbinder et al., 1990), and its formation is related to intracellular magnetite production in magnetotactic bacteria (Rahn-Lee and Komeil, 2013; Araujo et al., 2015). Biogenic magnetite crystals have also been reported to occur in botanical (grass) tissues (Gajdardziska-Josifovska et al., 2001). The inorganic precipitation of fine-grained magnetites from ferrihydrite and microbially mediated Fe^{3+} reduction (Maher and Taylor, 1988) is another possible way of formation. Maghemite is the other strongly magnetic mineral dominating the Fe mineralogy of soils from the tropics and subtropics (Taylor and Schwertmann, 1974; Goulard et al., 1998; Da Costa et al., 1999). Maghemite is also found to be the main source of magnetic enhancement in the Mediterranean as well as Chernozemic soils from the temperate climate belt (Torrent et al., 2006; Jordanova et al., 2010; Hu et al., 2013; Gorka-Kostrubiec et al., 2016).

Pedogenic Fe oxides in soils are characterized by small crystal sizes—between a few to several tens of nanometers. Their structural properties are strongly influenced by the common presence of impurities and substitutions in the crystal lattice, usually by aluminum because of the wide occurrence of Al ions in the soil solution released during the weathering of the primary silicate minerals. The highest degree of Al substitutions is found in soil goethites [up to Al/(Al + Fe) of 0.33], but various degrees of substitutions depend on the availability of Al sources (feldspars, micas, kaolinite, gibbsite), soil pH, organic matter, etc. (Cornell and Schwertmann, 2003).

1.1.3 THE ROLE OF WEATHERING PROCESSES IN PEDOGENESIS

The formation of pedogenic Fe oxides is triggered and fed primarily by the processes of weathering of the primary silicates of the parent rock. These primary minerals usually occur in the silt and sand fractions as residuals from the physical disintegration of the solid rock. Weathering is generally considered to proceed in water or a solution. Thus, the hydrolysis is the main mechanism of silicate weathering. The relative stability of different primary minerals with regard to alteration (weathering) follows the opposite order of the temperature of the crystallization of these minerals from magma. This is related to the fact that the higher the crystallization temperature, the bigger is the difference in the Earth's surface temperature, which drives the mineral out of its equilibrium state. According to Churchamn and Lowe (2012) after Goldich (1938), the relative order of stability against the weathering of the minerals is: volcanic glass = olivine < pyroxenes < amphyboles < biotite < K-feldspars < muscovite < quartz. The stabilities of plagioclase feldspars are lower than K-feldspars. The incorporation of Fe into the structure of primary silicate minerals plays an important role in their weathering intensity. Because of its presence in a divalent state (Fe^{2+}) within the silicates, iron is easily oxidized to Fe^{3+}, which causes a charge imbalance of the mineral and increases its dissolution through hydrolysis. Once released in the solution, Fe^{2+} is quickly oxidized to Fe^{3+} and subsequently hydrolyzed to form Fe oxyhydroxide. The processes of the weathering of the minerals in a soil environment are strongly dependent on the presence and properties of biota. Acidification of the uppermost soil horizons is realized through release of protons (H^+) from plant roots for a charge balance when the roots absorb more cations than anions. As a result, amorphous Fe and Al oxides from the soil around the roots can be dissolved (Calvaruso et al., 2009). Roots and fungal hyphae also exude organic acids into the soil. Another counteracting effect is the "nutrient uplift" or "biological pumping" of elements such as K and Si from lower layers toward the soil surface, which induces the retention and formation of new alumosilicate minerals (He et al., 2008). All these biological effects influence the degree of weathering of the primary minerals in a soil.

There are remarkable differences between the weathering rates of minerals determined in laboratory studies compared with the much slower rates measured in field studies (White and Brantley, 2003; Ganor et al., 2007). At the same time, different minerals have different reaction kinetics, which may limit the weathering rate of a particular mineral if its residence time in the weathering environment is longer than the timescale of the weathering reaction. Similarly, weathering intensity may be limited due to the shortage of a fresh mineral supply in the near surface environment due to low denudation rates (Hilley et al., 2010 and references therein). It was observed that the depletion of primary minerals and the neoformation of secondary clays and oxides in soil chronosequences progressively decrease with soil age (Taylor and Blum, 1995; Reeves and Rothman, 2013). This slowing down of chemical weathering has been explained by different factors inhibiting the rate of dissolution of primary alumosilicate minerals, such as a decrease in the reactive surface area of minerals due to physical occlusion by secondary precipitates and leached layers (Maher et al., 2009; Emmanuel and Ague, 2011); different fluid dynamics, which control the thermodynamic saturation of the primary dissolving phases (Maher, 2010); bacterial/fungal communities and their evolution over time (Moore et al., 2010); and biological activity (Calvaruso et al., 2009). Consequently, the soil's evolution and the corresponding development of the magnetic signature will also be influenced and will reflect the effect of all of these factors. Thus, the presence of Fe in the primary minerals of the parent rock is not a sufficient condition for its release and subsequent precipitation as a secondary pedogenic Fe oxide. The exact type of Fe oxide that will be formed after the hydrolysis of the Fe^{2+} ions released in the soil solution strongly depends on the relative amount of Fe^{2+} compared with Fe^{3+}, the oxidation rate, pH, and presence of organic acids (Cornell and Schwertmann, 2003; H. Liu et al., 2007; Colombo et al., 2015; Pedrosa et al., 2015). The role of organic acids in weathering reactions and pedogenesis at a profile scale has been modeled for the chronosequence of soils near Santa Cruz, CA (Lawrence et al., 2014). It was shown that profile evolution is sensitive to kaolinite precipitation and oxalate decomposition rates. Geochemical gradients along the profile's depth have been considered to determine the reactions within the system (Fig. 1.1.1).

Looking at the pore scale (Fig. 1.1.1a), interaction among the variety of processes can be observed: the input of organic acids, the complexation of organic matter on mineral surfaces, the dissolution of primary alumosilicate minerals, the precipitation of secondary minerals, and the transport of weathering products. As a result of the above-mentioned processes acting at the pore scale within the solum, macroscopic gradients can be defined along the soil depth (Fig. 1.1.1b). The concentration of organic carbon (organic matter) typically decreases with depth within the biotic zone. This biotic zone overlaps the underlying abiotic zone where weathering reactions occur and lead to changes in mineral abundances. The speed of the downward propagation of the mineral weathering front is defined as weathering velocity Vw. Another characteristic parameter is the equilibrium length scale Leq, which represents the depth interval where the equilibrium of the soil's solution with dissolving species is reached. It corresponds to the depth where the concentration of the secondary mineral phases reaches a

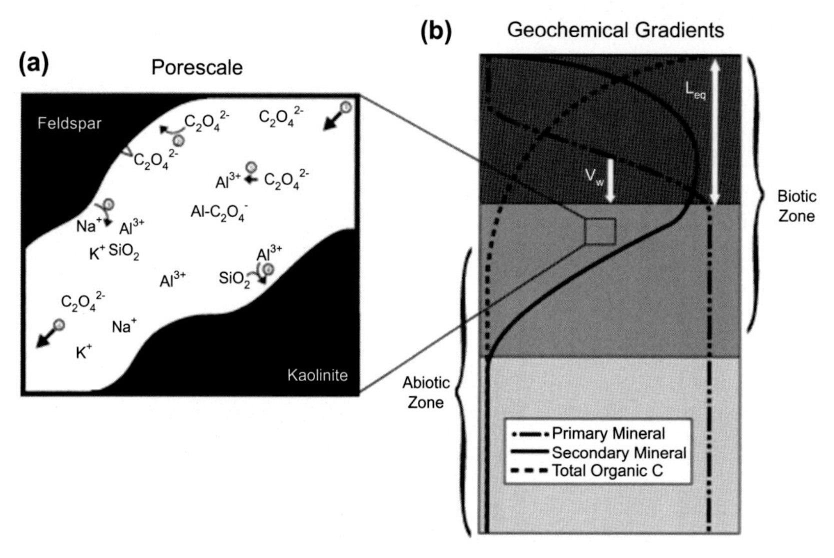

FIGURE 1.1.1

(a) Biogeochemical reactions related to mineral weathering in soils: 1—input of organic acids; 2—complexation of organic matter on mineral surfaces; 3—dissolution of primary minerals; 4—precipitation of secondary phases; 5—aqueous organic–metal complexation; 6—transport of secondary products. (b) Geochemical gradients arising from soil mineral weathering. The biotic zone is defined as the area where declined organic inputs are observed. The abiotic zone, where chemical weathering of primary minerals occurs and leads to changes in mineral abundancies, overlaps the biotic zone

Reprinted with permission from Lawrence, C., Harden, J., Maher, K., 2014. Modeling the influence of organic acids on soil weathering. Geochimica et Cosmochimica Acta 139, 487–507.

plateau (Fig. 1.1.1b). In the uppermost levels (the topsoil) of a profile without external sediment input during pedogenesis, the concentration of primary minerals approaches zero when all possible weathering reactions within this zone have taken place. This approach to studies of soil processes is also very useful in analyzing the behavior of magnetic characteristics along depth, which will be discussed later.

The soil formation process is a complex interplay between specific pedogenic processes, creating a set of solid-phase pedogenic features. In fact, the soil is an open system where biogeochemical reactions are constrained by the laws of thermodynamics and determine the existence of pedogenic thresholds. However, the heterogeneity of the matrix, coupled with the different reaction kinetics of the soil's constituents, leads to significant variability in the final pedogenic states (Chadwick and Chorover, 2001). During the pedogenesis, the soil's body exerts systematic transformations until a "mature" state of equilibrium with the fluxes of matter and energy in its environment is reached. This means that all pedogenic processes (including the slowest) are already terminated or are in dynamic equilibrium with the external environment (Targulian and Krasilnikov, 2007). It is well documented that different pedogenic processes have different characteristic times (rates) and are usually grouped into three main classes: rapid (10^1–10^2 years), medium rate (10^3–10^4 years), and slow (10^5–10^6 years). The processes of the formation of pedogenic Fe oxides fall within the group of medium-rate development, together with humification, lessivage, cheluviation, andosolization, etc. (Targulilan and Krasilnikov, 2007). However, the correct estimation of the age of pedogenic Fe oxides, as far as they form over the time of the soil's evolution, is still a real problem. Recently, new instrumental techniques have become available for dating the secondary pedogenic phases in a soil (Cornu et al., 2009). The age of a soil cannot be considered as equal to the degree of mineral weathering, as far as the primary minerals from the parent material enter the zone of soil formation when the weathering front propagates down to the corresponding depth (Yoo and Mudd, 2008). Thus, in a study of Mn oxide concretions from lateritic iron deposits in the Quadrilatero Ferifero (Brazil), Spier et al. (2006) found that the oldest grains of Mn oxides occur near the surface, while younger oxides are found deeper in the solum.

1.2 PEDOGENIC MODELS AND SOIL MAGNETIC ENHANCEMENT PHENOMENA—EXISTING THEORIES AND FINDINGS

1.2.1 MODELS OF PEDOGENESIS

The historic evolution of studies dealing with soil formation and development follows the advances of the natural sciences, through which the processes and pathways of pedogenesis can be satisfactorily described and explained. Detailed reviews on the existing pedogenic models are available in the relevant literature (Schaetzl and Anderson, 2009; Minasny et al., 2008). Here, we only briefly describe the main models of pedogenesis and their relevance to the magnetic signature of soils. The most well-known model of pedogenesis—at least to the geophysics community—is the state factor model by Hans Jenny (1941). He compiled a set of variables (state factors) that define the state of a soil system. He described a soil as a function of five major factors—climate (cl), organisms (o), relief (r), parent material (p), and time (t):

$$S = f\ (cl,\ o,\ r,\ p,\ t)$$

The role of each factor in the soil formation process can be quantitatively assessed by keeping the rest of the factors constant. The quantitative use of factorial models results in a definition of empirical relationships between soil-forming factors and soil attributes and influenced the development of soil classification systems (Bockheim et al., 2014). The state factors model is also the basis for the widely applied method for the estimation of palaeoprecipitation based on the magnetic susceptibility enhancement of palaeosols, usually within loess-palaeosol sequences (Maher and Thompson, 1992, 1995; Maher, 1998, 2011; Maher et al., 2003; Liu et al., 2012; Maher and Posolo, 2013; Maxbauer et al., 2016). Extending the set of magnetic characteristics used as a palaeoprecipitation proxy, Geiss and Zanner (2007), Geiss et al. (2008), Liu et al. (2013), and Hyland et al. (2015) suggested other sets of magnetic parameters based on measurements of magnetic remanences (e.g., anhysteretic remanence and isothermal remanence) to be more suitable as palaeoprecipitation proxies. Further discussion on this topic is presented in Chapter 10.

Another model considers the pedogenesis from the point of view of the energy of soil formation, developed in the works of Volobuyev (Minasny et al., 2008). Volobuyev's model considers the soil on a macro scale and explains soil formation in terms of the energies required to build a soil:

$$E = w_1 + w_2 + b_1 + b_2 + e_1 + e_2 + g + v,$$

where w_1 and w_2 are the energies of physical and chemical weathering; b_1 and b_2 are the energies of organic matter accumulation and transformation; e_1 and e_2 are the energies of evaporation and transpiration; g is energy loss due to leaching; and v is the energy of heat exchange between the soil and the atmosphere. Most prominent is the energy of evapotranspiration ($e_1 + e_2$), consisting of about 95–99.55% of the total energy balance, followed by the energy of biological activities and mineral weathering. The main factors influencing evapotranspiration are the mean annual temperature (MAT) and the moisture index $K = P/ETp$, where P is precipitation and ETp is the potential evapotranspiration. The dependence of different energy terms on the temperature and the K index is given in Fig. 1.2.1, presented in Minasny et al. (2008). The highest energy of soil formation is obtained for soils in the humid tropics, followed by soils under forests and steppes, and the lowest energy is seen in soil formation in tundras. The dominant role of evapotranspiration as a factor influencing the magnetic enhancement of soils is discussed in detail in Orgeira et al. (2011).

Other models of pedogenesis are the so-called soil mass–balance models. They judge the processes that lead to the formation of certain soil types starting from the parent rock and consider how soil changes over time (Minasny et al., 2008, 2015). Here, the soil profile models are dealing with the geochemistry of weathering and the vertical transport of matter as a result of pedogenic processes. Two contrasting means of soil development can be distinguished: (1) top-down pedogenesis, which is based on the assumption that soil develops downward without external input of new material, and (2) up-building pedogenesis, according to which the soil develops upward, transforming the incoming new material on the surface into soil (Johnson and Watson-Stegner, 1987; Schaetzl and Anderson, 2009). The relative importance of the pedogenic production rate versus the sedimentation of new material governs the direction of soil development—transgressive or regressive (Johnson and Watson-Stegner, 1987). When sediment accumulation is much faster than the rates of pedogenesis, a transgressive mode of soil evolution is realized, resulting in the weakening of pedogenic features. In cases of low to moderate input of new material, the constant input of fresh sediment prone to fast weathering, and the favorable high input of organics, the intensified processes of pedogenic Fe oxide formation and translocation lead to the development of thick illuvial horizons in soils with up-building pedogenesis (Lowe and Tonkin, 2010; Eger et al., 2012).

1.2.2 MODELS OF PEDOGENIC MAGNETIC ENHANCEMENT

The existing pedogenic hypotheses and models reveal the vital role of Fe oxides as sensitive indicators of soil formation and development and highlight the importance of pedogenic oxides for soil functioning (Churchman and Lowe, 2012). Consequently, the magnetic signature of a soil is the fingerprint of soil formation. The direct evidence of the pedogenic development of well-aerated and drained soils is the enhanced magnetic signal observed in the solum (Maher, 1986, 1998). A detailed review of the major hypotheses on the origin, geochemistry, and physics of magnetic enhancement of well-aerated

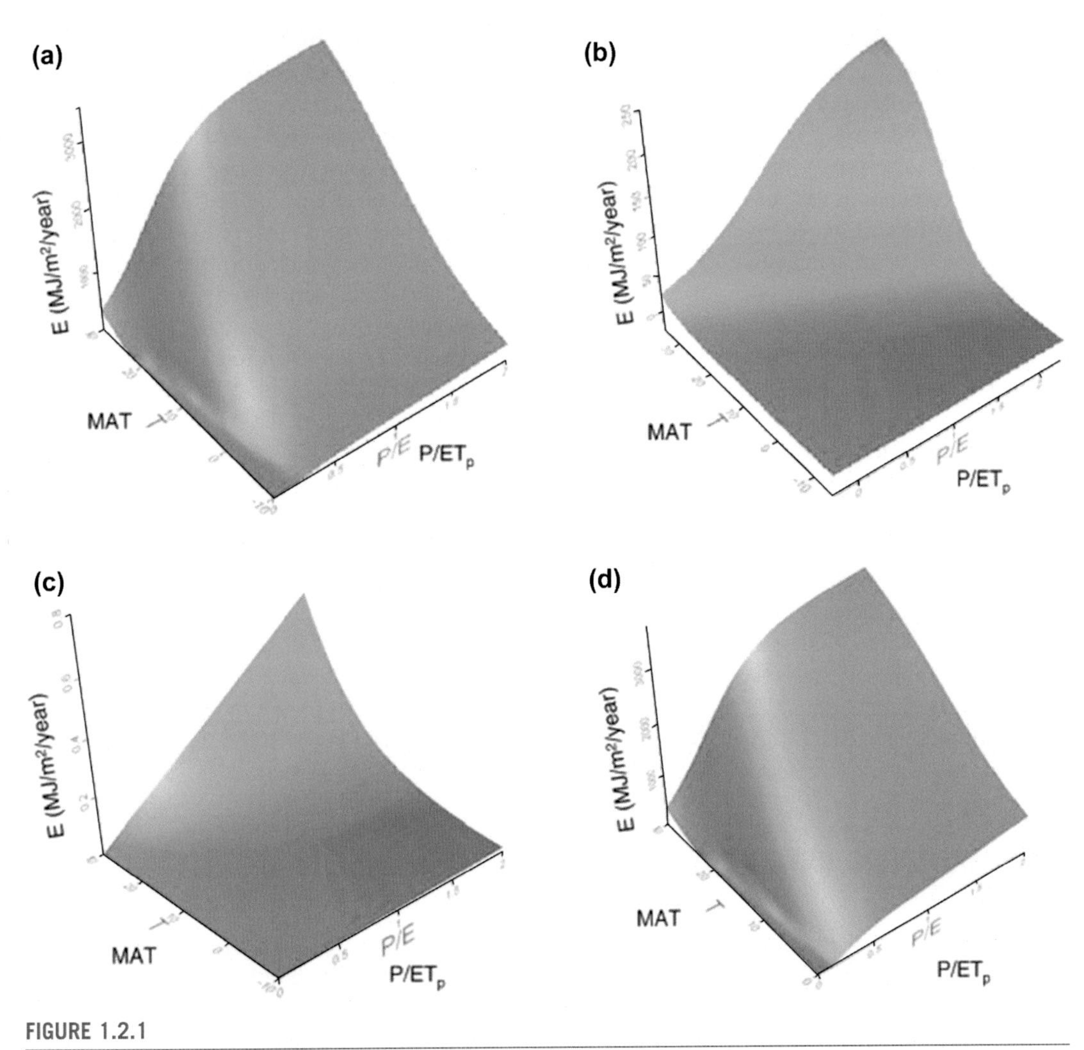

FIGURE 1.2.1

The dependence of different energy terms of soil formation on mean annual temperature (MAT) and the ratio of precipitation P over potential evapotranspiration (P/ETp): (a) energy of evapotranspiration; (b) energy of biomass production; (c) energy of mineral weathering; (d) total energy of soil formation

Reprinted with permission from Minasny, B., McBratney, A.B., Salvador-Blanes, S., 2008. Quantitative models for pedogenesis—A review. Geoderma 144, 140–157.

soils is given in Orgeira et al. (2011); Liu et al. (2012) and Maxbauer et al. (2016). We will briefly summarize the major points in each of the proposed models of magnetic enhancement.

The fermentation mechanism requires alterations of the reducing and oxidizing conditions at a micropore level in soils. During the anoxic (reduced) stage, the primary ferrihydrite is reduced, leading to the liberation of Fe^{2+} ions into the soil solution. This creates favorable conditions for the precipitation of new, strongly magnetic magnetite during the following period of oxidation upon drying at a near neutral pH (Maher, 1988, 1998). An alternative process for the creation of Fe^{2+} is the biotic pathway through the electron transfer during the reduction of organic matter by dissimilatory iron reducing bacteria, which results in the extracellular precipitation of nano-sized magnetite (Lovley et al., 1987). The purely biogenic origin of fine-grained magnetite (intracellular synthesis by magnetotactic bacteria) in soils has also been considered, but the amount of cells in the soil that are usually reported are far too low to explain the degree of magnetic enhancement observed (Maher, 1998).

The presence of maghemite in magnetically enhanced soils is explained in two different ways. According to laboratory studies, fine-grained (nanometer-sized) magnetite grains are prone to fast oxidation to maghemite in the Earth's ambient surface conditions (Rümenapp et al., 2015). Thus, the magnetite initially formed through the fermentation mechanism is supposed to transform to maghemite in a short time. The second genetically different pathway for maghemite formation in soils is proposed by Barrón and Torrent (2002) and further elucidated in Barrón et al. (2003) and Michel et al. (2010). It considers that the initial ferrihydrite formed as a result of chemical weathering in soils follows a transformation pathway to hydromaghemite (ferrimagnetic ferrihydrite) through maghemite to hematite in the presence of phosphate (or another ligand), which impedes the direct transformation of 2-line ferrihydrite into hematite or goethite (Barrón et al., 2003; Liu et al., 2010). Within such a chain of transformations, no magnetite step is possible.

Another hypothesis for the magnetic enhancement of soils discusses the effects of wildfires on the production of strongly magnetic phases (Le Borgne, 1955; Mullins, 1977; Kletetschka and Banerjee, 1995). It is supposed that nonmagnetic iron hydroxides like ferrihydrite and lepidocrocite transform into strongly magnetic maghemite during the thermal alteration under high temperatures (Hirt et al., 1993; Mitov et al., 2002; Cornell and Schwertmann, 2003; Machala et al., 2007). However, there are contradicting findings rejecting this hypothesis (Clement et al., 2010; Roman et al., 2013).

1.3 MAGNETIC PROXY PARAMETERS USED IN SOIL MAGNETISM STUDIES

The use of magnetic parameters in environmental studies is challenging for both researchers in natural sciences, on the one hand, and ecology and geography, on the other. The complexity of the soil-related processes that are governed by the laws and relationships of chemistry, physics, biology, and geomorphology requires a deep knowledge of all of the respective fields in order to obtain reliable environmental information from the soil. As far as the intention of this book is to show how soil magnetic signatures can be used to decipher soil genesis, evolution, and properties and how to apply it to resolving actual environmental problems, a detailed description of the physics of magnetism and magnetic parameters is not included here. Readers can find such information in a number of excellent works, such as Thompson and Oldfield (1986), Dunlop and Özdemir (1997), Maher and Thompson (1999), and Evans and Heller (2003) and recent reviews by Liu et al. (2012) and Maxbauer et al. (2016). Table 1.3.1 lists the main magnetic parameters and ratios used in soil magnetic studies, their general meaning, and corresponding reference work for further details.

Table 1.3.1 List of the most common magnetic parameters and ratios, utilized in soil magnetic studies

Parameter	Definition	Measurement units (SI)	Proxy for	References
Low field magnetic susceptibility (χ or χ_{lf})	$\chi = k/\rho$, where k is volume susceptibility; ρ is bulk density	m^3/kg	Concentration of strongly magnetic Fe oxides (magnetite, maghemite, titanomagnetite)	Hunt et al. (1995) Walden et al. (1999)
Frequency dependent magnetic susceptibility χ_{fd}	$\chi_{fd} = \chi_{lf} - \chi_{hf}$, where χ_{lf} and χ_{hf} are susceptibilities, measured at low- and high-frequency field (usually 0.47 and 4.7 kHz)	m^3/kg	Concentration of superparamagnetic strongly magnetic particles (magnetite, maghemite) within the grain size range (10–25 nm)	Eyre (1997) Worm (1998)
Percent frequency dependent magnetic susceptibility $\chi_{fd}\%$	$\chi_{fd}\% = 100^* (\chi_{lf} - \chi_{hf})/\chi_{lf}$	%	Relative proportion of the superparamagnetic fraction in the total magnetic susceptibility signal	Mullins and Tite (1977) Dearing et al. (1997)
High-field magnetic susceptibility (χ_{hf})	χ_{hf} is calculated from the high-field portion of the hysteresis loop	m^3/kg	Magnetic susceptibility of the paramagnetic minerals (e.g., clays) and high-coercivity antiferromagnetic Fe oxides (hematite, goethite)	Thompson and Oldfield (1986) Brachfield (2006)
Anhysteretic remanence (ARM)	ARM is acquired in the laboratory through simultaneous application of a weak dc-field (h) and an alternating magnetic field with decreasing amplitude	Am^2/kg	High values indicate higher concentration of magnetically stable single domain (SD) magnetite/maghemite grains	Maher (1988); Dunlop and Özdemir (1997); Dunlop and Argyle (1997); Egli and Lowrie (2002)
Anhysteretic susceptibility (χ_{arm})	$\chi_{arm} = ARM/h$	m^3/kg	The same as ARM but the dependence on the value of the inducing weak dc field is eliminated	Maher (1988); Peters and Dekkers (2003)

Continued

Table 1.3.1 List of the most common magnetic parameters and ratios, utilized in soil magnetic studies—cont'd

Parameter	Definition	Measurement units (SI)	Proxy for	References
Isothermal remanence (IRM, SIRM, M_{rs})	IRM is acquired in a strong dc field, usually of 1−2 T intensity. Saturation is reached for magnetite-like phases, while goethite and hematite need application of higher fields to be saturated	Am^2/kg	Concentration of remanence-carrying fraction of the magnetic minerals	Thompson and Oldfield (1986); Peters and Dekkers (2003)
S ratio	$S = -IRM_{300mT}/SIRM$	Dimensionless	Relative contribution of magnetically soft to magnetically hard remanence carrying mineral phases. Values close to 1 correspond to soft magnetite-like IRM carriers, lower values indicate progressive increase in the contribution of hard magnetic minerals (hematite + goethite)	Thompson and Oldfield (1986) Peters and Dekkers (2003)
χ_{ARM}/χ	Ratio of anhysteretic susceptibility to low field magnetic susceptibility	Dimensionless	Grain-size sensitive parameter. In case of uniform mineralogy, high values of χ_{ARM}/χ indicate relative enrichment with SD particles, while low values correspond to prevailing MD and/or SP particles	King et al. (1982) Liu et al. (2004) Liu et al. (2012)
$SIRM/\chi$	Ratio of saturation isothermal remanence to low field magnetic susceptibility	A/m	Grain size sensitive parameter. In case of uniform mineralogy and insignificant contribution of paramagnetic minerals to magnetic susceptibility, higher $SIRM/\chi$ indicates the presence of smaller (SD) grains	Thompson and Oldfield (1986) Peters and Dekkers (2003)

HIRM	"Hard" isothermal remanence HIRM = $0.5(IRM_{1T} + IRM_{-300mT})$	Am^2/kg	Concentration of the high-coercivity fraction	Robinson (1986) Thompson and Oldfield (1986)
χ_{fd}/HIRM	Ratio of frequency dependent susceptibility to the hard IRM	m/A	Relative importance of superparamagnetic pedogenic fraction versus pedogenic hematite, which is contributing to HIRM. In loess/palaeosol environmental applications is used as palaeoprecipitation proxy	Liu et al. (2013)
L ratio	$(SIRM + IRM_{-0.3T})/(SIRM + IRM_{-0.1T})$	Dimensionless	Reveals the effect of hematite grain size on the values of S ratio and HIRM. Uniform values indicate no change in the hematite grain size, which allows for conventional interpretation of the S ratio and HIRM	Q. Liu et al. (2007) Liu al. (2012)
M_s	Saturation magnetization	Am^2/kg	M_s depends only on the type of Fe oxide and its concentration in the material	Dunlop and Özdemir (1997)
B_c	Coercive force	mT	Mineral, grain size and structural dependence	Dunlop and Özdemir (1997) Peters and Dekkers (2003)
B_{cr}	Coercivity of remanence	mT	Mineral, grain size and structural dependence. For magnetite/maghemite Bc is in the range 20–30mT; for hematite—300–600 mT	Dunlop and Özdemir (1997) Peters and Dekkers (2003)
M_{rs}/M_s	Ratio of saturation remanence to saturation magnetization	Dimensionless	Grain size sensitive in case of magnetite like carrier of the remanence. Values close to 0.5 are typical for SD magnetites, lower values—for larger PSD and MD grains	Day et al. (1977) Dunlop and Özdemir (1997)

Continued

Table 1.3.1 List of the most common magnetic parameters and ratios, utilized in soil magnetic studies—cont'd

Parameter	Definition	Measurement units (SI)	Proxy for	References
B_{cr}/B_c	Ratio of coercivity of remanence to coercive force	Dimensionless	Grain size sensitive in case of magnetite like carrier of the remanence. Values up to 3 are typical for SD magnetites, larger values—for PSD and MD grains	Day et al. (1977) Dunlop and Özdemir (1997)
T_c (T_N)	Curie (Neel) temperature	Degrees Celsius (°C)	Mineral-specific value, depends only on the type of Fe oxide and the possible presence of substitutions in the crystal lattice. T_c of pure magnetite is 578°C; T_N of hematite—680°C	Dunlop and Özdemir (1997)

Reference to relevant publications, where more detailed information about the peculiarities and dependencies of the corresponding parameter could be found, is given in the last column.

1.4 SOIL SAMPLING METHODOLOGY FOR MAGNETIC STUDIES, LABORATORY INSTRUMENTATION, AND METHODS

1.4.1 METHODOLOGY FOR SAMPLING SOIL PROFILES

Most of the sampling sites of the soil profiles from Bulgaria are chosen at the locations of the representative soil profiles, as published in the "Soil Atlas of Bulgaria" by Koinov et al. (1998). Because of the lack of geographic coordinates for the sites in the Soil atlas, the exact geographic locations for sampling the soil profiles have been chosen according to the standard requirements for soil sampling in geochemistry (Darnley et al., 1995) and Food and Agriculture Organization recommendations (www.fao.org/soils-portal/). The soil profiles that were not described in the Soil atlas are sampled at locations considered according to the major soil type in the area and the characteristic features of the soil horizons. The vertical profiles were sampled in open pits. After careful cleaning, the profiles were described in a field book, and their geo-referenced coordinates and photographs were also taken. Loose bulk material of about 100 g weight was gathered at each 2-cm interval from the topsoil down to the parent rock material. Soil samples were sealed in plastic bags and air-dried in the laboratory. After gentle crushing in a mortar, the material was sieved through a 1-mm mesh and the resulting sample was used for the subsequent laboratory magnetic measurements and additional nonmagnetic analyses.

1.4.2 LABORATORY INSTRUMENTATION AND METHODS

Nonmagnetic analyses

Soil texture classes (i.e., clay $d < 2$ μm; silt 2 μm $< d <$ 50 μm, and sand $d >$ 50 μm) were determined through the wet sedimentation method, and the results are presented as percentages of the total composition. Soil preparation for wet sedimentation consisted of organic matter removal with 30% H_2O_2 and subsequent dispersion with sodium hexametaphosphate.

Soil pH was measured with a Hanna 213 pH-meter (HANNA Instruments, USA) in a 1:5 soil:H_2O proportion.

The **organic carbon** (C_{org}) content was determined for selected samples from the profiles using a modified Tyurin method (Kononova, 1966). A wet combustion technique was based on organic C oxidation by potassium dichromate $K_2Cr_2O_7$ (0.4N) in an acid solution (H_2SO_4:water = 1:1).

Chemical analyses were performed at the Laboratory of Environmental Chemistry at Swiss Federal Institute of Technology (ETH)-Zurich. Nano-crystalline pedogenic oxides were extracted by using a dithionite-citrate extraction (DCB extraction, Mehra and Jackson, 1960). Poorly crystalline and amorphous phases have been analyzed for material from the same levels using an acid-oxalate extraction, applying 4 h of shaking in the dark (McKeague and Day, 1966). The extracted amounts of Fe, Al, and Mn were measured by using an ICP-OES Varian Vista-MPX. For all extractions from each sample, two replicas were prepared in order to check the reproducibility of the results. Each value per sample is an average of the two replicas. Commonly, the deviation between the single determinations is in the range of 2–5%.

The **total element contents** were obtained through X-ray fluorescence spectrometry using an energy-dispersive XRF spectrometer, the Spectro-X-Lab 2000. The samples were prepared as pressed pellets using 4 g of soil and 0.9 g amide wax.

An analysis of **major and trace elements** as well as **rare earth elements** (REEs) in selected samples was performed with a laser ablation inductively coupled plasma mass spectrometer (LA-ICP-MS)

Perkin Elmer ELAN DRC-e. The samples were prepared in the following steps: (1) 1 g of soil material was heated to 450°C for 10 h to remove organic matter; (2) 0.6 g of the heated material was mixed with 2.4 g $B_4Li_2O_7$ and the homogenized mixture was heated for 10 min at 1100°C; (3) the melt was poured into a heated graphite crucible; and (4) after cooling, the samples for ICP-MS were cut. Each analysis is an average of three point determinations.

Soil mineralogy was investigated by using qualitative **XRD analysis** performed on bulk soil material from different horizons. XRD spectra were measured on a D2 Phaser-Bruker instrument using Ni-filtered Cu-Kα radiation in the interval 5–60° 2θ at 30 kV/10 mA. The powder diffraction file of the International Centre for Diffraction Data (ICDD) database was used for phase identification.

The **iron–manganese nodules** and concretions used for examination with a scanning electron microscope were mounted in epoxy resin, and polished cross sections were prepared. Scanning electron microscope observations coupled with EDX analyses were performed on a JEOL JSM6390 electron microscope. **Single magnetic particles** from the grinded bulk material were extracted by using a hand magnet and examined by scanning electron microscopy after carbon coating the surface.

The relative amounts of **goethite and hematite** at selected profile depths have been estimated by diffuse reflectance spectroscopy (DRS) (Schleinost et al., 1998). The spectra were measured on a Varian Cary 1E spectrophotometer equipped with a $BaSO_4$-coated integrating sphere and using $BaSO_4$ as the white standard. DR spectra were recorded from 380 nm to 710 nm in 0.5-nm steps. The measured reflectance values were transformed into the Kubelka–Munk remission function. The second derivative of this function was calculated by using a cubic spline procedure (software supplied by the instrument producer). The ratios of the intensities of the bands around 425 nm (goethite) and 530 nm (hematite) were used to estimate the proportion of the two minerals. The calibration curve from Torrent et al. (2007) was used for the estimation of the mass proportion of hematite and goethite.

Magnetic analyses

Low-field magnetic susceptibility and **frequency-dependent magnetic susceptibility** were measured with a Bartington MS2B kappameter (Bartington Ltd., UK) with a dual-frequency sensor at two frequencies (0.47 and 4.7 kHz). Weakly magnetic samples were measured using a multifunction kappabridge MFK-1FA (Agico, Brno, Czech Republic) at low (976 Hz) and high (15,616 Hz) frequencies. The mass specific magnetic susceptibility (χ) was calculated using the dry weight of the sample. For some soil profiles, magnetic susceptibility was measured after immersing soil samples in liquid nitrogen for about 2 h in order to reach liquid nitrogen temperature (77 K) and then immediately measured on the KLY-2 kappabridge (Agico, Brno, Czech Rep.). In this way, magnetic susceptibility χ_{77K} was obtained and compared with room-temperature susceptibility ($\chi = \chi_{293K}$).

For the measurements of laboratory magnetic remanences, 2 g of soil material from the corresponding sample was mixed with gypsum and a small amount of water in $2 \times 2 \times 2$ cm^3 plastic containers. The containers were removed after hardening.

If iron–manganese nodules were present in the profile, they were picked up from the bulk material before grinding. The grains were fixed in gypsum using the same procedure described here but using smaller plastic containers ($0.5 \times 0.5 \times 0.5$ cm^3) for remanence measurements.

Anhysteretic remanence (ARM) acquired in a 100-mT AF peak field and 0.1-mT DC bias field was imparted using a Molspin AF-tumbling demagnetizer, equipped with an ARM attachment (Molspin Ltd., UK).

Isothermal remanence (IRM) was induced using an ASC pulse magnetizer (ASC Scientific, USA). IRMs at 2 T, −0.3 T, and −0.1 T were imparted to all samples, and a field of 5 T was used to acquire IRM_{5T} in selected samples.

All remanence measurements were performed with a JR6a automatic spinner magnetometer with a sensitivity of 2×10^{-6} A/m (Agico, Czech Rep.). The residual magnetization from gypsum impurities was subtracted from all ARM and IRM measurements using equivalent measurements on a blank gypsum sample.

Pilot soil samples were subjected to stepwise IRM acquisition up to 5 T using the ASC pulse magnetizer. Some soil samples were magnetized up to 1 T using a vibrating sample magnetometer, EV9 VSM (DSM Magnetics, ADE Corp.). The IRM acquisition curves were used for the analysis of coercivity distributions using IRM-CLG 1.0 software by Kruiver et al. (2001).

The **composite IRM** was imparted in selected samples according to Lowrie's (1990) methodology. The DC fields of 300 mT (or 1 T in some cases) for the "soft" component and 2 T (or 5 T) for the "hard" components were imparted along perpendicular axes. The samples were subjected to stepwise thermal demagnetization in an MMTD shielded furnace (Magnetic Measurements Ltd., UK) from 20°C up to 700°C.

Thermomagnetic analyses of the magnetic susceptibility of bulk soil material were performed using a KLY-2 kappabridge with a CS-23 high-temperature furnace (Agico, Czech Rep.) applying a fast heating rate (11°C/min).

Hysteresis curves were measured for bulk samples on a Micromag 3900 (Princeton Measurements Corporation, USA) up to a 1-T saturation field and hysteresis parameters: saturation magnetization (M_s), saturation remanence (M_{rs}), and coercivity (B_c) were calculated after the subtraction of the paramagnetic contribution. Back-field DC demagnetization curves were measured to obtain the coercivity of remanence B_{cr}. The measurements were carried out at the palaeomagnetic laboratories at ETH Zurich and the University of Helsinki.

The **low-temperature dependence of an $IRM_{2.5T}$**, acquired at room temperature was measured in selected samples with a quantum design magnetic properties measurements system (MPMS), on cooling to 10 K and warming to 400 K in a zero field.

1.5 SAMPLING STRATEGIES IN SOIL SCIENCE AND IN ENVIRONMENTAL MAGNETISM

There are a vast number of papers dealing with the soil sampling strategies in pedology and soil-related disciplines such as ecology and hydrology (Pennock, 2004; Pennock et al., 2006). As pointed out by the latter authors, the primary determinant of the sampling design is a well-defined research problem, which can be resolved through an appropriate choice of scale of sampling in terms of support, spacing, and extent (Pennock, 2004). These terms are usually used in geostatistics. "Support" refers to the volume of the spatial sample; "spacing" describes the distance between the sampled units; and the "extent" is the length (or area) sampled in a given study. In classic soil profile studies, the aim is to define certain physical and geochemical soil properties that are typical for the sampled soil type.

Due to natural soil variability, composite samples are usually gathered from experimental plots. It is assumed that by using a composite sample, a valid estimate of the population mean of the measured property will be obtained (Pennock et al., 2006). However, different soil properties show spatial variations over different scales, depending on the physics and/or chemistry of the underlying processes. For example, short-range differences in parent material, drainage and biological activities result in significant differences in soil properties over short distances (Burrough, 1993). On the other hand, the variability in different soil properties (such as clay content, mineral C and N content) at different spatial scales shows various trends (increasing or decreasing), which indicates that, depending on the nature of the measured property, its spatial dependence may vary (Garten Jr. et al., 2007). Soil fungal activities and microbial communities show large variability at even centimeter and millimeter scales (Grundmann and Debouzie, 2000). Soil profiles are described, sampled, and characterized by genetic horizons according to the standardized methodology (Soil Survey Staff, 1999). Recent efforts for a more objective and computerized visualization, aggregation, and classification of soil profile collections propose, instead of defining "modal" profiles, an alternative concept of "representative depth functions" (Beaudette et al., 2013). Representative soil property depth functions defined in this way give users a continuous estimate of soil properties with depth.

The results provided by the magnetic studies of soils here are in line with these latter developments in soil science, as far as they deliver continuous depth functions of the magnetic signature. This is obtained though the applied methodology for soil sampling. Soil sampling of vertical profiles for environmental magnetic studies is designed and planned from a different perspective compared with classic sampling strategies in soil science. As far as the magnetic measurements do not require expensive laboratory reactives and consumables, time-consuming pretreatment, or sample preparation, continuous sampling with small depth increments can be realized. Typically, a 2-cm sampling interval is applied; for longer sequences, a 5-cm interval is considered plausible. The 2-cm interval has a basis with respect to the standard sample size used in commercial rock magnetometers for magnetic remanence measurements. It is a kind of restriction only in cases when nondisturbed samples, oriented with respect to geographic north, are necessary for obtaining information on the direction and intensity of the ancient Earth's magnetic field (Butler, 1992; Tauxe, 2003). In the case of a pure environmental study, loose powder samples are gathered. In special cases, even a smaller 0.5-cm spacing can be applied. The exact choice of the sampling interval depends on the aim of the planned study. However, in most cases, the magnetic signature of soil material gathered using a 2-cm interval is measured with an instrumental precision, comparable with the amplitude of the natural variation of the soil's magnetic property. Thus, denser sampling will not give a record with a higher depth resolution. A particular case is the magnetic study of fire-affected soils, where the soil magnetic signature varies strongly even at a millimeter scale in the uppermost fire-affected depth interval (see Chapter 10). Here, a denser 0.5-cm sampling interval is more appropriate for an environmental magnetic study. The standard amount of soil material taken from each depth interval is about 100 g. This amount is gathered from a slice of about 5 cm × 6 cm × 2 cm in dimension. As a comparison, the accepted standards in soil science studies designed for revealing soil genesis, classification, and mapping consider that a representative sample should be of a 1- to 1.5-kg soil material per horizon. However, if we compare the typical thickness of a soil horizon (e.g., about 30 cm for humic horizons) that should be sampled along its total length to get a representative sample of 1.5 kg for that horizon, then the amount sampled for environmental magnetic studies related to the particular sampling interval of 2 cm is of the same proportion related to the sampled thickness.

An important issue in the use of environmental data obtained from soil magnetic studies is the representativeness of the signature of the sampled profiles for the corresponding soil order—that is, whether the obtained relationships can be generalized for another soils of the same genesis and parent material at other locations. The approach of averaging enough single profiles to construct a "modal profile" would lead to an unnecessary loss of information because each individual magnetic record along a soil profile gives the dynamics of the pedogenic processes linked to the weathering intensity and formation of secondary Fe oxides under specific combination of soil-forming factors. Consequently, a comparison of depth variations in magnetic signal for several profiles sampled at different locations from the same soil type is better suited and reveals the dynamic effects of changing external conditions coupled with the fixed specific expressions of the pedogenesis for each particular soil type.

Due to these different approaches in sampling protocols in soil science and environmental magnetism, soil specialists sometimes raise the question of the effects of microvariability in soil properties and changing environmental conditions (seasonal and/or yearly) and the soil's magnetic properties as measured. The following paragraph aims to clear up these concerns.

1.5.1 SOIL MICROVARIABILITY

As was discussed earlier, a soil's variability is its intrinsic property spanning different scales from millimeter to kilometer and on a global scale (Heuvelink and Webster, 2001). Microvariability arises because of the natural variability in soil composition, texture, and biological activity in space (and time). A good example of the homogeneity of field magnetic susceptibility measurements is presented in Garbacea and Ioane (2010). The authors measured field magnetic susceptibility along 16 profiles of different soil types in southern Romania. They measured field (volume-specific) magnetic susceptibility K on vertical walls at a 10- to 20-cm distance for each locality and plotted maps of susceptibility variations on these vertical surfaces, showing a good consistency of the values measured per horizon.

Other experimental evidence regarding the reliability of mass-specific magnetic susceptibility variations along several soil profiles of Chernozems is presented next. Recent profiles of Chernozems from an area of about 40 km^2 near the town of Russe (NE Bulgaria) have been sampled for magnetic studies, using two parallel vertical sampling lines at a distance of about 30 cm. From each sampling depth for each of the two parallel sections, two or three single subsamples were measured, and the average value was taken to be representative for that depth. Four of the profiles are leached Chernozems and one is Luvic Phaeozem. In order to calculate the difference between the mean mass-specific magnetic susceptibility per sampled level, the percentage difference is calculated as: *delta* $\chi = 100*ABS(\chi_1 - \chi_2)/[(\chi_1 + \chi_2)/2]$, where χ_1 and χ_2 are the two mean susceptibility values from the same depth interval of the two parallel sections. The depth variations of this parameter are shown in Fig. 1.5.1.

The variations of *delta* χ observed are systematically lower than 20%; in the case of magnetically enhanced uppermost H and A (AB) horizons, this difference is below 10% (Fig. 1.5.1). The higher values obtained in the loess parent material are due to the possible effects of a disturbance in the signal of the unaltered loess due to crotovinas, carbonate concretions, roots, etc. Consequently, small-scale soil variability is reflected in the magnetic signal of the enhanced horizons up to 10% of its

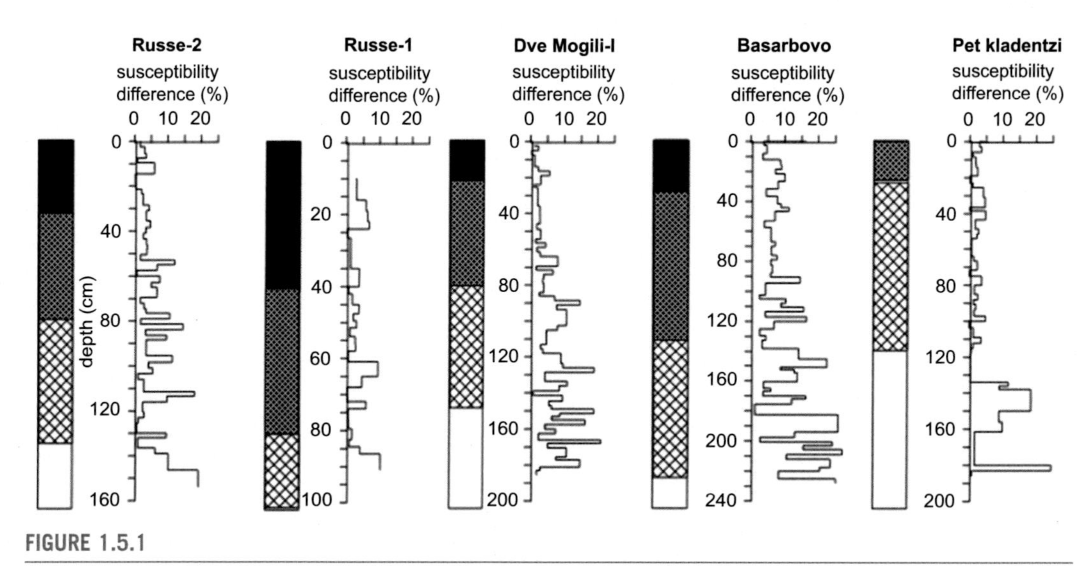

FIGURE 1.5.1

Depth variations of the percent difference in magnetic susceptibility (calculated per sampled level between two adjacent sampling profiles at about 30 cm distance on a cleaned vertical wall of a soil pit) for five profiles of Chernozem-like soils, developed on loess parent material (NE Bulgaria). Soil columns with distinction of different horizons are depicted on the left of each profile (black—H horizon; dark grey—A horizon; hatched—AB horizon, white—C horizon).

amplitude. At a profile scale, another estimate was performed. For the same set of five profiles, the mean values of magnetic susceptibility per horizon and the corresponding standard deviation were calculated. The data are presented in Table 1.5.1. and suggest the same order of uncertainty in the estimated mean values per horizon. At the same time, the origin of this uncertainty is rooted in the intrinsic within-horizon variations of the distribution of Fe oxides. The latter respond to changes in depth-dependent physical and chemical soil properties defined by the specific pedogenic evolution of each particular soil type.

The spatial locations of the studied profiles also permit us to come to a conclusion regarding the variability of the magnetic susceptibility of genetic horizons at plot scale (40–50 km^2). If we consider the soils according to their classification (typical and leached Chernozems and Luvic Phaeozem), the three profiles of leached Chernozems (Russe 1, Russe 2, and Dve Mogili I) show similar mean susceptibilities per horizon in the range of $116-129 \times 10^{-8} m^3/kg$ for H horizons, $103-118 \times 10^{-8} m^3/kg$ for A horizons, and $63-82 \times 10^{-8} m^3/kg$ for AB horizons. This again testifies to the magnetic signature of soils as being a reliable characteristic at different spatial scales (vertical and lateral). The obtained χ_{mean} values for the other two profiles are a reflection of a modified combination of environmental factors. Again, it shows the high sensitivity of the soil's magnetic properties to the exact combination of soil forming factors and its potential for detailed pedogenic studies.

Table 1.5.1 Calculated mean magnetic susceptibilities (in 10^{-8} m^3/kg) and the standard deviation per horizon for the studied two parallel depth profiles per each of the five locations. The two values per horizon are the means per each of the two profiles. The number of samples per horizon is given below the horizon designation column

area 25 x 20km							*CHERNOZEMS*							
Basarbovo		*typical*	Russe-1		*medium leached*	Russe-2		*medium leached*	Dve mogili I		*weakly leached*	5 kladentzi		*luvic phaeozem*
	Xmean	Xst dev		Xmean	Xst dev		Xmean	Xst dev		Xmean	Xst dev		Xmean	Xst dev
H	100.78	4.80	H	129.39	4.58	H	115.96	1.98	H	117.01	1.51			
12/13	107.12	4.95	9/12	127.85	8.73	15/14	115.30	3.04	7/6	117.92	2.33			
A	98.08	9.37	A	103.48	20.27	A	106.57	16.42	A	116.77	8.61	A	80.50	2.10
27/31	102.45	13.64	23/20	111.97	12.50	18/19	110.75	13.51	22/19	118.46	8.82	9/8	81.59	1.89
AB	46.36	9.07	AB	63.87	4.89	AB	81.77	7.98	AB	66.62	14.40	B	87.74	5.92
30/28	46.89	6.54	7/-			23/21	78.31	7.00	26/20	65.22	15.02	33/25	85.92	5.64
C	51.57	5.60	L1			C	79.37	6.59	C	49.00	5.34	C	44.67	8.49
7/-	57.70					12/14	74.25	5.73	17/17	48.88	3.38	23/10	44.04	10.57

H, *organic hor.*, A, *humic hor.*, AB, *transitional hor.*, C, *parent material (loess L1).*

1.5.2 THE TEMPORAL VARIABILITY OF SOIL'S MAGNETIC PROPERTIES

The other major aspect of a soil's magnetism variability is its time dependence. A valuable contribution in this respect is the publication by Geiss (2014), who studied the time and location dependence of the magnetic signature of five soil profiles from Nebraska. The profiles were sampled within a 6-year period from one sampling site at several meters' distance. Detailed magnetic susceptibility and remanence measurements along the profiles were compared to reveal the effect of seasonal and yearly changes in temperature and precipitation. The observed variations of the magnetic susceptibility and anhysteretic and isothermal remanences along the depth of the six cores (Fig. 1.5.2) are very consistent and do not show any bias in the magnetic proxy parameters used for palaeoprecipitation reconstructions.

The absence of any correlation between the magnetic parameters with seasonal and yearly precipitation values for the study site confirm that the soil's magnetic signal is built up and reaches an equilibrium state in much longer time periods, similar to the observed time-dependence of other soil physical parameters (clay content, organic content, etc.) (Schaetzl and Anderson, 2009).

Another critical methodological aspect of magnetic studies is the effect of soil moisture on magnetic susceptibility measurements. This issue was studied in detail by Maier et al. (2006) using laboratory and field experiments. The authors concluded that the influence of soil moisture and the resulting soil conductivity on the measured values of the topsoil's magnetic susceptibility is negligible compared with the effect of the contact between the measurement loop and the soil surface.

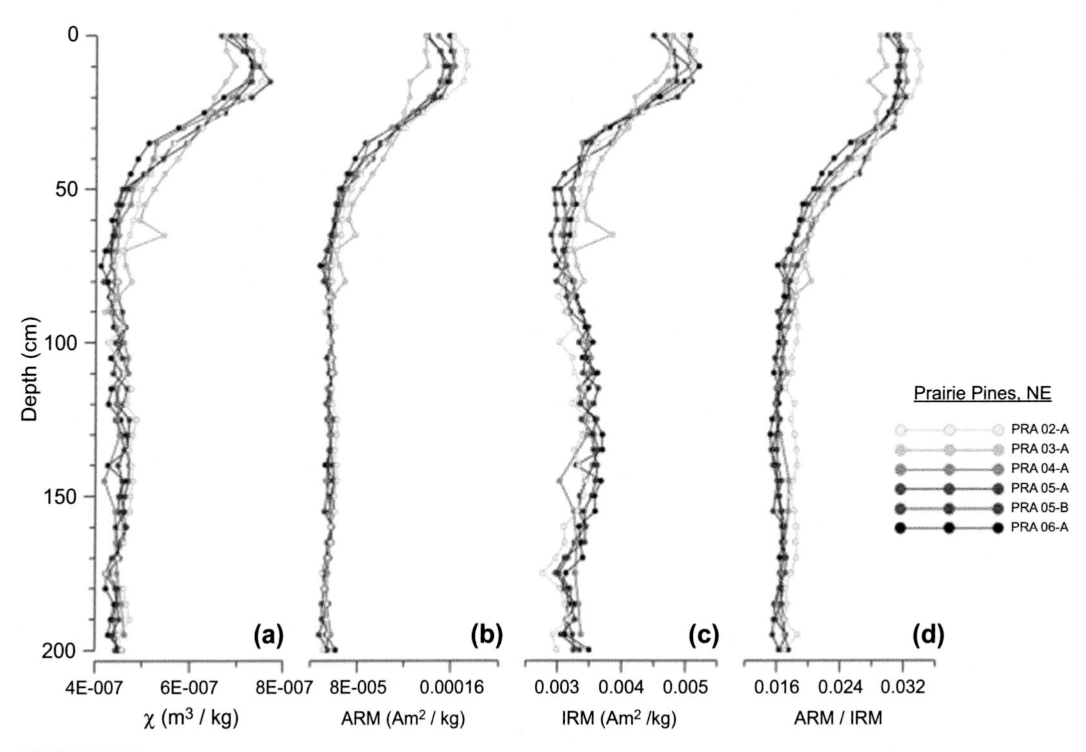

FIGURE 1.5.2

Magnetic properties of six cores of well drained Typic Argiudolls from Prairie Pines, sampled over the timespan of 5 years. *(Reprinted with permission from Geiss, C., 2014. Does timing or location matter? The influence of site variability and short-term variations in precipitation on magnetic enhancement in loessic soils. Geoderma 230–231, 280–287.)* (a) Magnetic susceptibility (χ); (b) Anhysteretic Remanence (ARM); (c) Isothermal Remanence (IRM) and (d) the ratio of ARM/IRM.

REFERENCES

Achat, D.L., Pousse, N., Nicolas, M., Bredoire, F., Augusto, L., 2016. Soil properties controlling inorganic phosphorus availability: general results from a national forest network and a global compilation of the literature. Biogeochemistry 127, 255–272.

Araujo, A.C.V., Abreu, F., Silva, K.T., Bazylinski, D.A., Lins, U., 2015. Magnetotactic bacteria as potential sources of bioproducts. Mar. Drugs 13, 389–430.

Arias, M., Barral, T., Diaz-Fierros, F., 1995. Effects of iron and aluminium oxides on the colloidal and surface properties of kaolin. Clays Clay Minerals 43 (4), 406–416.

Barrón, V., Torrent, J., 2002. Evidence for a simple pathway to maghemite in Earth and Mars soils. Geochim. Cosmochim. Acta 66, 2801–2806.

Barrón, V., Torrent, J., de Grave, E., 2003. Hydromaghemite, an intermediate in the hydrothermal transformation of 2-line ferrihydrite into hematite. Am. Mineral 88, 1679–1688.

Beaudette, D.E., Roudier, P., O'Geen, A.T., 2013. Algorithms for quantitative pedology: a toolkit for soil scientists. Comput. Geosci. 52, 258–268.

Bockheim, J.G., Gennadiyev, A.N., Hartemink, A.E., Brevik, E.C., 2014. Soil-forming factors and soil taxonomy. Geoderma 226–227, 231–237.

Brachfeld, S., 2006. High-field magnetic susceptibility (Xhf) as a proxy of biogenic sedimentation along the Antarctic Peninsula. Phys. Earth Planet. Inter 156, 274–282.

Burrough, P.A., 1993. Soil variability: a late 20th century view. Soils Fertil. 56, 529–562.

Butler, R.F., 1992. Paleomagnetism: Magnetic Domains to Geologic Terranes. Blackwell Sci. Publ., Boston, p. 319.

Calvaruso, E., Mareschal, L., Turpault, M.-P., Leclerc, E., 2009. Rapid clay weathering in the Rhizosphere of Norway Spruce and Oak in an acid forest ecosystem. Soil Sci. Soc. Am. J. 73, 331–338.

Chadwick, O.A., Chorover, J., 2001. The chemistry of pedogenic thresholds. Geoderma 100, 321–353.

Chikazumi, S., 2010. Physics of Ferromagnetism, second ed. Oxford University Press, Oxford, UK.

Clement, B.M., Javier, J., Sah, J.P., Ross, M.S., 2010. The effects of wildfires on the magnetic properties of soils in the Everglades. Earth Surf. Process. Landforms. http://dx.doi.org/10.1002/esp.2060.

Coey, J.M.D., 2009. Magnetism and Magnetic Materials. Cambridge University Press, Cambridge, USA.

Colombo, C., Palumbo, G., Sellitto, V.M., Cho, H.G., Amalfitano, C., Adamo, P., 2015. Stability of coprecipitated natural humic acid and ferrous iron under oxidative conditions. J. Geochem. Explor. 151, 50–56.

Cornell, R., Schwertmann, U., 2003. The Iron Oxides. Structure, Properties, Reactions, Occurrence and Uses. Weinheim, New York.

Cornu, S., Montagne, D., Vasconcelos, P.M., 2009. Dating constituent formation in soils to determine rates of soil processes: a review. Geoderma 153, 293–303.

Churchman, G.J., Lowe, D.J., 2012. Alteration, formation, and occurrence of minerals in soils. In: Huang, P.M., Li, Y., Sumner, M.E. (Eds.), Handbook of Soil Sciences, Properties and Processes, second ed., vol. 1. CRC Press (Taylor & Francis), Boka Raton, FL, pp. 20.1–20.72.

Da Costa, A.C.S., Bigham, J.M., Rhoton, F.E., Traina, S.J., 1999. Quantification and characterization of maghemite in soils derived from basaltic rocks in Southern Brazil. Clay Clay Minerals 47 (4), 466–473.

Darnley, A., Bjorklund, A., Bolviken, B., Gustavsson, N., Koval, P., Plant, J., Steenfelt, A., Tauchid, M., Xuejing, X., Garrett, R., Hall, G., 1995. A Global Geochemical Database for Environmental and Resource Management. Recommendations for International Geochemical Mapping. Final Report of IGCP Project 259, 19. UNESCO Publishing, Earth Sciences, ISBN 92-3-103085-X.

Day, R., Fuller, M., Schmidt, V.A., 1977. Hystereis properties of titanomagnetites – grain size and compositional dependence. Phys. Earth Planet. Interiors 13, 260–267.

Dearing, J., Bird, P., Dann, R., Benjamin, S., 1997. Secondary ferrimagnetic minerals in Welsh soils: a comparison of mineral magnetic detection methods and implications for mineral formation. Geophys. J. Int. 130, 727–736.

do Carmo Horta, M., Torrent, J., 2007. Phosphorus desorption kinetics in relation to phosphorus forms and sorption properties of Portuguese acid soils. Soil Sci. 172, 631–638.

Duiker, S.W., Rhoton, F.E., Torrent, J., Smeck, N.E., Lal, R., 2003. Iron (Hydr)Oxide crystallinity effects on soil aggregation. Soil Sci. Soc. Am. J. 67, 606–611.

Dunlop, D., Özdemir, O., 1997. Rock Magnetism. F undamentals and frontiers. In: Edwards, D. (Ed.), Cambridge Studies in Magnetism. Cambridge University Press.

Dunlop, D., Argyle, K., 1997. Thermoremanence, anhysteretic remanence and susceptibility of submicron magnetites: nonlinear field dependence and variation with grain size. J. Geophys. Res. 102 (B9), 20199–20210.

Eger, A., Almond, P.C., Condron, L.M., 2012. Upbuilding pedogenesis under active loess deposition in a super-humid, temperate climate — quantification of deposition rates, soil chemistry and pedogenic thresholds. Geoderma 189–190, 491–501.

Egli, R., Lowrie, W., 2002. Anhysteretic remanent magnetization of fine magnetic particles. J. Geophys. Res. 107, B10. http://dx.doi.org/10.1029/2001JB000671.

Emmanuel, S., Ague, J.J., 2011. Impact of nano-size weathering products on the dissolution rates of primary minerals. Chem. Geol. 282, 11–18.

Evans, M., Heller, F., 2003. Environmental Magnetism: Principles and Applications of Enviromagnetics. Academic Press, San Diego, CA.

Eyre, J.K., 1997. Frequency dependence of magnetic susceptibility for populations of single-domain grains. Geophys. J. Int. 129, 209–211.

Fassbinder, J.W.E., Stanjek, H., Vali, H., 1990. Occurrence of magnetic bacteria in soil. Nature 343, 161–163.

Fisher, H., Luster, J., Gehring, A., 2008. Magnetite weathering in a Vertisol with seasonal redox-dynamics. Geoderma 143, 41–48.

Gajdardziska-Josifovska, M., McClean, R.G., Schofield, M.A., Sommer, C.V., Kean, W.F., 2001. Eur. J. Mineral 13, 863–870.

Ganor, J., Lu, P., Zheng, Z., Zhu, C., 2007. Bridging the gap between laboratory measurements and field estimations of silicate weathering using simple calculations. Environ. Geol. http://dx.doi.org/10.1007/s00254-007-0675-0.

Garbacea, G.F., Ioane, D., 2010. Geophysical mapping of soils. New data on Romanian soils based on magnetic susceptibility. Rom. Geophys. J 54, 83–95.

Garten Jr., C.T., Kang, S., Brice, D.J., Schadt, C.W., Zhou, J., 2007. Variability in soil properties at different spatial scales (1 m—1 km) in a deciduous forest ecosystem. Soil Biol. Biochem. 39, 2621–2627.

Geiss, C.E., Zanner, C.W., 2007. Sediment magnetic signature of climate in modern loessic soils from the Great Plains. Quat. Int. 162–163, 97–110.

Geiss, C.E., Egli, R., Zanner, C.W., 2008. Direct estimates of pedogenic magnetite as a tool to reconstruct past climates from buried soils. J. Geophys. Res. 113, B11102. http://dx.doi.org/10.1029/2008JB005669.

Geiss, C., 2014. Does timing or location matter? the influence of site variability and short-term variations in precipitation on magnetic enhancement in loessic soils. Geoderma 230–231, 280–287.

Goldich, S.S., 1938. A study in rock weathering. J. Geol. 46, 17–58.

Górka-Kostrubiec, B., Teisseyre-Jeleńska, M., Dytłow, S.K., 2016. Magnetic properties as indicators of Chernozem soil development. Catena 138, 91–102.

Goulard, A.T., Fabris, J.D., de Jesus Filho, M.E., Coey, M.D., Da Costa, G.M., de Grave, E., 1998. Iron oxides in a soil developed from basalt. Clay Clay Minerals 46 (4), 369–378.

Grimley, D., Arruda, N., 2007. Observations of magnetite dissolution in poorly drained soils. Soil Sci. 172 (12), 968–982.

Grundmann, G.L., Debouzie, D., 2000. Geostatistical analysis of the distribution of NH_4^+ and NO_2^--oxidizing bacteria and serotypes at the millimeter scale along a soil transect. FEMS Microbiol. Ecol. 34, 57–62.

He, Y., Li, D.C., Velde, B., Yang, Y.F., Huang, C.M., Gong, Z.T., Zhang, G.L., 2008. Clay minerals in a soil chronosequence derived from basalt on Hainan Island, China and its implication for pedogenesis. Geoderma 148, 206–212.

Heuvelink, G.B.M., Webster, R., 2001. Modelling soil variation: past, present, and future. Geoderma 100, 269–301.

Hilley, G.E., Chamberlain, C.P., Moon, S., Porder, S., Willett, S.D., 2010. Competition between erosion and reaction kinetics in controlling silicate-weathering rates. Earth Planet. Sci. Lett. 293, 191–199.

Hirt, A., Banin, A., Gehring, A., 1993. Thermal generation of ferromagnetic minerals from iron-enriched smectites. Geophys. J. Int. 115 (3), 1161–1168.

Hu, P., Liu, Q., Torrent, J., Barron, V., Jin, C., 2013. Characterizing and quantifying iron oxides in Chinese loess/paleosols: implications for pedogenesis. Earth Planet. Sci. Lett. 369-370, 271–283.

Hunt, C.P., Moskowitz, B.M., Banerjee, S.K., 1995. Magnetic properties of rocks and minerals. In: Rock Physics and Phase Relations. A Handbook of Physical Constants. AGU Reference Shelf, 3, pp. 189–204.

Hyland, E.G., Sheldon, N.D., Van der Voo, R., Badgley, C., Abrajevitch, A., 2015. A new paleoprecipitation proxy based on soil magnetic properties: implications for expanding paleoclimate reconstructions. Geol. Soc. Am. Bull 127 (7/8), 975–981.

Ilbert, M., Bonnefoy, V., 2013. Insight into the evolution of the iron oxidation pathways. Biochim. Biophys. Acta 1827, 161–175.

Jenny, H., 1941. Factors of Soil Formation. A System of Quantitative Pedology. R. Amundson, Dover Publ. Inc, New York (Edition 1994).

Jordanova, D., Jordanova, N., Petrov, P., Tsacheva, T., 2010. Soil development of three Chernozem-like profiles from North Bulgaria revealed by magnetic studies. Catena 83 (2–3), 158–169.

Johnson, D.L., Watson-Stegner, D., 1987. Evolution model of pedogenesis. Soil Sci. 143, 349–366.

King, J., Banerjee, S.K., Marvin, J., Özdemir, O., 1982. A comparison of different magnetic methods for determining the relative grain size of magnetite in natural materials: some results from lake sediments. Earth Planet. Sci. Lett. 59, 404–419.

Kletetschka, G., Banerjee, S.K., 1995. Magnetic stratigraphy of Chinese loess as a record of natural fires. Geophys. Res. Lett. 22, 1241–1343.

Koinov, V., Kabakchiev, I., Boneva, K., 1998. Soil Atlas of Bulgaria. Zemizdat, Sofia, 320p.

Kononova, M.M., 1966. Soil Organic Matter: Its Nature, its Role in Soil Formation and in Soil Fertility, second ed. Pergamon Press, Oxford, England. 544 pp.

Kruiver, P., Dekkers, M., Heslop, D., 2001. Quantifcation of magnetic coercivity components by the analysis of acquisition curves of isothermal remanent magnetization. Earth Planet. Sci. Lett. 189, 269–276.

Lawrence, C., Harden, J., Maher, K., 2014. Modeling the influence of organic acids on soil weathering. Geochimica et Cosmochimica Acta 139, 487–507.

Le Borgne, E., 1955. Abnormal magnetic susceptibility of the top soil. Ann. Geophys. 11, 399–419.

Liu, H., Li, P., Zhu, M., Wei, Y., Sun, Y., 2007. Fe(II)-induced transformation from ferrihydrite to lepidocrocite and goethite. J. Solid State Chem. 180, 2121–2128.

Liu, Q., Banerjee, S., Jackson, M., Maher, B., Pan, Y., Zhu, R., Deng, C., Chen, F., 2004. Grain sizes, susceptibility and anhysteretic remanent magnetisation carriers in Chiness loess/paleosol sequences. J. Geophys. Res. 109, B03101. http://dx.doi.org/10.1029/2003JB002747.

Liu, Q.S., Roberts, A.P., Torrent, J., Horng, C.S., Larrasoaña, J.C., 2007. What do the HIRM and S-ratio really measure in environmental magnetism? Geochem. Geophys. Geosyst 8, Q09011. http://dx.doi.org/10.1029/2007GC001717.

Liu, Q., Hu, P., Torrent, J., Barrón, V., Zhao, X., Jiang, Z., Su, Y., 2010. Environmental magnetic study of a Xeralf chronosequence in northwestern Spain. Indications for pedogenesis. Palaeogeogr. Palaeoclimatol. Palaeoecol 293, 144–156.

Liu, Q., Roberts, A., Larrasoaña, J., Banerjee, S., Guyodo, Y., Tauxe, L., Oldfield, F., 2012. Environmental magnetism: principles and applications. Rev. Geophys. 50, RG4002.

Liu, Z., Liu, Q., Torrent, J., Barr?n, V., Hu, P., 2013. Testing the magnetic proxy *x*FD/HIRM for quantifying paleoprecipitation in modern soil profiles from Shaanxi Province, China. Glob. Planet. Change 110, 368–378.

Lovley, D.R., Stolz, J.F., Nord, G.L., Phillips, E.J.P., 1987. Anaerobic production of magnetite by a dissimilatory iron-reducing microorganism. Nature 330, 252–254.

Lowe, D.J., Tonkin, P.J., 2010. Unravelling upbuilding pedogenesis in tephra and loess sequences in New Zealand using tephrochronology. In: Gilkes, R.J., Prakongkep, N. (Eds.), Proceedings of the 19th World Congress of Soil Science "Soil Solutions for a Changing World", Brisbane, pp. 34–37.

Lowrie, W., 1990. Identification of ferromagnetic minerals in a rock by coercivity and unblocking temperature properties. Geophys. Res. Lett. 17, 159–162.

Machala, L., Zboril, R., Gedanken, A., 2007. Amorphous iron(III) oxides. A review. J. Phys. Chem. B 111, 4003–4018.
Maher, B., 1986. Characterization of soils by mineral magnetic measurements. Phys. Earth Planet. Inter 42, 76–92.
Maher, B., 1988. Magnetic properties of some synthetic sub-micron magnetites. Geophys. J. R. Astr. Soc. 94, 83–96.
Maher, B., Taylor, R., 1988. Formation of ultra fine-grained magnetite in soils. Nature 336, 368–370.
Maher, B.A., 1998. Magnetic properties of modern soils and Quaternary loessic paleosols: paleoclimatic implications. Palaeogeogr. Palaeoclimatol. Palaeoecol 137, 25–54.
Maher, B., Thompson, R., 1999. Quaternary Climates. Cambridge University Press, Cambridge, UK. Environments and Magnetism.
Maher, B.A., 2011. The magnetic properties of Quaternary aeolian dusts and sediments, and their palaeoclimatic significance. Aeolian Res. 3, 87–144.
Maher, B.A., Alekseev, A., Alekseeva, T., 2003. Magnetic mineralogy of soils across the Russian Steppe: climatic dependence of pedogenic magnetite formation. Palaeogeogr. Palaeoclimatol. Palaeoecol 201, 321–341.
Maher, B.A., Possolo, A., 2013. Statistical models for use of palaeosol magnetic properties as proxies of palaeorainfall. Glob. Planet. Change 111, 280–287.
Maher, B.A., Thompson, R., 1992. Paleoclimatic significance of the mineral magnetic record of the Chinese loess and paleosols. Quatern. Res. 37, 155–170.
Maher, B.A., Thompson, R., 1995. Paleorainfall reconstructions from pedogenic magnetic susceptibility variations in the Chinese loess and paleosols. Quat. Int. 44, 383–391.
Maher, K., 2010. The dependence of chemical weathering rates on fluid residence time. Earth Planet. Sci. Lett. 294, 101–110.
Maher, K., Steefel, C.I., White, A.F., Stonestrom, D.A., 2009. The role of reaction affinity and secondary minerals in regulating chemical weathering rates at the Santa Cruz Soil Chronosequence, California. Geochimica et Cosmochimica Acta 73, 2804–2831.
Maier, G., Scholger, R., Schon, J., 2006. The influence of soil moisture on magnetic susceptibility measurements. J. Appl. Geophys. 59, 162–175.
Marques, R., Waerenborgh, J.C., Prudêncio, M.I., Dias, M.I., Rocha, F., Ferreira da Silva, E., 2014. Iron speciation in volcanic topsoils from Fogo island (Cape Verde)—Iron oxide nanoparticles and trace elements concentrations. Catena 113, 95–106.
Maxbauer, D.P., Feinberg, J.M., Fox, D.L., 2016. Magnetic mineral assemblages in soils and paleosols as the basis for paleoprecipitation proxies: a review of magnetic methods and challenges. Earth Sci. Rev. 155, 28–48.
McKeague, J.A., Day, J.H., 1966. Dithionite and oxalate extractable Fe and Al as aids in differentiating various classes of soils. Can. J. Soil Sci. 46, 13–22.
Mehra, O., Jackson, M., 1960. Iron oxide removal from soils and clays by a dithionite-citrate system buffered with sodium bicarbonate. Clays Clay Minerals 7, 317–327.
Minasny, B., McBratney, A.B., Salvador-Blanes, S., 2008. Quantitative models for pedogenesis—A review. Geoderma 144, 140–157.
Minasny, B., Finke, P., Stockmann, U., Vanwalleghem, T., McBratney, A.B., 2015. Resolving the integral connection between pedogenesis and landscape evolution. Earth Sci. Rev. 150, 102–120.
Michel, F.M., Barrón, V., Torrent, J., Morales, M.P., Serna, C.J., Boily, J.-F., Liu, Q., Ambrosini, A., Cismasu, A.C., Brown, G.E., 2010. Ordered ferrimagnetic form of ferrihydrite reveals links among structure, composition, and magnetism. Proc. Natl. Acad. Sci. U.S.A 107, 2787–2792.
Mitov, I., Paneva, D., Kunev, B., 2002. Comparative study of the thermal decomposition of iron oxyhydroxides. Thermochimica Acta 386, 179–188.

Moore, J., Macalady, J.L., Schulz, M.S., White, A.F., Brantley, S.L., 2010. Shifting microbial community structure across a marine terrace grassland chronosequence, Santa Cruz, California. Soil Biol. Biochem. 42, 21–31.

Mullins, C.E., 1977. Magnetic susceptibility of the soil and its significance in soil science—a review. J. Soil Sci. 28, 223–246.

O'Reilly, W., 1984. Rock and Mineral Magnetism. Blakie, Glasgow. Chapman and Hall.

Orgeira, M.J., Egli, R., Comapgnucci, R.H., 2011. A quantitative model of magnetic enhancement in loessic soils. In: Petrovsky, E., Ivers, D., Harinarayana, T., Herrero-Bervera, E. (Eds.), The Earth's Magnetic Interior, IAGA Special Sopron Book Series, vol. 1. Springer, ISBN 978-94-007-0322-3.

Pedrosa, J., Costa, B.F.O., Portugal, A., Duraes, L., 2015. Controlled phase formation of nanocrystalline iron oxides/hydroxides in solution - an insight on the phase transformation mechanisms. Mater. Chem. Phys. 163, 88–98.

Pailhe, N., Wattiaux, A., Gaudon, M., Demourgues, A., 2008. Impact of structural features on pigment properties of alpha-Fe2O3 haematite. J. Solid State Chem. 181, 2697–2704.

Pennock, D., 2004. Designing field studies in soil science. Can. J. Soil Sci. 84, 1–10.

Pennock, D., Yates, Th, Braidek, J., 2006. Soil sampling designs. In: Carter, M.R., Gregorich, E.G. (Eds.), "Soil Sampling and Methods of Analysis, second Ed. Canadian Society of Soil Science, CRC Press Taylor & Francis Group.

Peters, C., Dekkers, M.J., 2003. Selected room temperature magnetic parameters as a function of mineralogy, concentration and grain size. Phys. Chem. Earth 28, 659–667.

Rahn-Lee, L., Komeil, A., 2013. The magnetosome model: insights into the mechanisms of bacterial biomineralization. Front. Microbiol. 4 http://dx.doi.org/10.3389/fmicb.2013.00352, 352.

Reeves, D., Rothman, D.H., 2013. Age dependence of mineral dissolution and precipitation rates. Glob. Biogeochem. Cycles 27, 906–919. http://dx.doi.org/10.1002/gbc.20082.

Robinson, S.G., 1986. The late Pleistocene paleoclimatic record of North Atlantic deep-sea sediments revealed by mineral-magnetic measurements. Phys. Earth Planet. Inter 42, 22–47.

Roman, S.A., Johnson, W.C., Geiss, C.E., 2013. Grass fires—an unlikely process to explain the magnetic properties of prairie soils. Geophys. J. Int. 195 (3), 1566–1575.

Rümenapp, C., Wagner, F.E., Gleich, B., 2015. Monitoring of the aging of magnetic nanoparticles using Mössbauer spectroscopy. J. Magn. Magn. Mater. 380, 241–245.

Schaetzl, R., Anderson, A., 2009. Soils. Genesis and Geomorphology. Cambridge University Press, UK, ISBN 978-0-521-81201-6.

Schwertmann, U., 1988. Occurrence and formation of iron oxides in various pedoenvironments. In: Stucki, J., Goodman, B., Schwertmann, U. (Eds.), Iron in Soils and Clay Minerals, NATO ASI Series, Series C: Mathematical and Physical Sciences, vol. 217. Reidel Publishing Company, pp. 267–308.

Scheinost, A., Chavernas, A., Barron, V., Torrent, J., 1998. Use and limitations of second-derivative Diffuse Reflectance spectroscopy in the visible to near-infrared range to identify and quantify Fe oxide minerals in soils. Clays Clay Minerals 46 (5), 528–536.

Soil Survey Staff, 1999. Soil Taxonomy: A Basic System of Soil Classification for Making and Interpreting Soil Surveys. Number 436 in Agricultural Handbook. USDA.

Spier, C.A., Vasconcelos, P.M., Oliveira, S.M.B., 2006. $^{40}Ar/^{39}Ar$ geochronological constraints on the evolution of lateritic iron deposits in the Quadrilátero Ferrífero, Minas Gerais, Brazil. Chem. Geol. 234, 79–104.

Targulian, V.O., Krasilnikov, P.V., 2007. Soil system and pedogenic processes: self- organization, time scales, and environmental significance. Catena 71, 373–381.

Tauxe, L., 2003. Palaeomagnetic Principles and Practice. Kluwer Academic Publishers, Dordrecht, ISBN 0-7923-5258-0.

Taylor, A., Blum, J.D., 1995. Relation between soil age and silicate weathering rates determined from the chemical evolution of a glacial chronosequence. Geology 23, 979–982.

Taylor, R., Schwertmann, U., 1974. Maghemite in soils and its origin. I. Properties and observations on soil maghemites. Clay Minerals 10, 289–298.

Thompson, R., Oldfield, F., 1986. Environmental Magnetism. Allen&Unwin, London.

Torrent, J., Barrón, V., Liu, Q., 2006. Magnetic enhancement is linked to and precedes hematite formation in aerobic soil. Geophys. Res. Lett. 33, L02401. http://dx.doi.org/10.29/2005GL024818.

Torrent, J., Liu, Q., Bloemendal, J., Barrón, V., 2007. Magnetic enhancement and iron oxides in the upper Luochuan loess–paleosol sequence, Chinese loess plateau. Soil Sci. Soc. Am. J. 71, 1570–1578. http://dx.doi.org/10.2136/sssaj2006.0328.

Van Breemen, N., 1988. Long-term chemical, mineralogical and morphological effects of iron-redox processes in periodically flooded soils. In: Stucki, J., Goodman, B., Schwertmann, U. (Eds.), Iron in Soils and ClayMinerals. D. Reidel Publishing company, NATO ASI Series.

Walden, J., Oldfield, F., Smith, J. (Eds.), 1999. Environmental Magnetism. A Practical Guide. Technical Guide No 6. Quaternary Research Association, London.

White, A.F., Brantley, S.L., 2003. The effect of time on the weathering of silicate minerals: why do weathering rates differ in the laboratory and field? Chem. Geol. 202, 479–506.

Worm, H.-U., 1998. On the superparamagnetic - stable single domain transition for magnetite, and frequency dependence of susceptibility. Geophys. J. Int. 133, 201–206.

Yoo, K., Mudd, S.M., 2008. Discrepancy between mineral residence time and soil age: implications for the interpretation of chemical weathering rates. Geology 36 (1), 35–38.

CHAPTER

MAGNETISM OF SOILS WITH A PRONOUNCED ACCUMULATION OF ORGANIC MATTER IN THE MINERAL TOPSOIL: CHERNOZEMS AND PHAEOZEMS

2

2.1 CHERNOZEMS AND PHAEOZEMS—MAIN CHARACTERISTICS, FORMATION PROCESSES, AND DISTRIBUTION

Some of the most fertile soils on Earth are the organic-rich soils of the temperate climate belt, classified as Chernozems in the Food and Agriculture Organization (FAO)−UNESCO system, Mollisols in the US Department of Agriculture system of soil classification, and black soils in China. The main characteristic of these soils is their dark-colored, humus-rich surface horizon with a high base saturation (Eckmeier et al., 2007). According to the classic genetic classification of Chernozems by Docuchaev, Chernozems in Russia are zonal soils developed under steppe vegetation and a dry continental to semiarid climate. However, an increasing number of studies also report Chernozem-like soils developed in forest-steppe and open forest vegetation cover (Eckmeier et al., 2007). The so-called anthropogenic pedogenesis of Chernozems is still a matter of debates, which considers that humans' shaping of the environment during historic times plays a role in the formation of this soil type (Gerlach et al., 2006; Lorz and Saile, 2011; Richter and Yaloon, 2012). Chernozems (Mollisols) are typically formed on aeolian or glacial till sediments, characterized by a high silica content and a prevailing silt fraction (Schaetzl and Anderson, 2009).

Chernozems' genesis is primarily related to the development of humus-enriched A horizons as a result of the accumulation and decomposition of xerophitic and mesophitic grasses and/or wood species. The rate of accumulation of organic matter is greater than decomposition, which results in humus accumulation (decomposed and resistant organic matter) in the surface horizons. A particular property of Chernozems is that between 5% and 30% of the soil organic carbon stock is represented by black carbon (BC) (Rodionov et al., 2010). The BC is a product of incomplete burning, suggesting that fires were often operating during Chernozem formation in the Holocene period. Chendev et al. (2010) claim that it is possible to discriminate long-term (interglacial) from short-term (within different climatic periods of the Holocene) trends in the development of the Chernozems from the East European Plain. The long-term trend consists of a systematic increase in the thickness of the humic horizon and an oscillating increase in the depth of the accumulation of carbonates. The short-term trend is related to Late Holocene local climatic changes and differences in the mineralogy of the parent rocks. The mineralogy and texture of the parent material also play a particular role in the formation of Chernozems. Easily weathered silt-size and

Soil Magnetism. http://dx.doi.org/10.1016/B978-0-12-809239-2.00002-4

clay-sized primary silicate minerals result in the formation of new pedogenic silicate clays with a high cation-exchange capacity (Hole and Nielsen, 1970; Pye, 1987). The accumulation of carbonates within 2 m from the surface is another major feature of their classification.

The distribution of Chernozems worldwide follows the temperate areas mainly in the northern hemisphere (central North America, central to southeastern Europe, central Asia, Russia, and China) and in the southern hemisphere—large areas in the Parana-La Plata basin of South America (Eckmeier et al., 2007; X. Liu et al., 2012). Fig. 2.1.1 depicts the distribution of Chernozems according to FAO

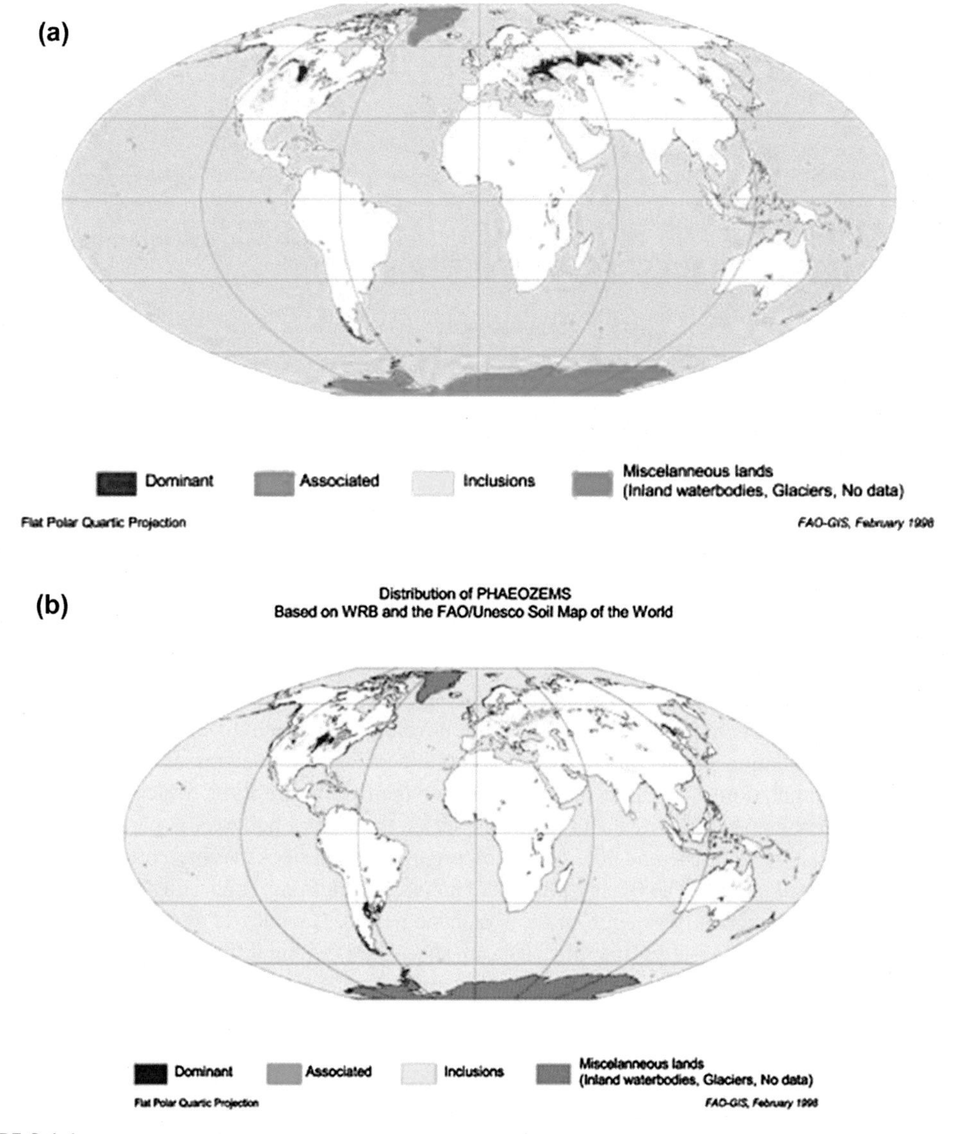

FIGURE 2.1.1

Spatial distribution of Chernozems (a) and Phaeozems (b) in the world.

FAO World Soil Resources Reports, 2001. Lecture Notes on the Major Soils of the World. ISSN:0532-0488. http://www.fao.org/docrep/003/y1899e/y1899e00.htm#toc.

soil maps (FAO, 2001). They are the major representative of the "tall grass steppe" soils, bordered by Phaeozems at steppe areas with increased precipitation.

Chernozems gradually transform into Phaeozems under a progressive increase in precipitation and temperature. In other national classifications, Phaeozems appear as Brunizems (Argentina, France), degraded Chernozems (former USSR), Parabraunerde-Tschernozems (Germany), and "Udolls" and "Aquolls" in the order of the Mollisols (Soil Taxonomy) (Spaargaren, 2008).

Increased temperature and precipitation lead to more intense leaching and differentiation of the profile (textural and geochemical), which represent characteristic genetic features of the Phaeozems group. The upper 2 m of the Phaeozems profile lacks carbonates.

Distribution and characteristics of Chernozems and Phaeozems in Bulgaria.

Chernozem-like soils in Bulgaria are found mainly in the lower Danube plain. The latter is blanketed by loess cover, which is continuous to the north and patchy to the south, formed during the glacial Pleistocene (Evlogiev, 2006; Haase et al., 2007; Jordanova et al., 2008). The gradual transition from Chernozems to Phaeozems is revealed southward from the Danube River, resulting from increased precipitation toward the Stara Planina mountain range and growing amount of the clay fraction in the loess material due to the increased distance from the dust source area (Evlogiev, 2006). The climate during Chernozems formation is of a temperate continental regime—a warm summer and cold winter with unsteady snow cover (Koinov et al., 1998). Epicalcic Chernozems are distributed immediately south from the Danube River, mainly in central North Bulgaria (Koinov et al., 1998; Ninov, 2002), and represent about 28% of the total area covered by Chernozems. Epicalcic Vermic Chernozems are characterized by a better developed humic horizon, and carbonates are found lower than 40–50 cm depth (Shishkov and Kolev, 2014). They cover 48% of the total area of the Chernozems in Bulgaria. Luvic Phaeozems are found in isolated areas at the foothills of the Stara Planina mountain range. They are formed under forest-steppe vegetation and have a prevailing heavy mechanical fraction.

2.2 DESCRIPTION OF THE PROFILES STUDIED

2.2.1 EPICALCIC CHERNOZEM (TYPICAL CHERNOZEM), PROFILE DESIGNATION: TB

Location: Situated close to Tsarev Brod village, Shumen district; northeast Bulgaria; N 43°20′06.4″, E 27°01′08.6″, elevation 234 m a.s.l.

Present-day climate conditions are characterized by a mean annual temperature (MAT) of 10.7°C and mean annual precipitation (MAP) of 546 mm. Vegetation is characterized by grasses and small bushes. A general view of the soil profile is presented in Fig. 2.2.1 and a closer view at different depths is given in Figs. 2.2.2–2.2.5.

Profile description

0–30 cm: A_k horizon (A_{kp} + A_k') humic horizon, dark brown, crumbly, rare carbonate concretions

30–50 cm: A_k horizon—humic horizon, brown, wealth of carbonate mycelia, crumbly

50–70 cm: AB_k horizon—carbonate rich horizon, light-brown specks, denser than upper levels

70–90 cm: BC_k horizon—transitional horizon, clayey, light brown, large $CaCO_3$ concretions

FIGURE 2.2.1

View of the profile of Calcic Chernozem (TB).

FIGURE 2.2.2

Profile TB. Ak horizon—detail.

FIGURE 2.2.3

Profile TB. ABk horizon—detail.

FIGURE 2.2.4

Profile TB. BCk horizon—detail.

FIGURE 2.2.5

Profile TB. Ck horizon—detail.

90–200 cm: C_k horizon—typical loess, light brown to yellow color, $CaCO_3$ concretions
Parent rock—typical loess

2.2.2 EPICALCIC CHERNOZEM (LEACHED CHERNOZEM), PROFILE DESIGNATION: OV

Location: N 43°48′18.4″, E 27°05′43.0″, elevation 188 m a.s.l.

Soil profile OV is situated in the Razgrad region (northeast Bulgaria) near the village of Oven. Present-day climate conditions are MAT = 10.5°C; MAP = 515 mm. Vegetation—grasses. Nearby, there is open forest at about 400 m to the southwest and northwest. A general view of the profile is presented in Fig. 2.2.6 and a closer view at different depths is given in Figs. 2.2.7–2.2.9.

Profile description
0–50 cm: A horizon, dark brown, prismatic structure, weakly clayey
50–112 cm: B_k horizon, illuvial, brown-orange, dense, clayey, $CaCO_3$ and Mn- concretions
112–126 cm: BC_k horizon, transitional, light brown–orange, dense, $CaCO_3$ concretions
126–196 cm: C_k horizon, light brown, sandy-clayey, $CaCO_3$ concretions
Parent rock—weathered loess

FIGURE 2.2.6

View of the profile of Epicalcic Chernozem (OV).

FIGURE 2.2.7

Profile OV. A horizon—detail.

FIGURE 2.2.8

Profile OV. B_k horizon—detail.

FIGURE 2.2.9

Profile OV. BC_k horizon—detail.

2.2.3 LUVIC PHAEOZEM (GRAY FOREST SOIL), PROFILE DESIGNATION: GF

Location: N 43°32′33.8″ E 27°13′31.5″, elevation 350 m a.s.l.

The profile is located near the village of Nikola Kozlevo. Present-day climate conditions: MAT = 11.3°C, MAP = 601 mm. Vegetation—forest, *Quercus ilex*.

A general view of the profile is presented in Fig. 2.2.10 and a closer view is given at different depths in Figs. 2.2.11–2.2.14.

Profile description

0–10 cm: A_h horizon, humic horizon, black to dark brown
10–40 cm: AE horizon, dark brown, dense
40–60 cm: AB horizon, light brown, heavy clayey, dense
60–160 cm: B_g horizon, brown, clayey, presence of black Fe-Mn concretions
160–200 cm: C_k horizon, clayey loess, light brown to yellow, sandy, $CaCO_3$ concretions

FIGURE 2.2.10

View of the profile of Luvic Phaeozem (GF).

FIGURE 2.2.11

Profile GF. A_h horizon—detail.

FIGURE 2.2.12

Profile GF. AB horizon—detail.

FIGURE 2.2.13

Profile GF. B_g horizon—detail.

FIGURE 2.2.14

Profile GF. C_k horizon—detail.

2.3 TEXTURE, SOIL REACTION, AND GEOCHEMISTRY

Soil reaction (pH) and the three mechanical classes (clay, silt, sand) show systematic depth variations along the three soil profiles studied (Fig. 2.3.1). There is a strongly alkaline pH along the TB profile (Fig. 2.3.1a) with well-expressed maxima in the solum caused by the presence of $CaCO_3$ concretions already at the soil's surface. Its preservation in the upper part of the soil suggests the persistence of dry climate conditions, which do not support carbonate dissolution and washing-out in depth. In contrast to

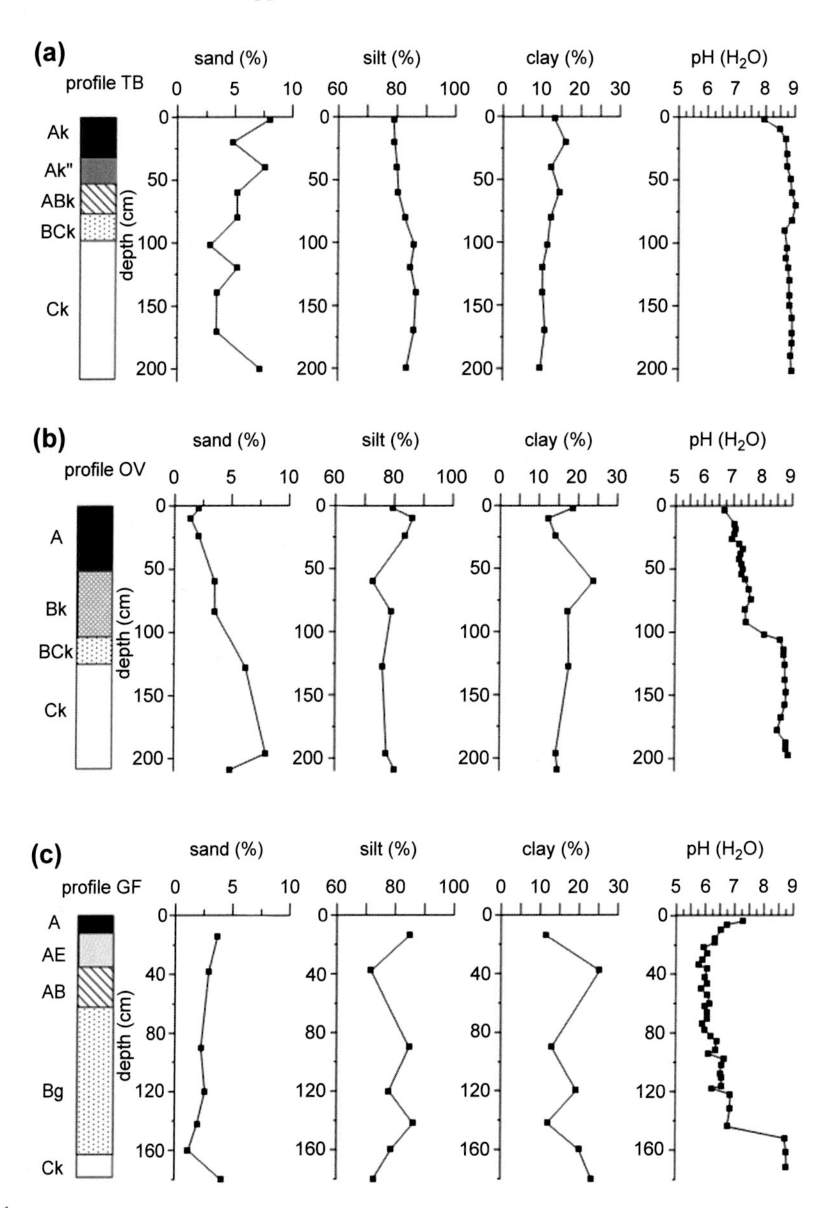

FIGURE 2.3.1

Mechanical fractions (sand, silt, clay) and soil reaction pH along the profiles of (a) Calcic Chernozem (TB), (b) Epicalcic Chernozem (OV), and (c) Luvic Phaeozem (GF).

this, the pH along the other Chernozem and Phaeozem profiles (OV and GF) is characterized by relatively less alkaline to neutral soil reaction, which becomes strongly alkaline in the C_k horizons due to higher $CaCO_3$ content caused by carbonate leaching and secondary precipitation. The contents of the clay and sand fractions along the three profiles show higher differentiation in the profile of Luvic Phaeozem (GF), while the TB profile reveals a smooth decrease in the clay fraction in depth, which is probably an expression of in situ weathering of primary minerals. The strongest textural horizon development in the GF profile leads to well-expressed variations in depth (Fig. 2.3.1c).

Major and trace element geochemistry for representative levels from the three profiles—two Chernozems and a Phaeozem—have been obtained by XRF analysis. The Chemical Index of Alteration

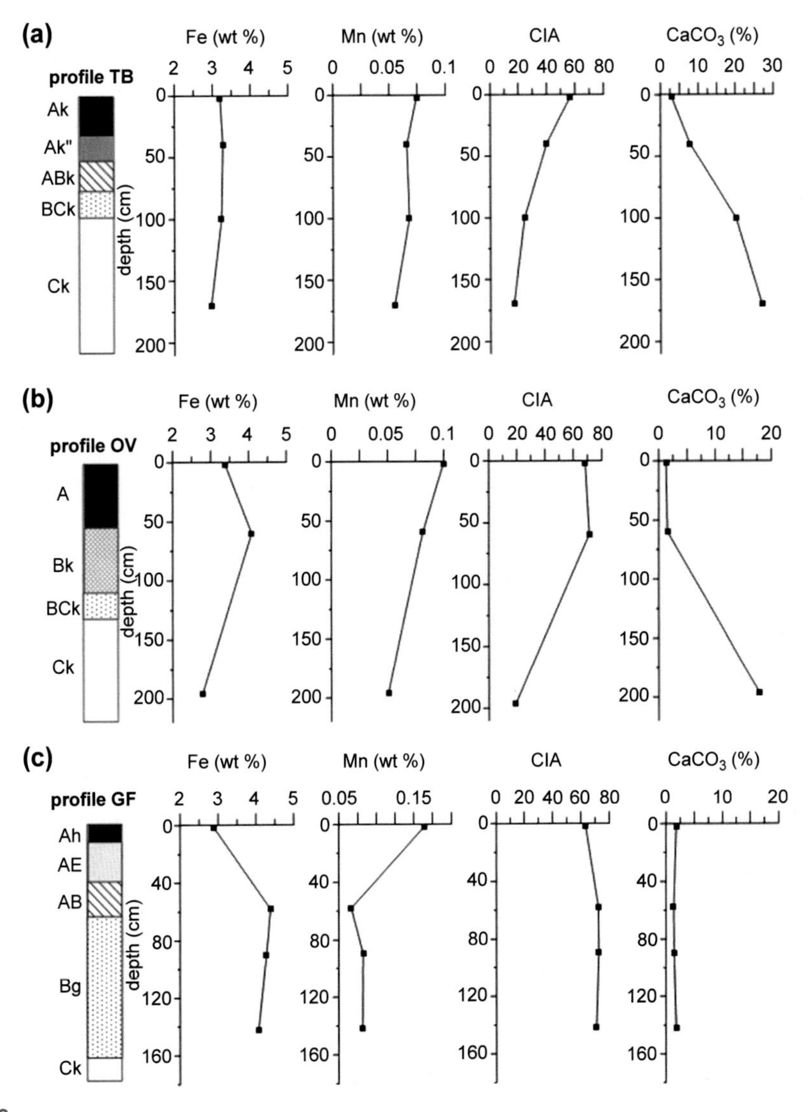

FIGURE 2.3.2

Content of iron (Fe), manganese (Mn), calcium carbonate ($CaCO_3$), and the Chemical Index of Alteration (CIA) along depth of the studied profiles: (a) profile TB, (b) profile OV, and (c) profile GF.

(CIA) is calculated as: CIA = [Al_2O_3/(Al_2O_3 + Na_2O + CaO* + K_2O)] x 100 (CaO only in silicates) according to Nesbitt and Young (1982). Additionally the content of $CaCO_3$ is calculated as:

$$\%CaCO_3 = \left(\%CaO / M_{(CaO)} - 1 * \left(\%S / M_{(S)}\right)\right) - [(5 * 2)/3] * \left(\%P_2O_5 / M_{(P_2O_5)}\right) * M_{(CaCO_3)}$$

where M(x) is the molar mass of compound X (Buggle et al., 2008). Depth variations of the two parameters, along with those of Mn and Fe contents, are given in Fig. 2.3.2. In the Epicalcic Chernozem TB, a smooth increase of $CaCO_3$ is seen with increasing depth up to about 26% in the C_k horizon. On the other hand, the calcium carbonate content in the C_k horizons of the profiles OV and GF is up to 15% (Fig. 2.3.2b,c). The difference is attributed to the process of enhanced leaching of carbonates out of the OV and GF profiles, also facilitated by the lower pH in the solum. The total iron content is similar in the three profiles but have different behavior along depths. The iron content is uniform along profile TB, while in the other two profiles, it shows a maximum in the upper part of the AB_k horizons. The manganese content reveals a systematic increase in the order TB → OV → GF. These phenomena are related to the strength of the pedogenic alteration of the parent material in the three profiles. The immobile elements Ti and Zr have analogous behavior, which reflects the uniformity of the parent rock mineralogy (loess sediments). The CIA shows a progressive decrease along depth of the TB profile (Fig. 2.3.2a) and systematically lower values compared with those for profiles OV and GF.

Maximum values of the CIA index for the profiles OV and GF are observed in deeper horizons (B_k and AB-B_g), implying that the most intense weathering of the primary minerals in these soils is occurring under enhanced pedogenic processes of leaching and illuviation. The low CIA values for the parent loess material are caused by high CaO content.

2.4 MAGNETIC PROPERTIES OF CHERNOZEMS AND PHAEOZEMS

The magnetic mineralogy of the soils studied is deduced from high-temperature thermal behavior of magnetic susceptibility (thermomagnetic analysis of magnetic susceptibility) of samples from different horizons (Fig. 2.4.1). The observed shapes and intensity of the peaks for the heating and cooling curves are similar to the ones typically found in organic-rich, strongly magnetic soils developed on loess sediments (Liu et al., 2005). The thermal behavior of magnetic susceptibility suggests that maghemite is the main pedogenic strongly magnetic mineral in the soil horizons. This supposition is based on the strong decrease observed in the signal in the temperature interval 300–420°C, which is the characteristic range of temperatures for the thermal inversion of maghemite to hematite (Dunlop and Özdemir, 1997). Strong thermal transformations occur in the topsoil samples (upper 2 cm), which are mediated by the reducing conditions maintained by burning organic matter (typically occurring at about 300°C (e.g. Hanesch et al., 2006)). The latter may cause the reduction of hematite into strongly magnetic magnetite (Evans and Heller, 2003). Further enhancement of this process is facilitated by the evolving gases from thermal transformations of clay minerals at higher temperatures (600–700°C) (Rowland, 1955). Thermal transformations in the samples from the parent loess (Fig. 2.4.1c) do not resemble that of maghemite decomposition, as seen by the different shape of the thermomagnetic curves. It suggests that only magnetite is present, as far as the observed Curie temperatures are at about 590°C. However, a significant increase in susceptibility after cooling to room temperature is evident for samples TB170 and GF142, which is related to the transformations of the instable clay minerals in the loess material.

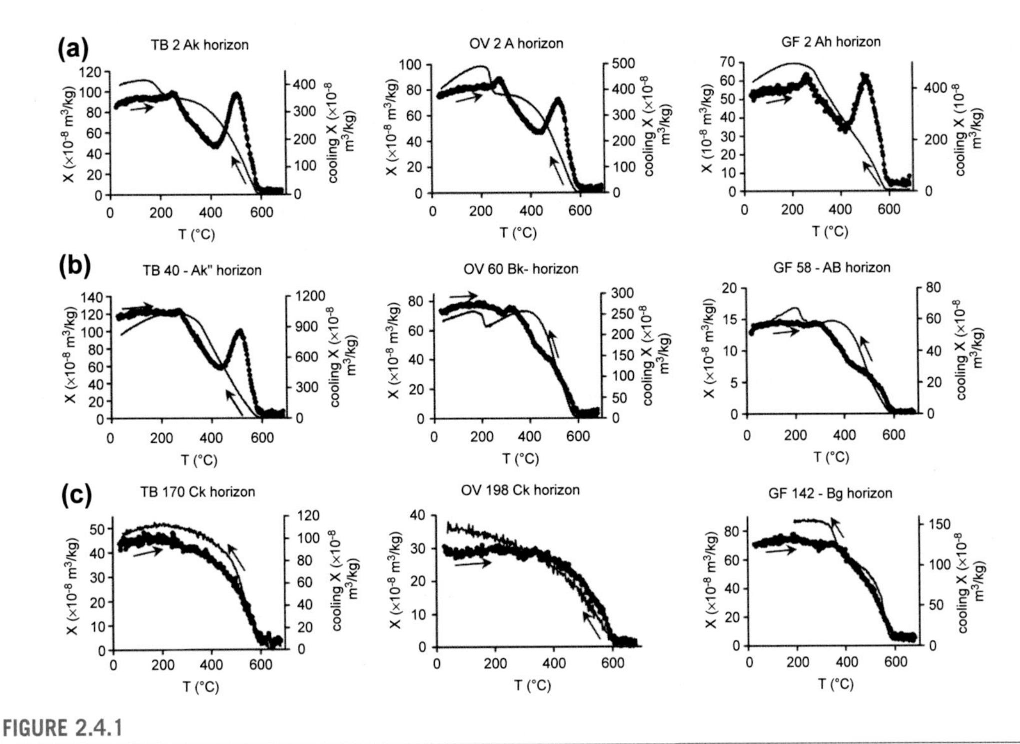

FIGURE 2.4.1

Thermomagnetic analyses of magnetic susceptibility for samples from different horizons of the three profiles. Arrows indicate heating and cooling runs. Heating in air, fast heating rate (11°C/min).

Redrawn from Jordanova, D., Jordanova, N., Petrov, P., Tsacheva, T., 2010. Soil development of three Chernozem-like profiles from North Bulgaria revealed by magnetic studies. Catena 83, 2–3, 158-169 with permission from Elsevier Ltd.

The mineralogical composition of selected samples from the three profiles is obtained through XRD analysis (Table 2.4.1). The main clay minerals are kaolinite and illite, in most cases present in amounts up to 10 wt%. These minerals show a characteristic endothermic peak at about 600°C on DTA curves (Grim and Rowland, 1942; Brown, 1998), reflecting a dehydroxylation reaction and formation of metakaolin (Bertolino et al., 2010). Liberated Fe^{2+} ions from the substitutions in the kaolinite structure are sources of a newly created strongly magnetic SP phase—magnetite (Djemai et al., 2001).

Typical morphologies of the coarse magnetic particles in Chernozems are irregular with strong evidence of surface weathering (cracks) (Fig. 2.4.2a,b). Smaller micrometer-size grains (Fig. 2.4.2c) appear more homogeneous.

The behavior of concentration-dependent magnetic parameters along the TB profile (exhibiting the highest concentration of carbonates in the solum) reflects the presence of a strongly magnetic mineral fraction in the upper part of the profile, a result of pedogenic "in situ" formation of fine magnetite/maghemite grains (Maher, 1986). A shift in the depth of the maxima of low-field susceptibility (χ) and saturation magnetization (M_s) in the uppermost topsoil levels relative to the higher depth of the maxima of magnetic remanences—isothermal (IRM) and anhysteretic (ARM)—is clearly observed (Fig. 2.4.3). It suggests that finer superparamagnetic grains dominate

Table 2.4.1 Mineralogical composition of bulk samples from the soil profiles TB, OV, and GF, obtained by XRD analysis

Profile	Sample	Quartz %	Albite %	Muscovite + Illite %	Kaolinite %	Montmorillonite	Calcite %	Hematite %	Microcline %
TB	TB 6	46	16	15	10		3		9
	TB 42	44	19	14	15		6		
OV	OV 64	48	21	16	10			4 (amorphous)	
	OV 182	29	14	20			16		14
GF	GF 56	47	16	18	14			3	
	GF 140	45	16	15	12	4		6 (amorphous)	

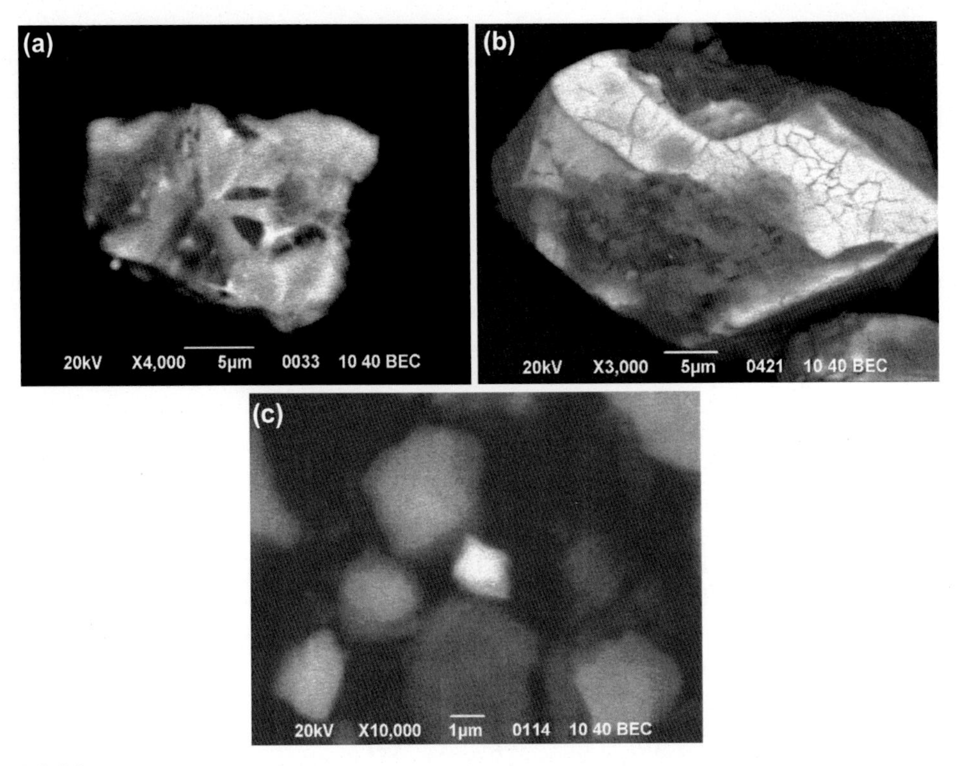

FIGURE 2.4.2

Microphotographs of single Fe-containing magnetic particles from the soil samples: (a) sample TB20, (b) sample OV22, and (c) sample OV22.

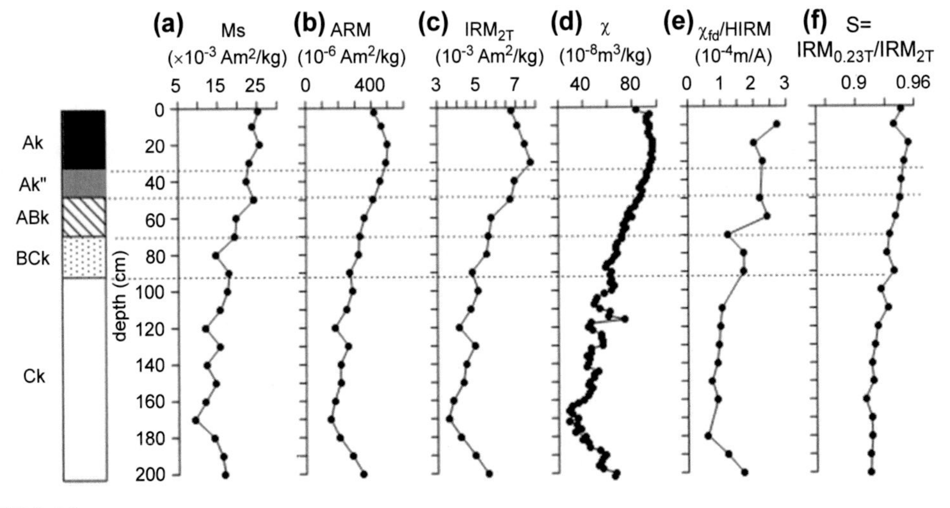

FIGURE 2.4.3

Soil profile TB. Depth variations of concentration-dependent magnetic characteristics and the ratios χ_{fd}/HIRM and S.

the uppermost approximately 10 cm, while downward more stable single-domain (SD)–like pedogenic grains give rise to the higher intensity of the remanences. Depth variations of the χ_{fd}/HIRM ratio reveal the relative changes in the proportion of pedogenic strongly magnetic fraction (maghemite) and pedogenic hematite (Q. Liu et al., 2012). Systematically higher values in the upper horizons (down to the bottom of AB_k) point to the prevailing contribution of the strongly magnetic pedogenic fraction. It decreases in the lower BC_k horizon at the expense of an increase in hematite content of the parent rock mineralogy. The same information is given by the S-ratio, revealing a small systematic decrease down the profile. Hysteresis ratios M_{rs}/M_s and B_{cr}/B_c do not show significant changes with depth, which signifies that the main carriers of the isothermal remanence have a similar grain-size distribution along the profile (Fig. 2.4.4). The ARM/IRM_{2T} and χ_{ARM}/χ ratios again show maxima at around a 30-cm depth, similar to the maximum in the concentration-dependent parameters ARM and IRM_{2T}, which, again, proves the enrichment of this level with more stable SD-like pedogenic grains.

The second Epicalcic Chernozem (profile OV) is characterized by higher carbonate leaching, expressed by the establishment of a structural B_k horizon. According to this stronger profile differentiation, the magnetic parameters along the OV profile show more distinct variations (Figs. 2.4.5 and 2.4.6). The magnetic susceptibility and saturation magnetization vary synchronously, showing maxima within the depth interval 30–60 cm, below a decline in the uppermost 25 cm. Such a decrease is not visible on the ARM and IRM_{2T} curves, implying an effect from the changes in the grain size of the remanence-carrying fraction. The behavior of the grain-size sensitive parameters M_{rs}/M_s and

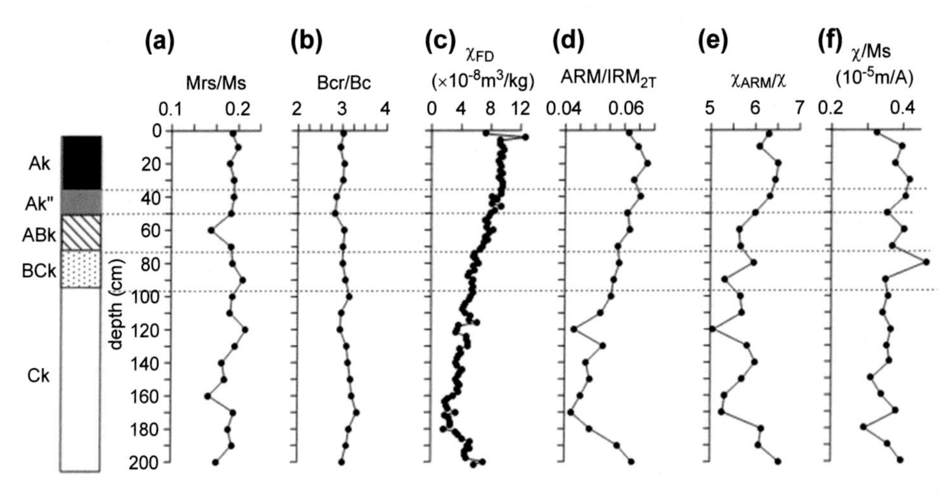

FIGURE 2.4.4

Soil profile TB. Depth variations of grain-size–sensitive magnetic characteristics.

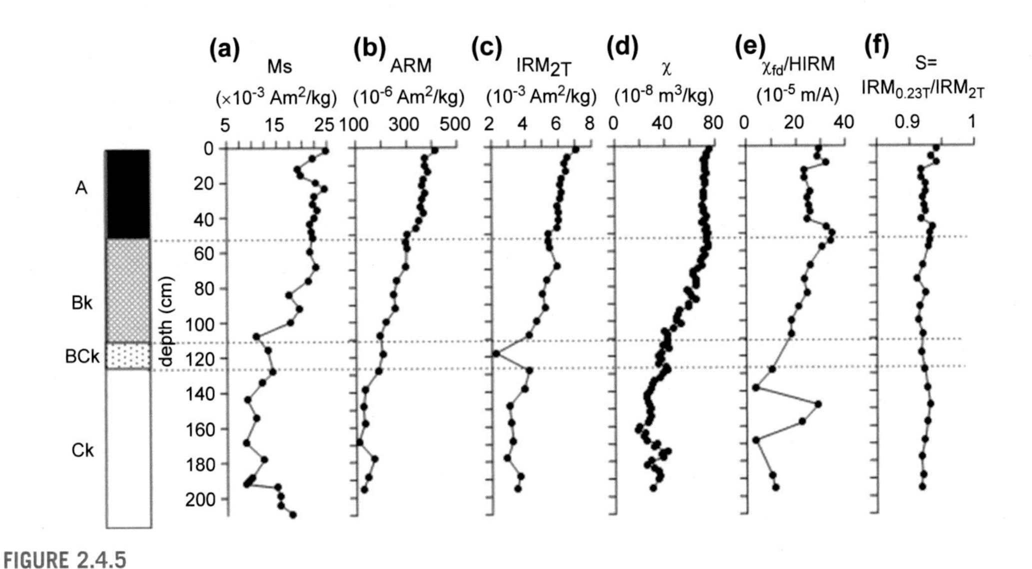

FIGURE 2.4.5

Soil profile OV. Depth variations of concentration-dependent magnetic characteristics and the ratios χ_{fd}/HIRM and S.

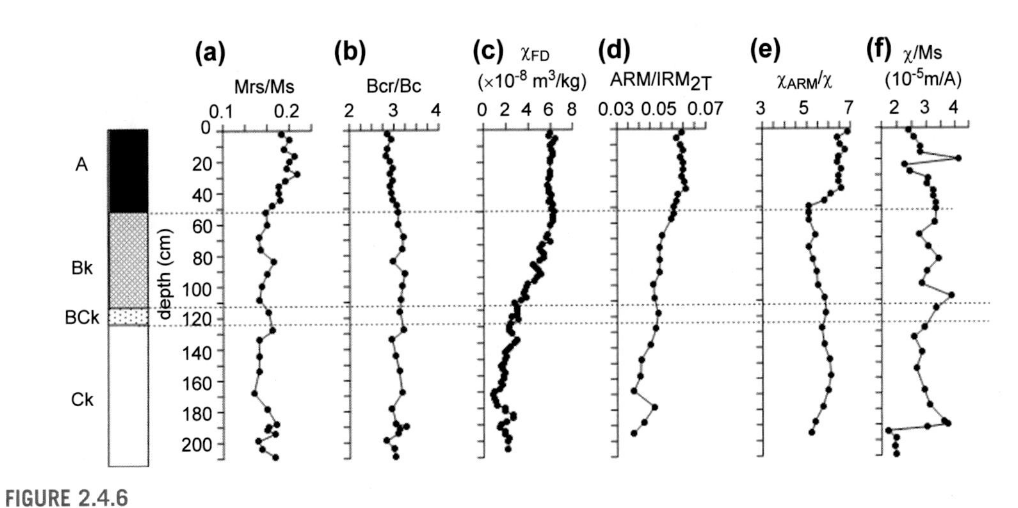

FIGURE 2.4.6

Soil profile OV. Depth variations of grain-size–sensitive magnetic characteristics.

χ_{ARM}/χ (Fig. 2.4.5a,e) point to the enhancement of the upper 30 cm with a stable SD-like fraction, which may be one of the reasons for the decrease in χ, as far as SD grains possess the lowest susceptibility (Dunlop and Özdemir, 1997). Using the parameters $\chi_{fd}/HIRM$ and χ_{fd} (Figs. 2.4.5e and 2.4.6c) as indicators for the dominance of superparamagnetic grains in the strongly magnetic pedogenic fraction (Q. Liu et al., 2012), we can infer that, at depths of 50–10 cm, we have a higher amount of SP grains compared with the upper levels. A mirrored behavior of the $\chi_{fd}/HIRM$ and χ_{ARM}/χ curves in the C_k horizon, where pedogenesis is negligible, suggests that the maximum on χ_{ARM}/χ in C_k is due to the change in the mineralogy of the samples; thus, the ratio cannot be interpreted as a grain-size indicator.

Depth variations of magnetic characteristics along the profile of Luvic Phaeozem (profile GF) are strongly bound to the observed horizons' designation (see profile's description). The uppermost A and AE horizons are magnetically depleted, as is obvious from the significantly lower values of the concentration-dependent parameters (M_s, χ, IRM_{2T}) (Fig. 2.4.7). The processes of leaching and translocation of substances down-profile of the Luvic Phaeozems result in the formation of the AB and B_g clayey horizons. They are characterized by increased magnetization and susceptibility values with a maximum in the AB and upper B_g levels. This phenomenon is related to the enhanced concentration of the pedogenic ferrimagnetic fraction in the above-mentioned levels. The magnetic grain-size proxy parameters for this profile (Fig. 2.4.8) show an enhanced concentration of stable

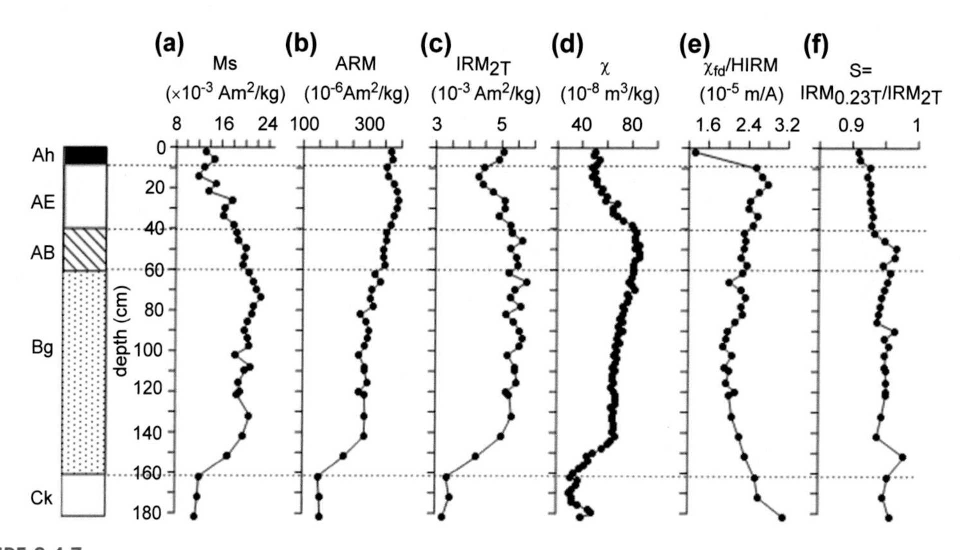

FIGURE 2.4.7

Soil profile GF. Depth variations of concentration-dependent magnetic characteristics and the ratios χ_{fd}/HIRM and S.

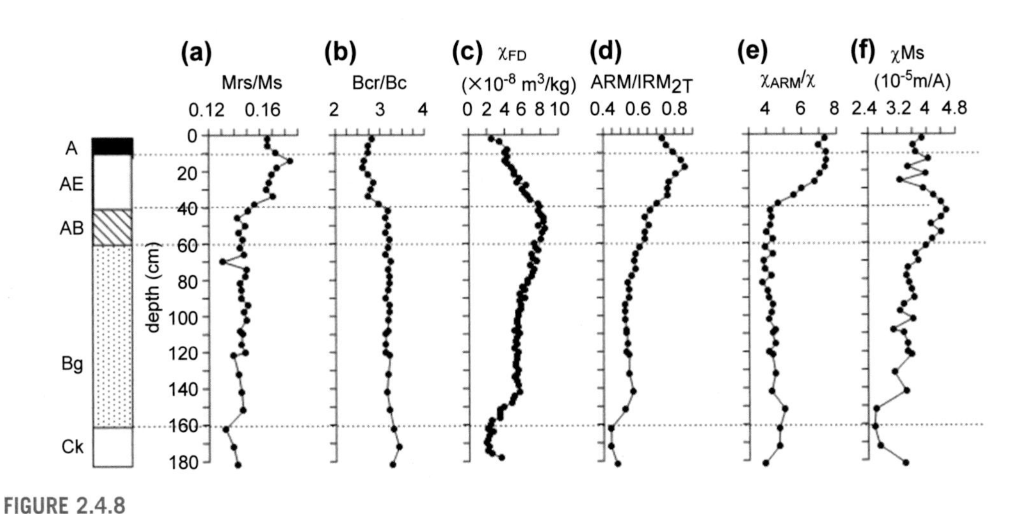

FIGURE 2.4.8

Soil profile GF. Depth variations of grain-size–sensitive magnetic characteristics.

SD-like particles in the AE horizon deduced from the observed maxima in M_{rs}/M_S, ARM/IRM_{2T}, and χ_{ARM}/χ and low B_{cr}/B_c. The existence of a clearly detectable depth interval of a superparamagnetic (SP) enrichment at a deeper level is evidenced by the maximum on the χ/M_s curve in the AB horizon.

2.5 LOW-TEMPERATURE HYSTERESIS MEASUREMENTS

Useful information about the mineralogy and prevailing magnetic grain size in the studied soil profiles can be obtained from the low temperature hysteresis measurements. The low temperature dependences of the ratio M_{rs}/M_S, saturation magnetization (M_S), and the coercive force (B_c) for samples from profiles TB and GF are shown in Fig. 2.5.1.

All samples analyzed reveal similar behavior. A specific feature on the M_{rs}/M_S temperature dependencies obtained is the observed maximum between 80 and 40K, depending on the sample and the increase in M_s at temperatures higher than 50K. At the same time, coercive force B_c shows only a slight systematic decrease upon heating to room temperature. Such variations are consistent with the result reported by Kosterov (2002) on low-temperature magnetism of partially oxidized magnetites. His results for sample BK5099, characterized by a smaller grain size and more advanced surface oxidation, fully comply with the present data for the Chernozems samples.

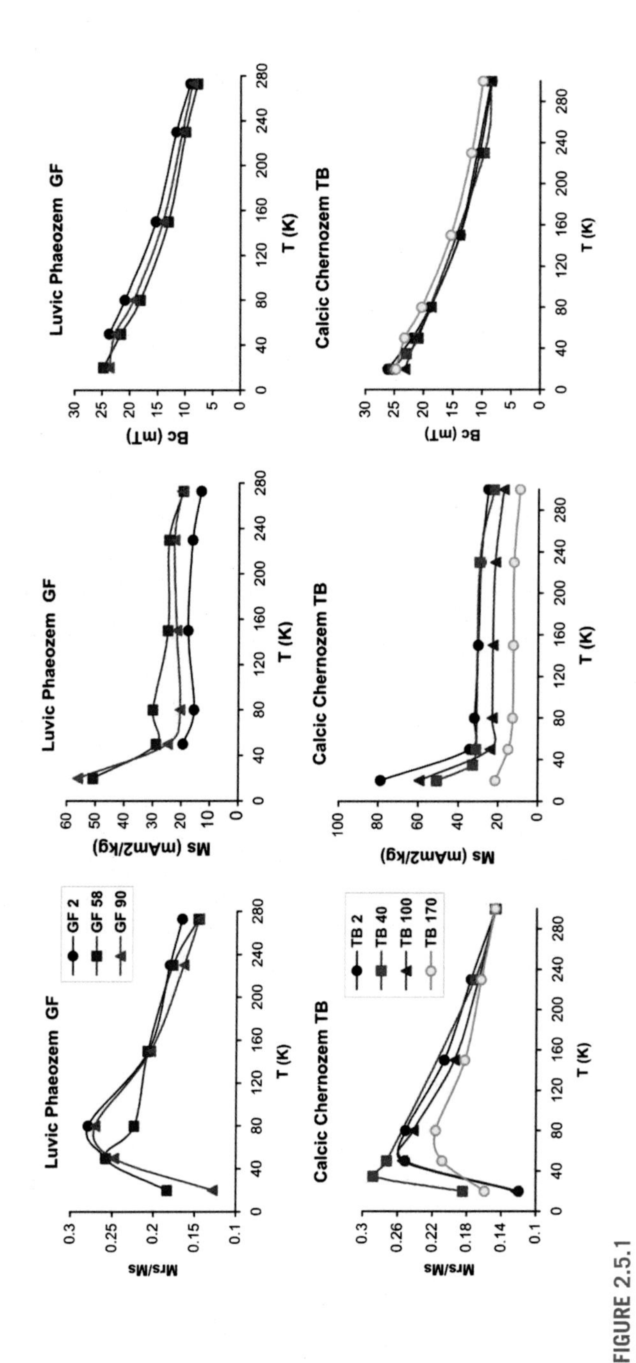

FIGURE 2.5.1

Low temperature dependence of the ratio M_{rs}/M_S, saturation magnetization M_S, and coercive force B_c for samples from the profiles GF and TB.

2.6 OXALATE AND DITHIONITE EXTRACTABLE IRON AND MAGNETIC CHARACTERISTICS OF THE PROFILES OF CHERNOZEM (OV) AND PHAEOZEM (GF)

The obtained results for the amounts of dithionite (Fe_d) and oxalate (Fe_o) extractable iron are shown in Table 2.6.1. It is observed that Fe_o increases in the more degraded profile of Luvic Phaeozem GF compared with the Epicalcic Chernozem OV. A systematically decreasing amount of oxalate-extractable Fe along the depth of the two profiles reflects the existence of a more intense pedogenic formation of amorphous Fe compounds in the topsoil levels in the compartment of active interaction of organic matter and soil mineral compounds. Measurements of the hysteresis loops for samples before and after dithionite extraction allow us to draw conclusions about the underlying processes that cause changes in the hysteresis parameters. Fig. 2.6.1a presents the obtained linear dependence between the change in coercive force B_c after DCB extraction and the frequency-dependent susceptibility χ_{fd}. A similar relationship is also observed between the amount of Fe_d and χ_{fd} (Fig. 2.6.1b).

As far as fine strongly magnetic pedogenic SP particles solely contribute to the χ_{fd} (Mullins and Tite, 1973; Worm, 1998), linearly increasing difference in B_c (delta B_c) as well as Fe_d with increased χ_{fd} values proves that dithionite extraction is effective in specifically removing the fine-grained pedogenic component in the soils.

Table 2.6.1 Content of dithionite-extractable (Fed) and oxalate-extractable (Fe_o) iron and the ratio Fe_o/Fe_d for selected levels of the profiles OV and GF

Depth (cm)	Fe_o (mg/g)	Fe_d (mg/g)	Fe_o/Fe_d
Profile GF			
2	4.04	7.07	0.57
58	2.82	11.54	0.24
90	1.91	10.96	0.17
142	1.50	9.86	0.15
Profile OV			
2	1.89	8.76	0.22
60	1.78	11.54	0.15
196	0.33	9.86	0.03

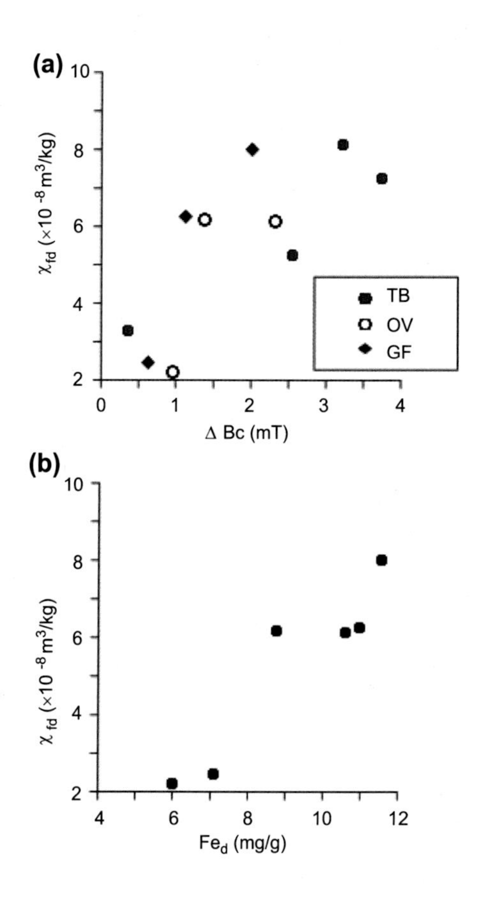

FIGURE 2.6.1

(a) Frequency-dependent magnetic susceptibility χ_{fd} as a function of the difference (ΔB_c) of the coercive force after CBD extraction minus initial B_c value; and (b) frequency-dependent magnetic susceptibility χ_{fd} as a function of the amount of dithionite—extractable iron (Fe_d).

2.7 MAGNETIC SIGNATURE OF PEDOGENESIS IN CHERNOZEMS AND PHAEOZEMS

Iron oxides are one of the most important soil constituents and determine soil color. The exact phase composition of the Fe oxides reflects in a subtle way any changes in the soil's microenvironment and soil age (Schwertmann, 1988; Cornell and Schwertmann, 2003). The large surface area of the pedogenic iron oxides makes them an important cementing agent. They often exist as a film on the surface of larger clay minerals, which changes the clay's properties and behavior (Cornell and Schwertmann, 2003). The magnetic mineralogy of Chernozems is uniformly reported as dominated by pedogenic maghemite and/or partially oxidized magnetite. The results from mineral magnetic measurements of Bulgarian Chernozems and Phaeozems (Jordanova et al., 1999, 2010) comply with the published data on modern (Holocene) Chernozem-like soils worldwide. Well-expressed phase transformations during thermomagnetic analysis of samples from the uppermost organic rich horizons (Fig. 2.4.1) are systematically observed in the studies of soils from Poland, Ukraine (Jelenska et al., 2004, 2010; Gorcka-Kostrubiec et al., 2016), and Chinese Loess Plateau (Liu et al., 2005). Differences are observed in the behavior of samples from different soil horizons, reflecting mainly the varying content of organic matter, which is the controlling factor of the effective atmosphere (reducing or oxidative) during laboratory experiments. Organic matter is known to play a major role in mineral transformations during heat treatment of thermally instable Fe oxides/hydroxides, which are usually present in the soil environment (Hanesch et al., 2006). Partial heating–cooling cycles (Jordanova et al., 2010) as well as Moessbauer analysis on material heated at different temperatures (Jelenska et al., 2010), suggest that the strongly magnetic phase initially present is maghemite, which is stable up to temperatures higher than 400°C. Such stability may be explained by the observed thermodynamic stability of nanometer-sized Fe oxides (Ye et al., 1998; Navrotsky et al., 2008) in contrast to their coarser analogues. Another reason for the thermal stability of maghemite could be the presence of Al substitutions in the maghemite structure (Özdemir and Banerjee, 1984; Özdemir and Dunlop, 1993; Dunlop and Özdemir, 1997; Gnanaprakash et al., 2006). Such a supposition leads to the conclusion that this maghemite is not a product of the low-temperature oxidation of magnetite, as far as magnetite synthesis is retarded by an Al presence in the solution (Cornell and Schwertmann, 2003). On the other hand, Al is often present in the soil solution, mainly in soils with a lower pH, which facilitates the chemical weathering of primary aluminosilicate minerals from the parent rock. The resulting "in situ" formation of pedogenic maghemite from ferrihydrite (Barrón and Torrent, 2002) inherits Al substitutions of the latter. A significant increase of susceptibility in the temperature range 400–500°C reflects the creation of a new strongly magnetic phase on heating (Fig. 2.4.1). Possible sources for the synthesis of such a phase are Fe ions migrating during the dehydroxylation of clay minerals outward. According to our XRD data (Table 2.4.1), kaolinite and illite are the main clay minerals in the Bulgarian loessic soils. These minerals dehydroxylate in the temperature range 450–500°C and new metakaolin is formed. Further, at temperatures greater than 650°C, new superparamagnetic magnetite is formed (Riccardi et al., 1999; Murad and Wagner, 1998; Djemai et al., 2001). The newly formed magnetite of superparamagnetic-single domain (SP-SD) size during heating experiments is also identified by Liu et al. (2005), Kopcewicz et al. (2005), and others. Another source for the observed susceptibility decrease in the interaval 300–420°C is the oxidation of already weathered (maghemitized) large lithogenic ferrimagnetic grains (Fig. 2.4.3). According to the mineral weathering classification of Grimley and Arruda (2007), the observed in Fig. 2.4.3. degree of cracking corresponds to weathering index 2. An enhanced chemical weathering is also indicated by the calculated values up to 73 of the CIA weathering index (Fig. 2.3.2). Lower CIA indexes along the profile TB may be either related to incomplete subtraction of Ca in carbonates through the applied calculations according to Buggle (2008) or indicate lower weathering of the primary silicate minerals due to aeolian sedimentation going on simultaneously with soil development. Such up-building pedogenesis is documented for Chinese and Midwestern US soils formed since the Holocene (Huang et al., 2009; Jacobs and Mason, 2005), as well as in a strong weathering environment (Eger et al., 2012). The available OSL-IRSL ages for the loess-soil deposits of the last glacial and the Holocene (Constantin et al., 2014; Timar-Gabor et al., 2011; Stevens et al., 2011; Fitzsimmons and Hambach, 2014; Markovic et al., 2014; Hatte et al., 2013) suggest that loess accumulation in Europe persisted well into the Early Holocene before the Chernozems started to develop. However, further information on the existence of dust accumulation during their formation is missing.

A characteristic feature of all depth variations of concentration-dependent magnetic parameters is the significant enhancement of the signal in the upper soil horizons. This behavior is reported for all studied Chernozem-like soil profiles worldwide. A summary of the published data is given in Table 2.7.1. Modern soils from the Great Plains (Midwestern United States) studied by Geiss and Zanner (2007), as well as Chernozems from Poland and Ukraine (Jelenska et al., 2008), show a dominant enhancement in the A horizons, while the downward B horizon is less magnetically enriched. If we consider the relative magnetic enhancement of magnetic susceptibility (the ratio of maximum susceptibility in the topsoil horizon divided by the minimum susceptibility of the loess parent material χ_{Amax}/χ_{Cmin}) as a measure of the degree of pedogenesis, it appears that there are generally two groups of sites—one with relatively low magnetic susceptibility of the parent loess (about 25–30 x 10^{-8} m^3/kg) for which the enhancement is about a factor of 3, and a second group of sites where magnetic susceptibility of the parent loess is high, in the interval of 40–120 x 10^{-8} m^3/kg, for which magnetic enhancement is low, in order of 1.2–2.0 (see Table 2.7.1). The latter sites are mainly soils from China, for which modern soil is considered to be formed after 1500 y BP, when prevailing loess accumulation during mid-Holocene ceased (Porter et al., 2001; Huang et al., 2006, 2009). The effect of widespread wildfires on the development of these soils is considered because of the presence of a significant amount of charcoal particles (fire remains). On the other hand, those Chinese profiles for which there is no sign in the magnetic signal for the presence of loess deposition

Table 2.7.1 Pedogenic enhancement of the recent Chernozem-like soils from around the world expressed as a ratio of the maximum magnetic susceptibility (χ_{max}) in the topsoil to the minimum susceptibility (χ_{min}) of the parent loess material. Soil type is indicated according to the authors' classifications

Site	χ_{max}/χ_{min}	χ_{max} (x10^{-8} m^3/kg)	χ_{min} (x10^{-8} m^3/kg)	Soil type	References
		Russian steppe, Poland, and Ukraine			
A99-6	5.4	27	5	Kastanozem	Maher et al. (2003)
A99-5	2.1	50	24	Kastanozem	
A99-10	2.8	75	27	Kastanozem	
A99-11	3.8	95	25	Kastanozem	
Abganerovo	2.3	46	20	Light Kastanozem	Alekseeva et al. (2007)
Peregruznoe	2.3	34	15	Kastanozem	
Kalmikya	3.2	51	16	Kastanozem	
Malyaevka	2.7	40	15	Kastanozem	
Avilov	3.7	59	16	Kastanozem	
MDZ, Poland	2.7	40	15	Nondegraded Chernozem	Gorka-Kostrubiec (2016)
VOD, Poland	4.0	60	15	Calcaro Haplic Chernozem	
KOL, Poland	5.3	80	15	Degraded Chernozem	

Table 2.7.1 Pedogenic enhancement of the recent Chernozem-like soils from around the world expressed as a ratio of the maximum magnetic susceptibility (χmax) in the topsoil to the minimum susceptibility (χmin) of the parent loess material. Soil type is indicated according to the authors' classifications —cont'd

Site	χ_{max}/χ_{min}	χ_{max} (x10^{-8} m^3/kg)	χ_{min} (x10^{-8} m^3/kg)	Soil type	References
H2, Ukraine	3.2	80	25	Typical Chernozem	
Southern Chernozem, Ukraine	1.9	87	45	Southern Chernozem	Menshov and Sukhorada (2012)
H1, Ukraine	4.1	81	20	Humic Chernozem	Jelenska et al. (2008)
A1, Ukraine	4.2	83	20	Saline Kastanozem	
MTS, Ukraine	2.1	52	25	Typical Chernozem	
Zemechy, Czech Republic	1.8	55	30	Brown soil	Hosek et al. (2015)
Great Plains (USA) and Canada					
Nebraska, Boone County	2.0	84	42	Mesic Typic Argiudoll	Geiss et al. (2004)
Miriam Cemetery (MIR 04-A)	1.5	75	50	Holdrege silt loam	Geiss and Zanner. (2007)
Prairie Pines (PRA 02-A)	1.4	63	45	Sharpsburg silt loam	
Mount Calvary Cemetery (MTC 03-A	2.1	75	35	Marshall silty clay loam	
Davisdale Conservation area (DAV 03-A)	2.5	50	20	Winfield silt loam	
Nebraska, Eustis sequence, modern soil	1.9	58	30		Rousseau et al. (2007)
Porcupine loess section, Wyoming, modern soil	1.7	125	75	Haplocryoll	Pierce et al. (2011)
Saskatchewan, Canada, site 9	3.25	65	20	Orthic dark-brown Chernozem	De Jong (2002)
Saskatchewan, Canada, site 10	1.57	58	37	Gleyed red–dark brown Chernozem	

Continued

Table 2.7.1 Pedogenic enhancement of the recent Chernozem-like soils from around the world expressed as a ratio of the maximum magnetic susceptibility (χmax) in the topsoil to the minimum susceptibility (χmin) of the parent loess material. Soil type is indicated according to the authors' classifications —cont'd

Site	χ_{max}/χ_{min}	χ_{max} (x10^{-8} m^3/kg)	χ_{min} (x10^{-8} m^3/kg)	Soil type	References
Chinese loess plateau					
ETC, China	1.3	100	75	Isohumisol	Huang et al. (2009)
XHC, China	1.2	110	90	Isohumisol	
JYC	1.2	140	120	Topsoil	Huang et al. (2006)
XJN	1.6	95	60	Topsoil	
DXF-S	3.3	100	30	Topsoil	
DXF-N	3.2	80	25	Topsoil	
Weinan, CLP	2.1	170	80	So	Dong et al. (2015)
Yaoxian	1.6	110	70	So	
Luochuan	3.2	130	40	So	
Jingchuan	2.0	80	40	So	
Yulin	6.0	30	5	So	
Huanxian	2.4	60	25	So	
Ganzi loess section, Tibetan plateau	3.3	100	30	Modern soil	Hu et al. (2015)
Argentina					
Veronica (undulating Pampa) profile R3	0.71	120	170	Argiudoll	Orgeira et al. (2008) Only well drained soils on loess
Zarate (undulating Pampa) profile SZ	1.4	147	105	Argiudoll	
CAS (undulating Pampa)	1.2	282	226	Typic Argiudoll	Liu et al. (2010)
GAO (undulating Pampa)	1.8	96	53	Vertic Argiudoll	
Loess sequence at Cordoba, Argentina	1.1	320	300		Rouzaut et al. (2012)
Siberia and Alaska					
Kurtak, Siberia, recent Chernozem	0.8	250	300	Chernozem	Evans and Heller (2001)
Halfway House, Alaska	0.5	100	200	Modern soil	Jensen et al. (2016)
Gold Hill, Alaska	0.3	50	150	Modern soil	

Table 2.7.1 Pedogenic enhancement of the recent Chernozem-like soils from around the world expressed as a ratio of the maximum magnetic susceptibility (χmax) in the topsoil to the minimum susceptibility (χmin) of the parent loess material. Soil type is indicated according to the authors' classifications —cont'd

Site	χ_{max}/χ_{min}	χ_{max} ($x10^{-8}$ m^3/kg)	χ_{min} ($x10^{-8}$ m^3/kg)	Soil type	References
			Bulgaria		
TB	3.5	97	28	Calcic Chernozem	Jordanova et al. (2010) and this study
OV	4.3	78	18	Epicalcic Chernozem	
GF	1.7	50	30	Luvic Phaeozem	
Dve mogili	2.5	118	47.5	Leached Chernozem	Jordanova (1996)
Russe, DZS	2.2	132	58.8	Calcic Chernozem	
Basarbovo	2.0	104	51.7	Epicalcic Chernozem	
Pet kladentzi	2.0	81	40.6	Epicalcic Chernozem	Jordanova et al. (1997)
Dve mogili II	2.4	95	39.4	Epicalcic Chernozem	
Harletz	1.9	65	33.4	Calcic Chernozem	Avramov et al. (2006)
Durankulak	2.0	58	28.1	Calcic Chernozem	
Orsoja	1.7	60	35	Calcic Chernozem	
Lubenovo	2.2	62	28.5	Bathicalcic Chernozem	Jordanova et al. (2007)
Viatovo	2.9	104	35.7	Luvic Chernozem	
Koriten	3.3	84	25.6	Leached Chernozem	Jordanova and Petersen (1999)
Kochava	3.6	100	27.6	Epicalcic Chernozem	Unpublished data
Gomotartzi	2.1	65	30.9	Epicalcic Chernozem	
G. Toshevo	3.4	85	25	Bathicalcic Chernozem	

during the mid-Holocene and low susceptibility of the parent material (Hu et al., 2015; profiles DXF-S and DXF-N from Huang et al., 2006) show higher magnetic enhancement. This pattern suggests that (as also mentioned by Geiss and Zanner, 2007) the enhancement is governed not only by climate factors but also by the properties of the parent material.

Another important question concerning the magnetism of Chernozems and Phaeozems is the role of biologically induced magnetic mineralization in the pedogenic fraction. The rare occurrence of magnetosomes of magnetotactic bacteria in the well-drained and oxic soils from China and the Russian steppe are reported (Maher and Thompson, 1995, 1999; Maher et al., 2003), which cannot account for the observed magnetic enhancement (see also Dearing et al., 1996). The obtained depth variations of the magnetic characteristics of the three profiles from Bulgaria (Figs. 2.4.3—2.4.8) provide additional detailed information about the high-resolution (each 2 cm) changes in the effective magnetic grain size of the strongly magnetic fraction. The experimental data suggest the following pattern: Calcic Chernozem (TB) has a prevailing SP contribution in the uppermost levels (the A_k horizon) followed by a higher SD contribution in the deeper levels within the A_k" horizon. The second profile (OV), showing a better-developed horizon differentiation and the presence of the B_k horizon, has a more SD-like magnetic fraction in the upper A horizon, while the SP fraction is more regularly distributed along the $A + B_k$ horizons but has a small maxima at the top of the B_k horizon (see marked $\chi_{fd}/HIRM$ maxima accompanied by χ_{arm}/χ minima in Figs. 2.4.6 and 2.4.7). The strongest grain-size differentiation in the depth distribution of the pedogenic strongly magnetic fraction is observed for the profile of Luvic Phaeozem (profile GF), which is characterized by the presence of a 30-cm-thick eluvial horizon (AE). Here, we have clearly dominating SP and SD fractions in different horizons. A magnetically stable SD-like fraction prevails in the upper A + AE horizons, deduced by the clear maxima of χ_{ARM}/χ, M_{rs}/M_s, and ARM/SIRM in Fig. 2.4.8. The lower AB horizon hosts the most significant concentration of superparamagnetic pedogenic grains, leading to the highest values of χ, χ_{fd} and χ/M_s (Fig. 2.5.1) along the GF profile. The lowermost B_g horizon is still magnetically enhanced but probably with larger PSD grains. Consequently, preferential synthesis and preservation of SP and SD fractions are genetically linked to the soil microenvironment and pedogenic processes in different horizons. In the profile of Calcic Chernozem, where no translocation of substances is evident, the SP fraction dominated the uppermost A horizon. With increasing degree of pedogenically induced processes of leaching and illuviation (degradation) within the soil profiles (e.g., TB → OV → GF), the maximum SP contribution moves downward the profile—in the upper part of the illuvial horizon—while the stable SD fraction is fixed to the upper A (AE) horizons. This observation may suggest a biologically mediated origin of the SD component, as also proposed by Geiss and Zanner (2007); and an inorganic origin of the SP fraction. Another explanation for the prevalence of the SP fraction in deeper B horizons of more-differentiated Phaeozem soils is the translocation of the already synthesized SP grains by clay illuviation, in the case of SP particle absorbance on the clay surfaces.

Further insight into the relative enhancement between the three profiles is provided by Fig. 2.7.1, showing depth variations of the parameters χ_{fd}, χ_{ARM}/χ, and χ_{ARM}/IRM_{100mT}. The latter ratio is considered as the most suitable to represent the contribution of the stable SD fraction, as far as the same ferrimagnetic fraction contributes to both the ARM and IRM remanences (see discussion in Geiss et al., 2008).

The absolute values of frequency-dependent magnetic susceptibility χ_{fd} along the three profiles show a clearly distinctive picture in the uppermost 50 cm of the profiles. As was discussed earlier, the strongest SP enhancement is observed in the TB profile, followed by the OV profile, and a depletion of

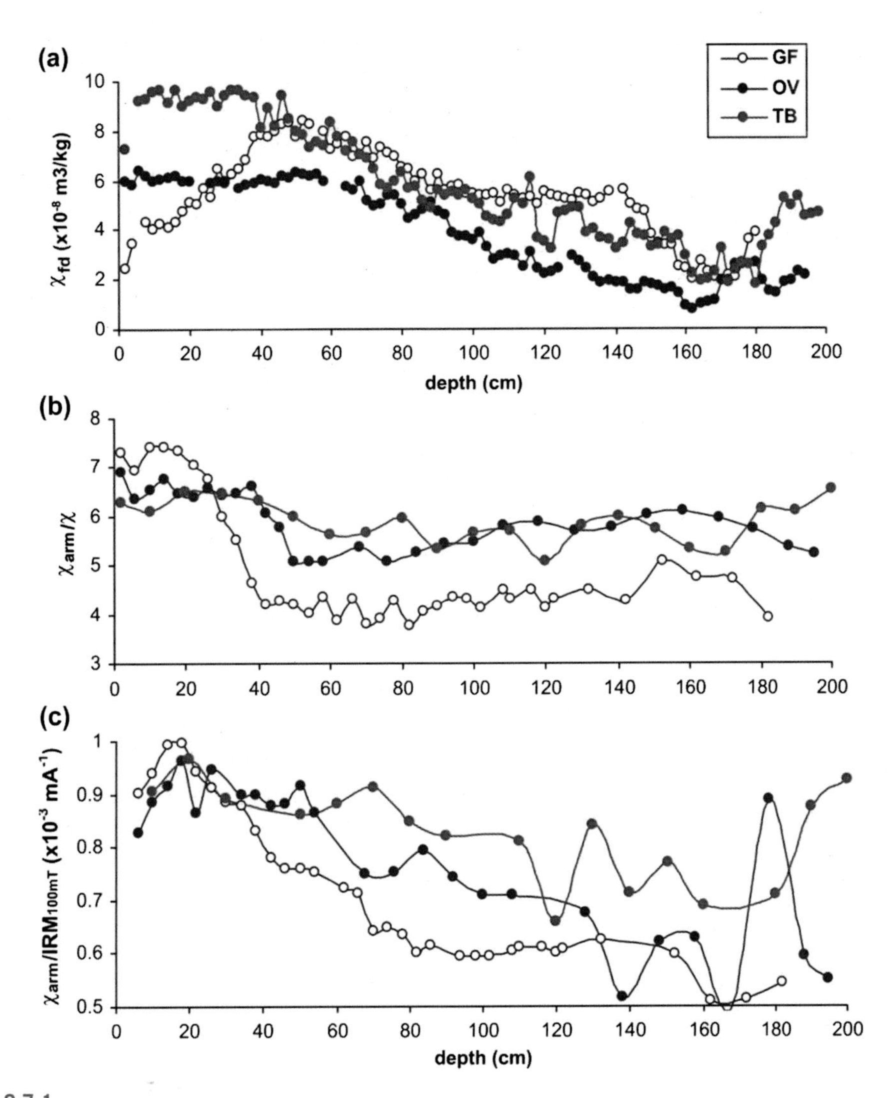

FIGURE 2.7.1

Comparison of the absolute magnetic enhancement of the three studied profiles (TB, OV, and GF): (a) depth variations of the content of superparamagnetic pedogenic fraction, deduced from χ_{fd} values; (b) depth variations of the content of stable SD-like pedogenic fraction, deduced by χ_{ARM}/χ proxy parameter; (c) depth variations of the ratio χ_{ARM}/IRM_{100mT} used by Geiss et al. (2008) as an indicator for the pedogenic magnetic enhancement.

the SP fraction relative to the others is typical for GF in the interval 0–50 cm. The content of SP particles decreases gradually in all profiles toward the parent loess material. The relative proportion of the stable SD fraction represented by the ratio χ_{ARM}/χ along the three profiles shows exactly the opposite enhancement—the highest values are obtained for the upper A + AE horizons of the GF profile, followed by SD enhancement of the relatively thicker (50–60 cm) interval in the OV and TB profiles. The ratio represents the relative abundance of SD (and MD) against the (SP + MD) fraction in the case of uniform mineralogy (Liu et al., 2004). On the other hand, the absolute values of the ARM/IRM_{100mT} ratio would provide an estimate unbiased by the SP presence of the real differences in the grain size of the ARM carriers (Geiss et al., 2008). The significantly higher values observed for the GF profile suggests that the stable pedogenic magnetic fraction dominates there. Similar values of the ratio χ_{ARM}/IRM_{100mT} for the pedogenic component in modern soils from the Midwestern United States is reported by Geiss et al. (2008) and interpreted as a rainfall indicator. Theoretical values of 1.4 x 10^{-3} mA^{-1} for χ_{ARM}/IRM_{100mT} in the case of a single-domain maghemite carrier (Egli and Lowrie, 2002; Geiss et al., 2008) are also consistent with the maximum of the ratio retained in the AE horizon of the profile of Luvic Phaeozem (GF). It should be mentioned that the obtained values for the ratio χ_{ARM}/IRM_{100mT} are related to the bulk material and not to a separated pedogenic magnetic component. Consequently, in the case of the GF profile, all remanence carriers in the upper 40 cm are represented solely by pedogenic single domain maghemite. For the two other Chernozem soils (TB and OV), pedogenic enhancement is more pronounced in the depth variations of the finest SP fraction (Figs. 2.4.4–2.4.6). Having in mind these considerations, we may draw the following conclusions about the magnetic expression of pedogenesis in Chernozems and Phaeozems:

1) The main strongly magnetic pedogenic fraction is represented by maghemite.
2) Magnetic enhancement is caused by "in situ" formation of superparamagnetic to single domain and pseudo–single domain fraction.
3) Profile degradation (e.g., progressive textural and mineralogical differentiation, leaching, and illuviation) is linked to gradual changes in the grain-size distribution of the pedogenic fraction along the depth of the soil profiles. Calcic Chernozems exhibit the strongest enhancement with SP grains compared with more degraded soil subgroups. The concentration of SP particles is the highest in the uppermost levels of A horizons. Going to more-leached soils, SP enhancement is seen preferentially in the bottom of A and the upper part of B_k horizons. Further degradation toward Phaeozems with Luvic features pushes SP enhancement to deeper soil levels—in the illuvial horizon—while at the same time the uppermost A (AE) horizons are enriched with stable SD pedogenic grains.

REFERENCES

Alekseeva, T., Alekseev, A., Maher, B.A., Demkin, V., 2007. Late Holocene climate reconstructions for the Russian steppe, based on mineralogical and magnetic properties of buried palaeosols. Palaeogeogr. Palaeoclimatol. Palaeoecol. 249, 103–127.

Avramov, V., Jordanova, D., Hoffmann, V., Roesler, W., 2006. The role of dust source area and pedogenesis in three loess-palaeosol sections from North Bulgaria: a mineral magnetic study. Studia Geoph. Geodaetica 50, 259–282.

Barrón, V., Torrent, J., 2002. Evidence for a simple pathway to maghemite in Earth and Mars soils. Geochimica et Cosmochimica Acta 66, 2801–2806.

Bertolino, L.C., Rossi, A.M., Scorzelli, R.B., Torem, M.L., 2010. Influence of iron on kaolin whiteness: an electron paramagnetic resonance study. Appl. Clay Sci. 49, 170–175.

Brown, M.E., 1998. Handbook of Thermal Analysis and Calorimetry. In: Principles and Practice, Vol. 1. Elsevier, The Netherlands, pp. 1–145.

Buggle, B., Glaser, B., Zöller, L., Hambach, U., Markovic, S., Glaser, I., Gerasimenko, N., 2008. Geochemical characterization and origin of Southeastern and eastern European loesses (Serbia, Romania, Ukraine). Quat. Sci. Rev. 27, 1058–1075.

Chendev Yu, G., Ivanov, I.V., Pesochina, L.S., 2010. Trends of the Natural Evolution of Chernozems on the East European plain. Eurasian Soil Sci. 43 (7), 728–736.

Constantin, D., Panaiotu, C., Necula, C., Codrea, V., Timar-Gabor, A., 2014. Sar-Osl Dating of Late Pleistocene Loess in Southern Romania Using Fine and Coarse-Quartz. LATE Pleistocene Holocene Climatic Variability Carpathian – Balkan Region 2014. Abstrtacts volume, 18 – 22.

Cornell, R., Schwertmann, U., 2003. The Iron Oxides. Structure, Properties, Reactions, Occurrence and Uses. Weinheim, New York.

Dearing, J.A., Dann, R.J.L., Hay, K., Lees, J.A., Loveland, P.J., Maher, B.A., O'Grady, K., 1996. Frequency-dependent susceptibility measurements of environmental materials. Geophys. J. Int. 127, 228–240.

Djemai, A., Balan, E., Morin, G., Hernandez, G., Labbe, J.C., Muller, J.P., 2001. Behaviour of paramagnetic iron during the thermal transformations of kaolinite. J. Am. Ceram. Soc. 84 (5), 1017–1024.

Dong, Y., Wu, N., Li, F., Huang, L., Wen, W., 2015. Time-transgressive nature of the magnetic susceptibility record across the Chinese loess plateau at the Pleistocene/Holocene transition. PLoS One 10 (7), e0133541. http://dx.doi.org/10.1371/journal.pone.0133541.

Dunlop, D., Özdemir, O., 1997. Rock Magnetism. Fundamentals and frontiers. In: Edwards, D. (Ed.), Cambridge Studies in Magnetism. Cambridge University Press.

Eckmeier, E., Gerlach, R., Gehrt, E., Schmidt, M.W.I., 2007. Pedogenesis of Chernozems in Central Europe—a review. Geoderma 139, 288–299.

Eger, A., Almond, P.C., Condron, L.M., 2012. Upbuilding pedogenesis under active loess deposition in a super-humid, temperate climate—quantification of deposition rates, soil chemistry and pedogenic thresholds. Geoderma 189–190, 491–501.

Egli, R., Lowrie, W., 2002. Anhysteretic remanent magnetization of fine magnetic particles. J. Geophys. Res. 107 (B10), 2209. http://dx.doi.org/10.1029/2001JB000671.

Evans, M.E., Heller, F., 2001. Magnetism of loess:palaeosol sequences: recent developments. Earth-Sci. Rev. 54, 129–144.

Evans, M., Heller, F., 2003. Environmental Magnetism: Principles and Applications of Enviromagnetics. Academic Press, San Diego, CA.

Evlogiev, J., 2006. Pleistocene and Holocene in the Danube Plain. DSc Dissertation. Geological Institute – Bulgarian Academy of Sciences (in Bulgarian), Sofia.

FAO World Soil Resources Reports, 2001. Lecture Notes on the Major Soils of the World. ISSN:0532-0488.

Fitzsimmons, K.E., Hambach, U., 2014. Loess accumulation during the last glacial maximum: evidence from Urluia, southeastern Romania. Quat. Int. 334-335, 74–85.

Górka-Kostrubiec, B., Teisseyre-Jeleńska, M., Dytłow, S.K., 2016. Magnetic properties as indicators of Chernozem soil development. Catena 138, 91–102.

Geiss, C.E., Zanner, C.W., Banerjee, S.K., Joanna, M., 2004. Signature of magnetic enhancement in a loessic soil in Nebraska, United States of America. Earth Planet. Sci. Lett. 28, 355–367.

Geiss, C.E., Zanner, C.W., 2007. Sediment magnetic signature of climate in modern loessic soils from the Great Plains. Quat. Int. 162– 163, 97–110.

Geiss, C.E., Egli, R., Zanner, C.W., 2008. Direct estimates of pedogenic magnetite as a tool to reconstruct past climates from buried soils. J. Geophys. Res. 113, B11102. http://dx.doi.org/10.1029/2008JB005669.

Gerlach, R., Baumewerd-Schmidt, H., van den Borg, K., Eckmeier, E., Schmidt, M., 2006. Prehistoric alteration of soil in the Lower Rhine Basin, Northwest Germany—archaeological, 14C and geochemical evidence. Geoderma 136, 38–50.

Grim, R., Rowland, R., 1942. Differential thermal analysis of clay minerals and other hydrous materials. Part 2 Am. Mineralogist 27, 801–818.

Grimley, D., Arruda, N., 2007. Observations of magnetite dissolution in poorly drained soils. Soil Sci. 172 (12), 968–982.

Gnanaprakash, G., Ayyappan, S., Jayakumar, T., Philip, J., Baldev, R., 2006. Magnetic nanoparticles with enhanced γ-Fe2O3 to α-Fe2O3 phase transition temperature. Nanotechnology 17, 5851–5857.

Haase, D., Fink, J., Haase, G., Ruske, R., Pecsi, M., Richter, H., Altermann, M., Jaeger, K., 2007. Loess in Europe—its spatial distribution based on a European loess map, scale 1:2, 500, 000. Quat. Sci.Rev 26, 1301–1312.

Hanesch, M., Stanjek, H., Petersen, N., 2006. Thermomagnetic measurements of soil iron minerals: the role of organic carbon. Geophys. J. Int. 165, 53–61.

Hatte, C., Gauthier, C., Rousseau, C.C., Antoine, P., Fuchs, M., Lagroix, F., Markovic, S., Moine, O., Sima, A., 2013. Excursions to C4 vegetation recorded in the Upper Pleistocene loess of Surduk (Northern Serbia): an organic isotope geochemistry study. Clim. Past 9, 1001–1014.

Hole, F.D., Nielsen, G.A., 1970. Soil genesis under prairie. In: Schramm, P. (Ed.), Proceedings of the Symposium on Prairie and Prairie Restoration, 3, pp. 28–34. Knox College, Galesberg, IL, Publ.

Hošek, J., Hambach, U., Lisá, L., Grygar, T.M., Horáček, I., Meszner, S., Knésl, I., 2015. An integrated rock-magnetic and geochemical approach to loess/paleosol sequences from Bohemia and Moravia (Czech Republic): implications for the Upper Pleistocene paleoenvironment in central Europe. Palaeogeogr. Palaeoclimatol. Palaeoecol. 418, 344–358.

Hu, P., Liu, Q., Heslop, D., Roberts, A.P., Jin, C., 2015. Soil moisture balance and magnetic enhancement in loess–paleosol sequences from the Tibetan Plateau and Chinese Loess Plateau. Earth Planet. Sci. Lett. 409, 120–132.

Huang, C.C., Pang, J.L., Chen, S.E., Su, H.X., Han, J., Cao, Y.F., Zhao, W.Y., Tan, Z.H., 2006. Charcoal records of wildfire history in the Holocene loess-soil sequences over the southern Loess Plateau of China. Palaeogeogr. Palaeoclimatol. Palaeoecol. 239, 28–44.

Huang, C.C., Pang, J., Su, H., Wang, L., Zhu, Y., 2009. The Ustic Isohumisol (Chernozem) distributed over the Chinese loess plateau: modern soil or palaeosol? Geoderma 150, 344–358.

de Jong, E., 2002. Magnetic susceptibility of Gleysolic and Chernozemic soils in Saskatchewan. Can. J. Soil Sci. 82, 191–199.

Jacobs, P.M., Mason, J.A., 2005. Impact of Holocene dust aggradation on A horizon characteristics and carbon storage in loess-derived Mollisols of the Great Plains, USA. Geoderma 125, 95–106.

Jelenska, M., Hasso-Agopsowicz, A., Kopcewicz, B., Sukhorada, A., Tyamina, K., Kadzialko-Hofmokl, M., Matviishina, Z., 2004. Magnetic properties of polluted and non-polluted soils. A case study from Ukraine. Geophys. J. Int. 159, 104–116.

Jeleńska, M., Hasso-Agopsowicz, A., Kądziałko-Hofmokl, M., Sukhorada, A., Bondar, K., Matviishina, Z., 2008. Magnetic iron oxides occurring in Chernozem soil from Ukraine and Poland as indicators of pedogenic processes. Stud. Geophys. Geod 52, 255–270.

Jelenska, M., Hasso-Agopsowicz, A., Kopcewicz, B., 2010. Thermally induced transformation of magnetic minerals in soil based on rock magnetic study and Mössbauer analysis. Phys. Earth Planetary Interiors 179, 164–177.

Jensen, B.J.L., Evans, M.E., Froese, D.G., Kravchinsky, V.A., 2016. 150,000 years of loess accumulation in central Alaska. Quat. Sci. Rev. 135, 1–23.

Jordanova, D., 1996. Magnetic Studies of Holocene Loess – Soil Sediments from Northeastern Bulgaria. Palaeoclimatic Significance of the Variations of Magnetic Susceptibility. PhD Dissertation. Sofia University "St. Kl. Ohridski", Sofia, in Bulgarian.

Jordanova, D., Petrovsky, E., Jordanova, N., Evlogiev, J., Butchvarova, V., 1997. Rockmagnetic properties of recent soils from North Eastern Bulgaria. Geophys J. Int. 128, 474–488.

Jordanova, D., Petersen, N., 1999. Palaeoclimatic record from loess-soil section in NE Bulgaria. Part I: rock-magnetic properties. Geophys. J. Int. 138, 520–532.

Jordanova, D., Hus, J., Geeraerts, R., 2007. Palaeoclimatic implications of the magnetic record from loess/palaeosol sequence Viatovo (NE Bulgaria). Geophys. J. Int. 171, 1036–1047.

Jordanova, D., Hus, J., Evlogiev, J., Geeraerts, R., 2008. Palaeomagnetism of the loess/palaeosol sequence in Viatovo (NE Bulgaria) in the Danube basin. Phys. Earth Planet. Inter. 167, 71–83.

Jordanova, D., Jordanova, N., Petrov, P., Tsacheva, T., 2010. Soil development of three Chernozem-like profiles from North Bulgaria revealed by magnetic studies. Catena 83 (2–3), 158–169.

Koinov, V., Kabakchiev, I., Boneva, K., 1998. Soil Atlas of Bulgaria. Zemizdat, Sofia, 320p.

Kopcewicz, B., Kopcewicz, M., Jelenska, M., Hasso-Acopsowicz, A., 2005. Moessbauer study of chemical transformations in soil samples during thermomagnetic measurements. Hyperfine Inter. 166 (1–4), 631–636.

Kosterov, A., 2002. Low-temperature magnetic hysteresis properties of partially oxidized magnetite. Geophys. J. Int. 149, 796–804.

Liu, Q., Banerjee, S.K., Jackson, M.J., Maher, B.A., Pan, Y., Zhu, R., Deng, C., Chen, F., 2004. Grain sizes of susceptibility and anhysteretic remanent magnetization carriers in Chinese loess/paleosol sequences. J. Geophys. Res. 109, B03101. http://dx.doi.org/10.1029/2003 JB002747.

Liu, Q., Deng, C., Yu, Y., Torrent, J., Jackson, M.I., Banerjee, S.K., Zhu, R., 2005. Temperature dependence of magnetic susceptibility in an argon environment: implications for pedogenesis of Chinese loess/palaeosols. Geophys. J. Int. 161, 102–112.

Liu, Q., Torrent, J., Morrás, H., Hong, A., Jiang, Z., Su, Y., 2010. Superparamagnetism of two modern soils from the northeastern Pampean region, Argentina and its paleoclimatic indications. Geophys. J. Int. 183, 695–705.

Liu, Q., Roberts, A., Larrasoaña, J., Banerjee, S., Guyodo, Y., Tauxe, L., Oldfield, F., 2012. Environmental magnetism: Principles and Applications. Rev. Geophys. 50, RG4002.

Liu, X., Burras, C.L., Kravchenko, Y.S., Duran, A., Huffman, T., Morras, H., Studdert, G., Zhang, X., Cruse, R.M., Yuan, X., 2012. Overview of Mollisols in the world: distribution, land use and management. Can. J. Soil Sci. 92, 383–402.

Lorz, C., Saile, T., 2011. Anthropogenic pedogenesis of Chernozems in Germany?—a critical review. Quat. Int. 243, 273–279.

Maher, B., 1986. Characterization of soils by mineral magnetic measurements. Phys. Earth Planet. Inter 42, 76–92.

Maher, B.A., Thompson, R., 1995. Paleorainfall reconstructions from pedogenic magnetic susceptibility variations in the Chinese loess and paleosols. Quat. Res. 44, 383–391. http://dx.doi.org/10.1006/qres.1995.1083.

Maher, B., Thompson, R., 1999. Palaeomonsoons I: the magnetic record of palaeoclimate in the terrestrial loess and palaeosol sequences. In: Maher, B., Thompson, R. (Eds.), Quaternary Climates, Environments and Magnetism. Cambridge University Press, Cambridge, UK, pp. 81–125.

Maher, B.A., Yu, H.M., Roberts, H.M., Wintle, A.G., 2003. Holocene loess accumulation and soil development at the western edge of the Chinese Loess Plateau: implications for magnetic proxies of palaeorainfall. Quat. Sci. Rev. 22, 445–451.

Markovic, S.B., Timar-Gabor, A., Stevens, T., Hambach, U., Popov, D., Tomic, N., Obreht, I., Jovanovic, M., Lehmkuhl, F., Kels, H., Markovic, R., Gavrilov, B., 2014. Environmental dynamics and luminescence chronology from the Orlovat loess–palaeosol sequence (Vojvodina, northern Serbia). J. Quat. Sci. 29, 189–199.

Menshov, A.I., Sukhorada, A.V., 2012. Soil magnetism in Ukraine (in Russian). Науковий вісник НГУ (1), 15–22, 2071–2227.

Mullins, C., Tite, M., 1973. Magnetic viscosity, quadrature susceptibility and frequency dependence of susceptibility in single-domain assemblies of magnetite and maghemite. J. Geophys. Res. 78 (5), 804–809.

Murad, E., Wagner, U., 1998. Clays and clay minerals: the firing process. Hyperfine Interact. 117, 337–356.

Navrotsky, A., Mazeina, L., Majzlan, J., 2008. Size-driven structural and thermodynamic complexity in iron oxides. Science 319, 1635–1638.

Nesbitt, H.W., Young, G.M., 1982. Early Proterozoic climates and plate motions inferred from major element chemistry of lutites. Nature 299, 715–717.

Ninov, N., 2002. Soils. In: Kopralev, I., Yordanova, M., Mladenov, C.H. (Eds.), Geography of Bulgaria. Inst. Geography – Bulgarian Academy of Sciences, ForKom, Sofia, pp. 277–303.

Orgeira, M.J., Pereyra, F.X., Vásquez, C., Castañeda, E., Compagnucci, R., 2008. Rock magnetism in modern soils, Buenos Aires province, Argentina. J. South Am. Earth Sci. 26, 217–224.

Özdemir, O., Banerjee, S., 1984. High temperature stability of maghemite (gamma Fe_2O_3). Geophys. Res. Lett. 11, 161–164.

Özdemir, O., Dunlop, D., 1993. Chemical remanent magnetization during gamma FeOOH phase transformations. J. Geophys. Res. 98, 4191–4198.

Pierce, K.L., Muhs, D.R., Fosberg, M.A., Mahan, C.A., Rosenbaum, J.G., Licciardi, J.M., Pavich, M.J., 2011. A loess–paleosol record of climate and glacial history over the past two glacial–interglacial cycles (~150 ka), southern Jackson Hole, Wyoming. Quat. Res. 76, 119–141.

Porter, S.C., Hallet, B., Wu, X., An, Z., 2001. Dependence of near-surface magnetic susceptibility on dust accumulation rate and precipitation on the Chinese loess plateau. Quat. Res. 55, 271–283.

Pye, K., 1987. Aeolian Dust and Dust Deposits. Academic Press. San Diego, California, 334 p. Riccardi.

Richter, M., Messiga, B., Duminuco, P., 1999. An approach to the dynamics of clay firing. Appl. Clay Sci. 15, 393–409.

Richter, D., de, B., Yaalon, D.H., 2012. "The changing Model of soil" Revisited. Soil Sci. Soc. Am. J. 76, 766–778. http://dx.doi.org/10.2136/sssaj2011.0407.

Rodionov, A., Amelung, W., Peinemann, N., Haumaier, L., Zhang, X., Kleber, M., Glaser, B., Urusevskaya, I., Zech, W., 2010. Black carbon in grassland ecosystems of the world. Glob. Biogeochem. Cycles 24, GB3013. http://dx.doi.org/10.1029/2009GB003669.

Rowland, R.A., 1955. Differential thermal analysis of clays and carbonates. In: Pask, G., Turner, J.M.W. (Eds.), Clays and Clay Technology, Bull, 169, pp. 151–163. California, SF.

Rousseau, D.-D., Antoine, P., Kunesch, S., Hatté, C., Rossignol, J., Packman, S., Lang, A., Gauthier, C., 2007. Evidence of cyclic dust deposition in the US Great plains during the last deglaciation from the high-resolution analysis of the Peoria Loess in the Eustis sequence (Nebraska, USA). Earth Planet. Sci. Lett. 262, 159–174.

Rouzaut, S., Orgeira, M.J., Vásquez, C., Ayala, R., Arguello, G.L., Tauber, A., Tófalo, R., Mansilla, L., Sanabria, J., 2012. Rock magnetism in two loess–paleosol sequences in Co'rdoba, Argentina. Environ. Earth Sci. http://dx.doi.org/10.1007/s12665-014-3855-8.

Schaetzl, R., Anderson, A., 2009. Soils. Genesis and Geomorphology. Cambridge University Press, UK, ISBN 978-0-521-81201-6.

Schwertmann, U., 1988. Occurrence and formation of iron oxides in various pedoenvironments. In: Stucki, J., Goodman, B., Schwertmann, U. (Eds.), Iron in Soils and Clay Minerals, NATO ASI Series, Series C: Mathematical and Physical Sciences, vol. 217. Reidel Publications Company, pp. 267–308.

Shishkov, T., Kolev, N., 2014. The Soils of Bulgaria. In: Hartemink, A.E. (Ed.), World Soils Book Series. Springer Science+Business Media Dordrecht, ISBN 978-94-007-7784-2, p. 205 (eBook).

Spaargaren, O., 2008. Phaeozems. In: Chesworth, W. (Ed.), Encyclopedia of Soil Science. Springer, ISBN 978-1-4020-3994-2, pp. 538–539.

Stevens, T., Marković, S.B., Zech, B., Hambach, U., Sümegi, P., 2011. Dust deposition and climate in the Carpathian basin over an independently dated last glacial–interglacial cycle. Quat. Sci. Rev. 30 (5), 662–681.

Timar-Gabor, A., Vandenberghe, D.A.G., Vasiliniuc, S., Panaoitu, C.E., Panaiotu, C.G., Dimofte, D., Cosma, C., 2011. Optical dating of Romanian loess: a comparison between silt-sized and sand-sized quartz. Quat. Int. 240, 62–70.

Worm, H.-U., 1998. On the superparamagnetic - stable single domain transition for magnetite, and frequency dependence of susceptibility. Geophys. J. Int. 133, 201–206.

Ye, X., Lin, D., Jiao, Z., Zhang, L., 1998. The thermal stability of nanocrystalline maghemite Fe2O3. J. Phys. D Appl. Phys. 31, 2739–2744.

CHAPTER

MAGNETISM OF SOILS WITH CLAY-ENRICHED SUBSOIL: LUVISOLS, ALISOLS, AND ACRISOLS

3

3.1 LUVISOLS: MAIN CHARACTERISTICS, FORMATION PROCESSES, AND DISTRIBUTION

The development of a clay-enriched subsoil horizon is the most characteristic feature of the Reference Soil Group Luvisols (IUSS Working Group WRB, 2014). During the pedogenesis, soil may become enriched in clay under the influence of several different processes, such as lessivage (downward leaching of clay particles in water suspension); the formation of new clay minerals in situ in the subsoil, the destruction of the sand and silt fractions, and the preferential erosion of fine particles from the upper soil horizons toward the illuvial (B_t) horizon (Schaetzl and Anderson, 2009; Quenard et al., 2011). In order for a soil to be classified as Luvisol, it is necessary to prove the presence of leaching of the clay fraction from upper soil to the subsoil and the development of an illuvial horizon. As a result of depletion of clay from the uppermost soil horizon, a bleached eluvial (E) horizon gradually develops bellow. Preferential weathering of most unstable primary silicate minerals occurs and the products are exported down, while only most immobile and coarse quartz and potassium feldspar grains remain in the eluviated part. Usually eluvial soil horizons are characterized by more acid soil reaction compared with the underlying illuvial horizons because of the loss of basic cations, as well the acidifying effect of the organics. During the lessivage, three processes take place: mobilization, transport, and deposition (Schaetzl and Anderson, 2009). Therefore, the clay should firstly be dispersed and involved in a water flow, then transported down, and finally settled down by some barrier to its movement. Usually, the enhanced concentration of basic cations (which are abundant in Ca-enriched parent material) is such a barrier. The development of illuvial horizons above Ca-rich parent rocks is due to the ability of carbonates to precipitate clay particles and the dissolution of carbonates in the upper soil horizons because of water penetration. During the initial stages of its pedogenic development, the illuvial horizon becomes increasingly thicker, redder and more clayey. In the climatic conditions of a temperate climate, B_t horizons continue growth until the beginning of the opposite process—degradation, because the surface horizon becomes more and more acidic and the leaching more intense. In such conditions, clay minerals in the upper part of B_t become unstable and transform into other clay minerals, which are usually transported and deposited at the bottom part of B_t.

The formation and transformations of iron oxides in soil proceed under the influence of a number of geochemical reactions. An initial reaction, involving Fe^{3+} ions into the pedogenic processes, is the hydrolysis and oxidation of lithogenic Fe-bearing minerals, mainly Fe^{2+}-silicates. Soils enriched in

Soil Magnetism. http://dx.doi.org/10.1016/B978-0-12-809239-2.00003-6

goethite and hematite develop in a warm and moist climate where well-aerated, oxidative conditions prevail (Schwertmann, 1988). Depending on the interrelation between temperature and moisture, one of these two minerals dominates. Both altitude and latitude influence the relative importance of goethite vs hematite in soils (Schwertmann, 1988). Increasing temperature and decreasing water activity lead to a dominating formation of hematite as a result of ferrihydrite dehydration (Cornell and Schwertmann, 2003). That is the reason why in low and middle geographical latitudes red-colored (hematitic) soils dominate, while yellow-orange (goethitic) soils are more abundant in northern latitudes, where the climate is cold and moist. Differentiation in the form of the iron oxides (goethite vs hematite) can be also found along the depth of a single soil profile, in which surface enriched organics strongly influence iron behavior. Organic matter often builds up humatic complexes with Fe ions, liberated during weathering, which inhibits hematite formation. That is why in leached soils hematite very often dominates in the illuvial horizon, while in the upper (eluvial) horizon goethite is found (Schwertmann, 1988; Cornell and Schwertmann, 2003). In a warm and moist climate (temperate continental and Mediterranean), the main magnetic mineral controlling the soil's magnetic properties is magnetite/maghemite, which, however, makes up only 1% to a few percent of the total mineral content in soil (Maher, 1998; Evans and Heller, 2003).

Worldwide, Luvisols are most widespread in the temperate continental belt—west/central Russia, United States, Europe (and especially the Mediterranean), south Australia, and China (Fig. 3.1.1). They are usually formed on flat/plane topographies with good drainage, in a climate

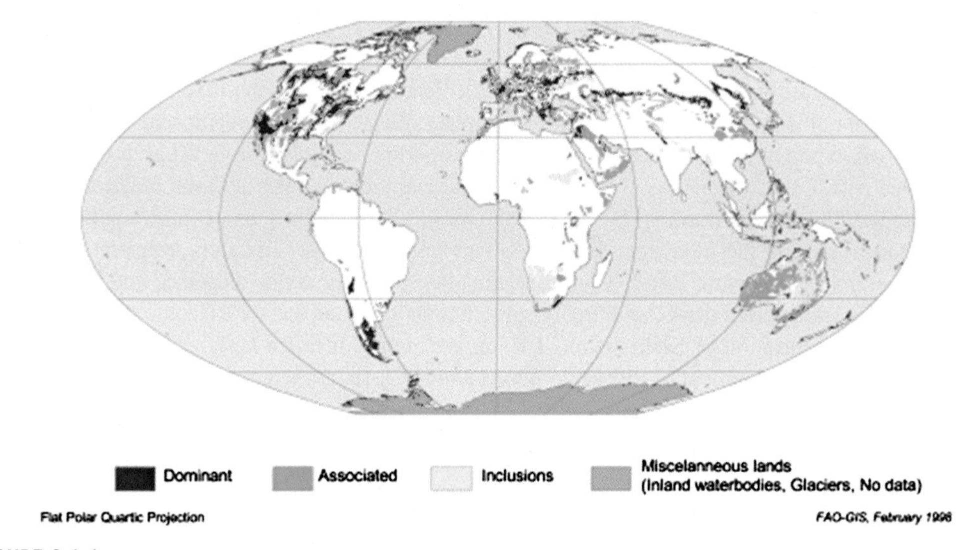

FIGURE 3.1.1

Spatial distribution of Luvisols worldwide (FAO, 2001). Lecture notes on the major soils of the world. http://www.fao.org/docrep/003/y1899e/y1899e00.htm#toc.

with a distinct seasonality of dry and wet periods. Luvisols develop on diverse unconsolidated parent materials, like aeolian, alluvial, colluvial, and glacial deposits (FAO World soil resources reports, 2001). Luvisols are common in Europe, apart from the most northern regions (Tóth et al., 2008; http://eusoils.jrc.ec.europa.eu/wrb/), and they represent the second most widespread soil type in the European Union.

In the Bulgarian soil classification system, Luvisols are typically correlated with cinnamonic forest soils (Shishkov, 2011; Shishkov and Kolev, 2014). They are found most frequently in cenral and southern Bulgaria (Koinov et al., 1998) and cover low undulating forms at the forefront of mountain ranges. Cinnamonic soils in Bulgaria are usually developed on Pliocene and old Quaternary deposits in regions influenced by the Mediterranean climate with alternating moist and dry seasons. Because of the characteristic pedogenic features of cinnamonic soils, they are considered as relict soils, formed during the Pliocene/Quaternary in a warmer and moister climate than today's climate. Climatic changes since their formation influence the pedogenesis of these cinnamonic soils. Three types of cinnamonic soils are distinguished in Bulgaria—typical, leached, and strongly leached (weakly podzolized). The most common is the leached variety, while the typical has a restricted distribution.

3.1.1 DESCRIPTION OF THE PROFILES STUDIED

Four profiles of Luvisols are presented here: two profiles of typical cinnamonic forest soil (Haplic Luvisol, Chromic), one profile of leached cinnamonic forest soil (Epicalcic Luvisol, Chromic), and strongly leached (weakly podzolized) cinnamonic forest soil (Bathycalcic Cutanic Luvisol, Chromic).

1. **Haplic Luvisol, Chromic** (typical cinnamonic forest soil) —**profile designation: CIN**
   Location: N 42°20′54.2″; E 25°08′20.7″, h = 274 m a.s.l.
   Present-day climate conditions: mean annual precipitation (MAP) = 529 mm; mean annual temperature (MAT) = 11.5°C, vegetation cover: broadleaf forest (hornbeam); relief: flat.
   Profile description: (general view and details from the horizons are shown in Figs. 3.1.2–3.1.5.)
   A_k horizon: 0–10 cm, humic horizon, brown-red color, dense, clayey, presence of rock fragments, vegetation roots and carbonate concretions
   $B_{tk}1$ horizon: 10–38 cm, illuvial horizon, red-brown color, heavy clayey, dense, presence of rock fragments, vegetation roots and carbonate concretions
   $B_{tk}2$ horizon: 38–60 cm, illuvial horizon, light brown color, clayey, carbonate concretions and rock fragments present, transition with $B_{tk}1$—sharp.
   BC_k horizon: 60–78 cm, transitional horizon, beige color with orange and grayish-green mottles, loose texture, abundant rock fragments
   C_k horizon: C horizon, light beige color, mottled, abundant rock fragments
   Parent rock: marls with chert

FIGURE 3.1.2

View of the profile of Haplic Luvisol (profile CIN).

FIGURE 3.1.3

Profile CIN. A_k horizon—detail.

FIGURE 3.1.4

Profile CIN. B_{tk} horizon—detail.

FIGURE 3.1.5

Profile CIN. C_k horizon—detail.

2. **Haplic Luvisol, Chromic** (typical cinnamonic forest soil) —**profile designation: PRV**
   Location: N 42°20′50.0″; E 25°07′48.0″, h = 306 m a.s.l.
   Present day climate conditions: MAP = 529 mm, MAT = 11.5°C, vegetation cover: broadleaf forest (hornbeam); relief: slight slope.
   Profile description: (general view and details from the horizons are shown in Figs. 3.1.6–3.1.10.)
   - A horizon: 0–18 cm, humic horizon, brown color, crumby structure, vegetation roots, and rock fragments
   - B_{tk} horizon: 18–80 cm, illuvial horizon, red-brown color, clayey, dense, rock fragments, and small (1- to 2-mm diameter) Fe-Mn concretions
   - BC_k horizon: 80–100 cm, transitional horizon, motley yellow-brown color, big rock pieces
   - C_k horizon: 80–100 cm, weathered parent rock
   - Parent rock: marls

FIGURE 3.1.6

View of the profile of Haplic Luvisol (profile PRV).

FIGURE 3.1.7

Profile PRV. A horizon—detail.

FIGURE 3.1.8

Profile PRV. $B_{tk}1$ horizon—detail.

FIGURE 3.1.9

Profile PRV. $B_{tk}2$ horizon—detail.

FIGURE 3.1.10

Profile PRV. C_k horizon—detail.

3. **Epicalcic Luvisol, Chromic** (leached cinnamonic forest soil) —**profile designation: F**
Location: N 42°12′43.1″; E 27°04′20.3″, h = 314 m a.s.l.
Present day climate conditions: MAP = 660 mm, MAT = 12.5°C, vegetation cover: broadleaf forest; relief: flat.
Profile description: (general view and details from the horizons are shown in Figs. 3.1.11–3.1.15.)
A_h horizon: 0–3 cm, humic horizon, black, loose, abundant organics
AE1 horizon: 3–10 cm, eluvial horizon, gray-brown, bleached, presence of small rock pieces
AE2 horizon: 10–29 cm, eluvial horizon, light beige, loose, small rock pieces
B_t horizon: 29–60 cm, illuvial horizon, dense, clayey, bright brown-red color, prismatic structure, vegetation roots, and carbonates
BC horizon: transitional horizon, dense, brown-red color, rock pieces, and carbonates
C_k horizon: 80–90 cm, C horizon, brown-red color, weathered rock, and rock fragments
Parent rock: metamorphosed Paleozoic sandstones

FIGURE 3.1.11

View of the profile of Epicalcic Luvisol (profile F).

FIGURE 3.1.12

Profile F. A horizon—detail.

FIGURE 3.1.13

Profile F. B_t horizon—detail.

FIGURE 3.1.14

Profile F. BC$_k$ horizon—detail.

FIGURE 3.1.15

Profile F. C$_k$ horizon—detail.

4. **Bathycalcic Cutanic Luvisol (Chromic)** (strongly leached cinnamonic forest soil) **profile designation: BGU**
   Location: N 42°59′20.0″; E 23°40′57.3″, h = 435 m a.s.l.
   Present-day climate conditions: MAP = 830 mm, MAT = 11°C, vegetation cover: broadleaf forest; relief: moderate slope
   Profile description: general view and details from the horizons are shown in Figs. 3.1.16–3.1.19.)
   AE horizon: 0–20 cm, humic-eluvial horizon, light beige color, bleached and loose, abundant Fe-Mn concretions
   B1 horizon: 20–80 cm, illuvial horizon, dark brown, clayey, prismatic structure, carbonates
   B2 horizon: 80–130 cm, illuvial horizon, dark brown, more carbonates than in B1, orange mottles, and rock fragments
   BC_k horizon: 130–170 cm, transitional horizon, clayey, mottled brown-orange color, abundant rock fragments
   Parent rock: granodiorites, lower carbon

FIGURE 3.1.16

View of the profile of Bathycalcic Cutanic Luvisol (profile BGU).

FIGURE 3.1.17

Profile BGU. AE horizon—detail.

FIGURE 3.1.18

Profile BGU. B1 horizon—detail.

FIGURE 3.1.19

Profile BGU. BC_k horizon—detail.

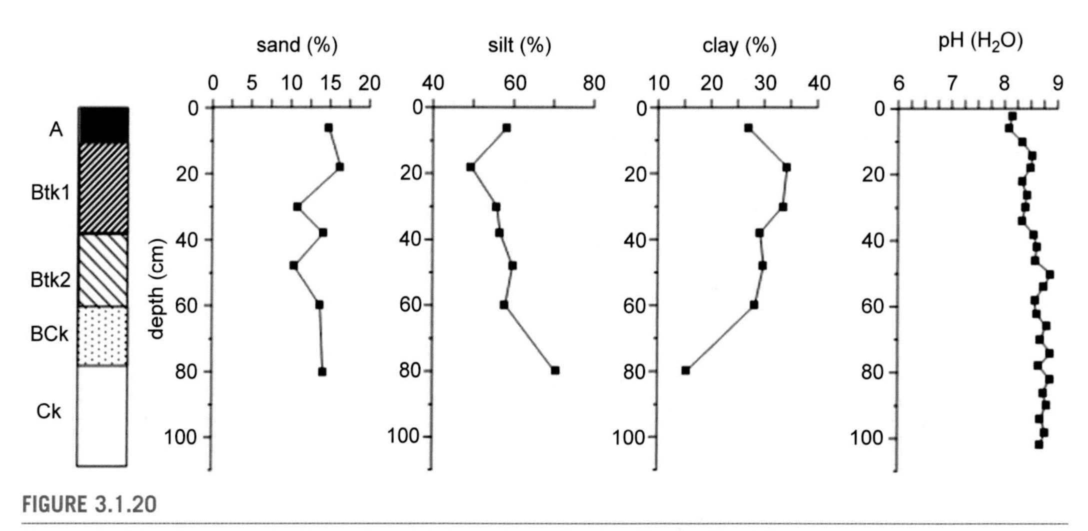

FIGURE 3.1.20

Mechanical fractions (sand, silt, clay) and soil reaction pH along the profile CIN.

3.1.2 TEXTURE, SOIL REACTION, AND MAJOR GEOCHEMISTRY OF LUVISOLS

Results from the texture analysis of selected samples from different genetic horizons of the CIN (Haplic Luvisol) and F (Epicalcic Luvisol) profiles are shown in Figs. 3.1.20 and 3.1.21. The most important diagnostic feature—clay enhancement in the illuvial horizons—is evidenced by the clay variations along both profiles. Clay depletion in eluvial horizons (AE1 and AE2) in the leached profile (F) is very well expressed, while in the Haplic Luvisol (CIN) clay content in the A horizon is only a little less than in the underlying B_t horizon (Fig. 3.1.20). Although clay variations along the profiles are well expressed, the maximum clay content is ~20–35% of the bulk mechanical composition. The silt fraction dominates the texture, making up 60–80% of the total composition. The uppermost clay-depleted horizon of the leached soil (profile F) is dominated by the silt fraction (Fig. 3.1.21), while for the CIN profile, the percentage of silt weakly increases in depth.

The sand fraction is of minor importance in both profiles, accounting for ~10–15% of the total composition, and does not vary along the depth. Therefore, the obtained texture variations in the profiles of Haplic Luvisol and Epicalcic Luvisol conform to the main classification features for Luvisols (Schaetzl and Anderson, 2009).

Soil reaction (pH) depends mainly on the parent material's nature but also on the type of the vegetation cover and its interaction with the mineral part of the soil, as well as possible podzolization processes. Soil reaction is alkaline (pH 8) in the upper part of the CIN profile (Fig. 3.1.20) and even continues to increase toward the C horizon. In contrast, pH along the leached Luvisol (profile F, Fig. 3.1.21) shows well-expressed minima (but reaching only near-neutral pH values), coinciding with the occurrence of the eluvial horizon, in accordance with the theories for eluviation/illuviation (Schaetzl and Anderson, 2009). A similar feature is underlined in the profile of weakly podzolized soil (profile BGU, Fig. 3.1.22), where the eluvial horizon is characterized by a sharp minimum in pH as well, reaching already acidic values (pH 4.5) in contrast to profile F. Afterward, it increases to pH 7 in depth.

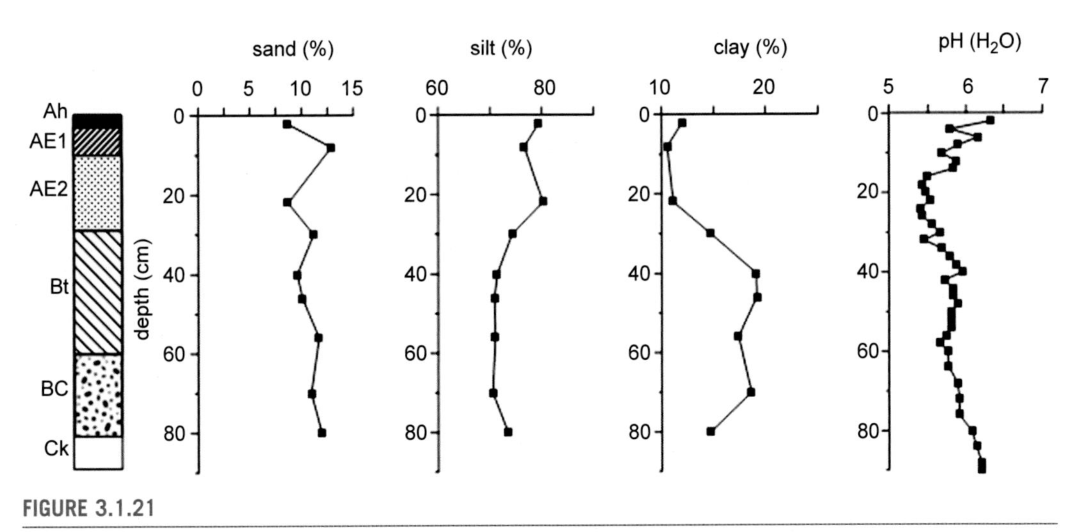

FIGURE 3.1.21

Mechanical fractions (sand, silt, clay) and soil reaction pH along the profile F.

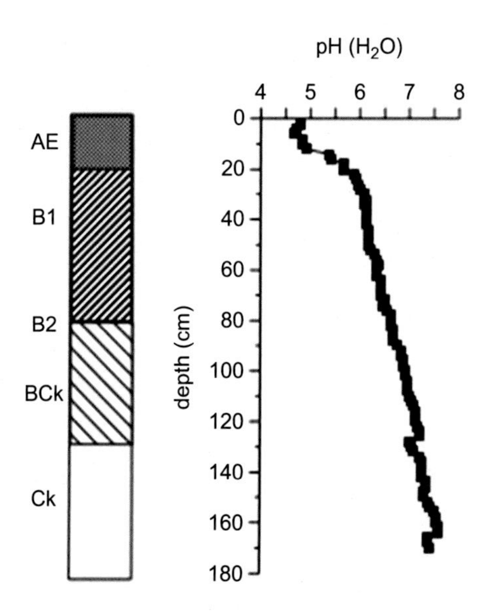

FIGURE 3.1.22

Soil reaction pH along the profile BGU.

Depth variations in soil texture and soil reaction for the Luvisols considered demonstrate progressively advancing processes of leaching, eluviation, and illuviation, which lead to the development of Haplic Luvisol, Epicalcic, and Bathycalcic Cutanic Luvisol.

For selected depths of the CIN (Haplic Luvisol) and BGU (Bathycalcic Cutanic Luvisol) profiles, the elemental composition is investigated through X-ray fluorescence (XRF) analysis. Figs. 3.1.23 and 3.1.24 illustrate variations of some of the measured elements.

The profile of Haplic Luvisol (CIN) (Fig. 3.1.23) is characterized by enhanced content of most of the major elements (Mg, K, Si, Al, Fe, Mn, Ti, and Zr) in A_k and $B_{tk}1$ horizons, which reveals the occurrence of significant pedogenic development of the upper part of the profile without eluviation. C_k horizon is enriched in Na, K, Ca, Si, and Al, thus reflecting the parent rock composition. Iron (Fe) and Al, which are also involved in oxide forms, show coherent variations along the depth, underlining the commonly observed link between both elements in iron oxide mineralogy of soils (Schwertmann, 1988). Usually Ti and Zr are considered as immobile elements and are used as indicators for the uniformity of the parent material (Fitzpatrick and Chittleborough, 2002; Stiles et al., 2003a,b). As can be seen in the CIN profile (Fig. 3.1.24.), both elements—Ti and Zr—are also enhanced in the upper horizons, while their ratio (Ti/Zr) retains a relatively constant value along the profile with a decrease only in the C_k horizon. Its constant value is regarded as an indication of the uniformity of the soil parent material (Reheis, 1990; Stiles et al., 2003a,b). It cannot be unambiguously stated that this behavior in the CIN profile is caused by differences in the parent material, as far as the data shown correspond to bulk soil analysis. It is, however, proved that often Ti and Zr can also reside in the fine-grained (pedogenic) fraction, not only in coarse lithogenic grains (Stiles et al., 2003a,b; Scheldon and Tabor, 2009).

Bathycalcic Cutanic Luvisol (profile BGU), having a well-developed eluvial horizon (see Figs. 3.1.18 and 3.1.22), exhibits an elemental composition, reflecting its main pedogenic characteristics. As it is evident from Fig. 3.1.24, the eluvial horizon is distinguished by a low content of certain

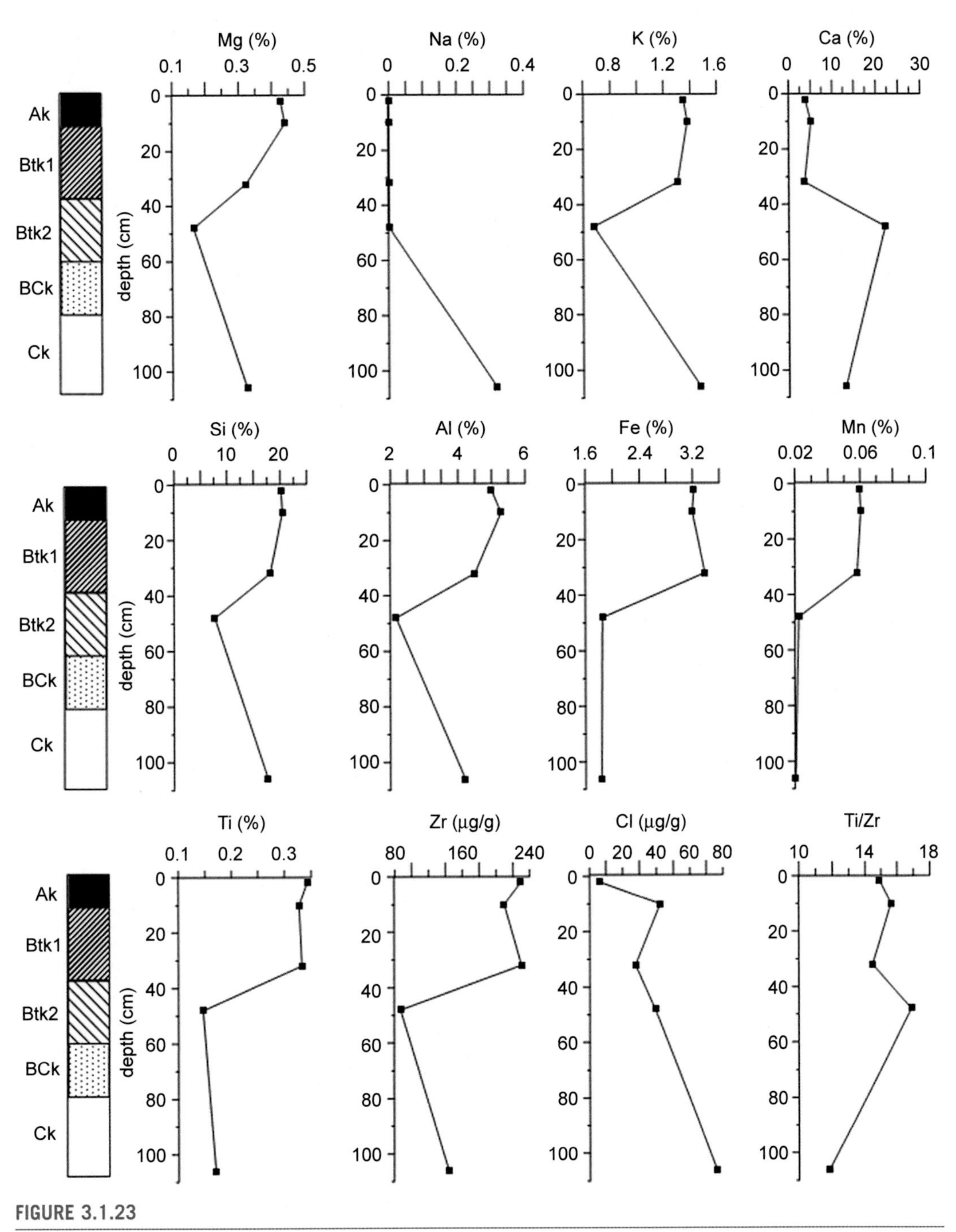

FIGURE 3.1.23

Content of some major and trace elements (Mg, Na, K, Ca, Si, Al, Fe, Mn, Ti, Zr, Cl) and the ratio Ti/Zr along depth of the profile of Haplic Luvisol (profile CIN) for selected samples.

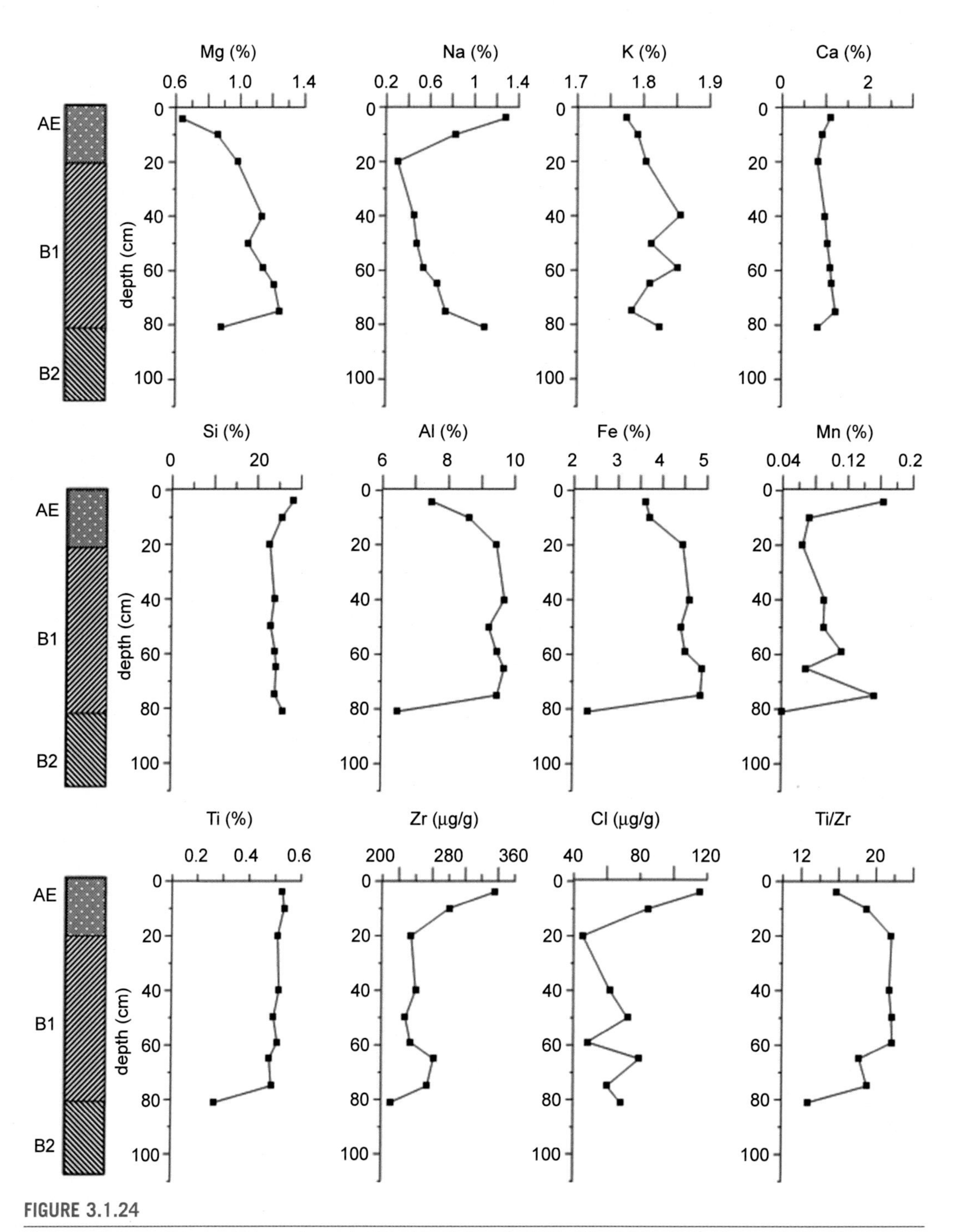

FIGURE 3.1.24

Content of some major and trace elements (Mg, Na, K, Ca, Si, Al, Fe, Mn, Ti, Zr, Cl) and the ratio Ti/Zr along depth of the profile of Bathycalcic Cutanic Luvisol (profile BGU) for selected samples.

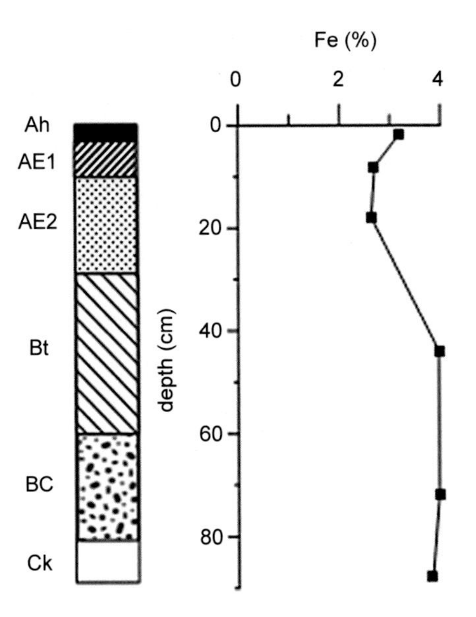

FIGURE 3.1.25

Content of iron (Fe%) along the profile of Epicalcic Luvisol (profile F), determined for selected samples.

elements—Mg, Al, and Fe—and higher contents of Na, Mn, Zr, and Cl. Depleted Fe and Al are signs of intense leaching processes. The latter leads to the destruction of the base minerals and their migration down-profile. Manganese enhancement in the AE horizon is due to the presence of abundant Mn concretions in this part of the profile (see the profile description). Variations in the ratio Ti/Zr cannot be uniquely assigned to the outer source of deposited material, since only Zr is enriched in the AE horizon.

For the leached Luvisol (profile F), only Fe content is determined for selected depths (Fig. 3.1.25). Its variations are also linked to the genetic soil horizons, showing a minimum in the eluvial horizons.

Chemical extractions in oxalate and dithionite (McKeague and Day, 1966; Mehra and Jackson, 1960) are fundamental methods in pedology for the determination of amorphous and crystalline forms of Fe, Al, and Mn (Cornell and Schwertmann, 2003). Table 3.1.1 presents the results for oxalate-

Table 3.1.1 Chemical extraction data for selected samples from CIN profile (Haplic Luvisol)

Sample depth (cm)	Al_o (g/kg)	Fe_o (g/kg)	Mn_o (g/kg)	Fe_d (g/kg)	Fe_{tot} (g/kg)	Fe_o/Fe_d	Fe_d/Fe_{tot}
2	1.583	1.119	0.328	8.559	32.21	0.131	0.267
10	1.439	1.107	0.330	8.890	31.97	0.124	0.278
32	1.613	1.113	0.358	9.106	33.85	0.122	0.269
48	0.276	0.000	0.022	4.100	18.52	0.00	0.221
106	0.494	0.000	0.037	2.131	18.37	0.00	0.116

Oxalate-extractable aluminum (Al_o), Iron (Fe_o) and manganese (Mn_o); dithionite-extractable iron (Fe_d), total Fe content (Fe_{tot}), and the ratios Fe_o/Fe_d and Fe_d/Fe_{tot}.

extractable Fe, Al, and Mn [oxalate-extractible iron (Fe_o), aluminum (Al_o), and manganese (Mn_o)] and dithionite-extractible Fe (Fe_d) for selected samples from profile CIN.

Amorphous forms of Al, Mn, and Fe are found only in the upper A_k and $B_{tk}1$ horizons, while no Fe_o is present in the deeper horizons and only minor amounts of Mn_o and Al_o are detected. Accordingly, better crystalline iron compounds, as reflected by the Fe_d content (Table 3.1.1), are in much higher amounts, compared with amorphous ones. The values of the ratios Fe_o/Fe_d and Fe_d/Fe_{tot}, used as indicators of soil pedogenic development and maturity in well-aerated climate conditions (Schwertmann, 1988) also indicate that the A_k and $B_{tk}1$ horizons are well developed and enhanced with crystalline iron oxides. An Fe_d/Fe_{tot} ratio of $\sim$0.27 is a typical value for soil developed in a temperate continental climate (Cornell and Schwertmann, 2003).

3.1.3 MAGNETIC PROPERTIES OF LUVISOLS

Haplic Luvisol, profile CIN

Thermomagnetic analysis (high-temperature behavior of magnetic susceptibility) is used for determining the kind of iron-containing minerals that dominate the magnetic signal. Fig. 3.1.26 shows

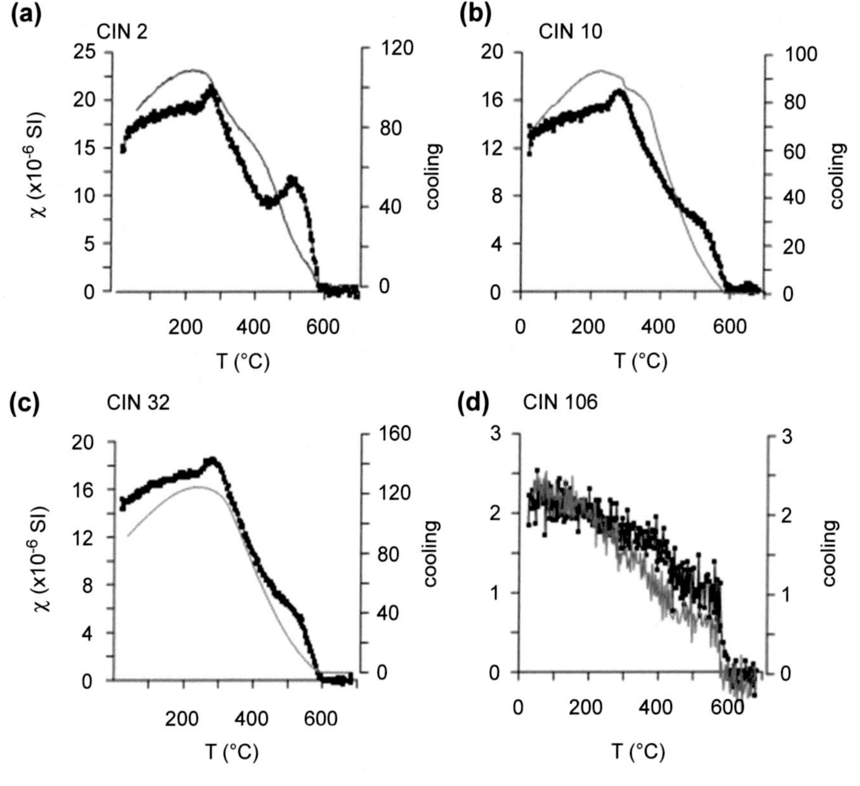

FIGURE 3.1.26

Thermomagnetic analyses of magnetic susceptibility for samples from different horizons of profile CIN. *Thick dotted line*—heating run, *thin line*—cooling run. Heating in air, fast heating rate (11°C/min). (a) Sample from A-horizon, (b and c) samples from $B_{tk}1$ horizon, and (d) sample from C_k horizon.

several examples of bulk soil analyses for selected depths. Magnetic susceptibility (χ) on heating shows similar behavior for samples from different soil depths, except for sample CIN 106, belonging to the C_k horizon (Fig. 3.1.26d). For soil samples from the upper horizons, $\chi(T)$ curves are characterized by an initial increase in χ up to ~300°C, followed by well-expressed small maximum at 300–320°C and a further sharp drop in χ until 400°C (Fig. 3.1.26a–c). After this significant drop in χ, a magnetite phase with T_c ~580°C is detected in all samples, more (sample CIN2) or less clearly expressed (samples CIN10 and CIN32). Such types of thermomagnetic heating curves are often obtained for soils, developed in the temperate continental climate in Europe (Liu et al., 2010; Chlupacova et al., 2010) and in China (Liu et al., 2005). One widely accepted interpretation of the observed variations in magnetic susceptibility is that pedogenic magnetite particles with a maghemite shell are initially present in the soil. During heating to 300°C, an effective increase in grain size because of stress release between the core and the shell (e.g., Van Velzen and Zijderveld, 1995) causes the observed χ increase, followed by the peak, corresponding to maghemite. A subsequent sharp χ decrease is linked to maghemite's inversion to hematite (Dunlop and Özdemir, 1997). A magnetite phase with $T_c = 580°C$ (Fig. 3.1.26) can be due to both (1) initially present magnetite (pedogenic and/or lithogenic) in the soil and (2) transformation-driven magnetite production as a result of reduction of hematite in the presence of organic matter or destruction of Fe-containing clay minerals (Zhang et al., 2012). It is possible to reduce this uncertainty by considering the thermomagnetic curve of the sample from the C horizon (Fig. 3.1.26d). It clearly shows the presence of magnetite with T_c of 580°C, implying that at least some part of the magnetite signal in upper soil depths is due to an initially existing lithogenic magnetite. On the other hand, taking into account the significantly stronger magnetic signal in the A_k and B_{tk} horizons, it could be supposed that pedogenic magnetite additionally contributes to this magnetic phase. Therefore, pedogenic magnetite and maghemite, as well as lithogenic magnetite, are the dominant magnetic minerals in the CIN profile.

Variations of different concentration-dependent magnetic parameters along the profile of Haplic Luvisol CIN are shown in Fig. 3.1.27. Magnetic susceptibility (χ), which depends mainly on the concentration of the strongly magnetic minerals, has high values in the upper A_k and $B_{tk}1$ horizons and gradually decreases in $B_{tk}2$ (Fig. 3.1.27). A similar behavior is exhibited by saturation magnetization (M_s) and remanences—anhysteretic remanence (ARM), isothermal remanence acquired in 2-T field (IRM_{2T}) high-field magnetic susceptibility (χ_{hf}), related mainly to the concentration of paramagnetic Fe-bearing clay minerals, varies coherently with the other concentration-dependent parameters. Therefore, it could be supposed that the concentration-dependent magnetic characteristics reflect pedogenic enhancement with iron oxides and clay minerals of the upper soil horizons.

As far as ARM is acquired most effectively by fine single-domain (SD) grains of magnetite/maghemite, while IRM_{2T} is acquired by coarser pseudo-SD (PSD) and multidomain (MD) grains (Maher, 1988; Dunlop and Özdemir, 1997; Evans and Heller, 2003), the variations in remanence parameters point to an enhancement of the A_k and $B_{tk}1$ horizons with SD, PSD, and MD magnetite/maghemite. Variations in the percent frequency-dependent magnetic susceptibility ($\chi_{fd}\%$) (Fig. 3.1.28c) reveal a significant portion of very fine superparamagnetic (SP) grains spread along the A_k, $B_{tk}1$, and $B_{tk}2$ horizons, while in BC_k it decreases to only 2%, indicating the presence of a very small amount of SP particles (Dearing et al., 1996). High values of $\chi_{fd}\%$ in the C_k horizon are probably due to measurement uncertainty regarding the Bartington MS2B dual-frequency sensor. Grain-size proxy ratios ARM/IRM_{2T} and χ_{ARM}/χ are used for detecting changes in the grain size of the SD-like remanence carriers, based on the established grain-size trends of ARM, IRM, and χ (e.g.,

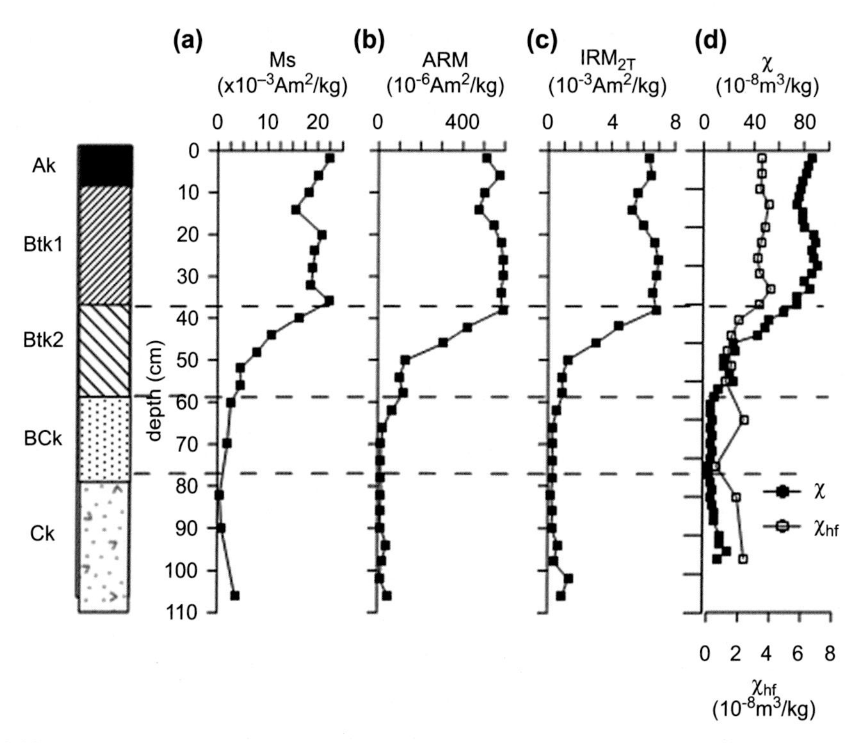

FIGURE 3.1.27

Soil profile CIN. Depth variations of concentration-dependent magnetic characteristics: (a) saturation magnetization (M_s), (b) anhysteretic remanence (ARM), (c) isothermal remanence, acquired in a field of 2 T (IRM_{2T}), (d) magnetic susceptibility (χ), and high-field magnetic susceptibility (χ_{hf}).

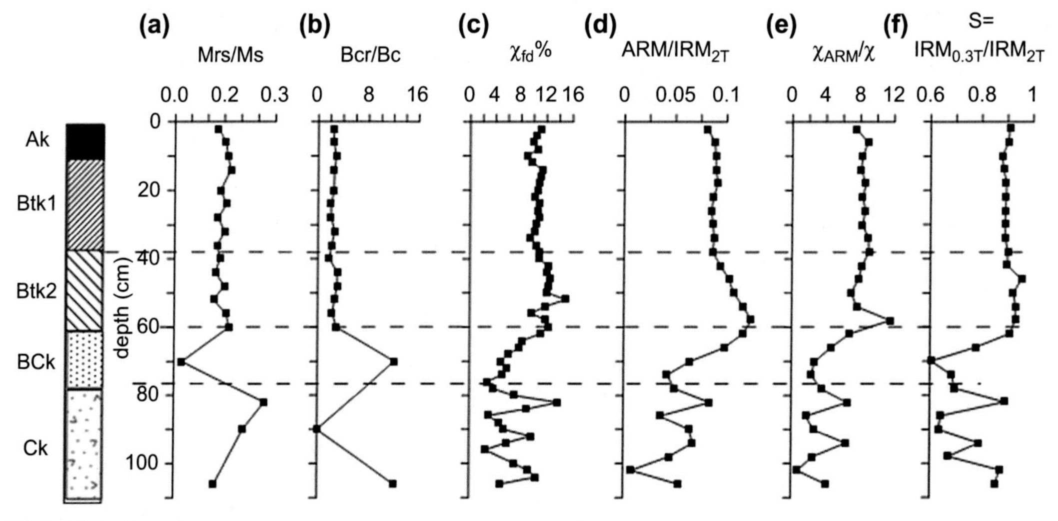

FIGURE 3.1.28

Soil profile CIN. Depth variations of grain-size–sensitive magnetic characteristics and ratios: (a) M_{rs}/M_s ratio, (b) B_{cr}/B_c ratio, (c) percent frequency dependent susceptibility $\chi_{fd}\%$, (d) ARM/IRM_{2T} ratio, (e) χ_{ARM}/χ ratio, and (f) S-ratio, calculated as: $IRM_{0.3T}/IRM_{2T}$.

Maher, 1988; Oldfield, 1999; Peters and Dekkers, 2003). Their behavior along the CIN profile (Fig. 3.1.28d and e) suggests that fine SD particles are abundant in the A_k, $B_{tk}1$, and $B_{tk}2$ horizons with the most SD-like grain size, linked to the bottom part of the $B_{tk}2$ horizon, where both ratios exhibit a maximum. Hysteresis ratios saturation remanent magnetization (M_{rs}) to saturation magnetization (M_s) (M_{rs}/M_s) and coercivity of remanence (B_{cr}) to coercive force (B_c) (B_{cr}/B_c) (Fig. 3.1.28a and b) point to uniform effective magnetic grain-size all along A_k, $B_{tk}1$ and $B_{tk}2$ and coarsening in the deeper parts of the profile. The S ratio (Fig. 3.1.28f) confirms a dominating magnetite-type mineral in the upper soil horizons (A_k, $B_{tk}1$, and $B_{tk}2$) and the appearance of high-coercivity minerals (probably hematite) in the transitional and the C_k horizons. This is also supported by the hysteresis ratios variations. Therefore, taking into account the variations of magnetic concentration and grain-size proxies, it could be concluded that pedogenic enhancement extends up to $B_{tk}2$ horizon. The finest pedogenic grains with significantly lower concentration are detected in $B_{tk}2$, as compared with the A_k and $B_{tk}1$ horizons.

Haplic Luvisol, profile PRV

Thermomagnetic analyses for selected samples from the second Haplic Luvisol—profile PRV (Fig. 3.1.29) show a general similarity with the curves, obtained for profile CIN (Fig. 3.1.26). The

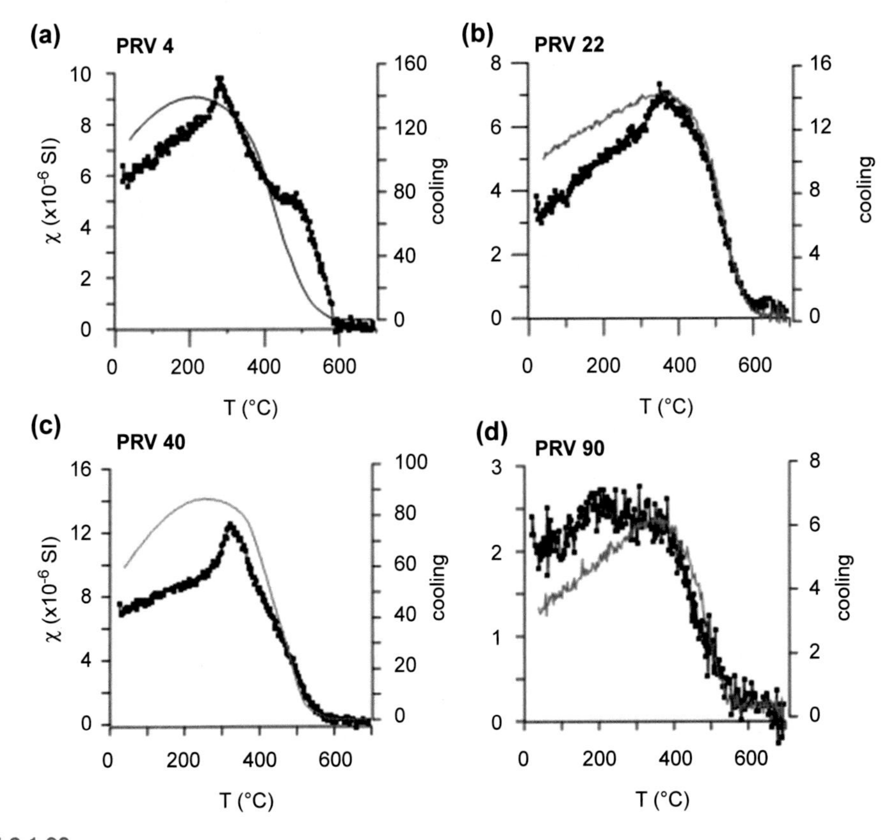

FIGURE 3.1.29

Thermomagnetic analyses of magnetic susceptibility for samples from different horizons of profile PRV. *Thick dotted line*—heating run, *thin line*—cooling run. Heating in air, fast heating rate (11°C/min). (a) Sample from A_k horizon, (b) sample from $B_{tk}1$ horizon, (c) sample from $B_{tk}2$ horizon, and (d) sample from C_k horizon.

"maghemite" peak at 300–320°C is sharper, compared with soil samples from the CIN profile. The magnetite component is clearly separated only on the χ(T) heating curve for the surface sample. Cooling curves again demonstrate intense thermal transformations in the magnetic mineralogy and the creation of a new strongly magnetic fraction on cooling. Thermomagnetic analysis for the sample from the BC_k horizon (Fig. 3.1.29d) reveals the presence of lithogenic magnetite as well as hematite with $T_N = 700$°C. Therefore, the thermomagnetic analyses show similar pedogenic magnetic minerals in both profiles of Haplic Luvisol, represented by magnetite and maghemite.

IRM stepwise acquisition up to the 5-T field has been carried out for selected samples from the profile (Fig. 3.1.30), and the obtained curves were analyzed using IRM-CLG1.0 software (Kruiver et al., 2001). Table 3.1.2 summarizes the obtained parameters of the separated coercivity components. The soft magnetic component dominates IRM and makes up 83–90% of the total IRM at 5 T. The coercivity of this component is relatively low (27–33 mT) and, considering the obtained DP values, corresponds to a wide grain-size distribution of the magnetic fraction. The high-coercivity component accounts for only 10–17% of IRM, and its coercivity corresponds to that of hematite (Table 3.1.2).

Magnetic parameters used as proxies for changes in the concentration of Fe oxides along the PRV profile display magnetic enhancement in the A horizon and the upper half of the B_{tk} horizon (Fig. 3.1.31a). The bottom part of B_{tk} is depleted in magnetic minerals and a minimum concentration is obtained in the BC_k and C_k horizons. Magnetic susceptibility varies in a similar way as in profile CIN, with a maximum value in the upper part of B_{tk}. Concentration of SD-like grains, registered through the intensity of ARM (Fig. 3.1.31b), shows a maximum at the bottom of A_k and the upper part of B_{tk}, while coarser grains, carrying IRM_{2T} more effectively, have a constant concentration along this depth interval (Fig. 3.1.31c). Since thermomagnetic analysis and IRM acquisition experiments showed the presence of hematite, a ratio χ_{fd}/HIRM, [hard isothermal remanent magnetization (HIRM) = (SIRM + $IRM_{-0.3T}$)/2 (Thompson and Oldfield, 1986)], is used for the estimation of

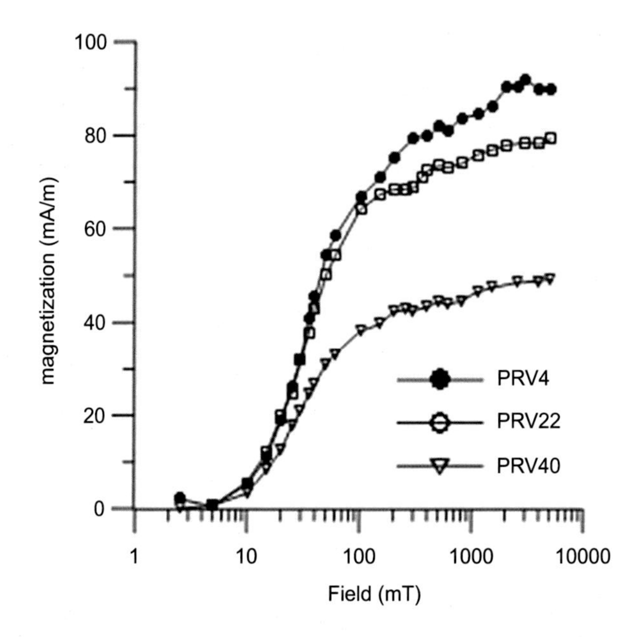

FIGURE 3.1.30

Stepwise acquisition of isothermal remanence (IRM) up to field of 5 T for selected samples from PRV profile.

Table 3.1.2 Summary parameters from coercivity analysis by fitting IRM acquisition curves with cumulative Gaussian functions, using IRM-CLG1.0 software (Kruiver et al., 2001) for selected samples from the PRV profile

Sample depth (cm)	IRM component 1				IRM component 2			
	Intensity (10^{-3} Am^2/kg)	% from IRM_{total}	$B_{1/2}$ (mT)	DP	Intensity (10^{-3} Am^2/kg)	% from IRM_{total}	$B_{1/2}$ (mT)	DP
Profile PRV, Haplic Luvisol								
4	5.792	84	33.10	0.35	1.513	16	457.1	0.42
22	5.441	90	33.10	0.35	1.093	10	457.1	0.42
40	3.337	83	27.50	0.35	1.093	17	457.1	0.42

The intensity, relative contribution (% from IRM_{total}), median acquisition field ($B_{1/2}$), and dispersion parameter (DP) for each coercivity component are shown.

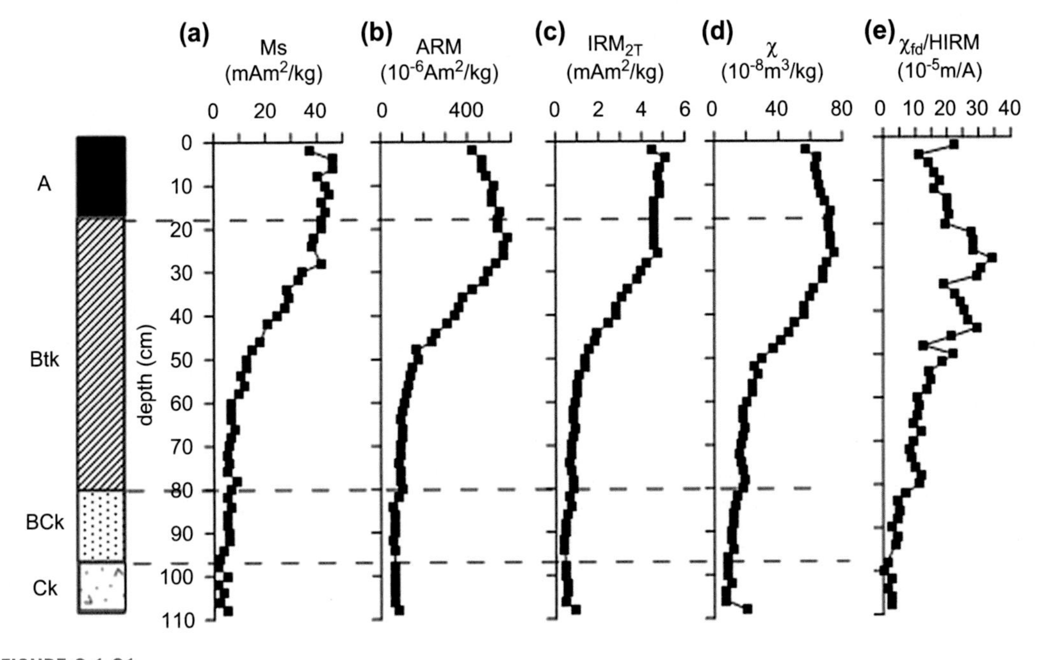

FIGURE 3.1.31

Soil profile PRV. Depth variations of concentration-dependent magnetic characteristics (a) saturation magnetization (Ms), (b) anhysteretic remanence (ARM), (c) isothermal remanence, acquired in a field of 2 T (IRM_{2T}), (d) magnetic susceptibility (χ), and (e) χ_{fd}/HIRM ratio.

pedogenic hematite in the profile (Liu et al., 2013). Variations in χ_{fd}/HIRM (Fig. 3.1.31e) indicate that pedogenic hematite is most probably present in the A and B_{tk} horizons with maximum concentration again at ~30-cm depth, similar to ARM. Hematite of both lithogenic and pedogenic origin is also evident from the variation in the S-ratio along the depth (Fig. 3.1.32f), which falls below ~0.6 in deeper soil horizons.

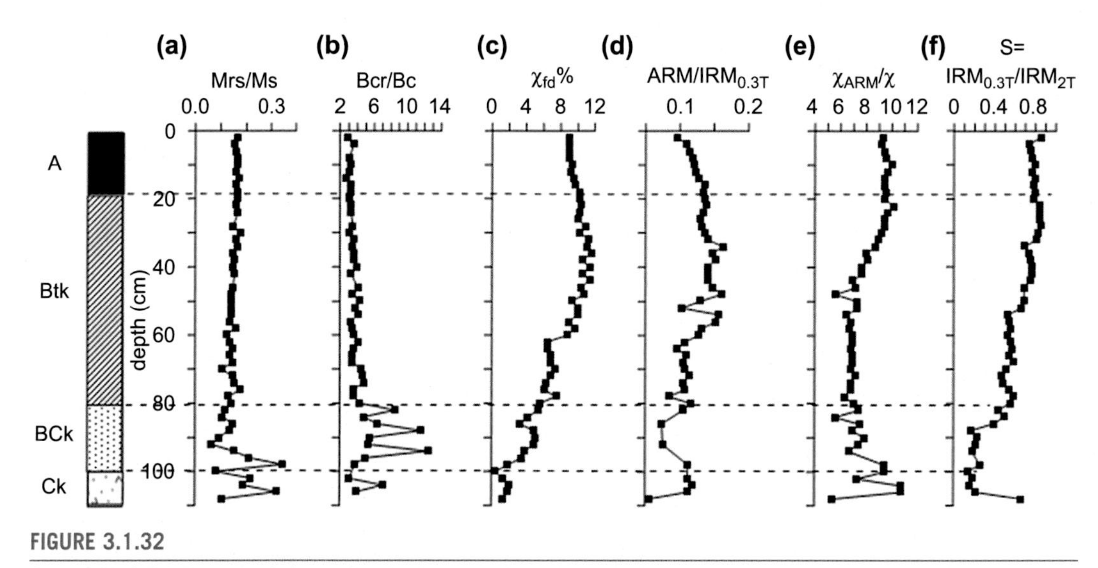

FIGURE 3.1.32

Soil profile PRV. Depth variations of grain size–sensitive magnetic characteristics and ratios: (a) M_{rs}/M_s ratio, (b) B_{cr}/B_c ratio, (c) percent frequency dependent susceptibility χ_{fd}%, (d) $ARM/IRM_{0.3T}$ ratio, (e) χ_{ARM}/χ ratio, and (f) S-ratio, calculated as: $IRM_{0.3}T/IRM_{2T}$.

Grain-size related parameters show depth variations, pointing to an enhanced concentration of SP grains in the upper part of the B_{tk} horizon (Fig. 3.1.32c) and a peak in the content of SD grains at the middle part of B_{tk}, according to the position of the maximum of the $ARM/IRM_{0.3T}$ ratio. The influence of lithogenic hematite is avoided by using $IRM_{0.3T}$ instead of IRM_{2T} for the calculation of the ARM/IRM ratio. The other parameter—χ_{ARM}/χ ratio—shows high values in A and the upper part of B_{tk}. The different behaviors of the two grain-size proxy ratios in the A horizon is due to the variations in IRM, which do not comply with that of χ (Fig. 3.1.31). The latter implies that the uppermost ~10 cm of the profile contains higher amount coarse magnetite particles, compared with the amount of SD and SP particles, while the finest fraction (SP + SD) is concentrated in the deeper levels. Therefore, magnetic susceptibility in the upper horizons of the PRV profile depends on both concentration and grain size, so that the ratio χ_{ARM}/χ cannot be used as a proxy for the content of SD grains. A more suitable parameter for this purpose is the $ARM/IRM_{0.3T}$ ratio. The bottom part of the profile—the BC_k and C_k horizons—is strongly influenced by the presence of lithogenic hematite, in addition to magnetite, which is evidenced by the hysteresis ratios (Fig. 3.1.32a and b), as well as by the S ratio.

Epicalcic Luvisol, profile F

Thermomagnetic analyses for selected samples from the leached Epicalcic Luvisol (profile F) show the dominant role of magnetite with clear T_c of 580°C (Fig. 3.1.33). The nearly rectangular shape of the heating $\chi(T)$ curve close to the T_c evidences the presence of coarse particles with a relatively narrow grain-size distribution. Thermomagnetic analysis for the sample from the parent rock (Fig. 3.1.33e) demonstrates that this magnetite is of lithogenic origin. The hematite phase with $T_N = 700$°C is also detected there. In addition, a sharp drop in $\chi(T)$ at ~120°C indicates the possible presence of

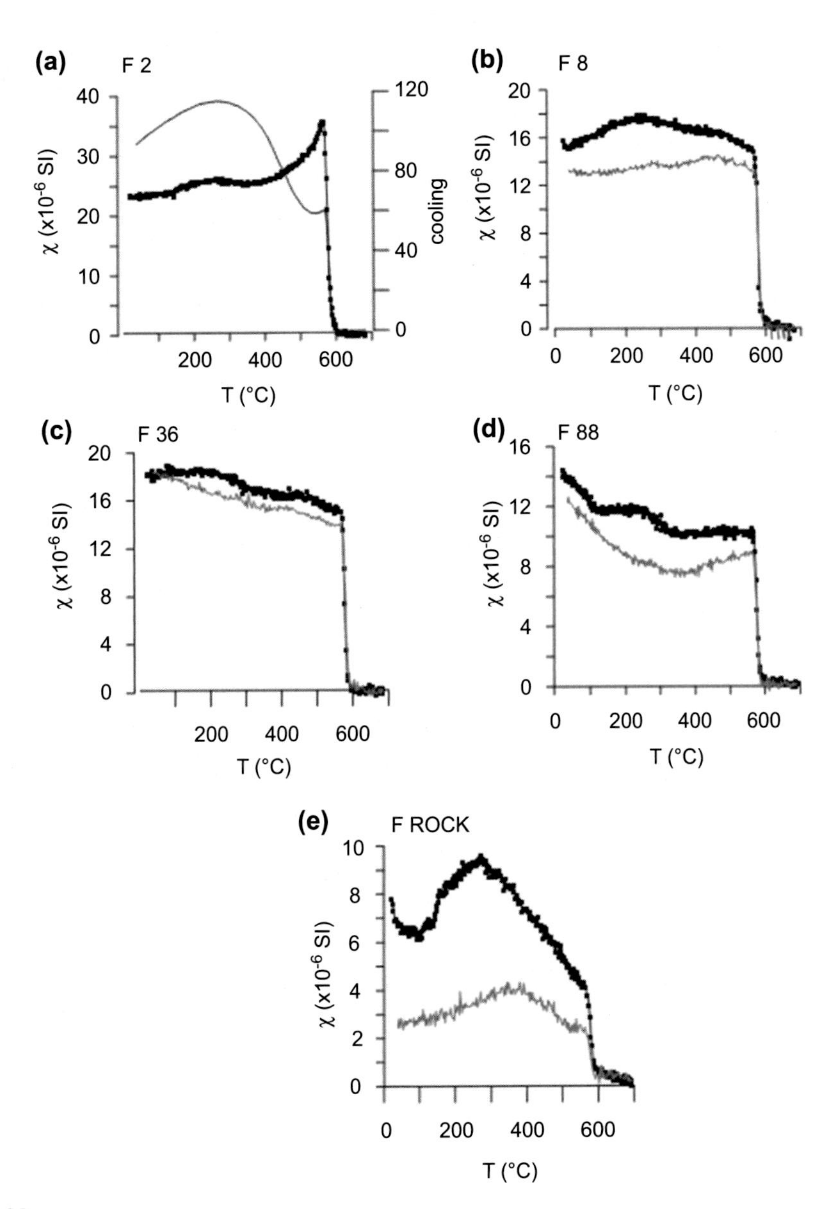

FIGURE 3.1.33

Thermomagnetic analyses of magnetic susceptibility for samples from different horizons of profile F. *Thick dotted line*—heating run, *thin line*—cooling run. Heating in air, fast heating rate (11°C/min). (a) sample from A_h horizon, (b) sample from AE1 horizon, (c) sample from B_t horizon, (d) sample from C_k horizon, and (e) sample from the parent rock.

lithogenic goethite as well. The latter can also be noticed on the heating curve for sample F88 from the C_k horizon (Fig. 3.1.33d).

For all samples analyzed, a bump in the $\chi(T)$ heating curve is observed at $\sim 300°C$, which probably reflects the presence of surface-oxidized magnetite grains. The surface sample (F2) is characterized by a sharp peak on the $\chi(T)$ curve before the T_c, corresponding to the so-called Hopkinson peak, an effect appearing in an assembly of SD particles, which become SP just below the T_c (O'Reilly, 1984: Dunlop and Özdemir, 1997; Dunlop, 2014). All cooling $\chi(T)$ curves, except that for sample F2, indicate little mineralogical transformations upon cooling.

Additional information about the magnetic mineral phases present in soil profile F is obtained from step-wise thermal demagnetization of IRM_{2T} (Fig. 3.1.34a–d). Demagnetization curves show similar behavior for all depths, characterized by three unblocking temperatures (T_{ub})—at 250°C, 600°C and 700°C. These temperatures could be ascribed to the unblocking of maghemitized magnetite grains, oxidized magnetite and hematite, respectively. Taking into account that these T_{ub} are found in all samples, it is supposed that they reflect a high concentration of different iron oxides in the parent material. On the other hand, diffuse reflectance (DRS) spectra (Fig. 3.1.34e) measured for soil samples from depths 4, 10, 22, 40 and 70 cm show the goethite presence only. The amplitude of the goethite peak (Scheinost et al., 1998) increases for samples from deeper levels (Fig. 3.1.34e). At the same time, no reliable indication for hematite presence in the spectra is observed.

Step-wise IRM acquisition up to the 1T field (Fig. 3.1.35) for several samples reveals only one coercivity component in the upper soil horizons. Only for samples F72 and F88, belonging to the BC_k and C_k horizons, a second weak high-coercivity component is separated (Table 3.1.3). According to the $B_{1/2}$ values obtained, it could be assigned to hematite. The dominating first component also has a relatively high coercivity (63–69 mT in the A_h, AE and B1 horizons and 91–95 mT in BC_k and C_k, respectively) (Table 3.1.3), which could indicate the important effect of maghemitization of the lithogenic magnetite grains. The fact that hematite is not identified in upper soil horizons, while thermal demagnetization suggests that hematite is present in all samples, could be explained by assuming that hematite in the upper part of the profile has a higher coercivity, compared with that in the parent material, and consequently is not magnetized in the magnetic field of 1 T (used for IRM acquisition).

In spite of the strong influence of parent lithology on the magnetic signal in profile F, the variations in concentration-dependent parameters clearly indicate an enhancement of the upper AE1, AE2 and B_t horizons with strongly magnetic minerals (Fig. 3.1.36). Magnetic enhancement gradually decreases along the lower part of the B_t horizon and has relative minima in the BC_k and C_k horizons. Magnetic susceptibility and saturation remanence show a very similar pattern (Fig. 3.1.36a and d), indicating that χ is governed by the concentration variations. Anhysteretic remanence (ARM) shows a different behavior in the near-surface horizons (Fig. 3.1.36b) with a gradual increase from A_h toward AE2, where its maximum is observed, and a further decrease, compatible with the variations in the other parameters. This suggests a maximum enhancement with SD magnetite grains in sub-soil depths (middle part of AE2). Isothermal remanence (IRM_{2T}) does not show such a depth trend. Instead, it has lower values in A_h and AE1 and slightly higher values in AE2 (Fig. 3.1.36c), followed by a rapid decrease at the middle part of B_t. Frequency-dependent magnetic susceptibility (χ_{fd}) shows a maximum concentration of SP fraction in AE2 and the upper part of B_t (10–40 cm depth) (Fig. 3.1.36e), while χ_{fd} is weaker in A_h and AE1. Therefore, clearly expressed variations in magnetic

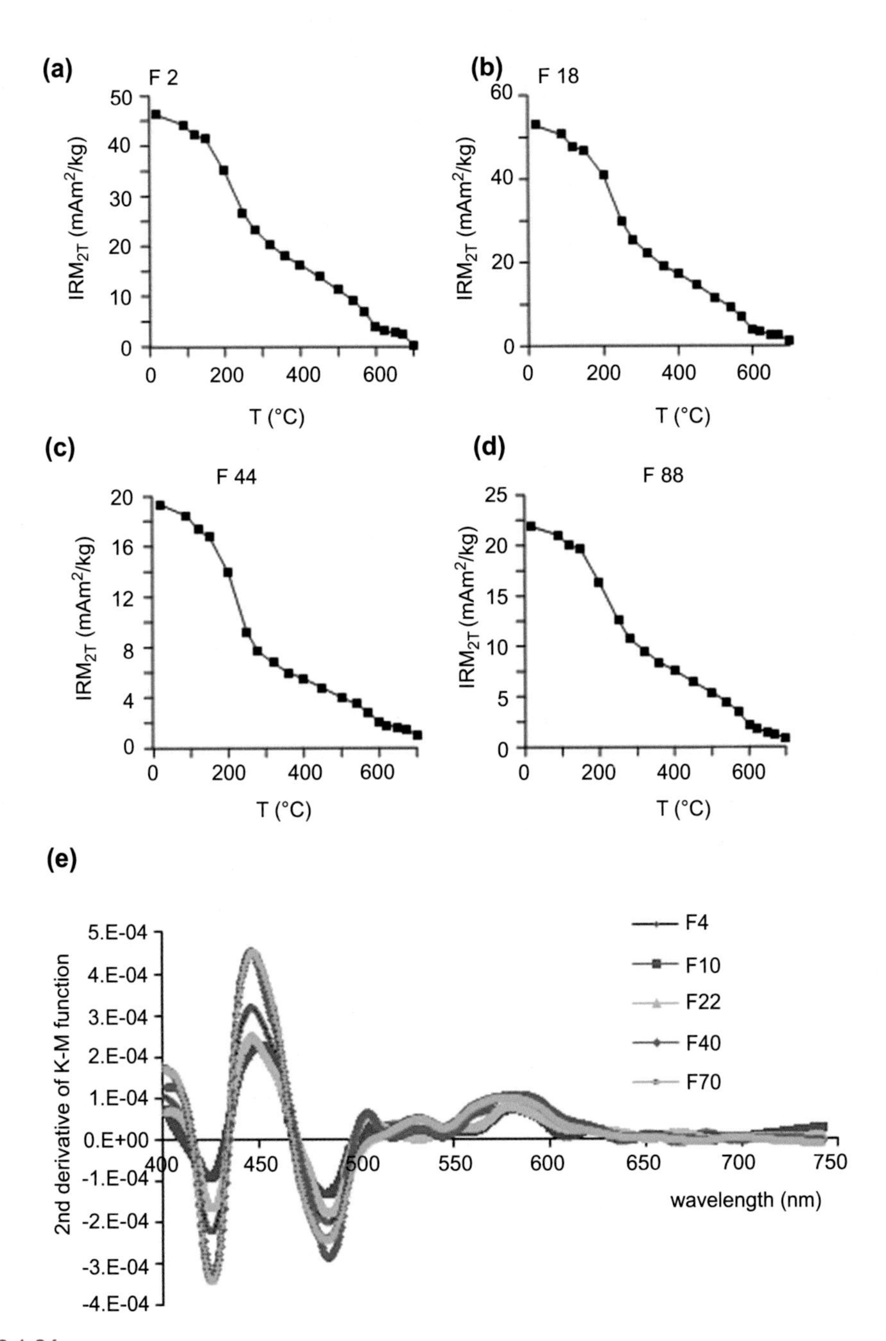

FIGURE 3.1.34

Thermal stepwise demagnetization of isothermal remanence (IRM_{2T}) for selected samples from profile F (a–d) and (e) second derivative of Kubelka-Munk remission spectra for pilot samples from F profile.

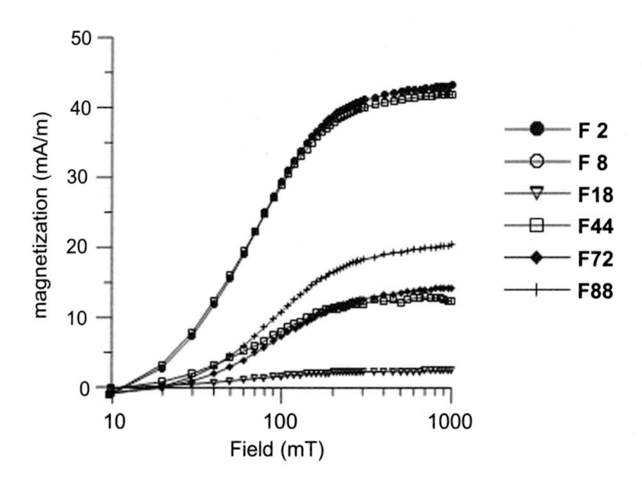

FIGURE 3.1.35

Stepwise acquisition of isothermal remanence (IRM) up to field of 1 T for selected samples from F profile.

Table 3.1.3 Summary parameters from coercivity analysis by fitting IRM acquisition curves with cumulative Gaussian functions, using IRM-CLG1.0 software (Kruiver et al., 2001) for selected samples from the F profile

Sample depth (cm)	IRM component 1				IRM component 2			
	Intensity (10^{-3} Am^2/kg)	% from IRM_{total}	$B_{1/2}$ (mT)	DP	Intensity (10^{-3} Am^2/kg)	% from IRM_{total}	$B_{1/2}$ (mT)	DP
Profile F, Epicalcic Luvisol								
2	43.0	100	63.1	0.38	–	–	–	–
8	42.0	100	63.1	0.37	–	–	–	–
18	2.3	100	63.1	0.37	–	–	–	–
44	12.5	100	69.2	0.35	–	–	–	–
72	14.0	87	95.5	0.37	0.5	3	794.3	0.20
88	20.0	97	91.2	0.37	0.6	3	794.3	0.20

The intensity, relative contribution (% from IRM_{total}), median acquisition field ($B_{1/2}$), and dispersion parameter (DP) for each coercivity component are shown.

grain size along the profile are linked to the genetic soil horizons. They are also underlined by the grain-size proxies (Fig. 3.1.37).

Because of the obviously large share of coarse magnetic grains in the iron oxide composition, $\chi_{fd}\%$ is low (between 0.5% and 2.5%), suggesting a relatively small contribution of the SP fraction (Dearing et al., 1996). Nevertheless, variations in $\chi_{fd}\%$ indicate maximum enhancement with SP grains at a ~10 cm depth (bottom of AE1) and a broad higher maximum at ~50-cm depth (middle part of B_t) (Fig. 3.1.37c). The SD-grain-size proxy ARM/IRM_{2T} shows the finest grain size in AE2 and the upper

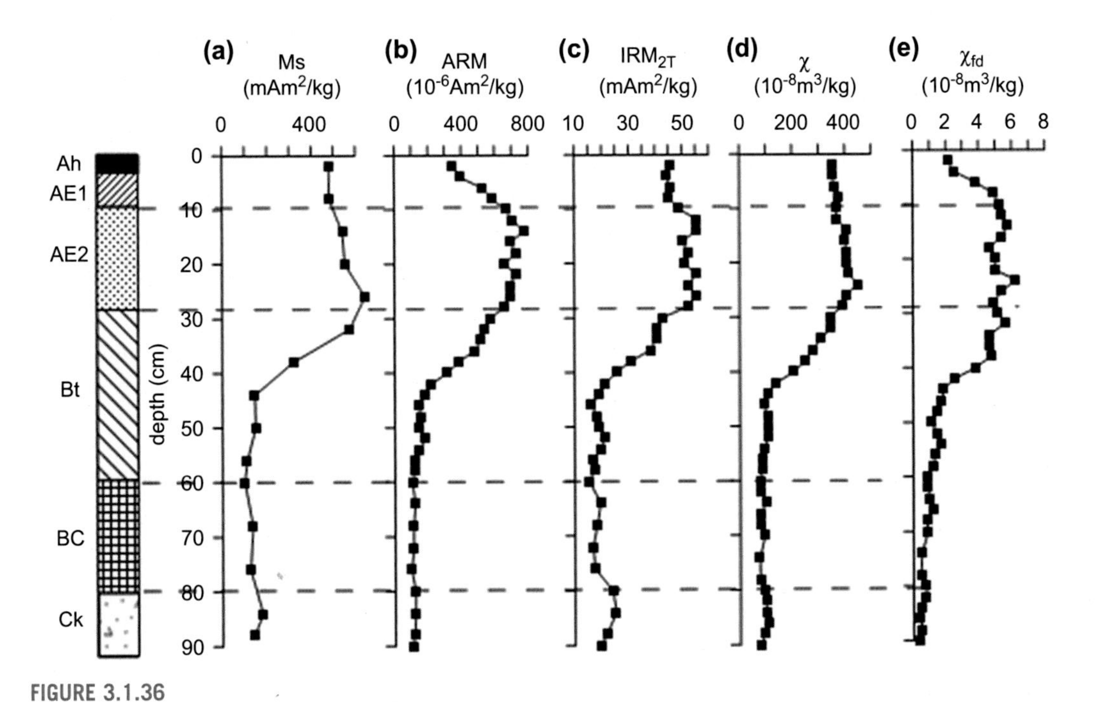

FIGURE 3.1.36

Soil profile F. Depth variations of concentration-dependent magnetic characteristics: (a) saturation magnetization (M_S), (b) anhysteretic remanence (ARM), (c) isothermal remanence, acquired in a field of 2 T (IRM_{2T}), (d) magnetic susceptibility (χ), and (e) frequency-dependent magnetic susceptibility (χ_{fd}).

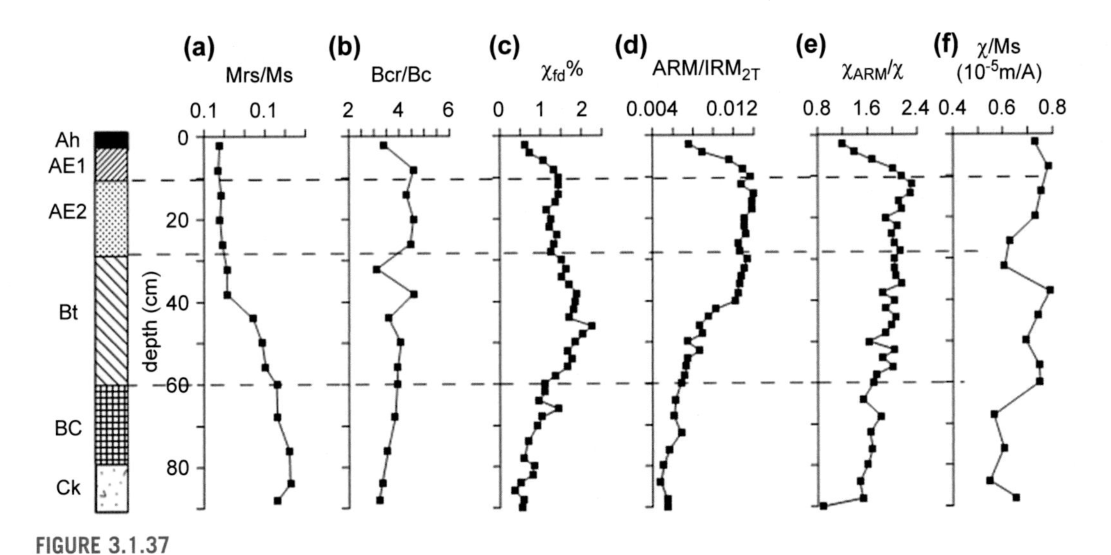

FIGURE 3.1.37

Soil profile F. Depth variations of grain size–sensitive magnetic characteristics and ratios: (a) M_{rs}/M_s ratio, (b) B_{cr}/B_c ratio, (c) percent frequency dependent susceptibility $\chi_{fd}\%$, (d) ARM/IRM_{2T} ratio, (e) χ_{ARM}/χ ratio, and (f) magnetic susceptibility to saturation magnetization (χ/M_s) ratio.

part of B_t (Fig. 3.1.37d). The minima in SP grains concentration in the surface levels could be due to depletion in such particles but also to the presence of particles smaller than the critical size, responding in the measuring frequency interval, which is ~0.0005 μm (Dearing et al., 1996). The latter hypothesis is supported by the obtained χ/Ms ratio (Fig. 3.1.37f), which shows high values in the A_h and AE2 horizons, and also confirms the broad maxima in B_t, obtained by $\chi_{fd}\%$. Hysteresis ratios (Fig. 3.1.37a and b) point to a sharp change in apparent grain size at a ~40-cm depth toward smaller grains, but this change is caused by a change in mineralogy, where the lithogenic fraction starts dominating the signal, as is clear from the concentration-dependent parameters (Fig. 3.1.36).

Bathycalcic Cutanic Luvisol (Chromic), profile BGU

Thermomagnetic analyses for selected samples from the BGU profile show that magnetic mineralogy is dominated by a magnetite phase (Fig. 3.1.38). In all samples, however, hematite's $T_N = 700°C$ is also present, suggesting that this mineral has a relatively important share in the magnetic mineral assembly, taking into account its weak magnetization (Dunlop and Özdemir, 1997). The sample from the surface level (Fig. 3.1.38a) shows behavior similar to surface samples from the other Luvisol profiles (CIN, PRV) with a bump at ~300°C, followed by a decrease and subsequent sharp peak at

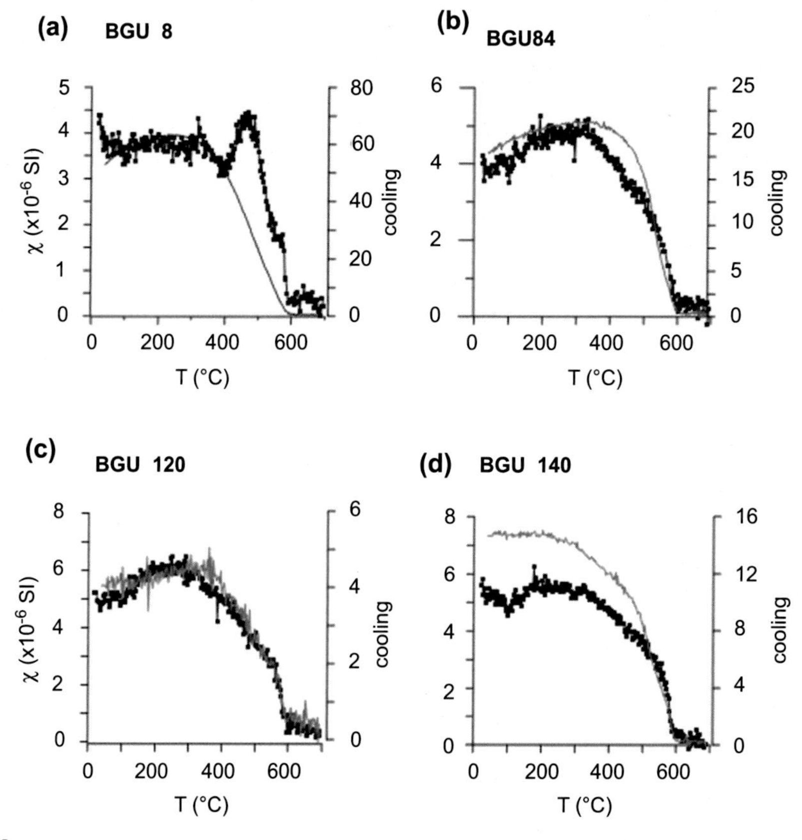

FIGURE 3.1.38

Thermomagnetic analyses of magnetic susceptibility for samples from different horizons of profile BGU. *Thick dotted line*—heating run, *thin line*—cooling run. Heating in air, fast heating rate (11°C/min). (a) Sample from AE horizon, (b and c) samples from B2 horizon, and (d) sample from BC_k horizon.

~500°C, suggesting the presence of pedogenic maghemite. Samples from the B2 and BC_k horizons (Fig. 3.1.38b and c) have similar heating curves, demonstrating the presence of magnetite and hematite. Sample BGU140 from the deepest part of the profile (Fig. 3.1.38d) also shows in addition to magnetite and hematite, $T_N = 100°C$, probably reflecting the presence of goethite. The strongest changes in magnetic mineralogy are observed in the surface sample (rich in organics), while for the deeper samples the susceptibility increase upon cooling is small. Therefore, the lithogenic magnetic fraction in profile BGU obviously plays an important role for the magnetism of soil.

In addition, the stepwise acquisition of isothermal remanence up to 5 T reveals the occurrence of a high-coercivity fraction in different soil depths (Fig. 3.1.39a). An analysis of coercivity components (Kruiver et al., 2001) shows the presence of two IRM components—low-coercivity, comprising 70–80% of the total IRM, and a high-coercivity component, accounting for 16–27% of IRM (Table 3.1.4). The obtained coercivities of the two components suggest that the first one is probably represented by magnetite/maghemite and the second one by goethite. No IRM component corresponding to hematite is obtained through coercivity analysis, pointing out that either it is in an SP state,

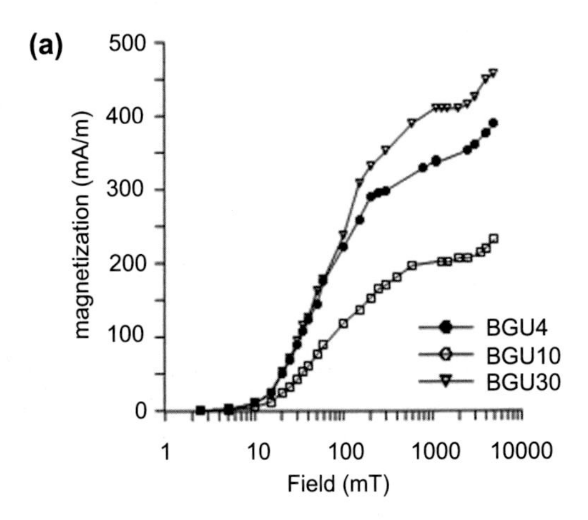

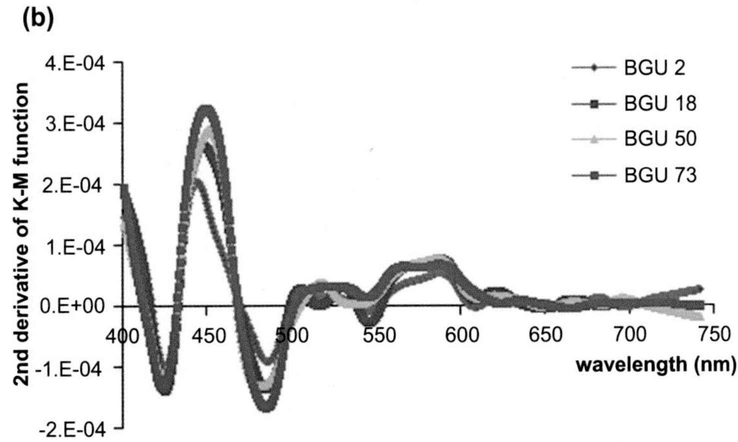

FIGURE 3.1.39

(a) Stepwise acquisition of isothermal remanence (IRM) up to field of 5 T for selected samples from profile BGU and (b) second derivative of Kubelka-Munk remission spectra for pilot samples from profile BGU.

Table 3.1.4 Summary parameters from coercivity analysis by fitting IRM acquisition curves with cumulative Gaussian functions, using IRM-CLG1.0 software (Kruiver et al., 2001) for selected samples from the BGU profile

Sample depth (cm)	IRM component 1				IRM component 2			
	Intensity (10^{-3} Am^2/kg)	% from IRM_{total}	$B_{1/2}$ (mT)	DP	Intensity (10^{-3} Am^2/kg)	% from IRM_{total}	$B_{1/2}$ (mT)	DP
Profile BGU, Bathycalcic Cutanic Luvisol								
4	16.0	81	55.0	0.41	3.75	19	3162.3	0.35
10	10.0	73	79.4	0.52	3.75	27	6309.6	0.35
30	20.5	84	74.1	0.53	4.00	16	5011.9	0.30

The intensity, relative contribution (% from IRM_{total}), median acquisition field ($B_{1/2}$), and dispersion parameter (DP) for each coercivity component are shown.

thus not contributing to IRM, or the hematite's coercivity spectra overlap with the coercivities of the other mineral phases and cannot be separated. In addition, DRS spectra, taken for samples from depths of 2, 18, 50, and 73 cm (Fig. 3.1.39b), reveal only the presence of goethite with increasing peak amplitude down-profile, while no hematite is detected.

The concentration of strongly magnetic iron compounds along the BGU profile shows clear variations linked to the described soil horizons (Fig. 3.1.40a–d). The most characteristic feature of all

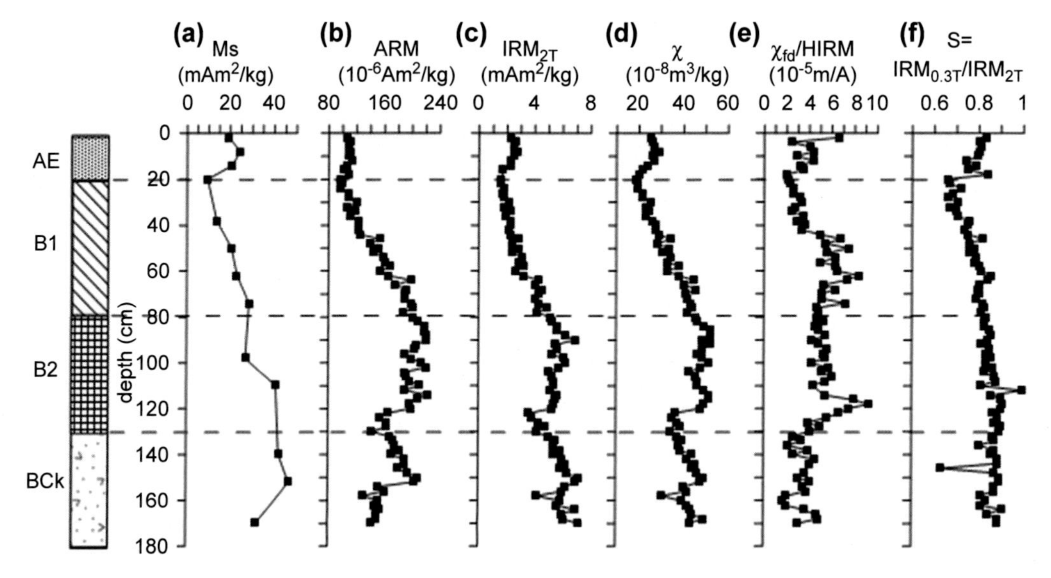

FIGURE 3.1.40

Soil profile BGU. Depth variations of concentration-dependent magnetic characteristics: (a) saturation magnetization (M_S), (b) anhysteretic remanence (ARM), (c) isothermal remanence, acquired in a field of 2 T (IRM_{2T}), (d) magnetic susceptibility (χ), and ratios χ_{fd}/HIRM (e) and S (f).

concentration-dependent parameters is their increase from the top toward the bottom of the profile, with the maximum in BC_k. Therefore, the significant influence of parent rock mineralogy on the soil's properties can be assumed. The task of separating pedogenic magnetic features in the bulk magnetic signal is challenging. The uppermost AE horizon is characterized by an increased concentration of SP, SD, as well as coarser grains in its middle part, where all parameters exhibit local maxima (Fig. 3.1.40). Along the B1 horizon, a gradual increase in the concentration of all grain-size fractions is revealed by the behavior of ARM, IRM_{2T}, χ, and M_s, while in the B2 horizon their concentration remains constant. The χ_{fd}/HIRM ratio, used as a proxy for the relative importance of pedogenic maghemite versus pedogenic hematite (Liu et al., 2013), suggests that in AE, the lower part of B1 and lower part of B2 pedogenic maghemite is more abundant, while the upper parts of the illuvial horizons B1 and B2 are enhanced with pedogenic hematite. The S ratio suggests that all along the profile, there is a high-coercivity fraction present, because S reaches only 0.8 in intervals where it has maxima (Fig. 3.1.40f). The minimum, corresponding to the depth interval at the limit between AE and B1, suggests the highest amount of high-coercivity minerals there. Taking into account, however, that the maximum field applied for IRM acquisition is 2 T, it is possible that the S ratio reflects changes in the hematite's content, rather than that of goethite, which is characterized by much higher coercivities (e.g., Dekkers, 1989).

Hysteresis ratios cannot be used to distinguish grain-size variations along the profile because of the presence of mineral mixtures with different coercivities and the interplay between lithogenic and pedogenic magnetic fractions. The percent frequency-dependent magnetic susceptibility (χ_{fd}%) shows relatively low values (2–6%) and points to local enhancement with SP grains at the middle of the AE horizon and a broad maximum, encompassing B1 and B2 (Fig. 3.1.41c). The finest SD grains are found

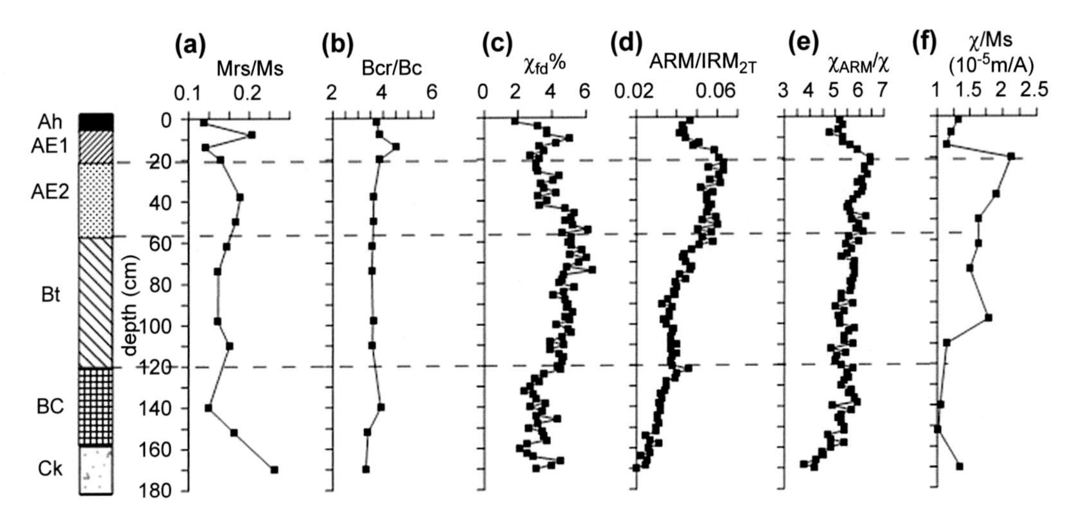

FIGURE 3.1.41

Soil profile BGU. Depth variations of grain size—sensitive magnetic characteristics and ratios: (a) M_{rs}/M_s ratio, (b) B_{cr}/B_c ratio, (c) percent frequency dependent susceptibility χ_{fd}%, (d) ARM/IRM_{2T} ratio, (e) χ_{ARM}/χ ratio, and (f) magnetic susceptibility to saturation magnetization (χ/M_s) ratio.

at the margin between AE and B1, according to the ratio ARM/IRM_{2T}, as well as the χ_{ARM}/χ ratio (Fig. 3.1.41d and e). Since χ_{ARM}/χ is influenced by a change in mineralogy (magnetite-hematite) along the profile, the other ratio—ARM/IRM_{2T} seems more appropriate as an SD-grain size proxy. Thus, considering ARM/IRM_{2T} variations, it could be stated that the finest SD grains are found in the B1 horizon, while in B2 coarser grains contribute to ARM (Fig. 3.1.41d). It is interesting to note that the ratio χ/M_s, usually regarded as a proxy for the SP fraction, in fact shows behavior similar to ARM/IRM_{2T} (Fig. 3.1.41f). It indicates that magnetic susceptibility is dominated by the SD fraction, and not by the SP part of the strongly magnetic assemblage.

3.1.4 PEDOGENESIS OF IRON OXIDES IN LUVISOLS, AS REFLECTED IN SOIL MAGNETISM

Magnetic susceptibility is a parameter which strongly depends on the concentration of the iron oxides in soil, of both lithogenic and pedogenic origin. Examples of Luvisols, considered above, come from different locations in Bulgaria—Haplic Luvisols (CIN and PRV) from south central Bulgaria, Epicalcic Luvisol (F) from southeast Bulgaria, and Bathycalcic Luvisol (BGU) from northwest Bulgaria. They are developed in different parent materials (e.g., a different initial lithology) of a different age. Thus, it is not possible to put directly side by side the magnetic characteristics obtained for these profiles in order to evaluate their pedogenic development and its magnetic expression. Usually, for an evaluation of magnetic enhancement of soils developed on loess, the difference between the magnetic susceptibility of upper soil horizons (χ_{soil}) and the susceptibility of the parent material (χ_C) (the least influenced by pedogenesis) is calculated (Maher and Thompson, 1995). However, in the case of different parent materials, it would be more appropriate to use the ratio (χ_{soil}/χ_C). χ_{soil} and $M_{s(soil)}$ are calculated as mean values for A + B horizons while χ_c and $M_{s(c)}$ are the mean values for the C-horizons. It comes out that for profiles CIN and PRV (Haplic Luvisols), the ratio χ_{soil}/χ_C is 8.0 and 4.9, respectively (Figs. 3.1.27 and 3.1.31). If the same ratio is calculated for Ms ($M_{s(soil)}/M_{s(c)}$), then the results show the opposite—3.7- and 7.7-times the difference between soil and parent material. For Epicalcic Luvisol (profile F), χ_{soil}/χ_C is 4.4, while $M_{s(soil)}/M_{s(C)}$ is 4.1. Similar values for magnetic susceptibility enhancement (~3–4 times) are obtained for Calcic Luvisols from Spain (Torrent et al., 2010a,b). Since the profile of Cutanic Luvisol (BGU) shows lower magnetic susceptibility in upper soil horizons compared with the subsoil, it is effectively magnetically depleted (Fig. 3.1.40) and the ratio $\chi_{soil}/\chi_C = 0.6$. A similar depletion of magnetic minerals content in the upper A_h and EB horizons in the Red Unit Luvisol Barranca Tlalpan in Mexico is found (Rivas et al., 2006).

In addition to parent material and climate, magnetic characteristics also depend on time. Different studies of chronosequences (i.e., soils developed on the same parent material and in the same climate but formed during different time periods) report increasing magnetic susceptibility with increasing time for pedogenic development (Singer and Fine, 1989; Singer et al., 1996; Torrent et al., 2010a). According to the most widely accepted hypotheses (Koinov et al., 1998; Ninov, 2002; Shishkov and Kolev, 2014), Luvisols from Bulgaria are considered as relict soils. Their formation and maturation took place in similar conditions during the Pliocene, when the climate was warmer and more humid than today's. Those most enhanced with SP magnetite/maghemite particles are the two Haplic Luvisols (CIN an PRV), where $\chi_{fd}\%$ reaches 10–12%, and according to the empirical models (Dearing et al., 1996), the superparamagnetic fraction predominates in the magnetic assemblage. Often, magnetic studies of Luvisols report a maximum concentration of strongly magnetic iron oxides in the illuvial horizon. Such a feature is obtained for Luvisols from Spain, Russia and Mexico (Rivas et al., 2006), as well as Calcic Luvisols from Bulgaria (profiles PRV and F, reported here). This magnetic signature could be linked to the process of illuviation and the leaching of substances down-profile, as it was explained at the beginning of this chapter.

The rock magnetic properties of different soils are strongly influenced by the parent rock lithology (e.g., Singer et al., 1996; Shenggao, 2000; Evans and Heller, 2003), as is also evidenced by the examples from Bulgaria. Strongly magnetic coarse lithogenic particles, present in large amounts in some profiles (for example, profiles F and BGU) lead to apparent very low "pedogenic" enhancement with SP grains, as deduced from the parameter $\chi_{fd}\%$. Despite this, variations in the absolute concentration of the frequency-dependent magnetic fraction (χ_{fd}) and the relative changes in the $\chi_{fd}\%$ in depth clearly indicate the intensity and extent of pedogenic processes.

Considering the variations in absolute concentration and relative contribution of different grain-size fractions of the "maghemite" pedogenic component, the following characteristic features in Luvisols can be identified:

1. In Luvisols without strong leaching (no eluvial horizons established), variations in the absolute content of SP and SD pedogenic maghemite (expressed by χ_{fd} and ARM, respectively) are coupled and these fractions are enhanced in A and the upper half of B horizons (see profiles CIN and PRV). The relative content of the SP strongly magnetic fraction ($\chi_{fd}\%$) is steadily increasing from the surface until the bottom part of B horizons. Similarly, the finest SD maghemite grains (maximum in ARM/IRM) are also found at the *bottom of B horizons.* Therefore, pedogenic enhancement with strongly magnetic maghemite grains proceeds through a concentration increase in the A horizons and grain growth in illuvial horizons (increasing from SP to SD grain size toward the surface).
2. Luvisols with a developed eluvial horizon are characterized by decoupled variations in SD and SP maghemite concentration and grain size in the eluvial and illuvial horizons. SD particles have minimum concentration in the eluvial horizon, while the SP fraction shows local maxima there and relative enhancement in the whole illuvial horizon (profile F). The most stable SD particles are found in the *upper part of the illuvial* horizon, while their grain size grows downward.
3. Strongly leached Luvisols (with signs of podzolization) (like profile BGU) show similar to moderately leached profiles (profile F) characteristic features in the concentration and grain-size variations in the pedogenic maghemite fraction along the profile but are more strongly expressed.
4. The coercivity analysis of IRM-acquisition curves reveals the dominant role of the magnetically soft fraction, which shows increasing coercivity (by means of $B_{1/2}$ parameter in Tables 2.2.2–2.2.4) in the order: Haplic Luvisol (PRV profile where $B_{1/2}$ is ~33 mT)—Calcic Luvisol (F profile where $B_{1/2}$ is ~63 mT)—Cutanic Luvisol (BGU profile where $B_{1/2}$ is ~76 mT). It could be supposed that the effective magnetic grain size of the pedogenic strongly magnetic fraction decreases with an increasing degree of lessivage and the development of eluvial-illuvial horizons in Luvisols.

5. Experimental data identifying the formation of pedogenic hematite are obtained only for the profile of Haplic Luvisol (CIN) through Visible Infrared Spectroscopy (VIS)—Diffuse Reflectance Spectroscopy (DRS) analysis, while for the profiles of Calcic Luvisol (F) and Cutanic Luvisol (BGU), a goethite signal (probably of lithogenic origin), is only evident from the DRS spectra.

3.1.5 RED MEDITERRANEAN SOILS

Red Mediterranean soils are spread mainly in the Mediterranean region, where they form under the influence of significant dust transport from the Sahara desert (Yaalon, 1997; Durn et al., 1999; Durn, 2003; Stuut et al., 2009). These specific soils develop on hard limestone and have a characteristic intense red/red-brown color, caused by the enhanced presence of ultrafine hematite grains. Often, red Mediterranean soils are referred to as *terra rossa*, and they are intensively studied because of their particular properties, mineralogy, and complex evolution under the influence of the Mediterranean climate and dust transport from Africa (Yaalon, 1997; Durn et al., 1999; Verheye and de la Rosa, 2005; Fedoroff and Courty, 2013). The most widely accepted view affirms that the development of the red Mediterranean soils was discontinuous, occurring during periods of environmental stability with dry and hot summers (like interglacials, when the climate was humid and precipitation exceeds evapotranspiration) (Fedoroff and Courty, 2013). However, the impact of glacial intervals on the development and properties of *terra rossa* soils is still not well studied. This specific soil is not a separate soil type in the classification systems, and *terra rossa* soils are mainly classified as Luvisols (Rhodic Luvisols/Rhodudalfs).

Bulgaria is situated in the northern periphery of the area, influenced by the present-day Mediterranean climate and only the most southeastern part of the country is characterized by such climatic conditions. Red-colored Mediterranean-type soils are found in the southeastern parts of the Strandja mountain. Their genesis and classification are still a matter of debate in Bulgarian pedology (Velizarova and Popov, 1999; Boyadzhiev, 1995; Boyadgiev, 1998; Ninov, 2002).

The soil profile of a Rhodic Luvisol is sampled in the Strandja mountain and studied in order to compare its properties with typical Mediterranean *terra rossa* soils. A detailed study is published in Jordanova et al. (2013).

Rhodic Luvisol, profile designation: RED

Location: N 42°13′10.8″ E 27°25′57.8″

Present-day climate: MAT = 12°C, MAP = 771 mm, altitude: 253 m a.s.l; vegetation: broadleaf deciduous forest—*Fagus orientalis* and *Rhododendron ponticum*

Description of soil profile: general view and details from the horizons are shown in Figs. 3.1.42—3.1.46).

AE horizon: 0—10 cm, humic horizon, light brown—red color, loose structure
B1 horizon: 10—50 cm, illuvial horizon, red-brown color, dense, clayey
B2 horizon: 50—60 cm, brown-red color, clayey, Fe-Mn concretions present
BC horizon: 60—90 cm, transitional horizon with signs of the initial rock texture
C horizon: 90—110 cm, parent material, gray-beige color, weathered rock fragments
Parent rock: Triassic metamorphosed limestones

FIGURE 3.1.42

View of the profile of Rhodic Luvisol (profile RED).

FIGURE 3.1.43

Profile RED. A horizon—detail.

FIGURE 3.1.44

Profile RED. B1 horizon—detail.

FIGURE 3.1.45

Profile RED. BC horizon—detail.

FIGURE 3.1.46

Profile RED. C horizon—detail.

Texture, soil reaction, and major geochemistry of Rhodic Luvisol (profile RED)

Similar to the other Luvisols presented earlier, the texture of the RED profile is dominated by the silt and clay fractions (Fig. 3.1.47b and c). The clay fraction is relatively enriched in the B1 and B2 horizons. The soil reaction is slightly alkaline to neutral, showing a decrease with increasing depth (Fig. 3.1.47d) and again an increase in the C horizon. The content of the major elements points to some interruption in the normal variations in the thin B2 horizon. A comparison of the elemental composition of soil samples with the composition of parent rock (limestone) (Fig. 3.1.47e–l) points to significant differences, as is also found for other red Mediterranean soils (Lucke et al., 2014). The observed significant enhancement of the soil with the major chemical elements cannot be explained solely by intense weathering and leaching out of the profile. Most recent geochemical investigations on *terra rossa* soils confirm the existence of a significant external input of dust/sediment particles to the solum (Stuut et al., 2009; Erel and Torrent, 2010; Muhs et al., 2010a,b; Lucke et al., 2014). In order to check if such an external dust source played a role in the RED profile, several soil samples were

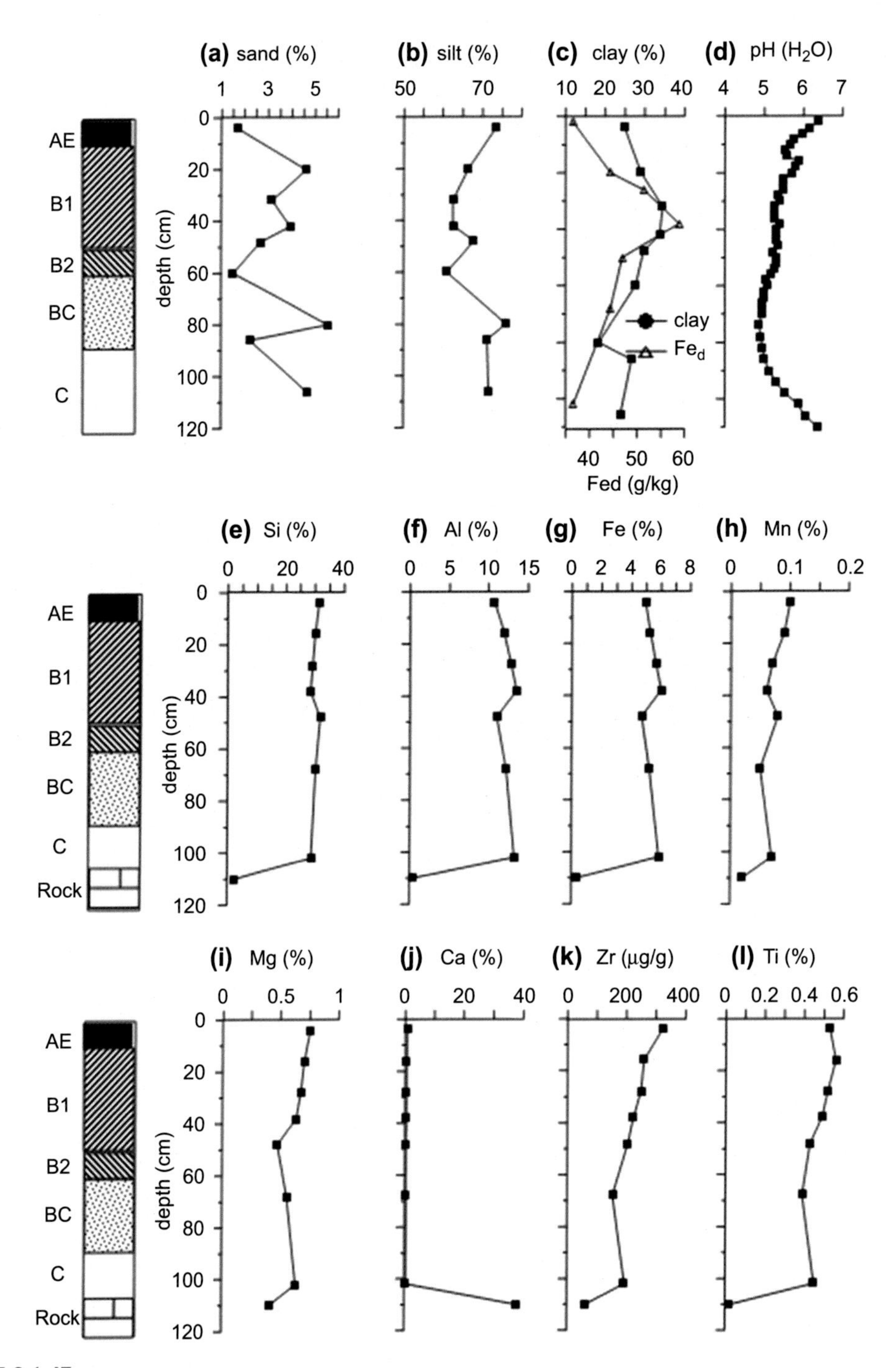

FIGURE 3.1.47

Mechanical fractions (sand, silt, clay), Fe_d content (a–c), and soil reaction pH (d) along the profile RED. Content of some major and trace elements (Si, Al, Fe, Mn, Mg, Ca, Ti, Zr) (e–l) along depth of profile of Rhodic Luvisol (profile RED) for selected samples.

Table 3.1.5 Chemical extraction data for selected samples from RED profile (Rhodic Luvisol)

Sample depth (cm)	Fe_o (g/kg)	Fe_d (g/kg)	Fe_{tot} (g/kg)	Fe_o/Fe_d	Fe_d/Fe_{tot}	Hematite (g/kg)	Goethite (g/kg)
4	3.62	36.46	63.30	0.10	0.58	28.62	7.85
18	5.90	44.44	67.81	0.13	0.66	35.60	8.84
28	7.08	51.35	72.84	0.14	0.70	39.57	11.77
38	6.93	58.77	77.28	0.12	0.76	47.08	11.69
48	9.25	46.88	60.32	0.20	0.78	38.86	8.02
68	6.58	44.38	67.11	0.15	0.66	23.70	20.68
102	5.02	36.56	75.77	0.14	0.48	13.75	22.81

Oxalate-extractable iron (Fe_o); dithionite-extractable iron (Fe_d), total Fe content (Fe_{tot}), and the ratios Fe_o/Fe_d and Fe_d/Fe_{tot} are shown. The hematite and goethite amounts, calculated from diffuse reflectance spectra (DRS), according to Torrent et al. (2007), are shown in the last two columns.

examined under an optical microscope and the observations showed an abundant presence of irregular quartz particles with magnetite inclusions and feldspar grains. The observed morphology suggests that these grains are wind-blown weathering products from the Paleozoic granitoids and Triassic metamorphic sediments outcropping in the region (Jordanova et al., 2013).

X-ray diffraction analysis on bulk material from the B1 horizon identified the presence of kaolinite, quartz, and mica, while in the C horizon kaolinite, quartz, mica, and K-feldspar are detected (Jordanova et al., 2013). Iron extraction data (Table 3.1.5.) show a low amount of oxalate-extractable Fe (Fe_o), compared with dithitonite-extractable Fe (Fe_d), pointing to a dominating role of better crystalline iron oxides in the soil. The ratio Fe_d/Fe_{tot}, used as an indicator of soil maturity (Cornell and Schwertmann, 2003) has relatively high values between 0.58 and 0.78 in upper soil horizons with a maximum in the B1 horizon. The clay content and Fe_d co-vary along the profile (Fig. 3.1.47c), indicating that crystalline iron oxides are linked mainly to the clay fraction. The obtained Fe_d/Fe_{tot} values, as well as the total iron content in the Bulgarian red soil are similar to the values reported for other Mediterranean *terra rossa* soils (Boero et al., 1992; Durn, 2003 Torrent et al., 2010a,b). The other widely used ratio—Fe_o/Fe_d, indicative about the relative contribution of the amorphous iron oxides in the total pedogenic iron oxides, shows low and almost constant values along the depth (Table 3.1.5). Despite the fact that a ferrimagnetic strongly magnetic phase dominates the magnetic signal, its contribution to the Fe_d amount in the upper soil horizons, is only a few percent. Thus, Fe_d could be approximated with the pedogenic hematite and goethite content in general. Results from the DRS analysis provide the ratio Hm/(Hm + Gt), which, in combination with Fe_d extraction data, is used to calculate the concentration of pedogenic hematite and goethite (according to the procedure described in Torrent et al. (2007)). In line with the data obtained for red Mediterranean soils from Spain (Torrent et al., 2010a,b), hematite is more abundant in the upper soil horizons of the RED profile, while the amount of goethite prevails in the deeper BC and C horizons.

Magnetic properties of Rhodic Luvisol

Thermomagnetic analysis of magnetic susceptibility was carried out using a slow and fast heating rate (~6.5°C and 11.5°C) in a KLY-2 CS23 furnace and using a low (~0.02 g) and a larger amount (~0.15 g) of material. As was shown in Jordanova and Jordanova (2016), using fast heating and more material in the analysis leads to the creation of local reducing conditions in the sample, which then experiences intense mineral transformations. Several examples of thermomagnetic analyses applying these two options (slow heating–low mass and fast heating–big mass) are shown in Fig. 3.1.48. Evidently, fast heating (Fig. 3.1.48a–d) leads to significant alterations in magnetic minerals on heating up to 700°C, which produces new strongly magnetic phases on cooling. In contrast, the slow heating of a small amount of material (Fig. 3.1.48e–h) shows that a magnetite-like phase dominates susceptibility signal in all samples and no significant mineralogical changes occur on cooling. In this case, as also pointed out by Jordanova and Jordanova (2016), thermomagnetic curves showing mineral transformations could be more informative about the real assemblage of Fe-containing phases through

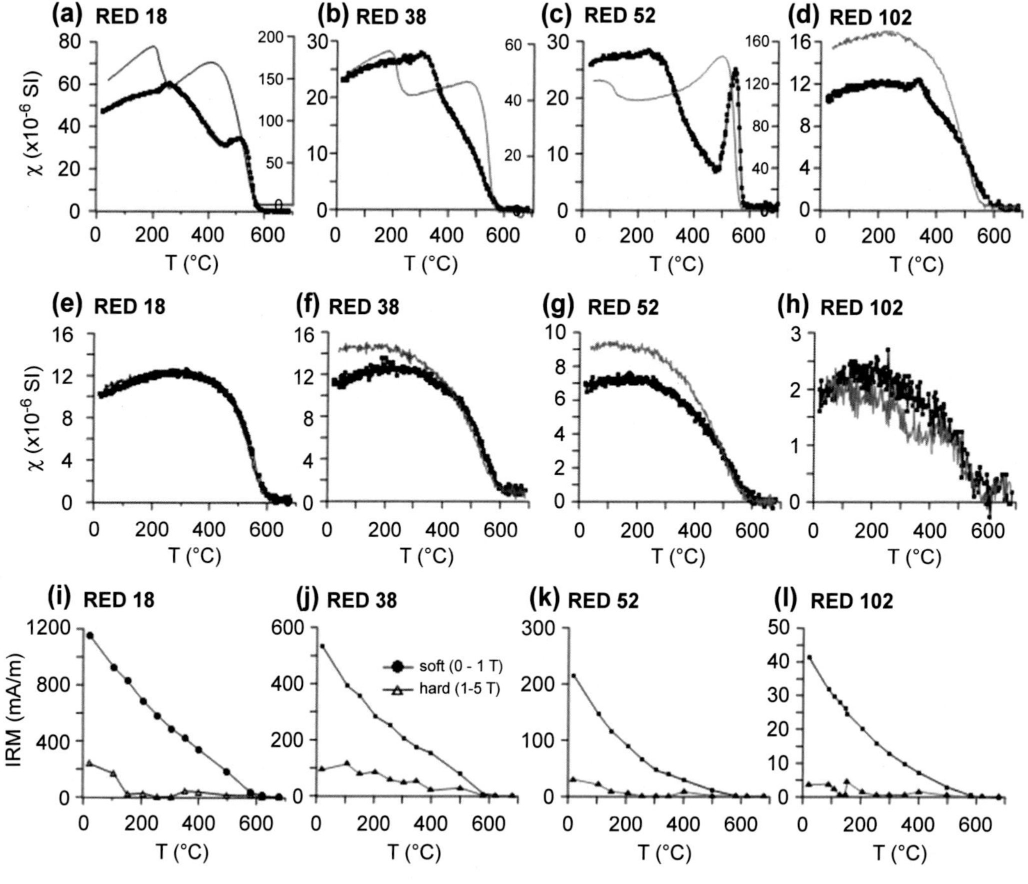

FIGURE 3.1.48

Thermomagnetic analyses of magnetic susceptibility for samples from different horizons of profile RED using fast heating rate (11°C/min) (a–d) and slow heating rate (6°C/min) (e–h). *Thick dotted line*—heating run, *thin line*—cooling run. Heating in air. Thermal stepwise demagnetization of two-component IRM [soft (0–1 T) and hard (1–5 T)] for selected samples from profile RED (i–l).

the noticeable alterations during heating. Considering the heating curves with a fast heating rate for samples from different soil horizons (Fig. 3.1.48a–c), it can be seen that χ increases slightly until 300°C, followed by a sharp decrease, a characteristic feature also observed in other Luvisols (see the profiles described earlier). The magnetite phase is also present in all samples but more (sample RED18 and RED52) or less well (sample RED38) expressed. Susceptibility behavior in the interval 20–400°C during fast heating experiments may be related to the presence of pedogenic maghemite. Magnetite could be both initially present and a secondary product from hematite reduction during heating. This nonuniqueness concerning the original presence of magnetite is resolved by using stepwise thermal demagnetization of isothermal remanence (Fig. 3.1.48i–l). The final unblocking of the soft magnetic component occurs at 580°C, which is evidence that magnetite is carrying this remanence (e.g., it is initially present in the soil samples). Wide demagnetization spectra and an often concave shape of the demagnetization curves of the soft component suggest that magnetite is of a larger grain size. Therefore, it is not of pedogenic origin but rather of lithogenic origin. The absence of maghemite's signature in the stepwise thermal demagnetization experiment could be either due to the stability of maghemite (which contradicts to the data from the fast $\chi(T)$ heating experiment) or to the fact that maghemite is superparamagnetic and does not carry remanence. Considering the microscopy observations, mentioned earlier, the magnetite fraction most probably comes from detrital external input from nearby outcropping metamorphic rocks, while maghemite is of pedogenic origin. Thermomagnetic analyses do not show the presence of hematite and/or goethite, suggesting that the strongly magnetic phase (magnetite plus maghemite) dominates the magnetic susceptibility. On the other hand, stepwise thermal demagnetization of the hard component of the composite IRM (Fig. 3.1.48i–l) demonstrates the existence of a high-coercivity phase (goethite) with T_{ub} ~100–120°C, especially strong in sample RED52, as well as hematite with T_{ub} ~700°C. Therefore, goethite and hematite with wide coercivity spectra are present in significant amounts in the soil, along with strongly magnetic maghemite/magnetite.

The coercivity distribution of the magnetic assemblage in different depth levels is also studied through stepwise IRM acquisition. As is evident from Fig. 3.1.49, a high-coercivity fraction in different relative amounts is found in all samples. Coercivity analysis by fitting IRM acquisition curves with cumulative Gaussian functions, using IRM-CLG1.0 software (Kruiver et al., 2001), reveals the presence of three IRM components in the studied samples (Table 3.1.6) characterized by median acquisition fields ($B_{1/2}$) of ~22 mT, ~0.25 T, and ~3.5 T, respectively. The low-coercivity component (IRM1) with $B_{1/2} = 22$ mT is compatible with the coercivity of pedogenic magnetite or MD magnetite (Egli, 2004). The second component could be ascribed to hematite, while the component with the highest coercivity $B_{1/2}$ of 3.5 T corresponds to goethite (Rochette et al., 2005; Dunlop and Özdemir, 1997).

Variations in concentration-dependent magnetic characteristics along the Rhodic Luvisol profile are shown in Fig. 3.1.50. The enhanced magnetic signal is characteristic for the upper part of the profile—the AE and B1 horizons, with maxima at ~20-cm depth (B1 horizon) (Fig. 3.1.50a–d). Changes in ARM and magnetic susceptibility are synchronous, while IRM_{2T} does not show lower values in the AE horizon. This means that IRM_{2T} in the surface levels is influenced by a magnetic mineral with low M_s, but contributing to IRM_{2T}, and not so much to ARM and χ. Thus, it could be tentatively ascribed to hematite. A small maximum in the concentration of the strongly magnetic fraction is also observed at the top of the B2 horizon, indicating local enhancement in this part of the profile.

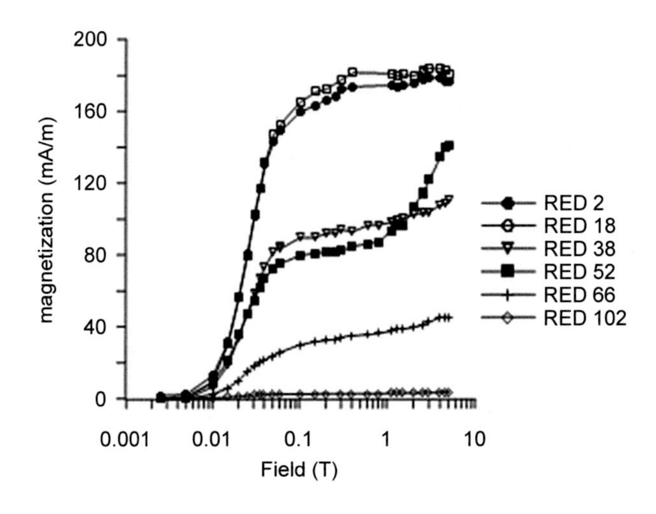

FIGURE 3.1.49

Stepwise acquisition of isothermal remanence (IRM) up to field of 5 T for selected samples from profile RED.

Table 3.1.6 Summary parameters from coercivity analysis by fitting IRM acquisition curves with cumulative Gaussian functions, using IRM-CLG1.0 software (Kruiver et al., 2001) for selected samples from the RED profile

Profile RED, Rhodic Luvisol									
	IRM component 1			IRM component 2			IRM component 3		
Depth (cm)	% from IRM_{total}	$B_{1/2}$ (mT)	DP	% from IRM_{total}	$B_{1/2}$ (mT)	DP	% from IRM_{total}	$B_{1/2}$ (mT)	DP
2	82	22.9	0.22	13	141.3	0.27	5	3020.0	0.28
28	89	24.5	0.22	9	162.2	0.35	2	3548.0	0.50
38	71	23.4	0.27	6	281.8	0.32	23	3981.1	0.40
52	51	22.9	0.22	5	316.2	0.20	44	3162.3	0.30
66	50	25.1	0.30	8	229.1	0.27	42	3981.1	0.48
102	59	20.9	0.28	4	251.2	0.55	37	3981.1	0.52

The relative intensity, median acquisition field ($B_{1/2}$), and dispersion parameter (DP) for each coercivity component are shown.

The S ratio calculated for all samples using Saturation Isothermal Remanent Magnetization (SIRM) acquired in the 2-T field and IRM = 100 mT for the low-coericivity fraction (Fig. 3.1.50f) shows high values in the AE, B1, and B2 horizons and slightly lower values in the BC and C horizons. However, using a field of 5 T for selected samples (Fig. 3.1.50e) in the calculation of the S ratio reveals very low S values all along the depth, but, again, higher S values around 0.6 are pertinent to AE and B1, while in BC it reaches S = 0.3. Therefore, the high-coercivity magnetic fraction that saturates mainly in DC fields between 2 and 5 T is detected in all horizons. The amount of this fraction generally increases toward the C horizon. Bearing in mind the high saturating fields, as well as the results from the

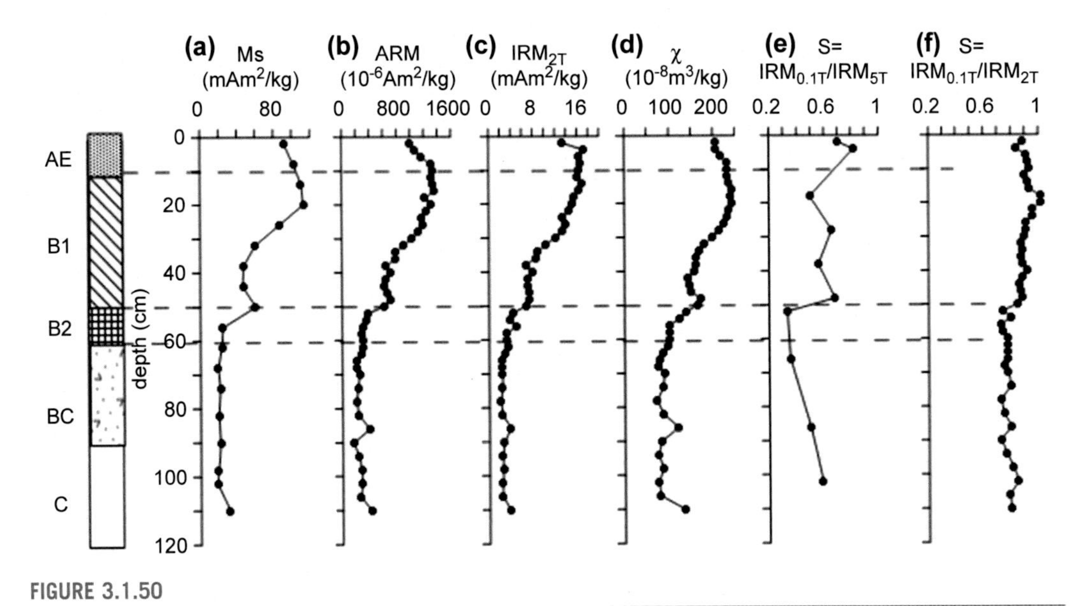

FIGURE 3.1.50

Soil profile RED. Depth variations of concentration-dependent magnetic characteristics: (a) saturation magnetization (M_s), (b) anhysteretic remanence (ARM), (c) isothermal remanence , acquired in a field of 2 T (IRM_{2T}), (d) magnetic susceptibility (χ), and S ratios, obtained using two maximum fields for SIRM (5 and 2 T) (e) and (f).

stepwise thermal demagnetization of the magnetically hard IRM component, it could be concluded that goethite is present in a lower amount in the upper part of the profile and is abundant in deeper levels. However, as shown by Liu et al. (2007), in environments with complex iron oxide mixtures involving strongly magnetic ferrimagnetic minerals as well as antiferromagnetic phases (hematite and goethite), the S ratio could be influenced not only by the concentration of high-coercivity minerals but also by the grain-size variations. In order to check for such a possibility, the behavior of different grain-size proxies along the solum of profile RED is plotted in Fig. 3.1.51.

Hysteresis parameters, usually interpreted as grain-size indicators for magnetite-bearing natural materials—M_{rs}/M_s and B_{cr}/B_c (Fig. 3.1.51a)—seem to indicate a finer grain size in AE plus B1 horizons and a coarser grain size in BC plus C horizons. However, as it became clear earlier, high-coercivity minerals (hematite and goethite) also contribute to the induced and remanent magnetization in the bulk soil samples. Thus, the ratios M_{rs}/M_s and B_{cr}/B_c are influenced more by the changing magnetic mineralogy down-profile, rather than a grain-size change in the strongly magnetic fraction. Frequency-dependent magnetic susceptibility (χ_{fd}) points to a pedogenic enhancement with very fine SP magnetite/maghemite particles in the upper AE-B1-B2 horizons (Fig. 3.1.51b), linked to soil formation. Therefore, variations in the percent frequency-dependent magnetic susceptibility (χ_{fd}%) (Fig. 3.1.51c) showing a progressive increase in χ_{fd}% in the AE-B1-B2 horizons, followed by constantly high ($\sim$12%) values in BC-C, could not be interpreted in a conventional way as being indicative for an increasing SP content with an increase in χ_{fd}%. This contradicts the fact that the absolute concentration of SP maghemite grains strongly decreases in the BC and C horizons (Fig. 3.1.51b).

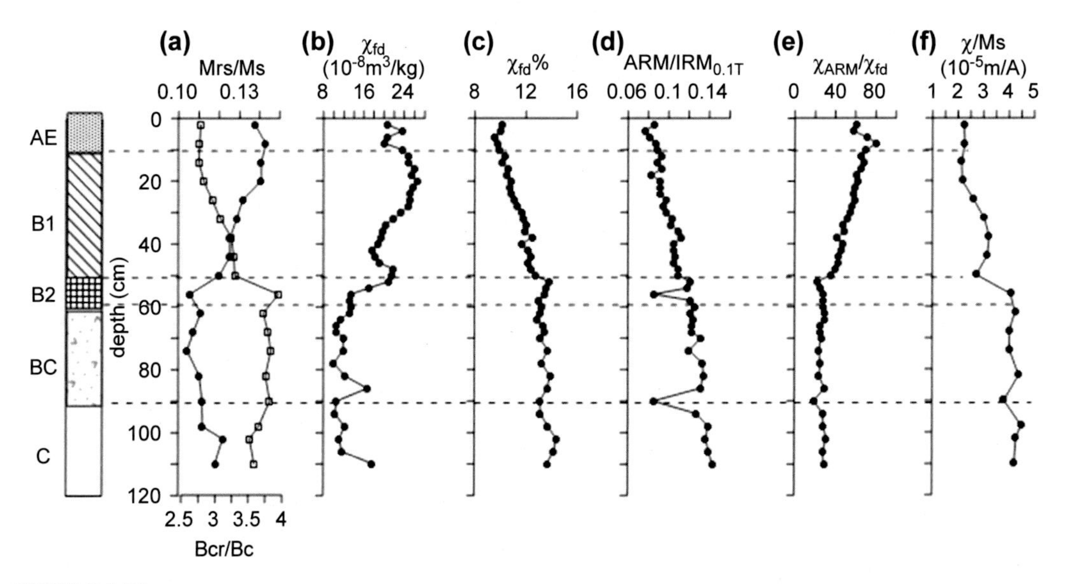

FIGURE 3.1.51

Soil profile RED. Depth variations of grain size–sensitive magnetic characteristics and ratios: (a) hysteresis ratios M_{rs}/M_s (*full dots*) and B_{cr}/B_c (*open dots*), (b) frequency dependent magnetic susceptibility χ_{fd}, (c) percent frequency dependent magnetic susceptibility $\chi_{fd}\%$, (d) $ARM/IRM_{0.1T}$ ratio, (e) χ_{ARM}/χ_{fd} ratio, and (f) χ/M_s ratio.

Grain-size variations in the low-coercivity ferrimagnetic component, linked to pedogenic maghemite, are studied through the parameters $ARM/IRM_{0.1T}$ and χ_{ARM}/χ_{fd} (Fig. 3.1.51d and e). A gradual increase in $ARM/IRM_{0.1T}$ until a 50-cm depth, pointing to an approach toward the SD threshold, is accompanied by a decrease in χ_{ARM}/χ_{fd} in the B1 horizon (suggesting a decreasing amount of the ARM-carrying fraction at the expense of the SP amount). Thus, magnetic grain-size proxies for the strongly magnetic fraction imply that the bottom part of B1 is enhanced with SP grains and small stable SD grains. Toward the surface, the concentration of the stable fraction increases (i.e., ARM has maxima there) and the SP fraction (χ_{fd}) increases (Fig. 3.1.51b). In the lower part of the profile, below ~60 cm, the ratio χ_{ARM}/χ_{fd} is constant, while $ARM/IRM_{0.1T}$ still increases. This means that $IRM_{0.1T}$ is increasing due to the contribution of nonpedogenic coarse (Ti) magnetite particles.

Based on the thermomagnetic analyses and coercivity distributions deduced from IRM-acquisition curves (Figs. 3.1.48 and 3.1.49), the following magnetic proxies for the content of maghemite, hematite, and goethite along the profile are proposed (Jordanova et al., 2013): (1) low-coercivity component $IRM_{0-0.15T}$ (isothermal remanence acquired in fields up to 0.15 mT) represents the contribution from pedogenic maghemite, lithogenic magnetite and coarse (titano)magnetite particles; (2) $IRM_{0.15-0.6T}$ reveals hematite content; and (3) the high-coercivity component $IRM_{0.6-5T}$ is reflecting the occurrence of pedogenic and lithogenic goethite. Variations in these three components are shown in Fig. 3.1.52, together with the parameters HIRM and L ratio. The HIRM is defined as 0.5 * (SIRM + $IRM_{300\ mT}$) (where IRM_{300mT} is the IRM induced in the opposite to SIRM direction and

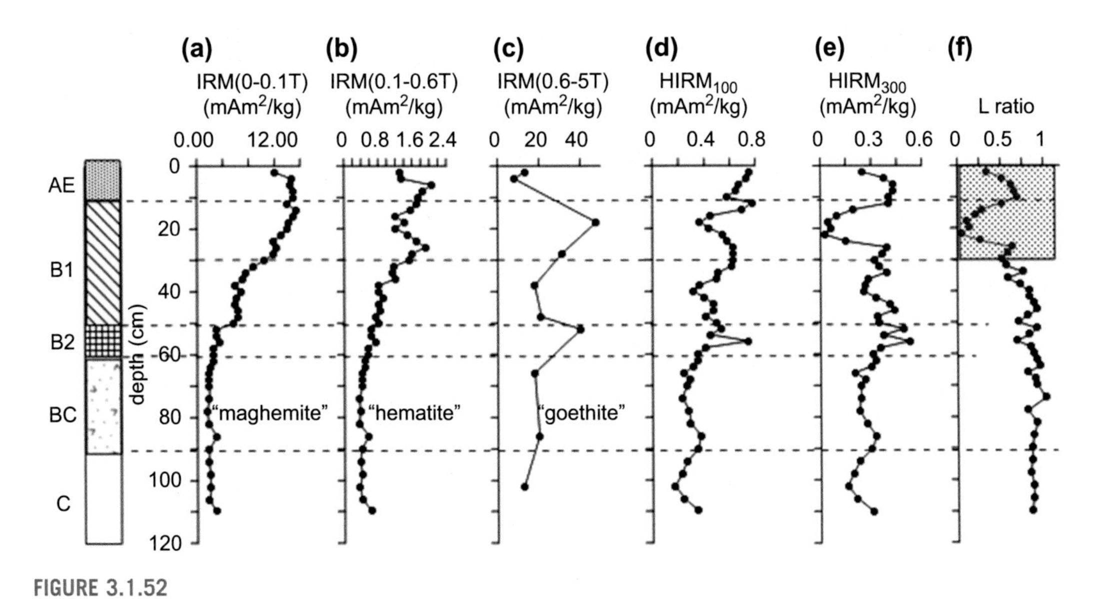

FIGURE 3.1.52

Soil profile RED. Variations of magnetic proxies for the content of maghemite, hematite, and goethite along the depth: (a) magnetic proxy for maghemite component, (b) magnetic proxy for hematite component, (c) magnetic proxy for goethite component, (d) "hard" isothermal remanence, calculated using a weak magnetic field of 100mT ($HIRM_{100}$), (e) "hard" isothermal remanence, calculated using a weak magnetic field of 300 mT ($HIRM_{300}$), and (f) L-ratio (definition in the text).

applying a field of 300 mT) and is commonly used to estimate the absolute concentration of hematite and/or goethite in mineral mixtures with magnetite/maghemite (Thompson and Oldfield, 1986; Liu et al., 2002). However, Liu et al. (2002) pointed out that such a straightforward interpretation of HIRM can be misleading when the remanence carried by hematite/goethite is masked by a strongly magnetic ferrimagnetic component and, in fact, HIRM has a similar magnitude to the measurement errors. In addition, in mixtures containing hematite and goethite, variations in HIRM could be due not only to changes in the relative abundance of these minerals but also to changes in grain size and/or isomorphic cation substitutions (Liu et al., 2007). The cited authors propose the use of a new parameter L $\{L = HIRM/[0.5 * (SIRM + IRM_{100\ mT})]\}$, (where IRM_{100mT} is the IRM induced in the opposite to SIRM direction and applying a field of 100 mT) which helps to determine whether samples contain hematite and goethite with variable coercivities in mixtures with strongly magnetic minerals. The "maghemite" component generally follows the behavior of magnetic susceptibility and reveals significant enhancement in the upper AE and B1 horizons with SP, SD, as well as coarser grains. The "hematite" component shows two local maxima in the AE and B1 horizons (Fig. 3.1.52b) with a gradually decreasing intensity down-profile. The "goethite" magnetic proxy has a maximum in the upper parts of B1 and B2. However, as demonstrated by Liu et al. (2007), when HIRM and L show linear dependence, then HIRM variations reflect changing coercivities in the hematite/goethite fraction. Such a relation is observed in the upper 30 cm of the profile (denoted by the red line in Fig. 3.1.52). Therefore, HIRM and L variations in this part of the profile are a sign of increasing coercivity of the hematite/goethite fraction. Considering this fact in combination with the variations in the "goethite" magnetic proxy, it could be supposed that in the interval of 10–24 cm, the goethite share in the high-coercivity fraction increases. Downwards, the L ratio and HIRM do not show a linear relation, which supposes that the maxima on $HIRM_{100}$ in the interval 25–35 cm (Fig. 3.1.52d) reflects an increased concentration of pedogenic hematite, as also supported by the "hematite" magnetic proxy (Fig. 3.1.52b). Local enhancement with hematite is also seen in the B2 horizon.

Magnetic proxies for the estimation of goethite abundance in the studied profile detect in fact only a part of the total goethite content, since the applied room-temperature measurements do not account for the para- and superparamagnetic goethite, as well as goethite with coercivities higher than 5 T. This is demonstrated by Jordanova et al. (2013), who compare the "magnetic proxy" goethite content and DRS-Fe_d-proxy for goethite (Fig. 3.1.53). The latter reveals constant high values of the ratio Hm/(Hm + Gt) of ~0.8 in the soil horizons and a gradual decrease toward the C horizon. In contrast, the "magnetic proxy" Hm/(Hm + Gt) ratio in the upper 0–30 cm is also constant, but the absolute value of the ratio is half that of the DRS-estimated ratio. In the interval 40–60 cm, virtually no goethite is detected by the magnetic ratio, meaning that it should be in a para- or SP state, while down-profile magnetic Hm/(Hm + Gt) is non-zero. Goethite in the upper soil horizon is supposed to be of pedogenic origin, while in the BC and C horizon, it is of lithogenic origin (Jordanova et al., 2013).

The hypothesis of significant African dust input into the solum of red Mediterranean soils (Muhs and Budahn, 2009; Stuut et al., 2009; Erel and Torrent, 2010; Maher et al., 2010; Muhs et al., 2010a,b; Muhs, 2013) is tested if it also applies to the Bulgarian Rhodic Luvisol. For this purpose, a rare earth

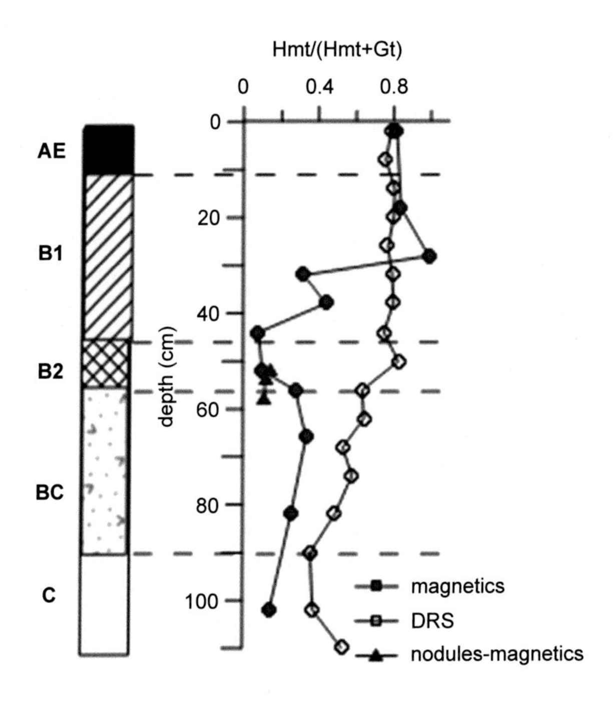

FIGURE 3.1.53

Profile of Hm/(Hm + Gt) ratio, obtained from DRS data and from magnetic proxies (explanations in the text).

Redrawn from Jordanova, N., Jordanova, D., Liu, Q., Hu, P., Petrov, P., Petrovský, E., 2013. Soil formation and mineralogy of a Rhodic Luvisol – insights from magnetic and geochemical studies. Glob. Planet. Change 110, 397–413, with permission from Elsevier Science.

elements (REE) analysis of selected soil samples and a limestone sample has been carried out, together with an analysis of the magnetic signature in the soil. The thin B2 horizon, characterized by a spike-like increase of concentration of the strongly magnetic SP and SD particles (Figs. 3.1.50 and 3.1.51), is also characterized by a different REE pattern, compared with the rest of the soil depths, as well as compared with the parent rock composition (Jordanova et al., 2013). It is proved that this depth interval represents increased dust input from Saharan/Sahel sources containing a high amount of maghemite and goethite.

One of the most disputable issues in the studies of red Mediterranean soils is the genetic relationship and proportions of pedogenic hematite, maghemite, and goethite. Experimental data concerning this problem are at the core of the theory that soil maghemite is an intermediate product in the chain transformation of ferrihydrite to hematite (Barrón and Torrent, 2002; Michel et al., 2010). According to the proposed model (Barrón and Torrent, 2002; Michel et al., 2010), in well-drained soils developed in a Mediterranean climate, an intermediate phase called "hydromaghemite" forms during the transformation of ferrihydrate into hematite. The important role of phosphate and/or other ligands in pore water directs the transformation toward maghemite formation. It is suggested that in the absence of such ligands, ferrihydrate transforms directly into hematite or goethite (Cornell and Schwertmann, 2003). Similar to red-colored soils from Spain, the main pedogenic minerals identified in the Rhodic Luvisol from Bulgaria through magnetic methods are maghemite, goethite, and hematite. Torrent et al. (2006) demonstrated that the plot of the χ_{fd} vs Hm content shows a linear trend, and the slope of the obtained regression depends on the climatic conditions (mainly precipitation). The authors postulate that for Mediterranean soils the Hm/χ_{fd} ratio is $> 5 \times 10^7$ g/m^3, while for soils and paleosols on loess in temperate areas (the Russian steppe, Chinese Loess Plateau paleosols), $Hm/\chi_{fd} < 5 \times 10^7$ g/m^3. For well-drained Brazilian Ferralsols, this ratio is $> 20 \times 10^7$ g/m^3 (Torrent et al., 2006). A compilation of data for χ_{fd} and the corresponding Hm content for samples from the Rhodic Luvisol (profile RED) and Haplic Luvisol (profile CIN) is shown in Fig. 3.1.54. A well-defined

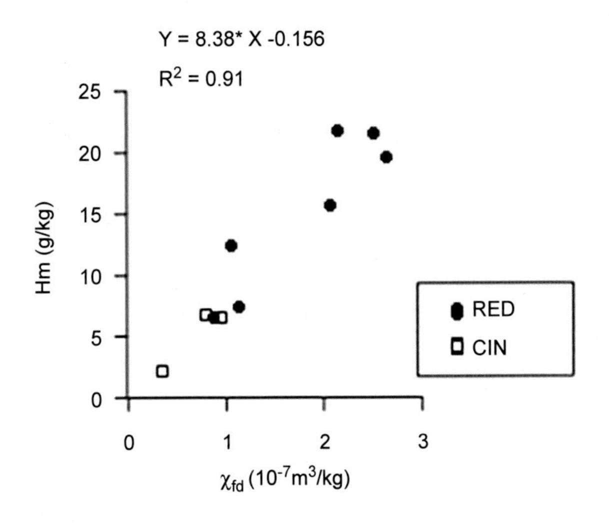

FIGURE 3.1.54

Dependence between the content of hematite (Hm), estimated from DRS measurements and chemical extraction data, and the frequency-dependent magnetic susceptibility (χ_{fd}) for samples from profiles of Haplic Luvisol (CIN) and Rhodic Luvisol (RED).

linear regression fits the relationship between χ_{fd} and Hm and shows a slope of 8.38. This value falls well into the range defined for the Mediterranean rubefied soils. Thus, data obtained from Bulgarian Luvisols support the model for pedogenic enhancement, proposed by Barrón and Torrent (2002).

Since only the two Luvisol profiles (CIN and RED) of the four examined showed the presence of pedogenic hematite (according to DRS results) and they both well fit a single trend, defined by the relation between Hm content and χ_{fd} (Fig. 3.1.54) for Mediterranean soils, it is worth comparing their magnetic signature. Profile CIN shows moderate magnetic enhancement ($\chi \sim 70–80 \times 10^{-8}$ m^3/kg) in the solum, and the grain-size magnetic proxies (Fig. 3.1.28) suggest both an increasing concentration and grain size of ferrimagnetic grains from the bottom of the illuvial horizon toward the surface. At the same time, no hematite was detected through magnetic methods—only through the analysis of DRS spectra. This suggests that hematite is in an amorphous state and does not contribute to the measured magnetic parameters. On the other hand, in the Rhodic Luvisol (profile RED, typical Mediterranean soil), hematite, along with maghemite and goethite, was detected (Figs. 3.1.51 and 3.1.52). Grain-size proxies for maghemite suggest that concentrations of SD and SP fractions do not co-vary in the surface horizon (AE). Therefore, in more mature Luvisol (RED), pedogenic development results in the coexistence of maghemite, hematite, and goethite (with hematite prevailing over goethite), while in the younger Haplic Luvisol (CIN), the ferrimagnetic fraction dominates the magnetic properties. Correspondingly, the concentration of Hm in CIN is lower compared with RED (Fig. 3.1.54). A similar conclusion, related to the degree of development and age, is drawn by Haidouti and Massas (1998) in a study of Haploxeralf and Rhodoxeralfs from Greece. The iron extraction data showing higher Feo/Fed in Palexeralfs as compared with Rhodoxeralf, as well as a similar relation in Mn-extraction data and the Redness Rating data, led the authors to the conclusion that Haploxeralfs are less-developed soils than Rhodoxeralfs.

Based on the available published data, some major characteristics of different Luvisols are summarized in Table 3.1.7. The soil profiles originate from Bulgaria, Spain, Iran, the United States, and Brazil. Obviously, the magnetic susceptibility (χ) of the soil horizons (humic and illuvial) varies across a wide range, which probably depends on the parent material and its composition, as well as the possible influence of micro-relief and local hydrological conditions. However, the other magnetic characteristic ($\chi_{fd}\%$) shows high values for most of the profiles—between 10% and 14%—except for profiles F and BGU, where lithogenic magnetic minerals dominate the magnetic signal (Table 3.1.7), and some of the soils from Iran located at higher altitudes. There is, however, no single relationship between the magnetic susceptibility (or frequency-dependent magnetic susceptibility) and the main climatic variables (MAP and MAT) (Table 3.1.7). Thus, a characteristic feature of most of the Luvisols is their strong enhancement with SP grains, which dominate the magnetic susceptibility. Similar to magnetic susceptibility, the ratio Hm/(Hm + Gt) also shows broad variations—from 0 to 0.8—implying that hematite is not always present in every Luvisol. It is remarkable that the two profiles from Bulgaria—F and PRV - which show medium to strong signs of leaching and eluviation, do not contain hematite, but lithogenic goethite prevails. A similar feature is obtained in profiles of some Typic and Ultic Palexeralfs from alluvial chronosequence in Spain (see Table 3.1.7), which are developed on higher terraces and influenced by a seasonal perched water table (Torrent et al., 2010a,b; Liu et al., 2010), cusing reductive dissolution of hematite. Magnetic susceptibility in these profiles is also low, pointing to the influence of reducing conditions on iron oxides (Maher, 1998; Cornell and Schwertmann, 2003).

Table 3.1.7 Compilation of magnetic susceptibility values (the average value in the most enhanced A and B horizons), the average value of the percent frequency-dependent magnetic susceptibility ($\chi_{fd}\%$) in magnetically enhanced horizons, the Hm/(Hm + Gt) ratio, the Fe_d/Fe_{tot} ratio (Fe_d/Fe_{tot}), and the weathering index CIA from published data on magnetic properties of Luvisol type soils in the world

Location	Soil type WRB 2014	Soil taxonomy USDA 2010	MAT (°C)	MAP (mm)	Elevation m a.s.l.	χ (10^{-8} m^3/kg)	$\chi_{fd}\%$	Hm/(Hm + Gt)	Fe_d/Fe_{tot}	CIA	References
Bulgaria	Haplic Luvisol, Chromic (CIN)	Haplic Palexeralf	11.5	529	274	80–90	10	0.71–0.76	0.27	37–42	This study
Bulgaria	Epicalcic Luvisol, (F)	Typic Palexeralf	12.5	660	314	320–400	1–2.5	0.0			This study
Bulgaria	Epicalcic Cutanic Luvisol (BGU)	Typic Paleustalf	11.0	830	435	20–60	2–5	0.0		71–77	This study
Bulgaria	Cutanic Luvisol (Belogradchik)	Typic Hapludalf	10	694	40 0	120–160	9–12				Grison et al. (2011)
Spain	Calcic Luvisol Chromic (RB)	Typic Palexeralf	17	600		30–40	13–14	0.6	0.38–0.5		Torrent et al. (2010b)
Spain	Calcic Luvisol Chromic (MO)	Typic Palexeralf	17	600		50–100	14–16	0.7–0.8	0.48–0.5		Torrent et al. (2010b)
Spain		Typic Haploxeralf	11.5	450	~800 average	34–68	11–13	0.17–0.35	0.7–0.76		Torrent et al. (2010a) and Liu et al. (2010)
		Typic Palexeralf									
		Ultic Palexeralf				48–68	11–12	0.09	0.62–0.79		
						60–136	10–13	0.3–0.4	0.71–0.83		
						30–50	7–11	0–0.12	0.66–0.86		
Bulgaria	Rhodic Luvisol (RED)	Typic Rhodustalfs	12	771	253	200–250	10–12	0.7–0.8	0.6–0.8	80–84	Jordanova et al. (2013) and this study
Iran, Kohgilouye province		Typic Haploxeralf	11	1009	2250	17–24	3–4		0.47–0.63		Owliaie et al. (2006)
Iran, Kohgilouye province		Calcic Rhodoxeralf	10	789	2280	50–59	5.2–5.3		0.54–0.72		

Continued

Table 3.1.7 Compilation of magnetic susceptibility values (the average value in the most enhanced A and B horizons), the average value of the percent frequency-dependent magnetic susceptibility ($\chi_{fd}\%$) in magnetically enhanced horizons, the Hm/(Hm + Gt) ratio, the Fe_d/Fe_{tot} ratio (Fe_d/Fe_{tot}), and the weathering index CIA from published data on magnetic properties of Luvisol type soils in the world—cont'd

Location	Soil type WRB 2014	Soil taxonomy USDA 2010	MAT (°C)	MAP (mm)	Elevation m a.s.l.	χ (10^{-8} m^3/kg)	$\chi_{fd}\%$	Hm/(Hm + Gt)	Fe_d/Fe_{tot}	CIA	References
Iran, Kohgilouye province		Calcic Haploxeralf	15	852	1800	27–31	3.3–3.9		0.44–0.52		
Iran, Kohgilouye province		Calcic Haplustalf	20	600	1000	43–48	6.8		0.35–0.38		
Iran, Kohgilouye province		Typic Haploxeralf	23	534	800	26–28	4.6–5.3		0.24		
Iran, Kohgilouye province		Calcic Haploxeralf	15	750	1300	76–126	8.7–10.7		0.49–0.65		
USA, coast range, King range chronosequence		Typic Haploxeralf	12–14	1000–1500		66	6–7				Fine et al. (1992) and Singer et al. (1992)
USA, coast range, King range chronosequence		Typic Hapludalt	12–14	1000–1500		250–300	10				
USA, Sacramento Valley seq.		Mollic Palexeralf	~10	~530		300–320	6–10				
USA, Sacramento Valley seq.		Ultic Palexeralf	~10	~530		400–500	6–12				
USA, California, Mendocino sequence		Ultic Haplustalf	~12.5	~1000		44–53					Fine et al. (1989)
Brazil, town of Catanduva (Sao Paulo state)		Typic Haplustalfs	22,4	1350	~520	108–374			0.46–0.77		Siqueira et al. (2010)

Soil types are indicated according to the authors' classification. Soil classification of the Bulgarian profiles, included in this study, is according to Shishkov (2011).

The Fe_d/Fe_{tot} ratio, used as an indicator for soil maturity, generally shows quite high values in Luvisols, indicating the prevailing role of better crystalline iron oxides in the total iron oxides content, compatible with relatively high values of the weathering index CIA calculated for the profiles BGU and RED (Table 3.1.7). Luvisols from the United States show the highest values of the magnetic susceptibility in the soil A/B horizons.

3.2 ALISOLS AND ACRISOLS

Alisols are a characteristic soil type for the humid tropics, subtropics, and warm temperate areas with active intense weathering. These soils show strong weathering of primary minerals, a loss of silica, and the neoformation of clay minerals as a result of transformations in weathering by-products (Schaetzl and Anderson, 2009). Often, Alisols are reported in association with Acrisols, but Alisols' clay content is of mixed mineralogy and they have high exchangeable aluminum content. With proceeding weathering, B_t horizons become enriched with clay, de-silication is advancing, and only quartz remains from the primary minerals in the profile (Spaargaren, 2008). Red and yellow colors dominate in tropical soils because of the abundance of Fe_2O_3 and Al_2O_3 minerals (Cornell and Schwertmann, 2003; Fritsch et al., 2005, 2007). In Alisols, secondary high-activity clays control the clay composition and release a considerable amount of Al. The soil profiles are characterized by an ochric surface horizon with a loose structure, which overlies a well-distinguished argic horizon. Alisols usually develop on old geomorphic surfaces. They are acid soils, rich in iron oxides and kaolinite (IUSS Working Group WRB, 2014; Schaetzl and Anderson, 2009; Jien et al., 2010).

Alisols are found mainly in Latin America (Ecuador, Nicaragua, Venezuela, Colombia, Peru, Brazil), West India, Africa, Southeast Asia, and northern Australia, but they also occur in subtropical and Mediterranean regions (China, Japan, the southeastern United States, the Iberian Peninsula, Greece) (Fig. 3.2.1a) and in restricted areas in Bulgaria and Romania (Tóth et al., 2008). Acrisols are characterized by a clay-enriched subsoil with low-activity clays and a low base saturation. Acrisols are also found in humid tropical, humid subtropical, and warm temperate regions. They are widespread in Southeastern Asia, the Amazon Basin, the southeastern United States, and Africa (Fig. 3.2.1b).

According to the Bulgarian soil classification system, the yellow podzolic soils are correlated to Alisols/Acrisols (Shishkov, 2011; Shishkov and Kolev, 2014; Ninov, 2002). Most of the yellow podzolic soils are considered as the oldest ones found in Bulgaria, but they do not correspond to the Russian yellow soils. There is still ongoing debate in Bulgarian pedology concerning the exact correlation of this specific soil to international classification systems. This is reflected in an apparent controversy: on the Soil Map of Europe 1:1,000,000, the soil is referred to as Stagnic Luvisol, while in the European Soil Database, it is Haplic Acrisol according to the FAO Revised Legend, 1990 (Shishkov and Kolev, 2014). In Bulgaria, Alisols/Acrisols have restricted distribution in the most southeastern parts of the country, influenced by the Mediterranean climate where the Rhodic Luvisols are found as well. They are formed under deciduous forests (beech, oak) as well as relict evergreen rhododendrons (*Rhododendron ponticum*), which is unique vegetation preserved since the Tertiary (Shishkov and Kolev, 2014).

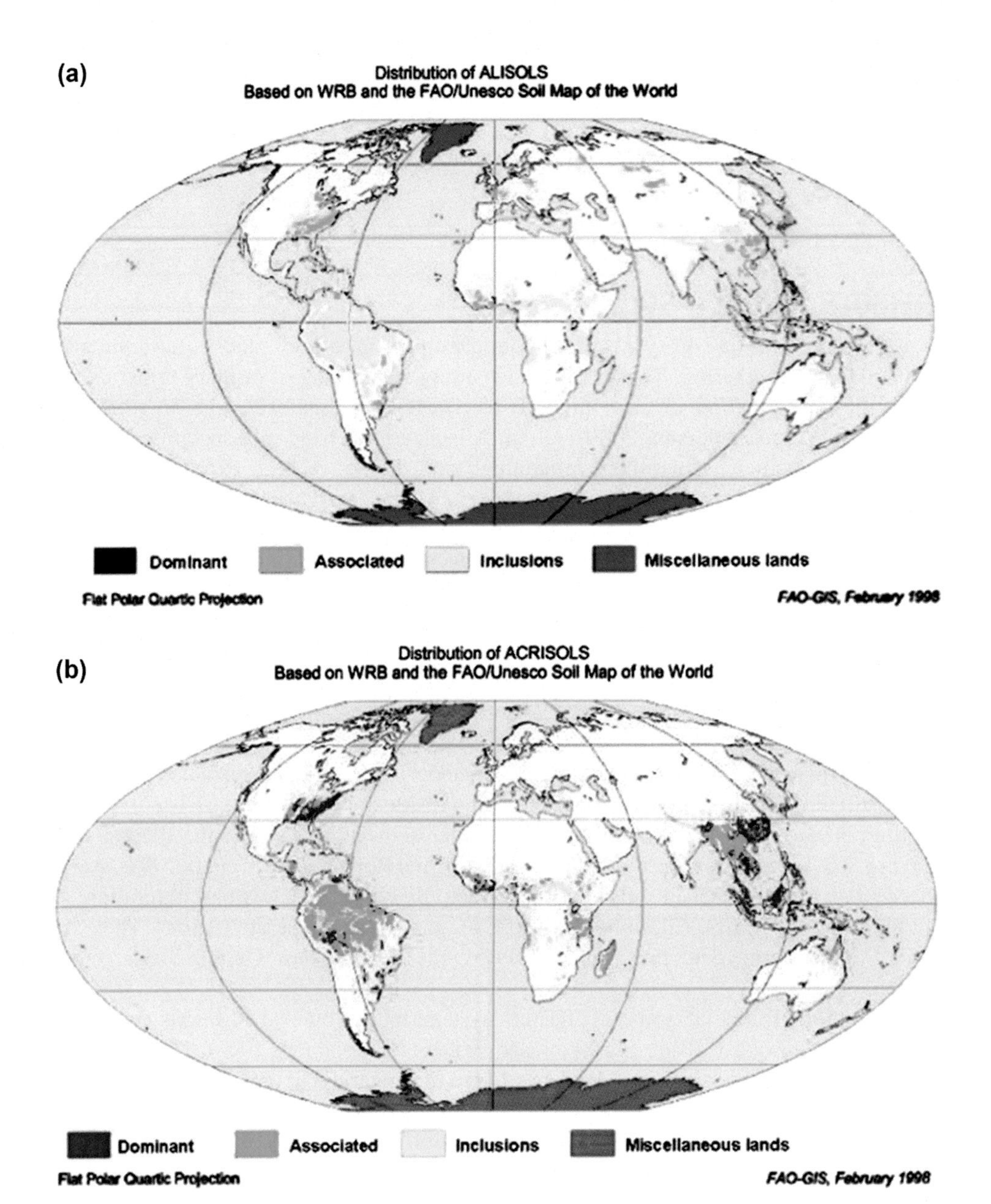

FIGURE 3.2.1

Spatial distribution of Alisols (a) and Acrisols (b) worldwide (FAO, 2001). Lecture notes on the major soils of the world. http://www.fao.org/docrep/003/y1899e/y1899e08a.htm#P304_38985).

3.2.1 PROFILE LOCATION AND DESCRIPTION

Alisol, *profile designation: YPS*

Location: N 42°29′16.7″ E 27°09′49.4″, altitude 250 m a.s.l.

Present-day climate: MAT = 13.1°C, MAP = 931 mm

Vegetation is represented predominantly by beech (*Fagus orientalis*), oak (*Quercus polycarpa*), and rhododendron (*Rhododendron ponticum*), relief: flat

Description of soil profile general view and details from the horizons are shown in Figs. 3.2.2–3.2.6

AE horizon: 0–2 cm, thin humic-eluvial horizon, dark gray-black color
E1 horizon: 2–17 cm, eluvial horizon, bleached, light yellow color, loose, sandy
E2 horizon: 17–27 cm, eluvial horizon, bleached, yellow color
$B_{tg}1$ horizon: 27–60 cm, illuvial horizon, red-brown color, prismatic structure, clayey, dense
$B_{tg}2$ horizon: 60–110 cm, illuvial horizon, brown color, clayey, Fe-Mn concretions present
C1 horizon: 110–130 cm, C horizon, light brown–beige color, orange mottles, amorphous Fe-Mn masses and gleyic spots visible in the bottom part
C2 horizon: C horizon, light beige, weathered shists

FIGURE 3.2.2

View of the profile of Alisol (profile YPS).

FIGURE 3.2.3

Profile YPS. A-E1-E2 horizons—detail.

FIGURE 3.2.4

Profile YPS. $B_{tg}1$ horizon—detail.

FIGURE 3.2.5

Profile YPS. $B_{tg}2$ horizon—detail.

FIGURE 3.2.6

Profile YPS. C1 horizon—detail.

3.2.2 TEXTURE, SOIL REACTION, AND MAJOR GEOCHEMISTRY OF ALISOL (YELLOW PODZOLIC SOIL; PROFILE YPS)

The mechanical composition of the yellow podzolic soil is dominated by the silt fraction, which is the highest in the upper eluvial horizons (Fig. 3.2.7a–c), while in the illuvial horizon, the clay is more enriched. The sand fraction comprises a small part of the texture, but it shows well-expressed maxima in the illuvial horizons. The soil is acidic (Fig. 3.2.7d) with minimum pH values in the bleached eluvial E1 horizon, where the pH ~4.5. A gradual increase in $B_{tg}1$ up to pH ~5.4 is observed, which remains constant in $B_{tg}2$. In the C horizon, the pH increases up to ~5.8. The strongly acidic pH obtained in the upper soil horizons addresses the enhanced weathering and destruction of the primary minerals. Iron

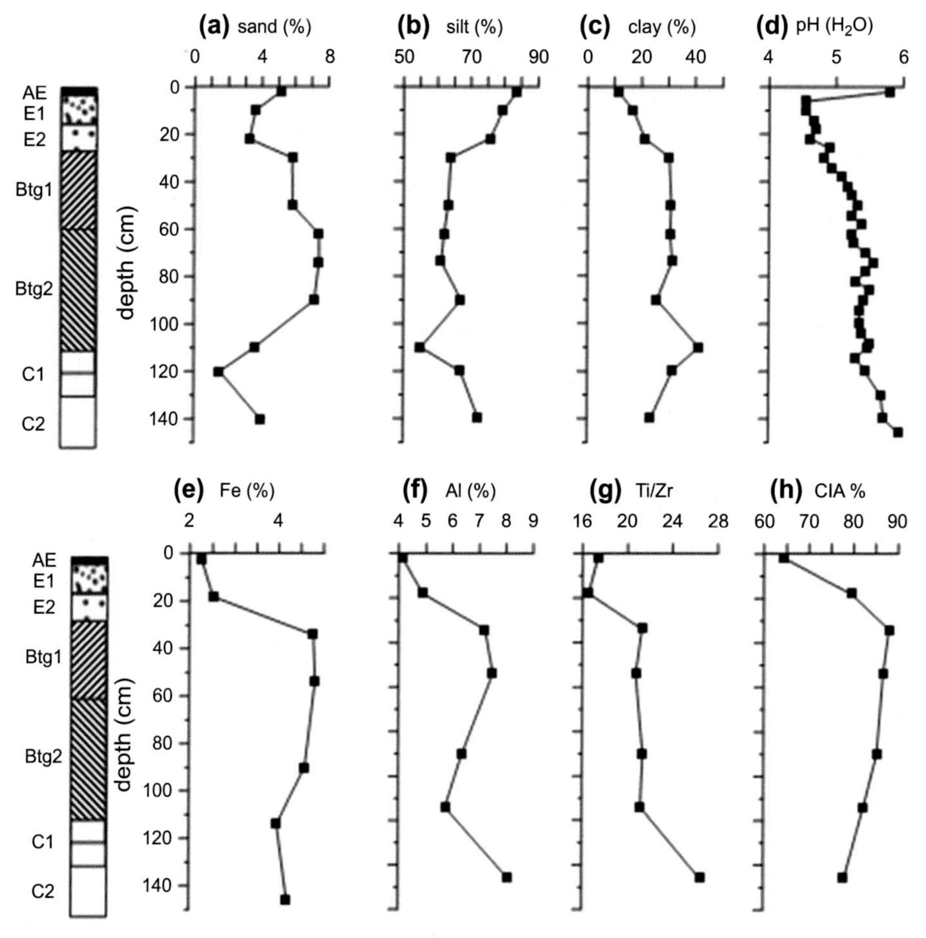

FIGURE 3.2.7

Mechanical fractions (sand, silt, clay) (a–c) and soil reaction pH (d) along the YPS profile. Content of aluminum (Al), iron (Fe) (e–f), ratio Ti/Zr, and the Chemical Index of Alteration (CIA) (g–h) along depth of profile.

Table 3.2.1 Chemical extraction data for selected samples from YPS profile (Alisol)

Sample depth (cm)	Al_o (g/kg)	Fe_o (g/kg)	Mn_o (g/kg)	Al_d (g/kg)	Mn_d (g/kg)	Fe_d (g/kg)	Fe_{tot} (g/kg)	Al_{tot} (g/kg)	Mn_{tot} (g/kg)	Fe_d/Fe_{tot}	Fe_o/Fe_d	C_{org} %
2	1.09	1.90	1.67	1.25	1.79	7.48	22.50	41.21	1.92	0.33	0.25	10.03
18	2.10	3.54	1.21	2.07	1.02	11.82	25.05	48.77	1.48	0.47	0.30	1.36
34	3.36	1.23	0.04	4.25	0.06	23.42	47.69	72.16	0.21	0.49	0.05	
54	3.14	1.13	0.03				48.06	75.08	0.16			
90	2.71	1.39	0.00	3.09	0.01	20.17	45.80	63.61	0.12	0.44	0.07	
114	2.27	1.72	0.00				39.57	57.35	0.11			1.64
146	1.18	1.29	0.96	1.14	0.76	8.72	41.91	80.97	1.07	0.21	0.15	

Oxalate-extractable aluminum (Al_o), iron (Fe_o), and manganese (Mn_o); dithionite-extractable aluminum (Al_d), iron (Fe_d), and manganese (Mn_d), total Al, Fe, Mn content (Al_{tot}, Fe_{tot}, Mn_{tot}), the ratios Fe_d/Fe_{tot} and Fe_o/Fe_d, and the organic carbon content (C_{org}) for selected samples are shown.

and aluminum are strongly bound to the illuvial horizons, while depletion is obvious in the eluvial horizons (Fig. 3.2.7e and f). An advanced weathering state is also evidenced by the high values of the chemical index of weathering CIA (Nesbitt and Young, 1982), reaching 90% in $B_{tg}2$ (Fig. 3.2.7h). Relatively weak variations in the Ti/Zr ratio suggest a uniformity of the parent material.

Chemical extraction data for oxalate- and dithionite-extractible Fe, Mn, and Al, as well as the total Fe content and determinations of organic carbon (C_{org}) for the selected samples, are summarized in Table 3.2.1. The relative importance of the free iron oxides expressed through the Fe_d/Fe_{tot} ratio shows relatively high values (0.3–0.5), notwithstanding the observed stagnic features in the profile. As a comparison, red-yellow soils from the upper Amazon basin (Fritsch et al., 2005) exhibit Fe_d/Fe_{tot} in the range 0.44–0.63. The second ratio—Fe_o/Fe_d—points to a significant amount of amorphous iron oxides in the uppermost soil horizons in the YPS profile (AE and E1), while downward, it decreases significantly (Table 3.2.1). The amount of oxalate extractable iron (Fe_o) is fairly high (1–3.4 g/kg) compared with the typical values obtained for tropical soils in the Amazon Basin (<1 g/kg) (Fritsch et al., 2007) as well as for red Ultisols (Kandiudults) from Thailand (Trakoonyingcharoen et al., 2006). However, regarding the depth variations in the Fe_o/Fe_d ratio, the obtained values for the profile YPS and tropical profiles from Amazonia (Fritsch et al., 2007) and Thailand (Trakoonyingcharoen et al., 2006) correlate very well, showing high values (0.2–0.3) in the humus-enriched uppermost horizons and almost zero in deeper soil levels (Table 3.2.1). It is also noticeable that almost the whole amount of Manganese is dithionite-extractable in these horizons, unlike in deeper soil horizons.

3.2.3 MAGNETIC PROPERTIES OF THE ALISOL PROFILE YPS

Thermomagnetic analyses of selected samples from the YPS profile display significant mineralogical transformations, occurring during the heating/cooling cycle (Fig. 3.2.8), expressed as sharp peaks. The topmost sample YPS2 has heating-cooling curves that resemble the characteristics of thermomagnetic curves for the surface organic-rich levels in Luvisols and Chernozems, suggesting the presence of maghemite and magnetite. The magnetite phase is well expressed for sample YPS2 with sharp T_c at 580°C. In contrast to this surface level, samples from the E1 (YPS 8) and E2 horizons (YPS 18) show

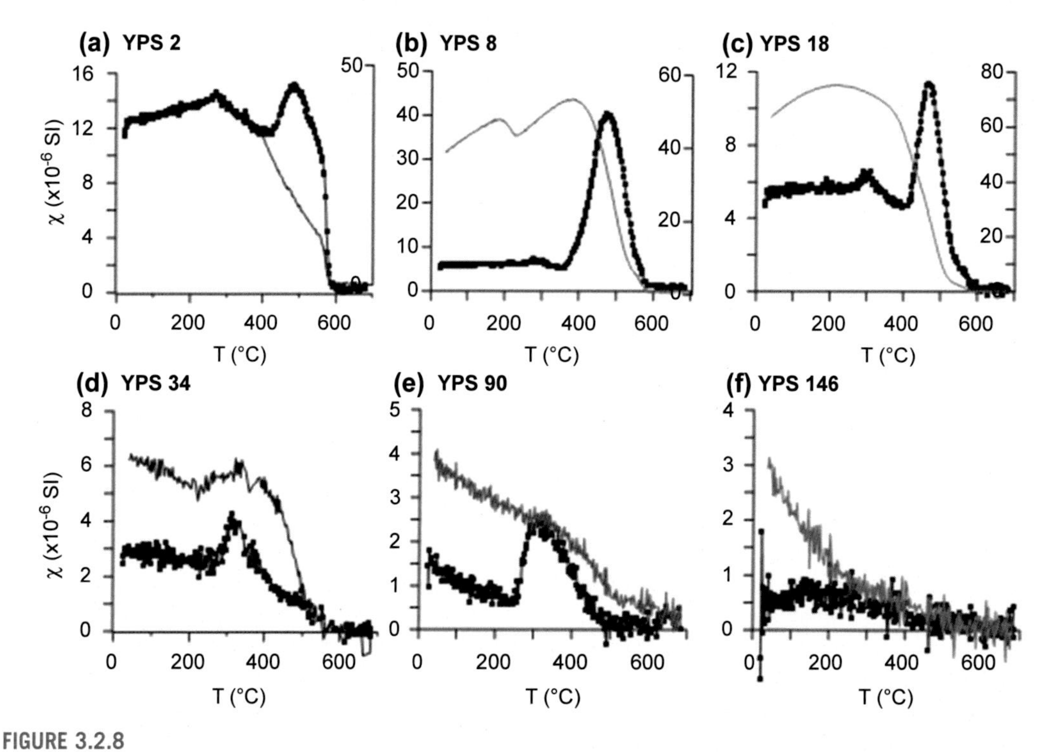

FIGURE 3.2.8

Thermomagnetic analyses of magnetic susceptibility for samples from different horizons of profile YPS. *Thick dotted line*—heating run, *thin line*—cooling run. Heating in air, fast heating rate (11°C/min).

transformation peaks, starting after ~400°C with maxima at ~480°C. A small bump at ~300°C highlights the presence of a certain amount of maghemite as well (Fig. 3.2.8b and c). Taking into account the very high intensity of the transformation peak and the established specific behavior of different iron oxyhydroxides in the presence of organic matter (Hanesch et al., 2006), it could be hypothesized that goethite in combination with organic matter are present in the eluvial horizons. Samples from the illuvial horizons (YPS 34 and YPS 90) show a different transformation picture with the formation of a new stronger magnetic phase after ~200°C with maxima at ~300°C and a smooth decrease until 600°C (Fig. 3.2.8d and e).

As demonstrated by Hanesch et al. (2006), two iron hydroxides start transforming into other phases at ~200°C—ferrihydrite and lepidocrocite. The lepidocrocite thermal transformation temperature depends on the crystallinity of the mineral, as well as on the substituting ions in the crystal lattice (Gehring and Hofmeister, 1994; Morris et al., 1998; Mitov et al., 2002). Lepidocrocite is usually found in soils with impeded drainage, like the pseudogley soils from New Zealand, South Africa, and Australia in a microclimate characterized by high precipitation (800–1200 mm) (Fitzpatrick et al., 1985). On the other hand, ferrihydrite is also common in soils (Cornell and Schewertmann, 2003) and especially in horizons experiencing fluctuating reduction conditions (van Breemen, 1988). During heating, ferrihydrite shows an endothermic peak at 220°C (Mitov et al., 2002), after which it transforms to maghemite (Eggleton and Fitzpatrick, 1988). Similar to goethite, the exact transformation temperature again depends on the crystallinity and foreign substitutions, especially Al (Schwertmann, 1988). Apart from the starting transformation temperature, the shape of the transformation peak in the presence of organics or without it is also diagnostic for the exact phase of the oxyhydroxide mineral (Hanesch et al., 2006). Considering the heating curves for samples YPS 34 and YPS 90 (Fig. 3.2.8d and e) in this respect, it seems more appropriate to assign the observed behavior to heating ferrihydrite without organics (this condition is met in the illuvial horizons).

Further information regarding the iron oxide forms in soil YPS is obtained through a component analysis of IRM-acquisition curves in fields up to 5 T. Only a small part of IRM is acquired above ~300 mT, as seen from the examples (Fig. 3.2.9). Coercivity analysis confirms the dominant role of a magnetically soft component (Table 3.2.2).

Three coercivity components are separated in both samples—maghemite-like with $B_{1/2}$ of 42 mT; hematite's IRM component 2 with $B_{1/2}$ of 280–350 mT; and a small goethite component with $B_{1/2}$ ~ 1990 mT (Table 3.2.2). Goethite and hematite are the main forms of iron oxides in tropical strongly weathered soils (Cornell and Schwertmann, 2003; Fritsch et al., 2005), but they are detected as a minor part of IRM because of the much stronger magnetic signal of the ferrimagnetic phase compared with antiferromagnetic ones (e.g., Oldfield, 1999). Also, the very fine grain size causes difficulties in the detection of hematite and goethite in soils (Cornell and Schwertmann, 2003), predetermining their paramagnetic behavior at room temperature (Banerjee, 2006; Barrero et al., 2006). The latter assumption is supported in the present study by the low-temperature hysteresis loops obtained for selected samples from different soil horizons. Hysteresis loops measured at room temperature (~300K), at 230K, and at 150K are shown in Fig. 3.2.10. Loops were also measured at 80K, but they

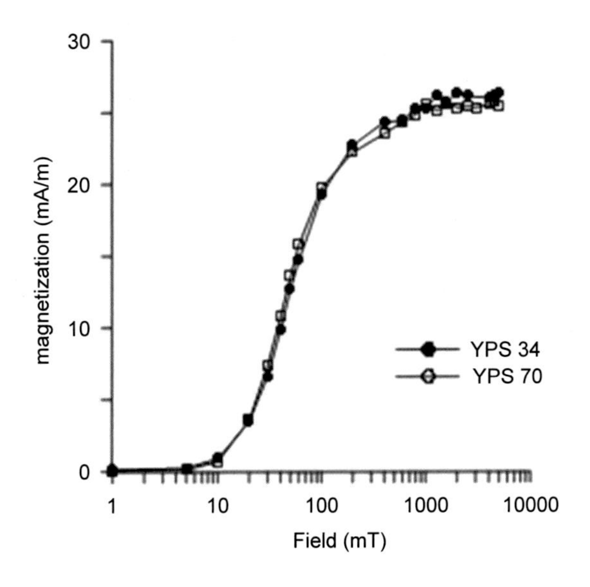

FIGURE 3.2.9

Stepwise acquisition of isothermal remanence (IRM) up to field of 5 T for selected samples from YPS profile.

Table 3.2.2 Summary parameters from coercivity analysis by fitting IRM acquisition curves with cumulative Gaussian functions, using IRM-CLG1.0 software (Kruiver et al., 2001) for selected samples from the YPS profile

Depth (cm)	IRM component 1			IRM component 2			IRM component 3		
	% from IRM_{total}	$B_{1/2}$ (mT)	DP	% from IRM_{total}	$B_{1/2}$ (mT)	DP	% from IRM_{total}	$B_{1/2}$ (mT)	DP
Profile YPS, Alisol/Acrisol									
34	84	42.7	0.31	15	354.8	0.40	1	1995.3	0.10
70	84	41.7	0.31	10	281.8	0.85	6	1995.3	0.35

The relative intensity, median acquisition field ($B_{1/2}$), and dispersion parameter (DP) for each coercivity component are shown.

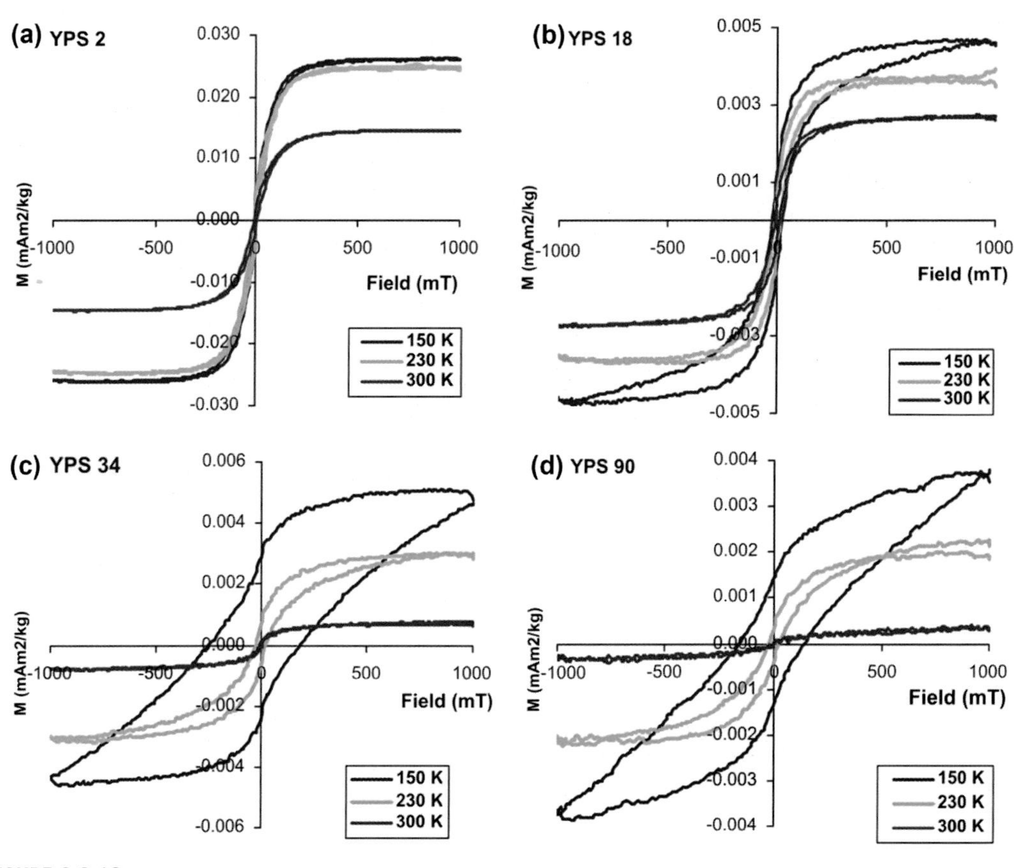

FIGURE 3.2.10

Hysteresis loops after correction for paramagnetic contribution, measured at room temperature (300K) and low temperatures (230 and 150K) for pilot samples from different horizons of YPS profile.

are not shown in the figure, since the curves are noisier although essentially the same as the ones measured at 150K.

Sample YPS 2 from the uppermost surface level (Fig. 3.2.10a) shows a continuous increase in magnetization and coercivity from room temperature toward low temperatures. This behavior could be ascribed to the temperature dependence of the SP–SD transition (Dunlop and Özdemir, 1997). Ferrimagnetic maghemite particles, which are in SP state at room temperature, become stable SD grains on cooling to 150K (Dunlop and Özdemir, 1997). Samples from E2 (YPS 18), $B_{tg}1$ (YPS 34), and $B_{tg}2$ (YPS 90) horizons show a constant rise in the contribution of the high-coercivity phase, which becomes visible only at temperatures below 230K and reaches full expansion at 150K. Room temperature hysteresis loops for the three samples mentioned above show a weak ferrimagnetic signal and completed saturation in fields of ~400–500 mT (Fig. 3.2.10b–d). Cooling to 230K causes the hysteresis loops to become wasp-waisted (Tauxe et al., 1996) but, again, seemingly saturated in 80 mT–1 T. Further cooling to 150K reveals nonsaturated loops in the maximum applied field of 1 T with much stronger coercivity in samples YPS 34 and YPS 90, while in sample YPS 18 the coercivity is the lowest (Fig. 3.2.10). The observed behavior could be explained by the low-temperature properties of goethite with varying degrees of Al substitutions in the lattice (Liu et al., 2006). As shown by these authors, goethite with high Al substitutions (~17%) exhibits a "paramagnetic"-like hysteresis loop at 300K, while at 20K the loop is strongly wasp-waisted with high coercivities. This is due to the decrease of goethite's T_N and T_b with increasing Al substitutions (Liu et al., 2004, 2006; Jiang et al., 2014a). In another study (Barrero et al., 2006), it has been observed that goethite's magnetic properties strongly depend on the adsorbed water content, causing paramagnetic behavior at room temperature, and above it, while at progressively lower temperatures the superparamagnetic/antiferomagnetic/weak ferromagnetic behavior is disclosed. Considering these findings, it could be supposed that a significant amount of Al-substituted goethite with T_N below room temperature is present in the samples studied, in a mixture with a small amount of maghemite (responsible for the weak ferrimagnetic properties at room temperature). Considering the observed shapes of the loops (Fig. 3.2.10c–e), a possible explanation is that below room temperature, already at 230K, Al-goethite becomes antiferromagnetic and, in a mixture with maghemite, causes the wasp-waistedness. The goethite signal at 150K in sample YPS 18 is weaker compared with the deeper samples, probably implying that a larger amount of maghemite is present there or, alternatively, that goethite has lower Al substitutions. Further low-temperature investigations (e.g., field cooling-zero field cooling, AC susceptibility, etc.) would be necessary to fully explain the observed particular hysteresis loops at low temperatures for these samples.

Depth variations of concentration-dependent magnetic parameters (Fig. 3.2.11) show an enhancement with strongly magnetic minerals of the upper part of the profile, including the A_h, E1, and E2 horizons. The uppermost thin A_h horizon reveals the highest magnetic enhancement, probably a result of existing local well-aerated conditions, favoring the pedogenic formation of strongly magnetic maghemite. Magnetic susceptibility, as well as remanent magnetizations (Fig. 3.2.11), points to a somewhat controversial (considering the Luvisol's magnetic properties) magnetic enrichment in the soil's eluvial horizons, subjected to intense weathering and bleaching (Schaetzl and Anderson, 2009). However, variations in high-field magnetic susceptibility (χ_{hf}), as well as the ratio of magnetic susceptibility measured at 77K to room temperature susceptibility (χ_{77}/χ) (Fig. 3.2.11d and e), suggest that deeper illuvial horizons are dominated by paramagnetic minerals (Wang and Løvlie, 2008). However, χ_{hf} may be influenced not only by paramagnetic minerals but also by nonsaturated high-

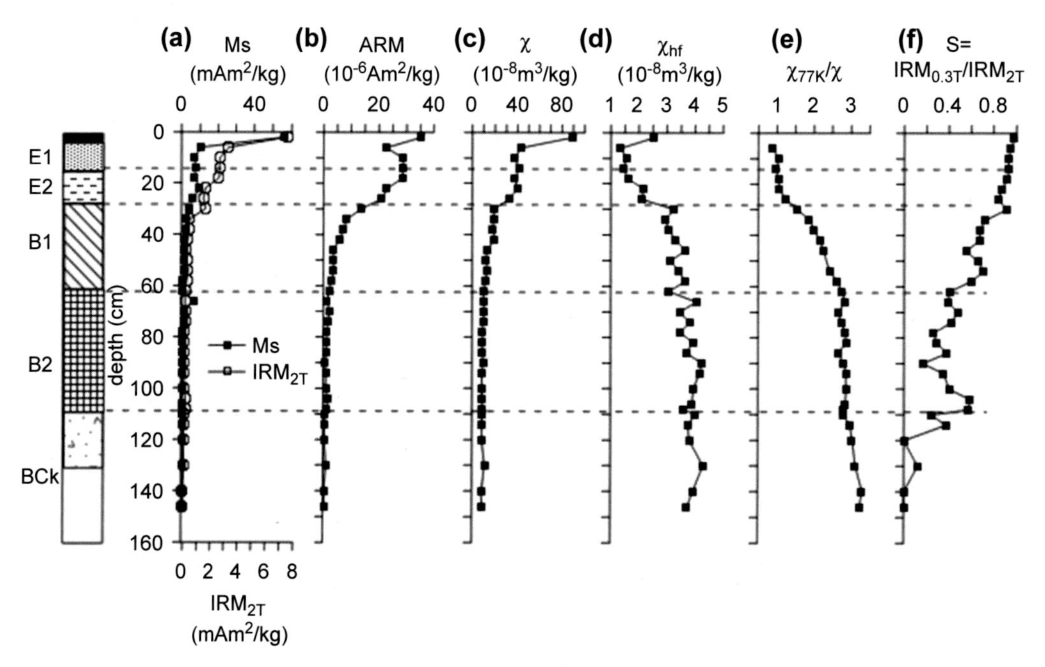

FIGURE 3.2.11

Soil profile YPS. Depth variations of concentration-dependent magnetic characteristics: (a) saturation magnetization (M_s) and isothermal remanent magnetization, acquired in a field of 2 T (IRM_{2T}), (b) anhysteretic remanence (ARM), (c) magnetic susceptibility (χ), (d) high-field susceptibility χ_{hf}, (e) ratio of magnetic susceptibility measured at 77K versus room temperature magnetic susceptibility (χ_{77K}/χ), and (f) the S ratio.

coercivity minerals, like hematite (Jiang et al., 2014a). It is probably not the case for the YPS profile studied, since the absolute value of χ_{hf} is very low, characteristic of clay minerals according to Jiang et al. (2014b). On the other hand, the S ratio (Fig. 3.2.11f) points to a strongly increasing high-coercivity fraction down. Taking into account that goethite is mostly paramagnetic and/or cannot be saturated in a field of 2 T (used for SIRM), the observed decrease in the S ratio along depth should be related to an increased hematite presence.

Diverse magnetic grain-size proxy parameters could be used for estimating the contribution of different fractions in the magnetic signal (Fig. 3.2.12). The concentration of the superparamagnetic fraction, deduced from χ_{fd} behavior (Fig. 3.2.12a), has a maximum in E2 and spreads downward to the bottom of $B_{tg}1$. The relative contribution of the SP fraction to the whole magnetic assembly ($\chi_{fd}\%$) shows a more extended pedogenic enhancement, compared with χ_{fd}, with broad maxima in E2 and a decreasing contribution until the middle part of $B_{tg}2$ (Fig. 3.2.12b). The relative change in the content of the SD-like fraction, approximated through ARM/IRM_{2T}, suggests that the most stable SD-like grains are present in the B1 horizon (Fig. 3.2.12d), while minimum values downward are related to larger sizes of the magnetite phase and an increasing contribution of hematite down-profile toward the C horizon. In fact, χ_{fd} and ARM/IRM_{2T} covary in the upper horizons, indicating a link between the SP and SD pedogenic fractions in the studied Alisol. The parameters HIRM and L ratio (Thompson

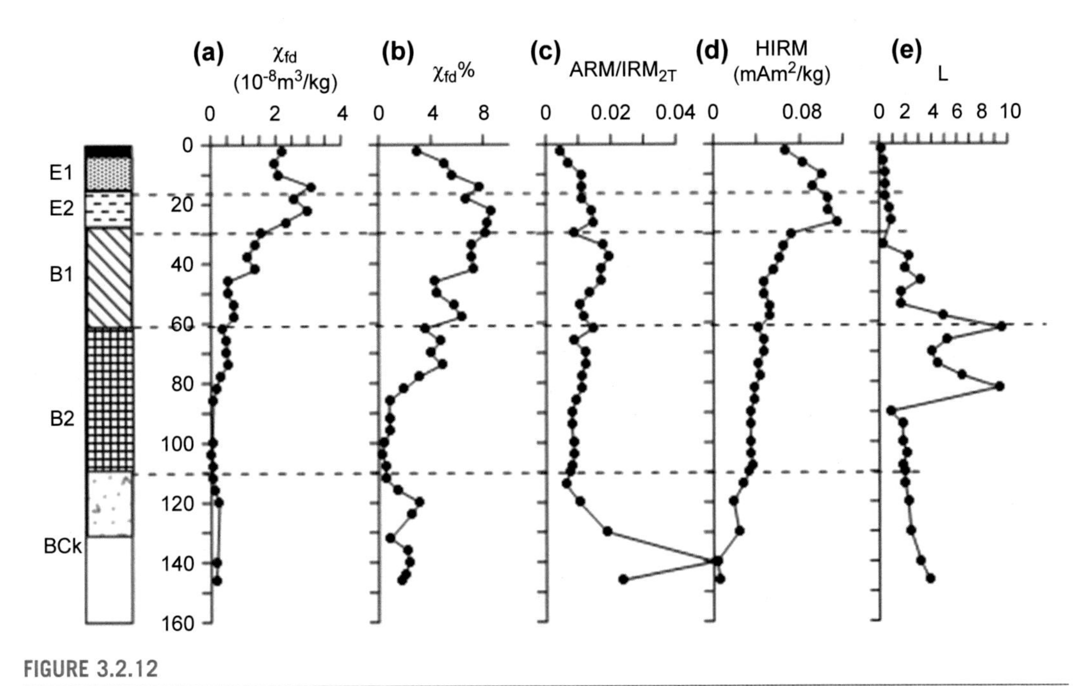

FIGURE 3.2.12

Soil profile YPS. Depth variations of grain size–sensitive magnetic characteristics and ratios: (a) frequency dependent magnetic susceptibility χ_{fd}, (b) percent frequency dependent magnetic susceptibility $\chi_{fd}\%$, (c) ARM/IRM_{2T} ratio, (d) "hard" IRM (HIRM), and (e) L-ratio.

and Oldfield, 1986; Liu et al., 2010) are used to explore the presence of high-coercivity minerals (mainly hematite) in the soil. They do not show a direct correlation and thus, HIRM could be used to infer changes in the concentration of hematite (Liu et al., 2010). HIRM variations suggest a maximum concentration of hematite in the E1-E2 horizons and decreasing to zero until the bottom of $B_{tg}2$ (Fig. 3.2.12d). Furthermore, the L ratio shows that hematite has the largest importance in the $B_{tg}2$ horizon, which is logical, taking into account the prevailing ferrimagnetic contribution in the upper horizons (Figs. 3.2.11 and 3.2.12b–d). Magnetic enhancement in the eluvial horizons of the studied Alisol, related to the presence of magnetite/maghemite (Figs. 3.2.8, 3.2.11 and 3.2.12) is concordant with the results of Fine et al. (1989), who studied magnetic enhancement in soil chronosequences, which also include four Acrisol profiles (two Typic Haplohumults, Plinthic Palehumult, and a Typic Haploxerult) from coastal regions of California (USA). The comparison between the magnetic susceptibility variations in the YPS profile (this study) and the above Acrisols from the United States is shown in Fig. 3.2.13a. Despite the much stronger susceptibility signal in the E horizons of the soils from the United States, compared with YPS soil, in all profiles maximum magnetic enhancement is linked to the eluvial horizons. The authors attribute this enhancement to the pedogenic processes, leading to neoformation of maghemite (Fine et al., 1989), which is easily removed by Dithionite-Citrate-Bicarbonate (DCB) treatment. The magnetic susceptibility before and after the DCB treatment for selected samples from the YPS profile is shown in Fig. 3.2.13b. As is evident, also for our profile samples from the eluvial horizon and the upper part of the B1 horizon show the most significant loss of magnetic susceptibility after DCB treatment. In contrast, the uppermost sample from the A_h layer shows almost no change in χ, suggesting that probably coarser maghemite grains are present

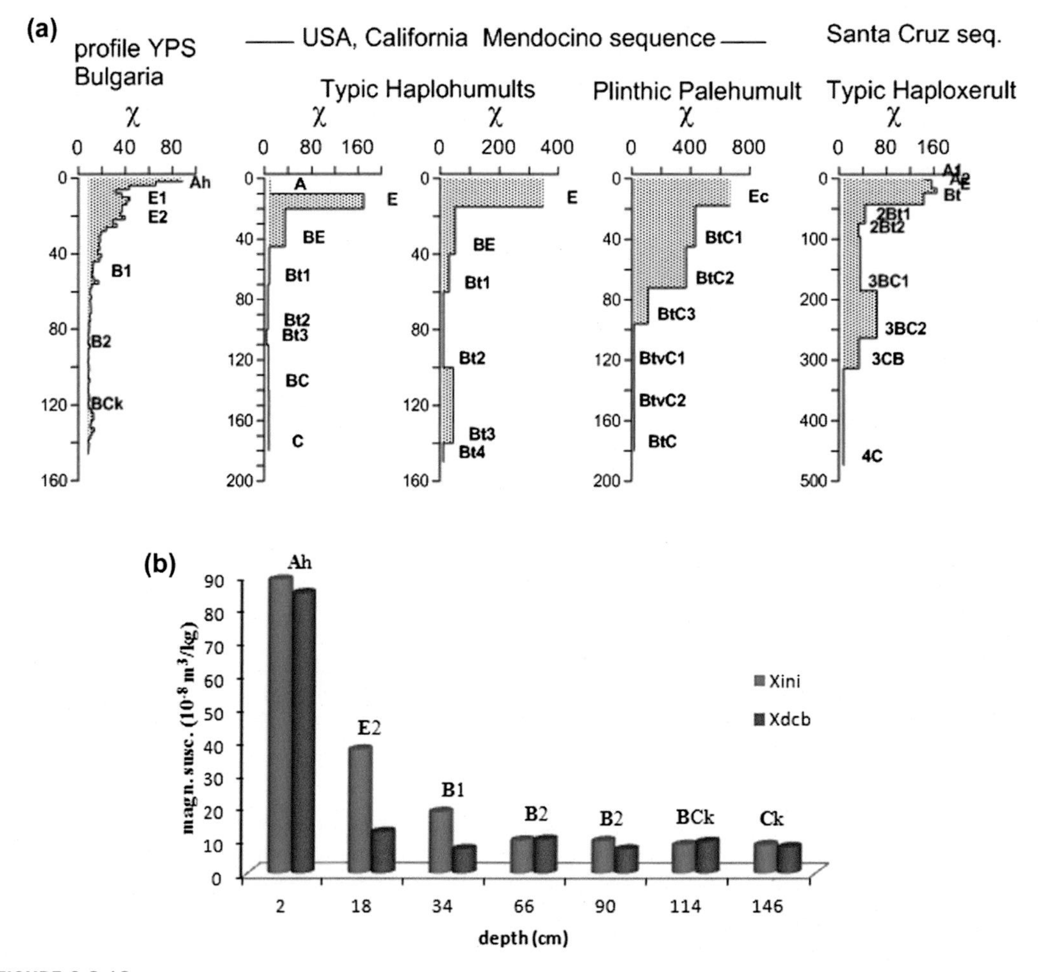

FIGURE 3.2.13

(a) Variations in magnetic susceptibility along the Alisol profile from Bulgaria (YPS) and several profiles of Acrisols from California (USA) *(Data from Fine, P., Singer, M., La Ven, R., Verosub, K., Southard, R., 1989. Role of pedogenesis in distribution of magnetic susceptibility in two California chronosequences. Geoderma 44, 287–306.)*; (b) comparison between initial magnetic susceptibility (χ_{ini}) and susceptibility after DCB-treatment (χ_{dcb}) for selected samples from different horizons of profile YPS.

there, which are not dissolved by dithionite. Furthermore, on the basis of magnetic susceptibilities of different grain-size fractions, Fine et al. (1989) concluded that the magnetic enhancement in the eluvial horizons is due not only to the finest grains but also coarser grain-size fractions that contribute to the χ. The authors attribute the observed magnetic enhancement of the eluvial horizons to both pedogenic formation of fine-grained maghemite and preferential accumulation of lithology-inherited magnetite grains.

Thus, the room temperature magnetic characteristics of the studied Alisol (YPS) imply that despite the intense processes of weathering, leaching, and transformations in the eluvial horizons E1 and E2, the latter are enhanced with ferrimagnetic SP and SD maghemite/magnetite (Figs. 3.2.11 and 3.2.12), as well as hematite (Fig. 3.2.12e). The illuvial horizons are enriched in highly substituted Al-goethite, which does not affect the room temperature magnetic characteristics, as demonstrated by the low-temperature hysteresis measurements. As is discussed in Fritsch et al. (2005), the development of yellow-colored horizons in tropical soils could be due to two different processes—(i) selective dissolution of iron oxides under reducing conditions, leading to the depletion of iron in these horizons and a yellowing of the soil, and (ii) in situ conversion of hematite to goethite without loss of iron. In the red-yellow soils of Amazonia, the latter mechanism has been shown to operate (Fritsch et al., 2005, 2007), since the iron pool in the soil horizons remained unchanged, while an intense formation of highly Al-substituted goethite in the upper part of the soil profiles is evidenced. In the Bulgarian Alisol studied (profile YPS), according to the data on Fe_{tot} and Fe_d (Table 3.2.1), as well as the Fe_2O_3/TiO_2 ratio calculated, it seems that the eluvial horizons are Fe-depleted with almost twice as low an amount of Fe_{tot}, compared with $B_{tg}1$-$B_{tg}2$. Therefore, most probably the first mechanism involving Fe oxide dissolution is responsible for the observed bleaching in the upper part of the profile, while in deeper illuvial horizons, highly Al-substituted very fine grained goethite is present. On the other hand, Jien et al. (2010) studied several Albic Alisols with stagnic features and report clear micromorphological, as well as geochemical difference between the eluvial and illuvial horizons, as well as an increase in pH from the E to the B horizons. Similar characteristics were obtained also for the YPS profile (Fig. 3.2.7), although the ratio Fe_o/Fe_d for the Alisols from Taiwan showed higher values (0.33–0.53, Jien et al., 2010), compared with the YPS profile (0.25–0.3, Table 3.2.1). Jien et al. (2010) also evidenced processes of Fe reduction and leaching from the A and E horizons, followed by oxidation and precipitation in the illuvial horizon. Similar features are observed in our YPS profile (Fig. 3.2.7 and Table 3.2.1). Thus, the Alisol profile considered in our study does not show distinctive features, established for the typical yellow-red soils from Amazonia (Fritsch et al., 2005, 2007) but rather for the Albic Alisols from subtropical mountain forest from Taiwan (Jien et al., 2010). Nevertheless, the obtained evidence for enrichment of the B1 and B2 horizons of YPS profile with highly Al-substituted goethite suggests that the soil has been formed in climate conditions, favoring intense weathering and formation of abundant pedogenic secondary goethite. An important point to be considered in the comparison between properties of the studied soil profile (YPS) and the tropical yellow-red soils and other Alisols from tropical/subtropical regions is the relict character of Alisols, found in Bulgaria. As pointed out by several authors (Fedoroff et al., 2010; Sauer et al., 2015), soils in contemporary temperate climate belt, older than the Holocene age, experienced cyclic changes in climatic conditions during the glacial-interglacial periods, which are reflected in a different degree in their properties.

REFERENCES

Banerjee, S., 2006. Environmental magnetism of nanophase iron minerals: testing the biomineralization pathway. Phys. Earth Planet. Inter. 154, 210–221.

Barrero, C., Betancur, J., Greneche, J., Goya, G., Berquó, T., 2006. Magnetism in non-stoichiometric goethite of varying total water content and surface area. Geophys. J. Int. 164, 331–339.

Barrón, V., Torrent, J., 2002. Evidence for a simple pathway to maghemite in Earth and Mars soils. Geochimica et Cosmochimica Acta 66, 2801–2806.

Boero, V., Premoli, A., Melis, P., Barberis, E., Arduino, E., 1992. Influence of climate on the iron oxide mineralogy of Terra Rossa. Clays Clay Miner. 40 (1), 8–13.

Boyadgiev, T., 1998. Red and yellow soils of Bulgaria. In: Sehgal, J., Blum, W.E., Gajbhiye, K.S., Balkema, A.A. (Eds.), Red and Lateritic Soils. Vol. 2. Red and Lateritic Soils of the World, ISBN 90 5410772 3. The Netherlands.

Boyadzhiyev, T., 1995. Characterisitics of soils with reddish Bt horizon in Bulgaria. Eurasian Soil Sci. 27 (4), 39–44.

Chlupacova, M., Hanak, J., Muller, P., 2010. Magnetic susceptibility of cambisol profiles in the vicinity of Vir Dam, Czech Republic. Studia Geophysika et Geodaetica 54, 153–184.

Cornell, R., Schwertmann, U., 2003. The Iron Oxides. Structure, Properties, Reactions, Occurrence and Uses (Weinheim, New York).

Dearing, J.A., Dann, R.J.L., Hay, K., Lees, J.A., Loveland, P.J., Maher, B.A., O'Grady, K., 1996. Frequency-dependent susceptibility measurements of environmental materials. Geophys. J. Int. 124, 228–240.

Dekkers, M., 1989. Magnetic properties of natural goethite-I. Grain-size dependence of some low- and high-field related rockmagnetic parameters measured at room temperature. Geophys. J. Int. 97 (2), 323–340.

Dunlop, D., Özdemir, Ö., 1997. Rock magnetism. Fundamentals and frontiers. In: Edwards, D. (Ed.), Cambridge Studies in Magnetism. Cambridge University Press.

Dunlop, D., 2014. High-temperature susceptibility of magnetite: a new pseudo-single-domain effect. Geophys. J. Int. 199, 707–716.

Durn, G., 2003. Terra rossa in the Mediterranean region: parent materials, composition and origin. Geol. Croat. 56/1, 83–100.

Durn, G., Ottner, F., Slovenec, D., 1999. Mineralogical and geochemical indicators of the polygenetic nature of terra rossa in Istria, Croatia. Geoderma 91, 125–150.

Eggleton, R.A., Fitzpatrick, R.W., 1988. New data and a revised structural model for ferrihydrite. Clays Clay Miner. 36, 111–124.

Egli, R., 2004. Characterization of individual rock magnetic components by analysis of remanence curves. 3. Bacterial magnetite and natural processes in lakes. Phys. Chem. Earth 29, 869–884.

Erel, Y., Torrent, J., 2010. Contribution of Saharan dust to Mediterranean soils assessed by sequential extraction and Pb and Sr isotopes. Chem. Geol. 275, 19–25.

Evans, M., Heller, F., 2003. Environmental Magnetism: Principles and Applications of Enviromagnetics. Academic Press, San Diego, CA.

FAO World soil resources reports, 2001. Lecture Notes on the Major Soils of the World. ISSN 0532–0488. http://www.fao.org/DOCREP/003/Y1899E/Y1899E00.HTM.

Fedoroff, N., Courty, M.-A., 2013. Revisiting the genesis of red Mediterranean soils. Turkish J. Earth Sci. 22, 359–375.

Fedoroff, N., Courty, M.-A., Guo, Z., 2010. Palaeosoils and relict soils. In: Stoops, G., Marcelino, V., Mees, F. (Eds.), Interpretation of Micromorphological Features of Soils and Regoliths. Elsevier, ISBN 978-0-444-53156-8, pp. 623–662.

Fine, P., Singer, M., La Ven, R., Verosub, K., Southard, R., 1989. Role of pedogenesis in distribution of magnetic susceptibility in two California chronosequences. Geoderma 44, 287–306.

Fine, P., Singer, M., Verosub, K., 1992. Use of magnetic-susceptibility measurements in assessing soil uniformity in chronosequence studies. Soil Sci. Soc. Am. J. 56, 1195–1199.

Fitzpatrick, R., Taylor, R., Schwertmann, U., Childs, C., 1985. Occurrence and properties of lepidocrocite in some soils of New Zealand, South Africa and Australia. Aust. J. Soil Res. 23, 543–567.

Fitzpatrick, R.W., Chittleborough, D.J., 2002. Titanium and zirconium minerals. In: Dixon, J.B., Schulze, D.G. (Eds.), Soil Mineralogy with Environmental Applications, Soil Sci. Soc. Amer. Book Series, 7, pp. 667–690.

Fritsch, E., Herbillon, A., Do Nascimento, N., Grimaldi, M., Melfi, A., 2007. From Plinthic Acrosols to Plinthosols and Gleysols: iron and groundwater dynamics in the tertiary sediments of the upper Amazon basin. Eur. J. Soil Sci. 58, 989–1006.

Fritsch, E., Morin, G., Bedidi, A., Bonnin, D., Balan, E., Caquineau, S., Calas, G., 2005. Transformation of haematite and Al-poor goethite to Al-rich goethite and associated yellowing in a ferralitic clay soil profile of the middle Amazon Basin (Manaus, Brazil). Eur. J. Soil Sci. 56, 575–588.

Gehring, A., Hofmeister, A., 1994. The transformation of lepidocrocite during heating: a magnetic and spectroscopic study. Clays Clay Miner. 42 (4), 409–415.

Grison, H., Petrovský, E., Jordanova, N., Kapička, A., 2011. Strongly magnetic soil developed on a non-magnetic rock basement: a case study from NW Bulgaria. Stud. Geophys. Geod. 55 (4), 697–716.

Haidouti, C., Massas, I., 1998. Distribution of iron and manganese oxides in Haploxeralfs and Rhodoxeralfs and their relation to the degree of soil development and soil colour. Z. Pfanzenernähr. Bodenk. 161, 141–145.

Hanesch, M., Stanjek, H., Petersen, N., 2006. Thermomagnetic measurements of soil iron minerals: the role of organic carbon. Geophys. J. Int. 165, 53–61.

IUSS Working Group WRB, 2014. World Reference Base for Soil Resources 2014. International Soil Classification System for Naming Soils and Creating Legends for Soil Maps. World Soil Resources Reports No. 106. FAO, Rome.

Jiang, Z., Liu, Q., Dekkers, M., Colombo, C., Yu, Y., Barrón, V., Torrent, J., 2014a. Ferro and antiferromagnetism of ultrafine-grained hematite. Geochem. Geophys. Geosyst. 15, 2699–2712. http://dx.doi.org/10.1002/2014GC005377.

Jiang, Z., Liu, Q., Colombo, C., Barrón, V., Torrent, J., Hu, P., 2014b. Quantification of Al-goethite from diffuse reflectance spectroscopy and magnetic methods. Geophys. J. Int. 196 (1), 131–144.

Jien, S.-H., Hseu, Z.-Y., Iizuka, Y., Chen, T.-H., Chiu, C.-Y., 2010. Geochemical characterization of placic horizons in subtropical montane forest soils, northeastern Taiwan. Eur. J. Soil Sci. 61, 319–332.

Jordanova, D., Jordanova, N., 2016. Thermomagnetic behaviour of magnetic susceptibility—heating rate and sample size effects. Front. Earth Sci. 3, 90. http://dx.doi.org/10.3389/feart.2015.00090.

Jordanova, N., Jordanova, D., Liu, Q., Hu, P., Petrov, P., Petrovský, E., 2013. Soil formation and mineralogy of a Rhodic Luvisol – insights from magnetic and geochemical studies. Glob. Planet. Change 110, 397–413.

Koinov, V., Kabakchiev, I., Boneva, K., 1998. In: Koinov, V., Boneva, K. (Eds.), Atlas of Soils in Bulgaria. Zemizdat, Sofia.

Kruiver, P., Dekkers, M., Heslop, D., 2001. Quantifcation of magnetic coercivity components by the analysis of acquisition curves of isothermal remanent magnetization. Earth Planet. Sci. Lett. 189, 269–276.

Liu, Q., Deng, C., Yu, Y., Torrent, J., Jackson, M.J., Banerjee, S., Zhu, R., 2005. Temperature dependence of magnetic susceptibility in an argon environment: implications for pedogenesis of Chinese loess/palaeosols. Geophys. J. Int. 161, 102–112.

Liu, Z., Liu, Q., Torrent, J., Barrón, V., Hu, P., 2013. Testing the magnetic proxy χFD/HIRM for quantifying paleoprecipitation in modern soil profiles from Shaanxi Province, China. Glob. Planet. Change 110, 368–378.

Liu, Q.S., Torrent, J., Yu, Y.J., Deng, C.L., 2004. Mechanism of the parasitic remanence of aluminous goethite [alpha-(Fe,Al)OOH]. J. Geophys. Res. 109, B12106. http://dx.doi.org/10.1029/2004JB003352.

Liu, Q., Banerjee, S., Jackson, M., Zhu, R., Pan, Y., 2002. A new method in mineral magnetism for the separation of weak antiferromagnetic signal from a strong ferrimagnetic background. Geophys. Res. Lett. 29 (12), 1565. http://dx.doi.org/10.1029/2002GL014699.

Liu, Q., Hu, P., Torrent, J., Barrón, V., Zhao, X., Jiang, Z., Su, Y., 2010. Environmental magnetic study of a Xeralf chronosequence in northwestern Spain: indications for pedogenesis. Palaeogeogr. Palaeoclimatol. Palaeoecol. 293, 144–156.

Liu, Q., Roberts, A., Torrent, J., Horng, Ch-Sh, Larrasoaña, J., 2007. What do the HIRM and S-ratio really measure in environmental magnetism? Geochem. Geophys. Geosyst. 8, Q09011. http://dx.doi.org/10.1029/2007GC001717.

Liu, Q., Yu, Y., Torrent, J., Roberts, A., Pan, Y., Zhu, R., 2006. Characteristic low-temperature magnetic properties of aluminous goethite [α-(Fe, Al)OOH] explained. J. Geophys. Res. 111, B12S34.

Lucke, B., Kemnitz, H., Bäumler, R., Schmidt, M., 2014. Red Mediterranean soils in Jordan: new insights in their origin, genesis, and role as environmental archives. Catena 112, 4–24.

Maher, B., Thompson, R., 1995. Paleorainfall reconstructions from pedogenic magnetic susceptibility variations in the Chinese loess and paleosols. Quat. Res. 44, 383–391.

Maher, B., 1988. Magnetic properties of some synthetic sub-micron magnetites. Geophys. J. R. Astr. Soc. 94, 83–96.

Maher, B., 1998. Magnetic properties of modern soils and Quaternary loessic paleosols: paleoclimatic implications. Palaeogeogr. Palaeoclimat. Palaeoecol. 137, 25–54.

Maher, B., Prospero, J., Mackie, D., Gaiero, D., Hesse, P., Balkanski, Y., 2010. Global connections between aeolian dust, climate and ocean biogeochemistry at the present day and at the last glacial maximum. Earth-Science Rev. 99, 61–97.

McKeague, J.A., Day, J.H., 1966. Dithionite and oxalate extractable Fe and Al as aids in differentiating various classes of soils. Can. J. Soil Sci. 46, 13–22.

Mehra, O., Jackson, M., 1960. Iron oxide removal from soils and clays by a dithionite-citrate system buffered with sodium bicarbonate. Clays Clay Miner. 7, 317–327.

Michel, F., Barron, V., Torrent, J., Morales, M., Serna, C., Boily, J.-F., Liu, Q., Ambrosini, A., Cismasu, C., Brown Jr., G., 2010. Ordered ferromagnetic form of ferrihydrite reveals links among structure, composition, and magnetism. PNAS 107, 2787–2792.

Mitov, I., Paneva, D., Kunev, B., 2002. Comparative study of the thermal decomposition of iron oxyhydroxides. Thermochim. Acta 386, 179–188.

Morris, R., Golden, D., Shelfer, T., Lauer, H., 1998. Lepidocrocite to maghemite to hematite: a pathway to magnetic and hematitic Martian soil. Meteorit. Planet. Sci. 33, 743–775.

Muhs, D., Budahn, J., Avila, A., Skipp, G., Freeman, J., Patterson, D.A., 2010a. The role of African dust in the formation of Quaternary soils on Mallorca, Spain and implications for the genesis of Red Mediterranean soils. Quaternary Sci. Rev. 29, 2518–2543.

Muhs, D., Budahn, J., Skipp, G., Prospero, J., Patterson, D.A., Arthur Bettis III, E., 2010b. Geochemical and mineralogical evidence for Sahara and Sahel dust additions to Quaternary soils on Lanzarote, eastern canary Islands, Spain. Terra Nova 22 (No. 6), 399–410.

Muhs, D., Budahn, J., 2009. Geochemical evidence for African dust and volcanic ash inputs to terra rossa soils on carbonate reef terraces, northern Jamaica, West Indies. Quat. Int. 196, 13–35.

Muhs, D.R., 2013. The geologic records of dust in the Quaternary. Aeolian Res. 9, 3–48.

Nesbitt, H.W., Young, G.M., 1982. Early Proterozoic climates and plate motions inferred from major element chemistry of lutites. Nature 199, 715–717.

Ninov, N., 2002. Soils. In: Kopralev, I., Yordanova, M., Mladenov, C. (Eds.), Geography of Bulgaria. Institute of Geography – Bulg. Acad. Sci. ForCom, Sofia, pp. 277–317.

O'Reilly, W., 1984. Rock and Mineral Magnetism. Blakie. Chapman and Hall, Glasgow.

Oldfield, F., 1999. The rock magnetic identification of magnetic mineral and magnetic grain size assemblages. In: Walden, J., Oldfield, F., Smith, J. (Eds.), Environmental Magnetism: A Practical Guide. Quaternary Research Association, Technical Guide No. 6, London, ISBN 0 907780 42 3, pp. 98–112, 0264-9241.

Owliaie, H., Heck, R., Abtahi, A., 2006. The magnetic susceptibility of soils in Kohgilouye, Iran. Can. J. Soil Sci. 86, 97–107.

Peters, C., Dekkers, M.J., 2003. Selected room temperature magnetic parameters as a function of mineralogy, concentration and grain size. Phys. Chem. Earth 28, 659–667.

Quénard, L., Samouëlian, A., Laroche, B., Cornu, S., 2011. Lessivage as a major process of soil formation: a revisitation of existing data. Geoderma 167–168, 135–147.

Reheis, M., 1990. Influence of climate and eolian dust on the major-element chemistry and clay mineralogy of soils in the northern Bighorn Basin, U.S.A. Catena 17, 219–248.

Rivas, J., Ortega, B., Sedov, S., Solleiro, E., Sychera, S., 2006. Rock magnetism and pedogenic processes in Luvisol profiles: examples from Central Russia and Central Mexico. Quat. Int. 15 (6–157), 212–223.

Rochette, P., Mathé, P.-E., Esteban, L., Rakoto, H., Bouchez, J.-L., Liu, Q., Torrent, J., 2005. Non-saturation of the defect moment of goethite and fine-grained hematite up to 57 Teslas. Geophys. Res. Lett. 32, 1–4, L22309.

Sauer, D., Schülli-Maurer, I., Wagner, S., Scarciglia, F., Sperstad, R., Svendgård-Stokke, S., Sørensen, R., Schellmann, G., 2015. Soil development over millennial timescales – a comparison of soil chronosequences of different climates and lithologies. In: IOP Conf. Series: Earth and Environmental Science, 25, p. 012009. http://dx.doi.org/10.1088/1755-1315/25/1/012009.

Schaetzl, R., Anderson, A., 2009. Soils. Genesis and Geomorphology. Cambridge Univ. Press, UK, ISBN 978-0-521-81201-6.

Scheinost, A., Chavernas, A., Barrón, V., Torrent, J., 1998. Use and limitations of second-derivative diffuse reflectance spectroscopy in the visible to near-infrared range to identify and quantify Fe oxide minerals in soils. Clays Clay Miner. 46 (No. 5), 528–536.

Schwertmann, U., 1988. Occurrence and formation of iron oxides in various pedoenvironments. In: Stucki, J.;, Goodman, B., Schwertmann, U. (Eds.), Iron in Soils and Clay Minerals, NATO ASI Series, Series C: Mathematical and Physical Sciences, vol. 217. Reidel Publ. Company, pp. 267–308.

Sheldon, N., Tabor, N., 2009. Quantitative paleoenvironmental and paleoclimatic reconstruction using paleosols. Earth-Science Rev. 95, 1–52.

Shenggao, L., 2000. Lithological factors affecting magnetic susceptibility of subtropical soils, Zhejiang Province, China. Catena 40, 359–373.

Shishkov, T., Kolev, N., 2014. The soils of Bulgaria. World soils book series. Series. In: Hartemink, A.E. (Ed.), Springer Science+Business Media Dordrecht 2014, ISBN 978-94-007-7784-2, p. 205 (eBook).

Shishkov, T., 2011. Implication of the world reference base and soil taxonomy within the framework of Bulgarian soil classification. Soil Sci. Agrochem. Ecol. XLV (3), 9–26.

Singer, M., Fine, P., Verosub, K., Chadwick, O., 1992. Time dependence of magnetic susceptibility of soil chronosequences on the California Coast. Quat. Res. 37, 323–332.

Singer, M., Verosub, K., Fine, P., TenPass, J., 1996. A conceptual model for the enhancement of magnetic susceptibility in soils. Quat. Int. 34–36, 243–248.

Singer, M.J., Fine, P., 1989. Pedogenic factors affecting magnetic susceptibility of Northern Californian soils. Soil Sci. Soc. Am. J. 53, 1119–1127.

Siqueira, D., Marques Jr., J., Matias, S., Barrón, V., Torrent, J., Baffa, O., Oliveira, L., 2010. Correlation of properties of Brazilian Haplustalfs with magnetic susceptibility measurements. Soil Use Management. 26, 425–431.

Spaargaren, O., 2008. Alisols. In: Chesworth, W. (Ed.), Encyclopedia of Soil Science. Springer Dodrecht, Berlin, Heidelberg, New York, ISBN 978-1-4020-3994-2, pp. 35–37.

Stiles, C., Mora, C., Driese, S., 2003a. Pedogenic processes and domain boundaries in a Vertisol climosequence: evidence from titanium and zirconium distribution and morphology. Geoderma 116, 279–299.

Stiles, C., Mora, C., Driese, S., Robinson, A., 2003b. Distinguishing climate and time in the soil record: mass-balance trends in Vertisols from the Texas coastal prarie. Geology 31 (4), 331–334.

Stuut, J.-B., Smalley, I., O'Hara-Dhand, K., 2009. Aeolian dust in Europe: African sources and European deposits. Quat. Int. 198, 234–245.

Tauxe, L., Mullender, T.A.T., Pick, T., 1996. Potbellies, wasp-waists, and superparamagnetism in magnetic hysteresis. J. Geophys. Res. 101 (B1), 571–583.

Thompson, R., Oldfield, F., 1986. Environmental Magnetism. Allen & Unwin, London.

Torrent, J., Barrón, V., Liu, Q., 2006. Magnetic enhancement is linked to and precedes hematite formation in aerobic soil. Geophys. Res. Lett. 33, L02401. http://dx.doi.org/10.1029/2005GL024818.

Torrent, J., Liu, Q., Bloemendal, J., Barron, V., 2007. Magnetic enhancement and iron oxides in the upper Luochuan loess–Paleosol sequence, Chinese loess plateau. Soil Sci. Soc. Am. J. 71, 1570–1578.

Torrent, J., Liu, Q.S., Barrón, V., 2010a. Magnetic susceptibility changes in relation to pedogenesis in a Xeralf chronosequence in northwestern Spain. Eur. J. Soil Sci. 61, 161–173.

Torrent, J., Liu, Q., Barrón, V., 2010b. Magnetic minerals in Calcic Luvisols (Chromic) developed in warm Mediterranean region of Spain: origin and paleoenvironmental significance. Geoderma 154 (3–4), 465–472.

Tóth, G., Montanarella, L., Máté, S.F., Bódis, K., Jones, A., Panagos, P., Van Liedekerke, M., 2008. Soils of the European Union.

Trakoonyingcharoen, P., Kheoruenromne, I., Suddhiprakarn, A., Gilkes, R., 2006. Properties of iron oxides in red Oxisols and red Ultisols as affected by rainfall and soil parent material. Austr. J. Soil Res. 44, 63–70.

Van Breemen, N., 1988. Long-term chemical, mineralogical and morphological effects of iron-redox processes in periodically flooded soils. In: Stucki, J., Goodman, B., Schwertmann, U. (Eds.), Iron in Soils and Clay Minerals. NATO ASI Series. D. Reidel Publishing company.

van Velzen, A.J., Zijderveld, J.D.A., 1995. Effects of weathering on single domain magnetite in Early Pliocene marine marls. Geophys. J. Int. 121, 267–278.

Velizarova, E., Popov, G., 1999. Study of the mineralogical composition of Cinnamonic forest soils from South-Eastern Bulgaria. Bulg. J. Agricult. Sci. 5, 55–59.

Verheye, W., de la Rosa, D., 2005. Mediterranean soils. In: Land Use and Land Cover, from Encyclopedia of Life Support Systems (EOLSS), Developed under the Auspices of the UNESCO. Eolss Publishers, Oxford, UK. http://www.eolss.net.

Wang, R., Løvlie, R., 2008. SP-grain production during thermal demagnetization of some Chinese loess/palaeosol. Geophys. J. Int. 172, 504–512.

Yaalon, D., 1997. Soils in the Mediterranean region: what makes them different? Catena 28, 157–169.

Zhang, Ch, Paterson, G., Liu, Q., 2012. A new mechanism for the magnetic enhancement of hematite during heating: the role of clay minerals. Stud. Geophys. Geod. 56, 845–860.

CHAPTER

THE MAGNETISM OF SOILS DISTINGUISHED BY IRON/ ALUMINUM CHEMISTRY: PLANOSOLS, POZDOLS, ANDOSOLS, FERRALSOLS, AND GLEYSOLS

4

4.1 PLANOSOLS: MAIN CHARACTERISTICS, FORMATION PROCESSES, AND DISTRIBUTION

Water stagnation is an essential environmental factor that strongly influences the redox-sensitive chemical elements in soil, such as Fe, Mn, and Al. Planosols are soils that are strongly affected by such hydrological control. The most important characteristics of the Planosols include: (1) the presence of a surface bleached eluvial horizon, showing signs of periodic water-logging and a sandy mechanical composition; and (2) a sharp textural contrast with the underlying illuvial horizon, containing a significantly higher amount of clay minerals. Planosols are a Reference Soil Group in the WRB classification, falling into the great group of Stagnosols (IUSS Working Group WRB, 2014).

According to the most recent views, several possible mechanisms for Planosol formation exist: (1) geogenic processes [e.g., coarse-grained sediment deposition over clayey material, colluvial deposition on clayey material, selective surface erosion of the finest mechanical fraction (Schaetzl and Anderson, 2009; Van Ranst et al., 2011)]; (2) physical pedogenic processes, such as eluviation and illuviation inside the solum with an unstable structure; and (3) chemical pedogenic processes—ferrolysis (Brinkman, 1970; Schaetzl and Anderson, 2009). It was precisely the Planosols and their properties that led Brinkman (1970) to develop the theory of ferrolysis. The latter consists of several key processes including the cyclic reduction and oxidation of iron, which makes Fe become mobile. During the reduction, Fe^{3+} ions are reduced to Fe^{2+}, which then substitute for the exchangeable cations on the silicates with surface charges. During the oxidation, Fe^{2+} oxidizes to iron hydroxide and H^+ ions are dissociated. They cause an acid soil reaction that further assists the dissolution of the clay minerals. During each reduction—oxidation cycle, soluble cations and part of the Al are exported from the site of the ferrolysis, part of the crystal lattice of clay minerals is destroyed, and the overall texture of the profile becomes strongly differentiated (van Breemen, 1988a; Van Ranst et al., 2011). The ferrolysis theory was widely accepted in the 1970's and 1980's of the 20th century, but a number of more recent studies question its importance in Planosol development (Montagne et al., 2008; Favre et al., 2002; Van Ranst and De Coninck, 2002; Van Ranst et al., 2011). These authors favor

Soil Magnetism. http://dx.doi.org/10.1016/B978-0-12-809239-2.00004-8

mostly erosional – sedimentary pedogenesis or an eluviation–illuviation bundle, which could also produce the specific characteristics observed in Planosols.

Due to the presence of a sandy eluvial horizon overlying the strongly clayey illuvial (B_t) horizon, at the boundary between these two horizons, good conditions for side water flow appear. As a result, the eluvial horizon is strongly leached and depleted of clay. Often, Fe-Mn concretions form at its base. Water-logging in the upper part of the illuvial horizon favors the processes of translocation and degradation because of the established reducing conditions (van Breemen, 1988a; Schaetzl and Anderson, 2009). The morphological features indicative of a periodic change of reducing–oxidizing conditions are concretions, nodules, amorphous masses on the surfaces of aggragates, etc. During the anaerobic periods, Fe and Mn become mobile as a result of their reduction, and can freely move along the existing concentration gradients or together with the soil solution. During the oxidative periods, Fe is oxidized and precipitates in the form of insoluble ferrihydrite or $Fe(OH)_3$, visualized as red and orange precipitates or black Fe-Mn concretions (Schaetzl and Anderson, 2009). Numerous repetitions of the reduction–oxidation cycles cause an increase in the concretions' size. Soils subjected to reduction–oxidation cycles are usually characterized by a mottled appearance reflecting the iron segregation processes.

The Planosols are found mostly in subtropical areas with alternating dry and wet seasons—Latin America (southern Brazil, Paraguay, and Argentina), Africa (the Sahelian zone, eastern and southern Africa), the eastern part of the United States, southeast Asia (Bangladesh and Thailand), and Australia (Fig. 4.1.1). Planosols are identified in Europe as well but show a relatively restricted distribution.

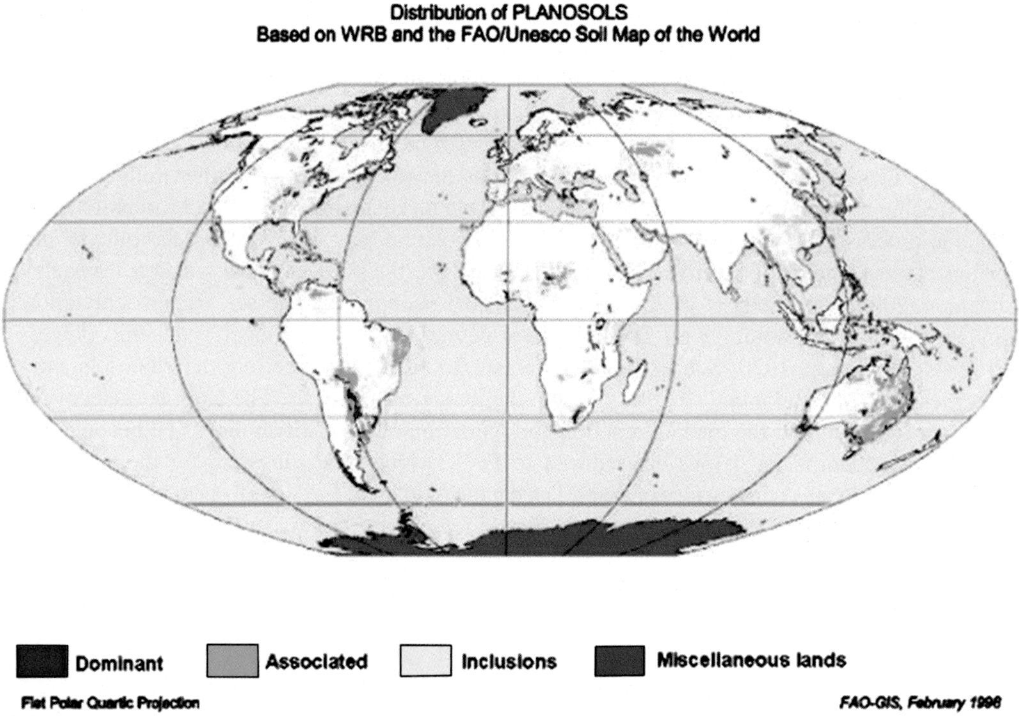

Distribution of PLANOSOLS
Based on WRB and the FAO/Unesco Soil Map of the World

Dominant
Associated
Inclusions
Miscellaneous lands
Flat Polar Quartic Projection
FAO-GIS, February 1998

FIGURE 4.1.1

Spatial distribution of Planosols worldwide (FAO, 2001). Lecture notes on the major soils of the world: http://www.fao.org/docrep/003/y1899e/y1899e00.htm#toc.

They are most characteristic for Austria and Bulgaria, less abundant in Estonia, and in some areas in southern England and have a minor appearance in other countries (Tóth et al., 2008). The parent materials are usually alluvial and colluvial deposits.

The Planosols are widespread in Bulgaria in both the northern and southern parts of the country and occupy ~9.75% of the territory (Shishkov and Kolev, 2014). According to the Bulgarian soil classification system, they are defined as "pseudopodzolic soils." Planosols commonly develop under mixed oak forests. The parent rocks are noncarbonaceous—granites, gneisses, sandstones, acid andesites, and the redeposited weathering products of these rocks. It is assumed that the Planosols in Bulgaria are genetically old soils with relict signs that developed on old geomorphic forms (Ninov, 2002; Shishkov and Kolev, 2014). The degree of their textural differentiation depends on the parent material—it is lower in Planosols developed on granites and their weathering products, and is stronger in Planosols developed on andesitic rocks. The main clay minerals are illite and kaolinite, while montmorillonite is present only in the illuvial horizon. A considerable differentiation in the chemical composition of the soil horizons is also observed—the surface humic-eluvial horizon is depleted in iron and enriched in Silica, while the illuvial horizon is Fe- enriched (Jokova, 1999; Jokova and Boyadjiev, 1993; Shishkov and Kolev, 2014). The soil reaction pH is acid.

4.2 DESCRIPTION OF THE PROFILES STUDIED

4.2.1 EUTRIC PLANOSOL (STRONGLY LEACHED CINNAMONIC FOREST SOIL), PROFILE DESIGNATION: OK

Situated close to the village of Otec Kirilovo, Plovdiv district (South Bulgaria)

Location: N 42°21′14.5″; E 24°59′56.1″, h = 250 m a.s.l.

Present-day climate conditions: mean annual temperature (MAT) = 12.1°C; mean annual precipitation (MAP) = 565 mm, vegetation cover: broadleaf forest; relief: flat.

Profile description: general view and details from the horizons are shown in Figs. 4.2.1–4.2.5.

0–3 cm: AE horizon, black, rich in organics
3–23 cm: E1 eluvial horizon, bleached, sandy, light beige color, Fe-Mn concretions
23–30 cm: E2 eluvial horizon, mottled beige-orange-brown, sandy
30–41 cm: B_t1 illuvial horizon, red-brown, dense, heavy clayey, Fe-Mn concretions
41–100 cm: B_t2 illuvial horizon, dark brown, prismatic structure, heavy clayey, Fe-Mn concretions
100–130 cm: C_k1 horizon, light-brown, clayey, large carbonate concretions and carbonate diffuse masses on the ped surfaces, Fe-Mn concretions
130–152 cm: C_k2 horizon, red-brown, carbonate concretions, loamy-sandy
152–182 cm: C_k3 horizon, weathered parent rock, sandy, dense
Parent material: loamy-sandy deposits of weathered south-Bulgarian Palaeozoic granites

FIGURE 4.2.1

View of the profile of Eutric Planosol (profile OK).

FIGURE 4.2.2

Profile OK. AE-E1 horizons—detail.

FIGURE 4.2.3

Profile OK. E2-B_t1 horizons—detail.

FIGURE 4.2.4

Profile OK. B_t2 horizon—detail.

FIGURE 4.2.5

Profile OK. C_k2 horizon—detail.

4.2.2 LUVIC ENDOGLEYC PLANOSOL (CINNAMONIC PSEUDO-PODZOLIC SOIL), PROFILE DESIGNATION: PR

Situated close to the village of Primorsko (Bourgas district), South Bulgaria.

Location: N 42°16′59.2″; E 27°44′15.4″, h = 50 m a.s.l.

Present-day climate conditions: MAT = 13.2°C; MAP = 692 mm, vegetation cover: broadleaf forest; relief: flat.

Profile description: general view and details from the horizons are shown in Figs. 4.2.6–4.2.11.

0–2 cm: AE horizon, rich in organics, black color
2–38 cm: E1 eluvial horizon, sandy, light-beige, bleached, vegetation roots abundant
38–60 cm: E2 eluvial horizon, mottled yellowish-brown, iron segregation spots
60–90 cm: $B_{tg}1$ horizon, orange-gray, heavy clayey, dense, gley spots visible
90–110 cm: $B_{tg}2$ horizon, brown-orange, heavy clayey, dense, iron segregation spots
110–145 cm: $B_{tg}3$ horizon, dark brown, clayey, dense, Fe-Mn concretions
145–180 cm: C_k horizon, mottled brown, sandy, weathered rock fragments present, iron segregation spots
Parent material: weathered pyroclastic deposits

FIGURE 4.2.6

View of the profile of Luvic Endogleyc Planosol (profile PR).

FIGURE 4.2.7

Profile PR. AE-E1 horizons—detail.

FIGURE 4.2.8

Profile PR. E1 horizon—detail.

FIGURE 4.2.9

Profile PR. $B_{tg}1$ horizon—detail.

FIGURE 4.2.10

Profile PR. C_k horizon—detail.

FIGURE 4.2.11

Luvic Endogleyc Planosol in the area of Primorsko, southeast Bulgaria. View of the location sampled.

4.2.3 LUVIC ENDOGLEYC PLANOSOL (CINNAMONIC PSEUDO-PODZOLIC SOIL), PROFILE DESIGNATION: ZL

Situated close to the village of Zlatosel (Plovdiv district), South Bulgaria.

Location: N 42°24′29.0″; E 25°01′08.6″, h = 350 m a.s.l.

Present-day climate conditions: MAT = 12.1°C; MAP = 865 mm, vegetation cover: broadleaf forest; relief: flat

Profile description: general view and details from the horizons are shown in Figs. 4.2.12–4.2.15.

0–2 cm: AE horizon, black, rich in decayed organics and vegetation
2–13 cm: E1 eluvial horizon, light-beige, bleached, sandy, vegetation roots present
13–24 cm: E2 eluvial horizon, beige-yellowish, sandy-loamy
24–38 cm: E3 horizon, mottled, gley spots, sandy
38–60 cm: $B_{tg}1$ illuvial horizon, red-brown, heavy clayey, dense, Fe-Mn concretions
60–76 cm: $B_{tg}2$ horizon, mottled, brown-beige, heavy clayey, Fe-Mn concretions, and Ca concretions
76–100 cm: C_k horizon, mottled beige-brown-grayish, carbonate concretions, and rock fragments
Parent rock: loamy-sandy deposits of weathered south-Bulgarian Palaeozoic granites

FIGURE 4.2.12

View of the profile of Luvic Endogleyc Planosol (profile ZL).

FIGURE 4.2.13

Profile ZL. AE-E1 horizons—detail.

FIGURE 4.2.14

Profile ZL. $B_{tg}1$-$B_{tg}2$ horizons—detail.

FIGURE 4.2.15

Profile ZL. C_k horizon—detail.

4.3 TEXTURE, SOIL REACTION, AND MAJOR GEOCHEMISTRY OF PLANOSOLS

The depth variations of magnetic susceptibility and some rock-magnetic mineral diagnostic analyses of the three Planosols described above are presented in Jordanova et al. (2011). The texture and the soil reaction changes along the profiles of the Planosols studied reveal features typical of this soil type (Figs. 4.3.1, 4.3.2, and 4.3.3). The uppermost eluvial horizons are characterized by a strongly acid pH, varying between 4.5 and 5.0, while downward, a systematic increase in pH reaching highly alkaline values is observed for all the three profiles (Figs. 4.3.1d, 4.3.2d, and 4.3.3d). The soil reaction in the OK and PR profiles starts increasing from the E1 horizon downward and shows a well-expressed fast

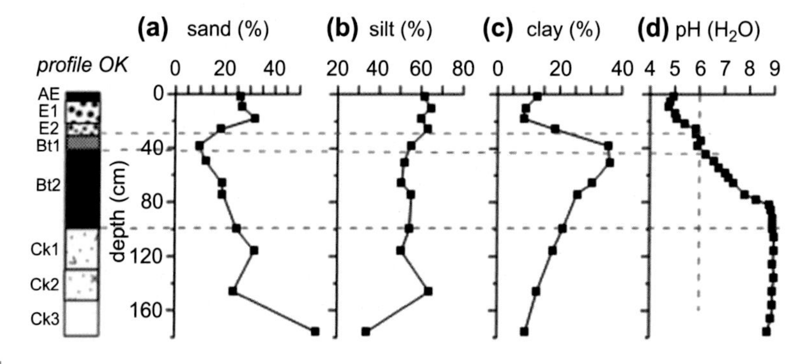

FIGURE 4.3.1

Mechanical fractions: (a) sand, (b) silt, (c) clay, and (d) soil reaction pH along the profile OK.

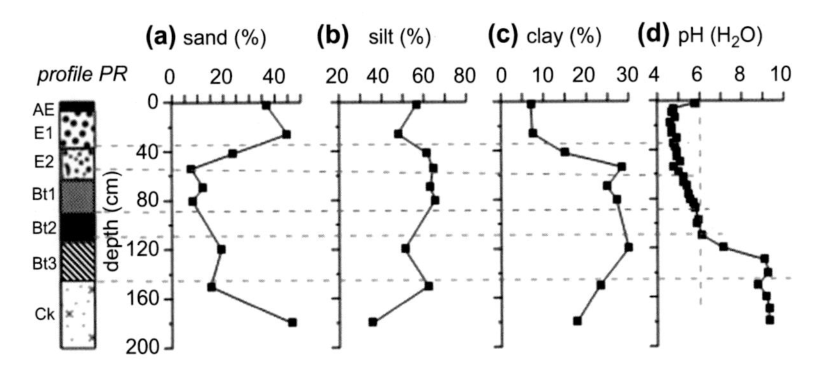

FIGURE 4.3.2

Mechanical fractions: (a) sand, (b) silt, (c) clay, and (d) soil reaction pH along the profile PR.

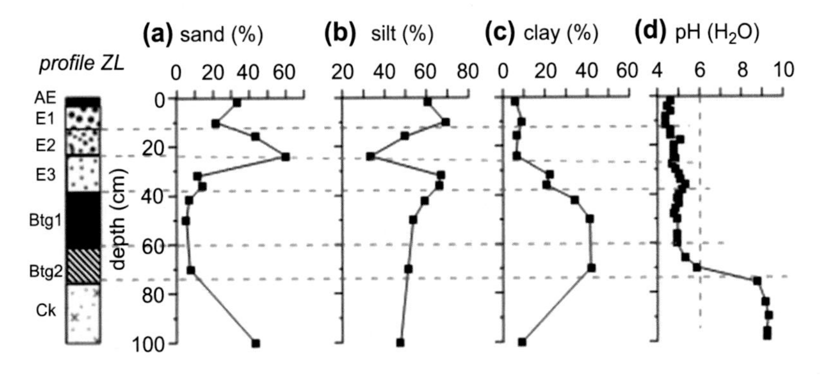

FIGURE 4.3.3

Mechanical fractions: (a) sand, (b) silt, (c) clay, and (d) soil reaction pH along the profile ZL.

rise within the illuvial horizons. For the other profile (ZL), the pH increase is sharper and starts at the bottom of $B_{tg}2$ (Fig. 4.3.3d). The soil texture is strongly differentiated, dominated by the sand/silt fractions in the eluvial horizons and by the clay fraction in the underlying illuvial horizons (Figs. 4.3.1a–c, 4.3.2a–c, and 4.3.3a–c). The clay content reaches 30–40% of the bulk mechanical composition. Sand and clay contents show opposite variations, while silt is more uniformly distributed along the depth, except for the ZL profile (Figs. 4.3.1c, 4.3.2c, and 4.3.3c). Analogous results for the Planosol texture are reported in Jokova and Boyadjiev (1993) for other locations from Bulgaria. Thus, taking into account the observed regularities in the variations of the three mechanical fractions in the Planosols, it could be assumed that the textural contrast in the profiles is due to internal pedogenic processes (either physical or chemical) but not to external sediment input, as reported in some studies (Dumon et al., 2014).

The content of some of the major chemical elements in selected soil samples from the three profiles is shown in Fig. 4.3.4. The data indicate that the eluvial horizons are enriched in Si and Mn and depleted in the other major elements. In contrast, the illuvial B_t horizons are rich in Fe and Al, while Si is at a minimum (Fig. 4.3.4). It is also notable that the maximum iron concentration is always located at the boundary between the deepest eluvial horizon and the underlying illuvial horizon. This is in agreement with the published data on other Planosols in Bulgaria (Jokova and Boyadjiev, 1993; Shishkov and Kolev, 2014). Calcium content is negligible in the upper part of the profiles, while it

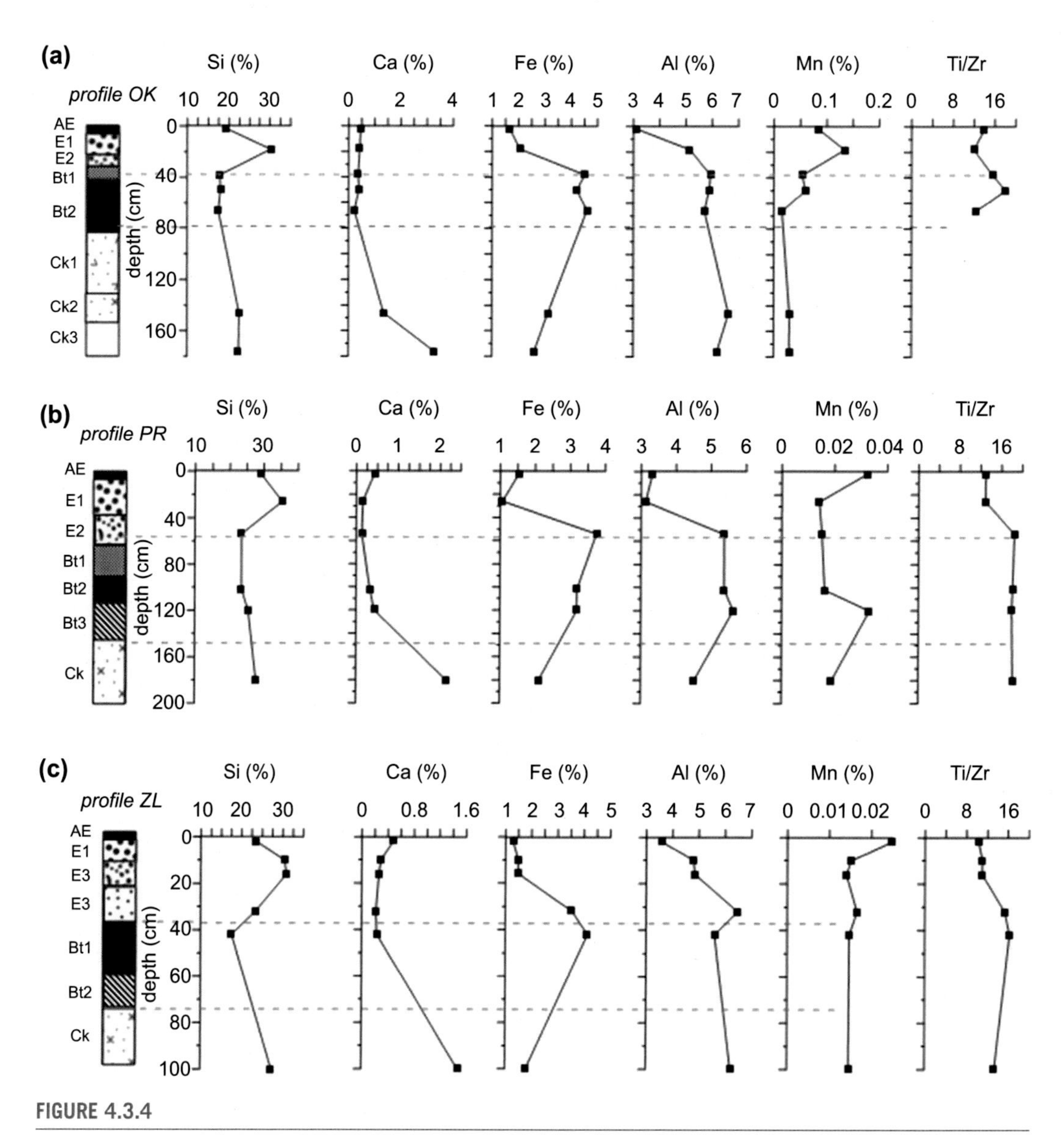

FIGURE 4.3.4

Content of some major and trace elements (Si, Ca, Fe, Al, Mn) and the ratio Ti/Zr for selected samples along the depth of profiles of (a) Eutric Planosol (OK), (b) Luvic Endogleyc Planosol (PR), and (c) Luvic Endogleyc Planosol (ZL).

increases in the C horizons, in line with the major concepts of Planosol pedogenesis (Schaetzl and Anderson, 2009). Iron and aluminum covary along the depth, except in the OK profile (Fig. 4.3.4). In the latter, however, the data for the parent material come from the C_k2 and C_k3 horizons, while most probably C_k1 is the real parent material for the above soil.

Manganese content is enhanced mostly in the uppermost soil depths where Fe-Mn concretions are also present, while downward its content decreases rapidly. The Titanium—Zircon ratio (Ti/Zr ratio), used as an indication of parent material uniformity (Stiles et al., 2003), shows similar values for the three Planosols, varying slightly between 10 and 16 (Fig. 4.3.4), and the minimum is bound to the

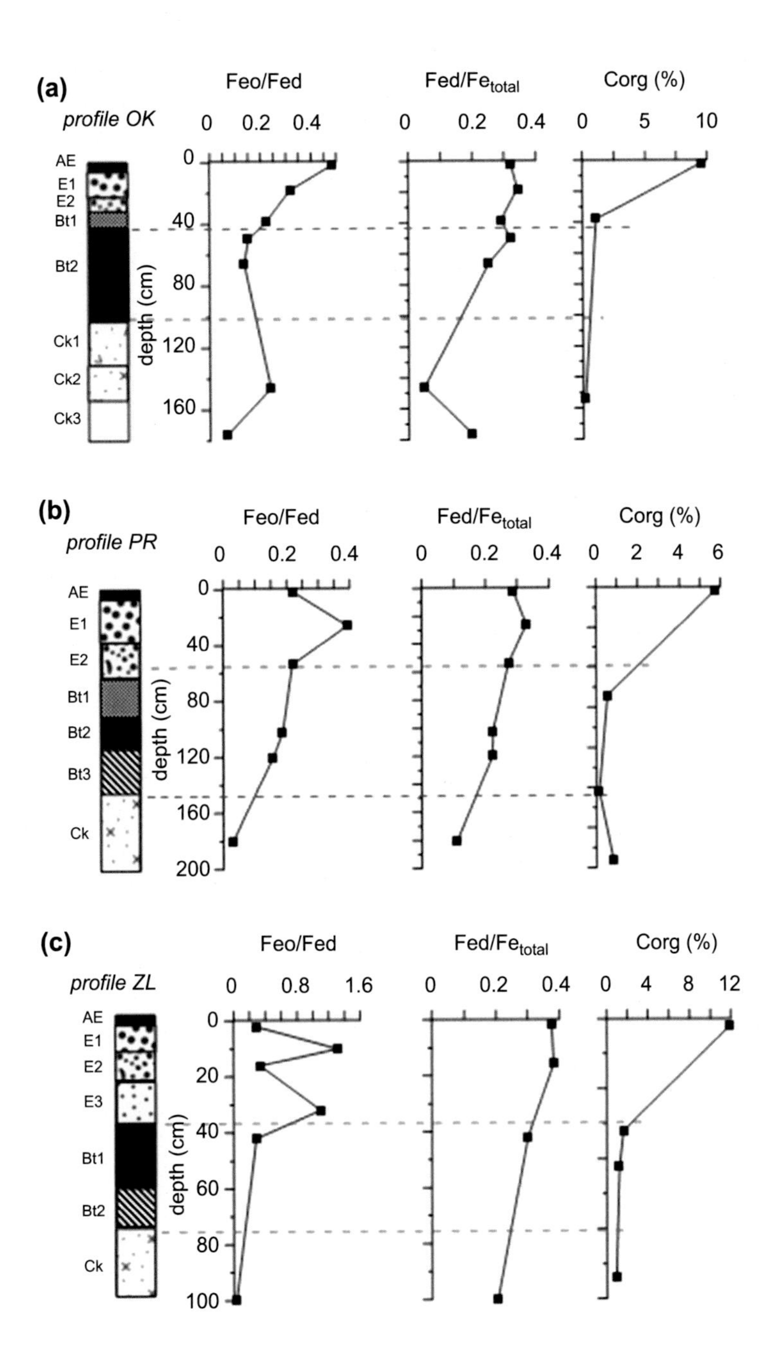

FIGURE 4.3.5

Variations in the ratio of oxalate-extractable iron to dithionite-extractable iron (Fe_o/Fe_d), dithionite-extractable iron to total iron (Fe_d/Fe_{tot}) and the organic carbon content (C_{org}) along the depth of (a) Eutric Planosol (OK), (b) Luvic Endogleyc Planosol (PR), and (c) Luvic Endogleyc Planosol (ZL).

eluvial horizons. Therefore, it further evidences that the profiles studied did not experience external sediment input during their development, which could be also a reason for the observed textural contrast (Van Ranst et al., 2011).

Dithionite and oxalate extraction data are reported in detail in Jordanova et al. (2011), while in Fig. 4.3.5, variations in the Fe_o/Fe_d and Fe_d/Fe_{tot} ratios as well the organic carbon content (C_{org} %) are shown. As becomes evident from Fig. 4.3.5, the Fe_o/Fe_d ratio (which is indicative of the presence of poorly crystalline iron phases, e.g., Schwertmann, 1988; Cornell and Schwertmann, 2003) points to a maximum amount of secondary amorphous Fe phases in the eluvial horizons and its sharp decrease in the B horizons of the three Planosols. The values of 0.2–0.4 of the ratio Fe_o/Fe_d in the E horizons indicate a significant share of the amorphous iron oxides in these parts of the profiles. The second ratio – Fe_d/Fe_{tot}—does not display similar sharp variations but rather a gradual decrease from the surface toward the bottom of the profiles. The obtained Fe_d/Fe_{tot} maximum of ~0.4 (Fig. 4.3.5) suggests the occurrence of a moderate amount of crystalline iron (oxy)hydroxides in the profiles as well. The C_{org} content determined for the selected samples reveals a high concentration in the uppermost thin AE horizons, while down-profile C_{org} is very low (Fig. 4.3.5).

4.4 MAGNETIC PROPERTIES OF PLANOSOLS

The magnetic mineralogy of the Planosols is studied through thermomagnetic analyses of magnetic susceptibility. Heating and cooling curves of $\chi(T)$ for selected samples from the three profiles are shown in Fig. 4.4.1. As seen from the figure, all samples are characterized by intense chemical transformations starting above 200°C. Only the samples from the deepest levels taken from the parent material horizons do not exhibit transformation peaks in their heating curves (Fig. 4.4.1a–e). The well-observed regularity of a decrease in the intensity of the second (higher temperature) transformation peak with increasing depth for the samples from the OK and ZL profiles (Fig. 4.4.1a and e) suggests the presence of iron oxyhydroxide that transforms under the decreasing influence of organic matter content with depth. The most probable mineral responsible for the observed behavior is ferrihydrite, which transforms into two steps when heated in the presence of organics (Hanesch et al., 2006). The decreasing amount of organics causes a less-reducing atmosphere at higher temperatures, diminishing the intensity of the peak at ~500°C (Fig. 4.4.1a and e). In the PR profile, only the uppermost sample (PR2) shows behavior similar to that described earlier, while for the deeper sample (PR26 from the E1 horizon) the transformation starts at ~300°C and a second broad maxima is expressed at 500°C.

The sample from the E2 horizon (PR54) shows only the first (~300°C) transformation (Fig. 4.4.1c). As shown in Hanesch et al. (2006), lepidocrocite exhibits similar one-stage transformation behavior (with or without organics) as well as a mixture of ferrihydrite, organic matter, and calcium carbonate. Therefore, lepidocrocite could also be present in the E2-B horizons in the PR profile, but additional research is needed to confidently confirm this hypothesis. Thermomagnetic curves for the samples from parent material from the three profiles generally indicate the presence of a very small amount of magnetite as well as a paramagnetic contribution at lower temperatures (especially in PR and ZL profiles) (Fig. 4.4.1). All cooling curves reveal the appearance of a large amount of magnetite-like new, strongly magnetic phases, possibly with different substituting elements, responsible for the lowering of the T_c in some of the samples (Fig. 4.4.1b, d, and f).

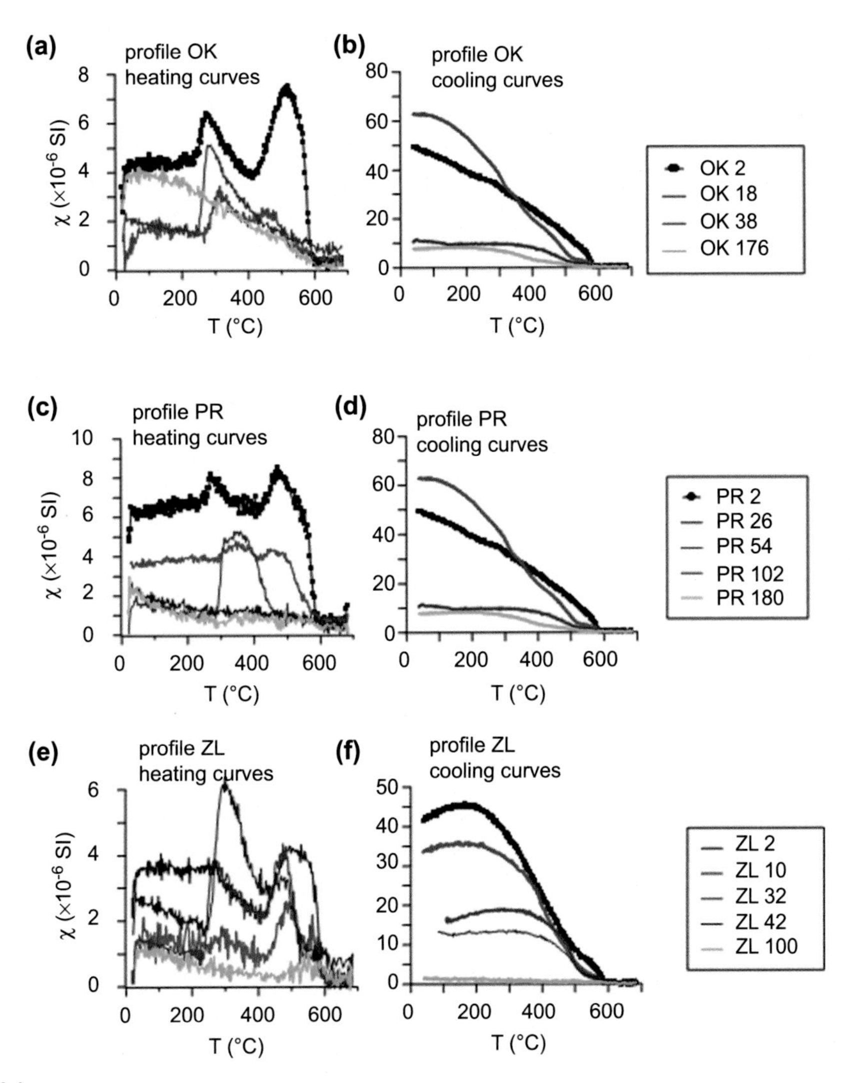

FIGURE 4.4.1

Thermomagnetic analysis of magnetic susceptibility for selected samples from different horizons of: OK Planosol profile: (a) heating curves and (b) cooling curves; PR Planosol profile: (c) heating curves and (d) cooling curves; ZL Planosol profile: (e) heating curves and (f) cooling curves. Heating and cooling runs for each sample are shown in equivalent color. Heating in air, fast heating rate (11°C/min).

The stable remanence-carrying magnetic minerals are studied through stepwise thermal demagnetization of the three-component isothermal remanence acquired at room temperature (Lowrie, 1990), using saturating fields of 0.23, 0.6, and 2 T. Representative examples for the three Planosols are shown in Fig. 4.4.2. In all the samples, the magnetically soft component acquired in the 0.23- T applied field is

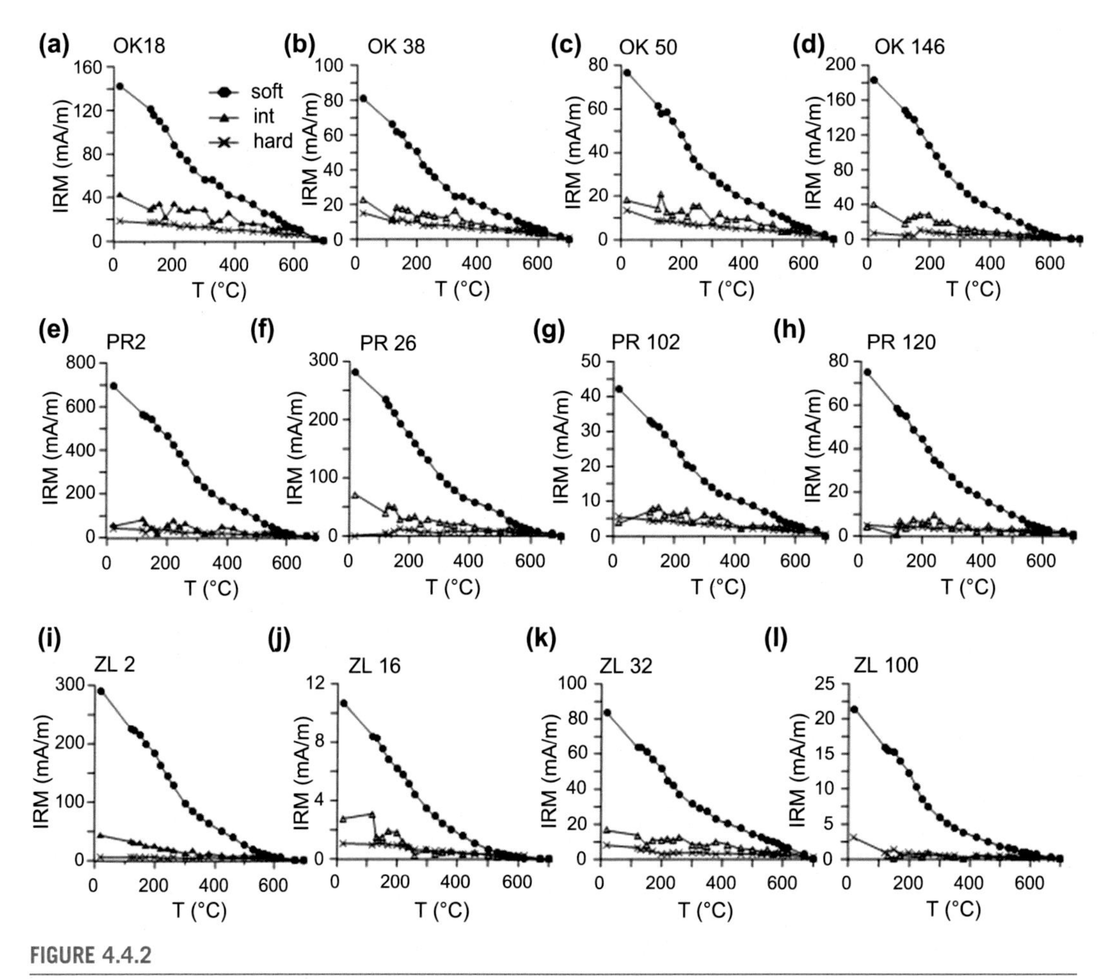

FIGURE 4.4.2

Examples of stepwise thermal demagnetization of composite IRM (soft fraction, acquired in a field 0.23 T; intermediate fraction, acquired between 0.23 and 0.46 T; hard fraction, acquired between 0.46 and 2.0 T). Examples a–d: OK profile; examples e–h: PR profile; examples i–l: ZL profile.

dominant in the total remanence. Two unblocking temperatures (T_{ub}) are present in the demagnetization curves of this soft component. For the samples from the PR and ZL profiles, these are at ~250°C and ~600°C (Fig. 4.4.2e–l), while for the OK profile, T_{ub1} is at ~250°C and the second (final) T_{ub2} is observed at 700°C (Fig. 4.4.2a–d). The intermediate and hard components all show the unblocking temperature of hematite at ~700°C. The hard component is the strongest (compared with the other profiles) in the upper horizons of the OK profile, while in the parent material it is negligible.

The lower T_{ub} of ~250°C observed could be related to the occurrence of unstable maghemite in soils (Dunlop and Özdemir, 1997) or, alternatively, to weakly crystalline hematite with Al substitutions (Schwertmann et al., 2000), while T_{ub} ~600°C most probably indicates the existence of oxidized magnetite. A small amount of hematite with T_{ub}~700°C is seen in all samples.

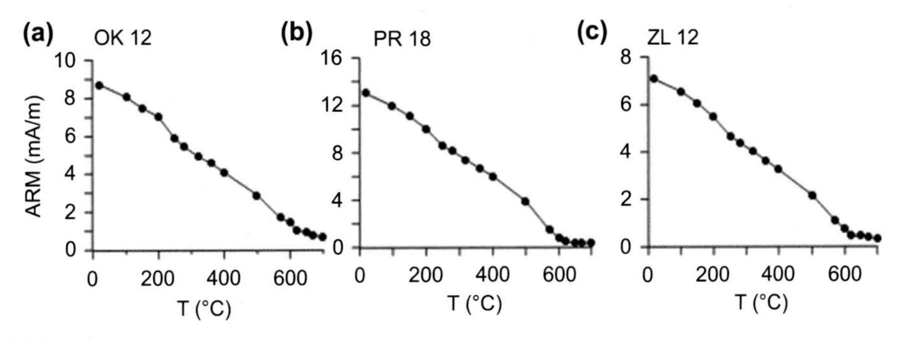

FIGURE 4.4.3

Examples of stepwise thermal demagnetization of anhysteretic remanence (ARM) for soil samples from the eluvial horizons of Plansols OK (a), PR (b), and ZL (C).

Stepwise thermal demagnetization of anhysteretic remanence (ARM) is carried out for samples from the eluvial horizons of the three Planosols in order to check if a stable single domain (SD)-like magnetite fraction is present. The obtained demagnetization curves are depicted in Fig. 4.4.3.

The convex shape of the curves is typical for SD-like ARM-carriers with a relatively narrow grain-size distribution (Dunlop and Argile, 1997). A small kink is seen at ~220°C, which again could be related to maghemite's presence. The origin of this magnetite fraction will be the subject of detailed discussion later in this chapter.

Further elucidation of the magnetic mineralogy is attained by a stepwise Isothermal Remanent Magnetization (IRM) acquisition in steady fields up to 5 T (Fig. 4.4.4). Samples from the illuvial horizons of the Planosols show a gradual increase in IRM, which reaches saturation above the 1- T

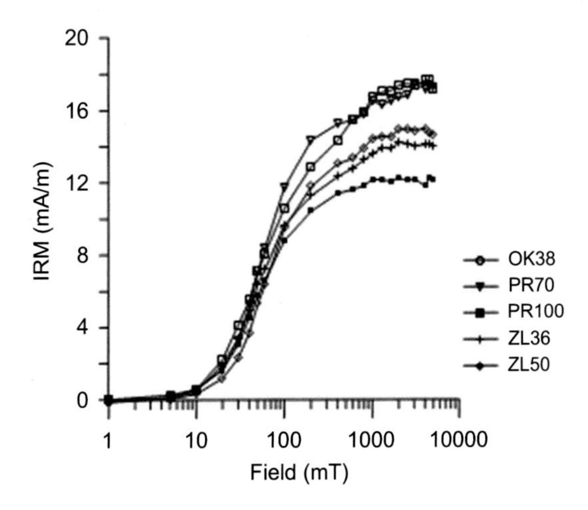

FIGURE 4.4.4

Stepwise acquisition of isothermal remanence (IRM) up to field of 5 T for selected samples from the Planosol profiles.

Table 4.4.1 Summary parameters from coercivity analysis by fitting IRM acquisition curves with cumulative Gaussian functions, using IRM-CLG1.0 software (Kruiver et al., 2001) for selected samples from the Planosol profiles OK, PR, and ZL

	IRM component 1				IRM component 2			
Sample depth (cm)	Intensity (10^{-6} Am^2/kg)	% from IRM_{total}	$B_{1/2}$ (mT)	DP	Intensity (10^{-6} Am^2/kg)	% from IRM_{total}	$B_{1/2}$ (mT)	DP
Profile OK								
2	0.586	91	41.70	0.28	0.059	9	398.10	0.37
38	0.675	77	47.90	0.40	0.205	23	631.00	0.33
Profile PR								
70	0.705	83	46.80	0.30	0.145	17	501.20	0.45
100	0.515	85	43.70	0.35	0.091	15	398.10	0.35
Profile ZL								
10	0.282	81	49.00	0.38	0.066	19	562.30	0.36
36	0.500	70	38.90	0.30	0.210	30	402.70	0.44
50	0.525	70	46.80	0.28	0.230	30	389.00	0.43

The intensity, relative contribution (% from IRM_{total}), median acquisition field ($B_{1/2}$), and dispersion parameter (DP) for each coercivity component are shown.

field. Coercivity analysis using the IRM-CLG1.0 software (Kriuver et al., 2001) reveals the presence of two IRM components. The soft IRM (IRM1 component) shows uniform coercivity for all samples expressed by the $B_{1/2}$ parameter generally in the range (40–46) mT (Table 4.4.1), as well as a similar width of coercivity distribution (the DP parameter). The intensity of the IRM1 component is similar among the three profiles and comprises between 70% and 80% of the total IRM (Table 4.4.1).

Consequently, the IRM1 is carried by identical mineral phases in all Planosols. The obtained coercivity of the soft component is rather high for pure SD magnetite (e.g., Egli, 2004); thus, it could indicate a maghemitized magnetite and/or maghemite as a mineral carrier. The second IRM component (IRM2) is characterized by a significantly weaker intensity, comprising a fairly low share in the total IRM (between 9% and 30%) and possessing higher coercivities in the interval of (389–631 mT) (Table 4.4.1). Such values could be ascribed to hematite of a different grain size and/or a degree of Al substitutions (Wells et al., 1999).

4.4.1 DEPTH VARIATIONS IN THE MAGNETIC CHARACTERISTICS OF PLANOSOLS

Magnetic susceptibility (χ) variations along the Planosol OK are shown in Fig. 4.4.5a. It is characterized by lower values in the upper part of the profile (eluvial and illuvial horizons) and a relative enhancement in the C horizons. On the other hand, the high-field susceptibility (χ_{hf}) exhibits very different behavior showing a minimum in the E horizons and a sharp increase in the illuvial horizons (Fig. 4.4.5b), coherent with the clay content variations (Fig. 4.3.1). Thus, the χ_{hf} reflects the changes in the content of the paramagnetic Fe-rich clay minerals in the illuvial horizons (B_t1 and B_t2) of the profile. The ferrimagnetic susceptibility calculated as a difference between the low-field susceptibility

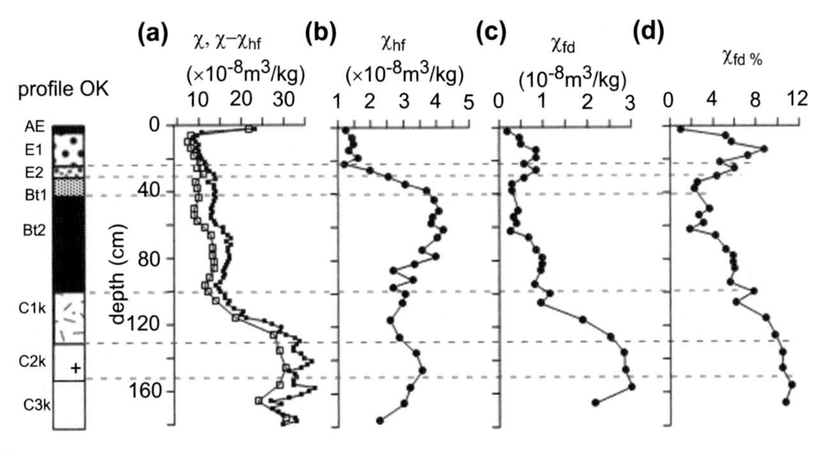

FIGURE 4.4.5

Profile OK. Depth variations of (a) magnetic susceptibility (χ) (*full dots*) and ferrimagnetic susceptibility ($\chi - \chi_{hf}$) (*open dots*); (b) high-field magnetic susceptibility (χ_{hf}); (c) frequency-dependent magnetic susceptibility (χ_{fd}), and (d) percent frequency-dependent magnetic susceptibility (χ_{fd}%).

(χ) and the high-field susceptibility (χ_{hf}) (Fig. 4.4.5a) again suggests a relative depletion in the strongly magnetic minerals in the upper soil part, compared with the parent material. The obtained χ values are very low in the upper soil horizons ($\chi \sim 10 \times 10^{-8}$ m^3/kg) and about twice as high in the C horizon. Nevertheless, frequency-dependent magnetic susceptibility (χ_{fd}) and the percent frequency-dependent magnetic susceptibility (χ_{fd}%) show well-constrained depth variations (Fig. 4.4.5c and d) with two maxima—a weaker one in the eluvial horizon E1 and a more pronounced one in the C1 horizon. The high values of χ_{fd}% (8–10%) obtained at the depth of the observed maxima reveal an important contribution of the nano-sized strong ferrimagnetic particles in the magnetic signal of the OK Planosol. Concerning the absolute amount of this SP fraction as revealed by the χ_{fd} (Fig. 4.4.5c), it appears that it is more abundant in the deeper part of the profile. The parent material of the profile is loamy-sandy re-deposited material from weathered granites (see the profile description earlier). Thus, one possible hypothesis is that the enhanced magnetism registered in the C horizons is due to strongly magnetic lithogenic iron oxides. However, a problem appears as how to explain the high concentration of the SP fraction in such deposits formed by weathered granite, because usually granites contain coarse magnetite/titanomagnetite grains (e.g., Henry et al., 2012). In addition to the SP fraction, the variations in the SD grain-size–sensitive anhysteretic remanence (ARM) shows enhanced values at the bottom of the OK profile (Fig. 4.4.5c).

Concentration-dependent magnetic parameters saturation magnetization M_S, isothermal remanence acquired in 2 T field IRM_{2T}, and anhysteretic remanence ARM (Fig. 4.4.6a–c) confirm the earlier considerations as well. Furthermore, the S ratio reflecting the relative contribution of high-coercivity minerals in the IRM (King and Channel, 1991) exhibits clear horizon-related variations—it strongly decreases from the surface through the eluvial horizons toward the bottom of the illuvial horizon Bt1 where it reaches a minimum, and again increases to high values ($\sim$0.8) at the bottom. Such behavior suggests an important role of a high-coercivity mineral component, most probably hematite, as deduced from the thermal demagnetization results (Fig. 4.4.2).

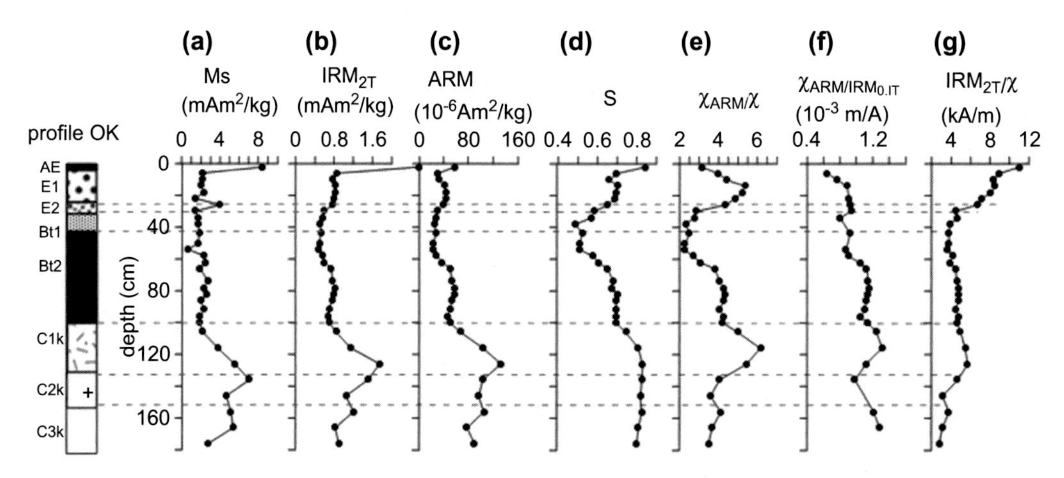

FIGURE 4.4.6

Depth variations of magnetic parameters and ratios along the OK Planosol profile: (a) saturation magnetization (M_S); (b) isothermal remanence acquired in a field of 2 T (IRM_{2T}); (c) anhysteretic remanence (ARM); (d) S ratio ($S = IRM_{0.3T}/IRM_{2T}$); (e) ratio between anhysteretic susceptibility to magnetic susceptibility (χ_{ARM}/χ); (f) ratio between anhysteretic susceptibility and isothermal remanence, acquired in 0.1 T field ($\chi_{ARM}/IRM_{0.1T}$); (g) ratio between IRM_{2T} and magnetic susceptibility (IRM_{2T}/χ).

Grain size related magnetic proxies display apparently controversial results. The χ_{ARM}/χ ratio, used as an SD grain-size proxy, shows in fact changes coherent with the S ratio along the depth except in the uppermost levels (Fig. 4.4.6d and e). Thus, its real grain-size dependence is questionable because magnetic susceptibility is probably strongly influenced by the hematite's contribution. In order to avoid the biasing effect of the mixtures in iron oxide mineralogy, the ratio $\chi_{ARM}/IRM_{0.1T}$ is used (Fig. 4.4.6f). Its variations reveal a coarser grain size of the magnetite-like (magnetically soft) fraction in the upper part of the solum, where the ratio has a value of ~0.4, followed by an overall gradual increase up to ~1.2, suggesting the dominance of SD-like grains (Geiss et al., 2008) in the C horizon. This conclusion is supported by another ratio—IRM_{2T}/χ (Fig. 4.4.6g) – which clearly separates the upper eluvial horizons containing a considerable fraction of magnetically stable grains (e.g., a maximum in the ratio) from the bottom parts, where despite the high χ values, the IRM_{2T}/χ is low. This observation could be explained by an SP fraction having a significant share in the magnetic susceptibility of the material from the C horizons, as also evidenced by the χ_{fd} variations (Fig. 4.4.5).

The overall behavior of the magnetic characteristics for the OK profile suggests an enhanced content of coarse magnetically stable grains in the eluvial horizons with maximum concentration at the surface. In addition, a small fraction of SP magnetite grains detected through χ_{fd} variations is linked to the middle part of the eluvial zone. The upper part of the illuvial zone (the B_t1 horizon) is enriched in high coercivity hematite, which has a content that diminishes downward. The bottom part of the B_t2 contains finer magnetite-like grains, as seen by the increase in $\chi_{ARM}/IRM_{0.1T}$ there, but the true SD grains, as well as a high amount of SP grains, are related to the C1 horizon.

The second Planosol profile, PR, shows a different character for the susceptibility variations in depth (Fig. 4.4.7a)—the upper eluvial horizon E1 is magnetically most enhanced, followed by a

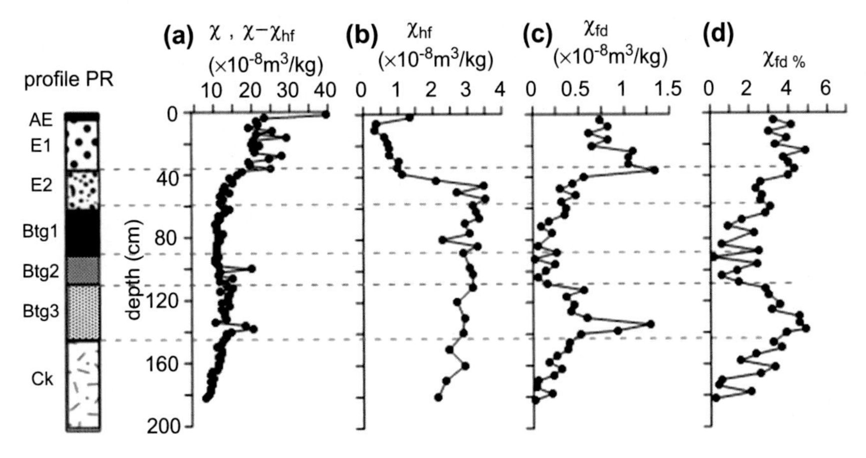

FIGURE 4.4.7

Profile PR. Depth variations of (a) magnetic susceptibility (χ) (full dots) and ferrimagnetic susceptibility ($\chi-\chi_{hf}$) (*open dots*); (b) high-field magnetic susceptibility (χ_{hf}); (c) frequency-dependent magnetic susceptibility (χ_{fd}); and (d) percent frequency-dependent magnetic susceptibility (χ_{fd}%).

gradual decrease in E2 and a constant χ along the illuvial horizons. A further χ decrease is registered until the bottom of the profile. Thus, despite the intense eluviation processes, the E1 horizon is magnetically enhanced with ferrimagnetic iron oxides, as evidenced by the χ_{ferri} ($\chi-\chi_{hf}$) variation. The high-field susceptibility (χ_{hf}) again sensitively locates clay-enriched illuvial horizons (Fig. 4.4.7b). However, considering the variations in the frequency-dependent magnetic susceptibility (Fig. 4.4.7c and d), similararities with the OK Planosol are revealed—there are again two maxima in the amount of the SP fraction – one linked to the middle of the eluvial zone and a second one linked to the bottom part ($B_{tg}3$ in the PR profile). The χ_{fd}% reaches lower values in the PR profile (~4%) compared with OK, suggesting a weaker SP contribution to the bulk low-field magnetic susceptibility, as also supported by the χ trend.

Depth variations in the concentration-related remanence characteristics along the PR profile corroborate the conclusions drawn from the susceptibility changes (Fig. 4.4.8a–c). The enhancement with coarse (IRM_{2T} maxima) and more stable SD-like (ARM maxima) particles in the eluvial horizons is well underlined. On the other hand, despite the differences between the susceptibility pattern along the OK and PR profiles, magnetic bi-parametric ratios S, χ_{ARM}/χ, and $\chi_{ARM}/IRM_{0.1T}$ (Fig. 4.4.8d, e, and g) demonstrate a remarkable similarity. Similar to the OK profile, a high coercivity fraction is identified in the illuvial horizon and, probably in this case, χ_{ARM}/χ is not reliable as an SD grain-size proxy because of its covariance with the S ratio. The IRM_{2T}/χ variations again suggest an enhancement with coarse stable grains in the upper eluvial horizons and a finer magnetic grain size dominated by an SD fraction at the bottom of the profile. The grain size of the magnetite-like fraction deduced from the behaviour of $\chi_{ARM}/IRM_{0.1T}$ ratio (Fig. 4.4.8f) decreases from the surface toward the C horizon, where it reaches values characteristic for the true SD magnetite (Geiss et al., 2008).

The magnetic susceptibility of the ZL Planosol profile closely resembles the pattern observed in the OK profile (Fig. 4.4.9a) with the strongest maximum associated with the deepest part. The magnetically depleted upper soil horizons exhibit low χ in the order of 10×10^{-8} m^3/kg, similar to the OK

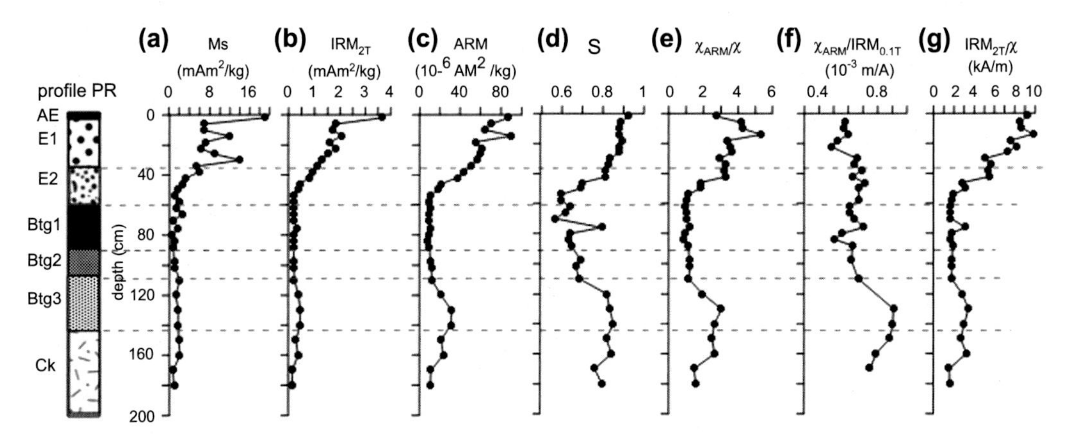

FIGURE 4.4.8

Depth variations of magnetic parameters and ratios along the PR Planosol profile: (a) saturation magnetization (M_s); (b) isothermal remanence acquired in a field of 2 T (IRM_{2T}); (c) anhysteretic remanence (ARM); (d) S ratio ($S = IRM_{0.3T}/IRM_{2T}$); (e) ratio between anhysteretic susceptibility to magnetic susceptibility (χ_{ARM}/χ); (f) ratio between anhysteretic susceptibility and isothermal remanence, acquired in 0.1 T field ($\chi_{ARM}/IRM_{0.1T}$); (g) ratio between IRM_{2T} and magnetic susceptibility (IRM_{2T}/χ).

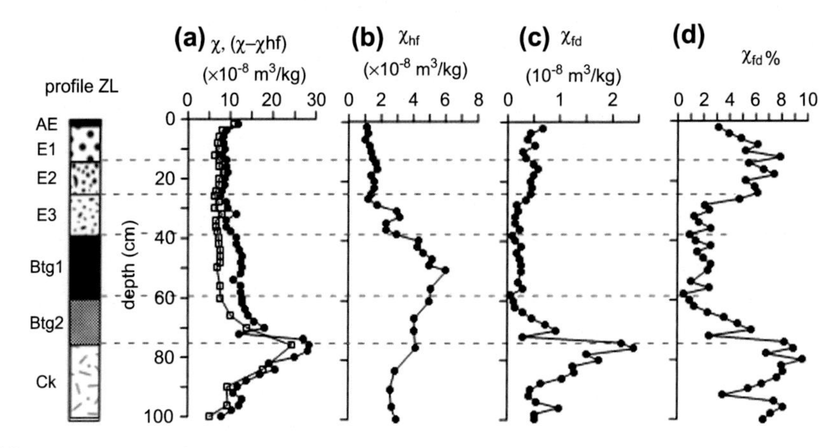

FIGURE 4.4.9

Profile ZL. Depth variations of: (a) magnetic susceptibility (χ) (*full dots*) and ferrimagnetic susceptibility ($\chi - \chi_{hf}$) (*open dots*); (b) high-field magnetic susceptibility (χ_{hf}); (c) frequency-dependent magnetic susceptibility (χ_{fd}); and (d) percent frequency-dependent magnetic susceptibility ($\chi_{fd}\%$).

profile. The high-field susceptibility is again enhanced in the illuvial horizons (Fig. 4.4.9b) and decreases in the C horizon, pointing to a link with the clay fraction. Comparing Figs. 4.4.5c–d and 4.4.9c–d, it is evident that χ_{fd} variations in both the OK and ZL profiles are identical, indicating a weaker SP enhancement in the eluvial horizons and a high amount of the SP fraction in the $B_{tg}2$-C_k horizons.

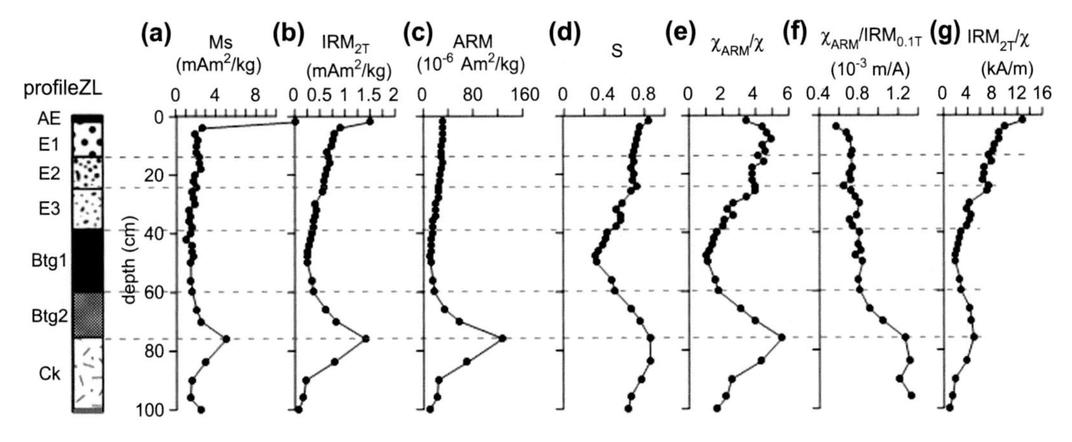

FIGURE 4.4.10

Depth variations of magnetic parameters and ratios along the ZL Planosol profile: (a) saturation magnetization (M_S); (b) isothermal remanence acquired in a field of 2 T (IRM_{2T}); (c) anhysteretic remanence (ARM); (d) S ratio ($S = IRM_{0.3T}/IRM_{2T}$); (e) ratio between anhysteretic susceptibility to magnetic susceptibility (χ_{ARM}/χ); (f) ratio between anhysteretic susceptibility and isothermal remanence, acquired in 0.1 T field ($\chi_{ARM}/IRM_{0.1T}$); (g) ratio between IRM_{2T} and magnetic susceptibility (IRM_{2T}/χ).

Analogously, variations in the saturation magnetization M_S and the remanent magnetizations (IRM_{2T}, ARM), as well as the different ratios (Fig. 4.4.10) along the ZL Planosol, closely match the respective characteristics obtained for the OK profile. The behavior of the $\chi_{ARM}/IRM_{0.1T}$ ratio in the ZL also points to an SD magnetite present at the bottom of the profile (Fig. 4.4.10f). Thus, the earlier considerations related to the meaning of the variations observed in the concentration and grain-size proxy parameters are identical with the other two Planosol profiles.

4.5 PEDOGENESIS OF IRON OXIDES IN PLANOSOLS, AS REFLECTED IN SOIL MAGNETISM

Planosols develop on different parent materials and in different climates, but all of them are characterized by identical properties related to their genesis. Aside from the "lithogenic" Planosols (those in which textural differentiation is due to external coarse textured sedimentation on an ancient surface of clay-enriched medium), the "pedogenic" Planosols are considered to be a result of several processes that become enhanced with advancing soil development. According to Baize (1989) physical macroporosity is created in the initially clayey parent material as a result of weathering, accompanied by lateral water circulation at the base of the topmost layer. Due to weathering, (1) Fe ions are released, clay is synthesised and, in the case of calcareous parent material, decarbonization proceeds; (2) these processes advance, surface water-logging conditions develop, and base desaturation starts; (3) reduction–oxidation cycles in the surface layer alternate, hydromorphic features develop at a shallow depth, and aluminization begins; (4) water-logging and mineral acidity increase and secondary illuviation expands; and (5) ferrolysis of some clay minerals begins with advancing acidity and water-logging; podzolization is also possible in the surface layer. As a result, three zones with different hydrodynamic conditions can be distinguished in the Planosols—(i) the uppermost eluvial horizon with a dynamic cyclic change in reduction-oxidation periods; (ii) the illuvial horizon, commonly enriched in illuviated clay in its upper surface and otherwise clay-enriched as a result of on-site clay formation. This part of the Planosol solum is regarded by some authors as predominantly well aerated (van Breemen, 1988a), but others (e.g., Baize, 1989) suppose that it contains a large amount of retained water and reducing conditions prevail; (iii) the clayey C horizon, containing weathered primary minerals, which could also be influenced by groundwater-level fluctuations.

Fig. 4.5.1 outlines the information on magnetic mineralogy and tendencies in grain-size proxies in the three Planosols studied. As demonstrated in Section 4.4 of this chapter, the E horizons, despite being bleached and clay depleted, are magnetically enhanced with coarser magnetite grains and a small amount of fine superparamagnetic grains. The coarser grains cannot be regarded as lithogenic, since the horizon has experienced intense redox changes and all primary Fe-containing minerals should have been dissolved, leaving only skeletal quartz grains (van Breemen, 1988a; Grimley and Arruda, 2007).

The dynamic interchange of the redox conditions in the soil pushes it to an unstable state because Fe and Mn released into the soil solution during reduction are mobilized and transported with the draining water, invoking an effective Fe^{2+} flow through the soil. An oxidative phase favors the precipitation of poorly crystalline ferrihydrite. During the next reduction–oxidation cycle, it could interact with the Fe^{2+} in the solution. According to van Breemen (1988a), the precipitation of better-crystalline Fe(III) oxides is a slow process, while the reduction is commonly a rapid process. A number of geochemical studies deal with possible mineralization pathways of ferrihydrite driven by Fe^{2+} concentration, pH, flow rate (Hansel et al., 2003, 2005; Tufano et al., 2009), the effect of Al substitutions in ferrihydrite (Hansel et al., 2011), Si and P content (Zachara et al., 2011), microbial activity

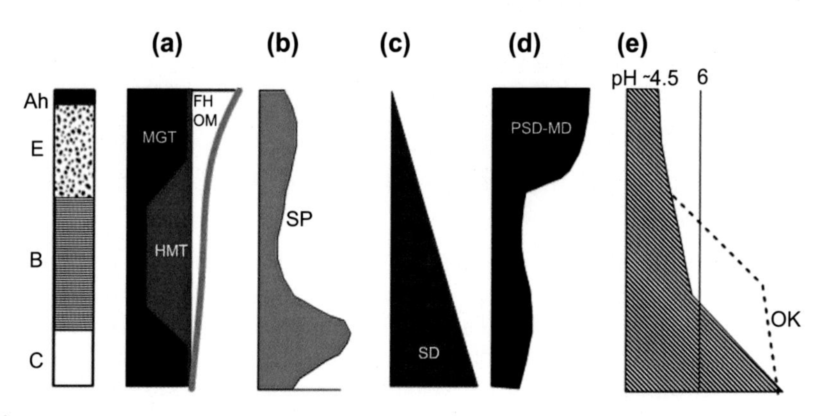

FIGURE 4.5.1

Schematic illustration of the major trends and relations among the magnetic mineralogy and magnetic grain size variations along the Planosol profiles, as deduced from the magnetic measurements: (a) magnetite (MGT)-hematite (HMT) and ferrihydrite/organic matter (FH, OM) distribution along depth; (b) variations in concentration of the superparamagnetic (SP) fraction along depth; (c) variations in concentration of the single-domain (SD) fraction along depth; (d) variations in concentration of the PSD-MD fractions along depth; (e) pH changes along profiles' depth.

(Hansel et al., 2003; Zachara et al., 2002; Usman et al., 2012), etc. The major findings suggest that, at low Fe^{2+} concentrations in the soil solution, the main secondary transformation product of ferrihydrite mineralization is goethite or lepidocrocite, while a high amount of Fe^{2+} and/or microbial activity triggers the fast precipitation of magnetite (Hansel et al., 2003, 2005). Considering the earlier findings, it could be hypothesized that the enhanced content of stable coarser magnetite grains obtained in the eluvial horizons of Planosols (Figs. 4.4.6–4.4.10 and 4.5.1) is a result of ferrihydrite mineralization under changing reducing–oxidative cycles. In addition, Thompson et al. (2006), in a laboratory-controlled experiment, show that, as a result of oscillating redox changes when the pH fluctuates between 4 and 5.5, the crystallinity of the precipitated products increases. This might be the reason for the well-expressed enhancement of E horizons with stable coarse grains (e.g., IRM_{2T}/χ maxima), while the amount of SP grains (as reflected in χ_{fd} variations) is low. Thus, grain growth is dominant over the grain nucleation (producing finer [SP] grains) in this part of the solum. The pedogenic origin of the magnetic enhancement of the eluvial soil horizons is also suggested by Fine et al. (1989) in a study of a chronosequence including Haplustalfs, Haplohumults, and Palehumults in the United States, but the authors do not reject a contribution from inherited lithogenic magnetite as well.

Further on, the magnetic properties of the Planosols studied suggest that the illuvial horizons are enriched with a high-coercivity magnetic mineral identified as hematite (Figs. 4.4.2–4.4.4). The theories and laboratory experiments on the reductive dissolution of ferrihydrite, referred to earlier, do not show the formation of hematite in any of the possible pathways. Thus, the observed magnetic signature in the B horizons cannot be assigned to such a process. The presence of hematite suggests the existence of oxidizing conditions in this part of the profiles, in accordance with van Breemen's view (van Breemen, 1988a). The author hypothesizes that the combination of a low pH and seasonally dry conditions provokes the nonreductive dissolution of goethite, followed by the precipitation of hematite. In the illuvial horizons of the Planosols, a very strong increase in pH, generally starting at the bottom of the deepest B horizon, is observed (Figs. 4.3.2, 4.3.3, and 4.5.1). This part of the profiles is also distinguished by sharp peaks in the concentration of fine-grained magnetic oxides (Fig. 4.5.1). The increase in the pH is exclusively sharp in the OK profile (Fig. 4.3.1), where it already starts from the eluvial horizons and spans ~3 pH units. The existence of a very fine-grained (SP and SD) magnetite fraction in the deepest parts of the profiles could not be related to lithogenic grains, since they are characterized by much larger grain sizes. At the same time, the ~SD magnetite occurrence in all three Planosols sampled in different locations (e.g., with different parent materials), would imply a strong influence by chemical weathering in the bottom parts of the three profiles. This alteration, however, is obviously linked to the Planosol's genesis, since a similar phenomenon is not observed in other soil types. The importance of the fine-grained, presumably pedogenic, iron oxides in the Planosols is also evidenced in Fig. 4.5.2, showing the ratio of the magnetic susceptibility of soil samples after the DCB extraction (χ_{DCB}) and its initial value (χ_{ini}).

As seen from Fig. 4.5.2, the largest part of χ is "erased" from the illuvial horizons in the middle parts of the profiles. However, ~60% of initial magnetic susceptibility is lost upon DCB extraction in the C horizons as well. One possible explanation of the observed enhancement with a fine SP-SD fraction at the bottom of the profiles is its formation under the influence of groundwater fluctuations, also causing the reductive dissolution of the primary minerals and further precipitation of a secondary magnetite phase, as already discussed. Since a high pH strongly favors magnetite precipitation as a transformation product of ferrihydrite (e.g., Hansel et al., 2005), its formation in the bottom parts of the Planosols is suggested.

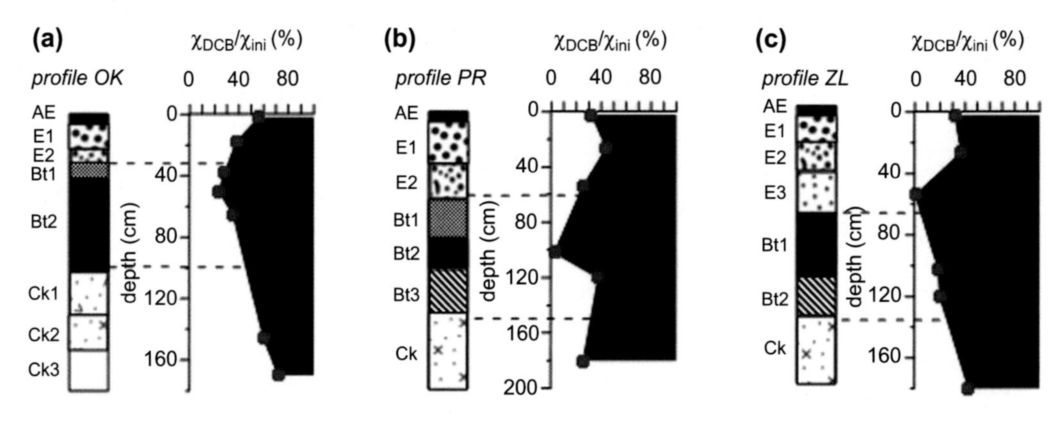

FIGURE 4.5.2

Changes in the ratio of magnetic susceptibility of soil samples, measured after DCB extraction (χ_{DCB}) and the initial magnetic susceptibility (χ_{ini}), expressed as a percent, for Planosols OK (a), PR (b), and ZL (c). Magnetic susceptibility part "erased" through the DCB is denoted by the *black-filled area* below the 100% value, while the DCB-resistant part of the χ is denoted by the *white area*. Depths of samples subjected to DCB extraction are shown by *dots*.

An important question regarding the mode of pedogenic iron oxide formation is the link between this magnetic fraction and the mechanical fractions in the soil. Usually, a good correlation is observed between the amount of pedogenic iron oxides and the percentage of a clay fraction in well-aerated soils such as Cambisols (Maher, 1986) and Luvisols (e.g., Torrent et al., 2010). In an attempt to establish the possible association of the strongly magnetic pedogenic fraction in Planosols to a certain mechanical grain-size fraction, we have found a good linear relation between frequency-dependent magnetic susceptibility (χ_{fd}) and the percentage of the sand fraction (Fig. 4.5.3).

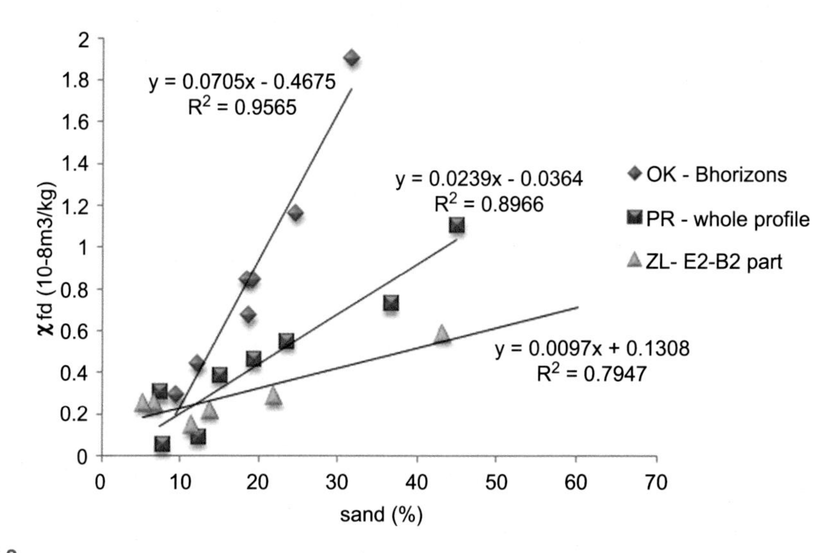

FIGURE 4.5.3

Linear regressions between frequency-dependent magnetic susceptibility (χ_{fd}) and the sand content for the three Planosol profiles.

The highest correlation coefficient (R^2) of the regression equation is obtained for the illuvial horizons of the OK profile, although the two other Planosols exhibit a good correlation between χ_{fd} and *sand* for almost the whole profiles. The obtained link between the finest pedogenic magnetite fraction and the sand fraction could be explained if we suppose that the pedogenic fraction has precipitated as a thin film on the surface of the coarse lithogenic grains. Such a situation is often encountered in different soils (e.g., Fine et al., 1989) and is observed where pedogenic minerals form under the influence of flow conditions. Rather, when it is bound to the clay fraction, the pedogenic iron oxides nucleate and grow in situ.

The magnetic signature of Planosols demonstrates a large heterogeneity in the distribution of the different magnetic grain-size fractions along the depth, most probably related to the specific hydrologic regime in this particular soil type. Mineral transformations occurring under the influence of oscillating reductive–oxidative conditions effectively cause pedogenic depletion rather than an enhancement of the soils. At the same time, secondary transformations lead to the appearance of a specific "doublet" in the depth distribution of the finest SP fraction. The middle parts of the eluvial horizons, as well as the deeper (bottom) part of the illuvial/parent material horizons, become relatively enriched in ultrafine SP magnetite grains. Although relatively depleted in Fe, the eluvial horizons become moderately enriched with stable (coarser) magnetic grains, either formed as a result of secondary mineralization of ferrihydrite or inherited from the parent lithology, or both. Unfortunately, no other published environmental magnetic study on Planosols is available in order to make better-constrained conclusions and to compare soil properties on a larger geographical scale.

4.6 MICROSCOPIC AND MAGNETIC STUDIES ON CONCRETIONS AND NODULES FROM PLANOSOLS

Nodules and concretions from different soil types are widely studied because they serve as a sink for a number of heavy and trace elements (Palumbo et al., 2001; Liu et al., 2002; Cornu et al., 2005). They can form via abiotic redox processes during wetting–drying cycles, as well as via biotic redox processes mediated by microorganisms (Jien et al., 2010b; Timofeeva et al., 2014; Zhang et al., 2014). Nodules are a characteristic feature for Planosols because of the strong textural contrast between the eluvial and illuvial horizons, where, along with seasonal water-logging, a lateral water flow is often created (Baize, 1989; van Breemen, 1988a; Schaetzl and Anderson, 2009) and the conditions for iron and manganese segregation are favorable.

Nodules and concretions are also abundant in the three Planosols from Bulgaria (see the profiles' descriptions). Several concretions from two of the profiles (OK and PR) are studied through scanning electron microscopy (SEM), and part of the data obtained are published in Jordanova et al. (2011). The concretions are cut diametrically, polished sections are prepared and observed under SEM with EDX analysis in selected spots.

SEM images in the back-scattered electrons mode of the three concretions from the OK profile, extracted from the soil material from an 82- to 84-cm depth, and two nodules from the PR profile, extracted from a 62- to 64-cm depth, are shown in Fig. 4.6.1. The concretions from the OK Planosol exhibit a differentiated layer-concentric structure (Fig. 4.6.1a–d) with a clearly seen "core" (commonly formed by a large soil particle) and numerous well-laminated layers around it. The higher-magnification images (not shown) also reveal the presence of small soil particles involved among the

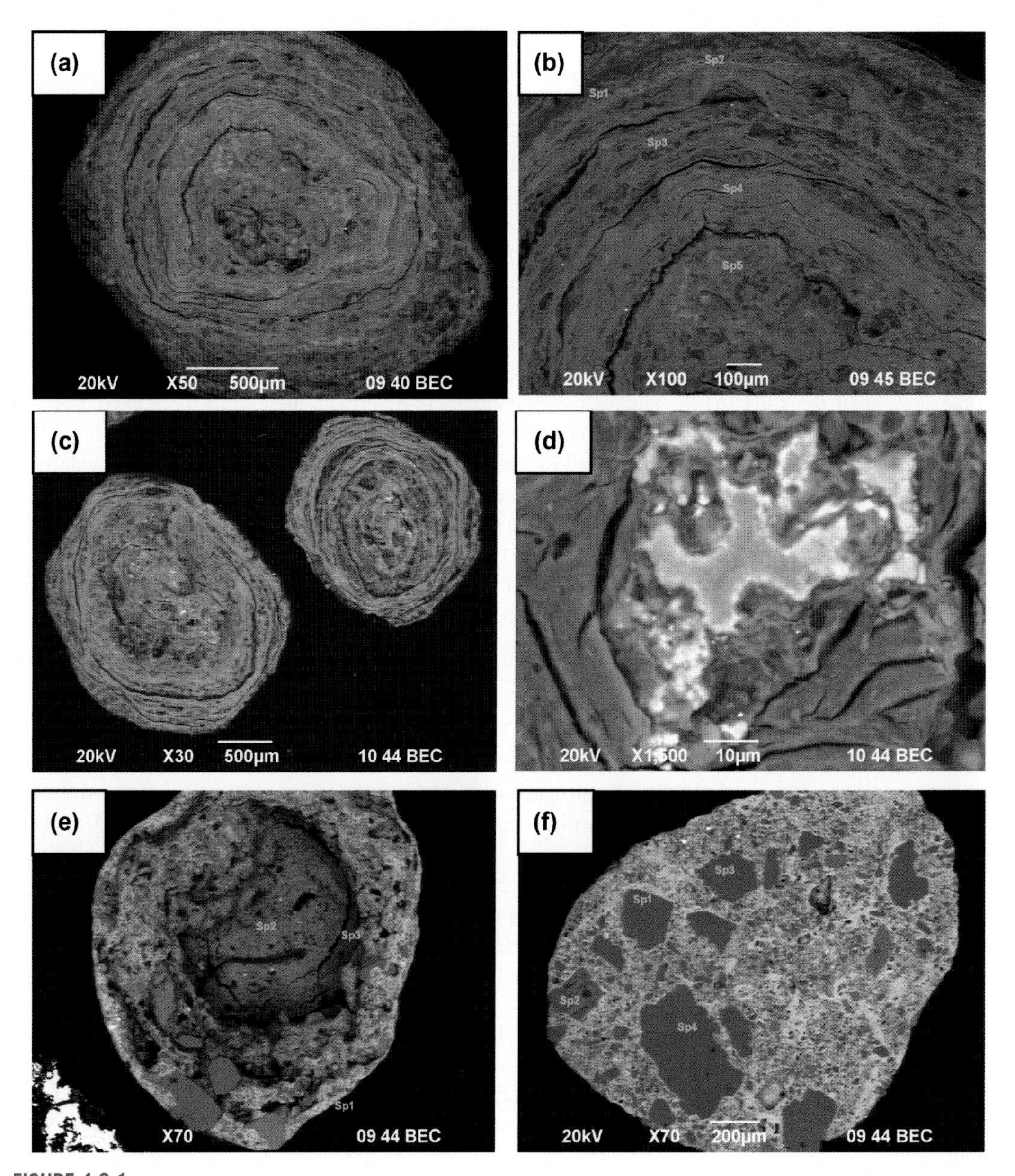

FIGURE 4.6.1

Scanning electron microscopy images of cross-sections of three concretions from the OK Planosol (a–d) and two nodules from the PR Planosol (e–f). A more-detailed image of the layered structure of the concretion from the OK profile displayed in (a) is shown in (b). Concentric-layered structure of two more concretions from OK profile is demonstrated in (c) as well. Detailed image of the central part of the smaller (right-hand) concretion from the OK profile in (c) is shown in (d).

Table 4.6.1 EDS spectra in selected spots of concretion 1 shown in the SEM image in Fig. 4.6.1a,b from the OK Planosol profile and "Integral Spectrum" (Int. Sp) obtained for the whole diametric surface and in the core of the concretion (Sp5-Int.Sp. CORE)

OK concr-1	O	Mg	Al	Si	K	Ca	Ti	Mn	Fe	Ba
Int. Sp.	56.46		6.96	13.28	1.48	1.16	0.65	9.04	10.98	—
Sp-2	53.12	0.91	3.97	6.39	0.43	0.95	—	4.43	29.80	—
Sp-3	52.41	0.95	8.05	14.38	1.36	1.02	0.38	9.53	11.01	0.91
Sp-3 narrow	46.17	1.33	4.17	4.76	0.52	0.99	—	29.51	7.08	3.91
Sp-4	49.14	0.99	7.32	11.63	1.47	1.43	—	16.12	9.69	2.14
Sp-4 bright	43.26	0.80	3.60	4.37	0.74	2.03	—	36.43	4.95	3.83
Sp-5-IntSp CORE	53.01	1.16	7.45	14.0	1.52	1.14	0.41	11.68	9.63	—

Elements shown in weight%.

different concretion layers. The elemental EDX analyses in selected spots, denoted on the SEM images, are summarized in Tables 4.6.1–4.6.4.

The elemental composition data shown in Tables 4.6.1–4.6.4 demonstrate enhanced Fe and Mn contents that are especially pronounced in the nodules from the PR Planosol. The major soil-forming elements such as Al, Mg, Si, Ti, and Ca are also present in the concretions, as well as Ba. Similar to the data for the concretions from other soil types (e.g., Vertisols and saline/sodic soils—see Chapter 5), a large heterogeneity in the elements' distribution inside the nodule is also observed in Planosols

Table 4.6.2 EDS spectra in two selected spots of concretion 2 shown in the SEM image in Fig. 4.6.1c–d from the OK Planosol profile and "Integral Spectrum" (Int. Sp) obtained for the whole diametric surface

OK concr-2	O	Mg	Al	Si	K	Ca	Ti	Mn	Fe	Ba	Ni	Ce
Int. Sp.	54.73	0.77	7.59	16.38	1.60	0.85	0.6	8.10	9.40	—	—	—
Sp-1	50.22	1.07	8.83	14.09	1.66	1.06	—	10.33	11.5	1.23	—	—
Sp-2	41.14	1.06	3.46	1.14	—	1.56	—	38.31	3.71	5.55	1.19	2.87

Elements shown in weight%.

Table 4.6.3 EDS spectra in a spot of the nodule shown in the SEM image in Fig. 4.6.1e from the PR Planosol profile

PR nodule 3	O	Na	Mg	Al	Si	K	Ca	Ti	Mn	Fe	Ba	Ni	Ce
Sp-1	46.67	—	—	4.9	12.24	0.49	1.10	—	—	34.61	—	—	—

Elements shown in weight%.

Table 4.6.4 EDS spectra in two selected spots of concretion shown in the SEM image in Fig. 4.6.1f from the PR Planosol profile and "Integral Spectrum" (Int. Sp) obtained for the whole diametric surface

PR dense nodule	O	Na	Mg	Al	Si	K	Ca	Ti	Mn	Fe	Ba	Ni	Cr
Int. Sp-1	43.81	—	—	2.96	14.29	0.62	—	—	—	37.62	—	—	0.71
Int. Sp.-2	45.36	—	—	2.67	17.00	0.77	—	—	—	34.21	—	—	—
Sp-1	57.86	—	—	—	42.14	—	—	—	—	—	—	—	—
Sp-2	51.89	—	—	9.22	25.85	13.05	—	—	—	—	—	—	—
Sp-3	37.83	—	—	1.11	1.08	—	—	—	0.70	59.29	—	—	—

Elements shown in weight%.

(Liu et al., 2002; Cornu et al., 2005; Gasparatos, 2013). The occurrence of a concentric laminated structure in the concretions is usually related to their formation in distinct stages during alternations of dry and wet periods (Palumbo et al., 2001; Cornu et al., 2005; Laveuf et al., 2012; Yu et al., 2015). Such a well-developed laminated structure is observed in all the three concretions from the OK Planosol (Fig. 4.6.1a–d), while in the PR profile both concretions (Fig. 4.6.1e) and nodules (Fig. 4.6.1f) are present. The concretions from PR does not show such a well-established layered structure as in the OK profile, although the inside "core" is also visible. The elemental composition of the concretions in the two profiles also differs—the iron content in PR is higher, while in Mn it is lower compared with the concretions from the OK profile (Tables 4.6.1–4.6.4). As also will be discussed in the Vertisols properties (Chapter 5), the layered concretions are thought to form in more loamy horizons with a frequent change in reduction–oxidation conditions, while dense nodules missing an ordered internal structure form in heavy clayey illuvial horizons (Zhang and Karathanasis, 1997).

The mineralogy of soil concretions is investigated in numerous studies, and most of them recognize goethite and/or hematite as the dominant crystalline iron (hydr)oxide (Zhang and Karathanasis, 1997; Cornu et al., 2005; Gasparatos et al., 2005; Chan et al., 2007). Magnetic nodules are also identified, mainly in Ferralsols (Mitsuchi, 1976) and lateritic soils from Australia (Singh and Gilkes, 1996), containing maghemite as a major iron oxide. The magnetic mineralogy of the Planosol concretions from Bulgaria is studied through a thermomagnetic analysis of magnetic susceptibility (Fig. 4.6.2). The heating curve shows mineral transformations above ~300°C with the creation of a new strongly magnetic phase. As already discussed in other chapters, such thermomagnetic behavior is typical for maghemite, which is unstable on heating (Dunlop and Özdemir, 1997). Thus, it could be supposed that concretions from the Plansols studied contain maghemite as a strongly magnetic mineral.

In order to verify the possible presence of high coercivity iron oxide minerals in the concretions, material from globules extracted from different depth levels from the B_t2 and C_k1 horizons of the OK Planosol are used for magnetic remanence measurements. A stepwise acquisition of IRM is carried out for material from ten depth levels. The IRM acquisition curves are depicted in Fig. 4.6.3.

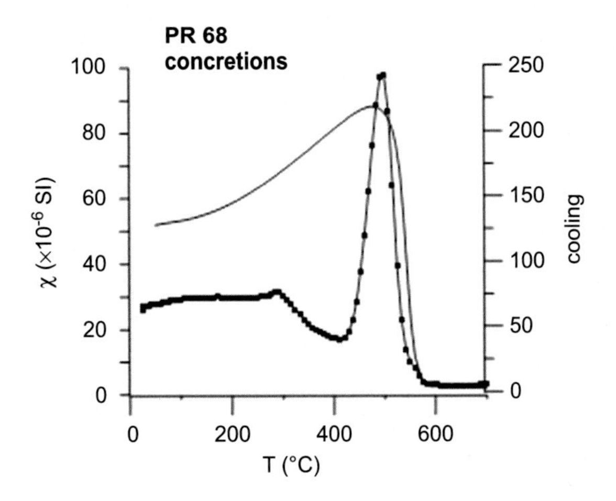

FIGURE 4.6.2

Thermomagnetic analysis of magnetic susceptibility of material from concretions/nodules extracted from depth 66–68 cm (B_t1 horizon) from the PR Planosol. Heating in air, fast heating rate (11°/min).

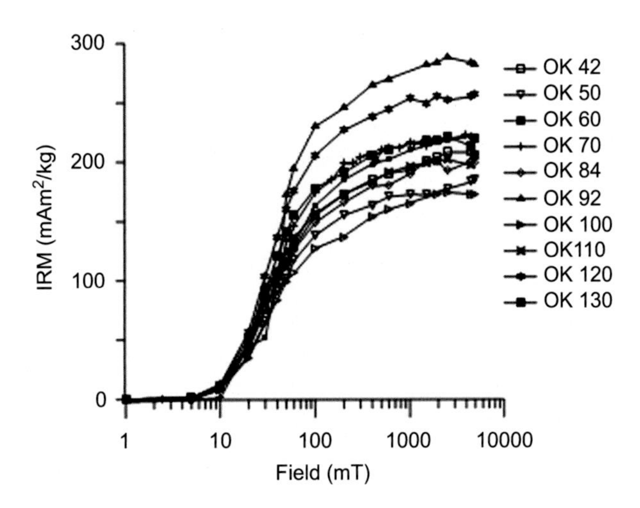

FIGURE 4.6.3

Stepwise acquisition of isothermal remanence (IRM) up to field of 5 T for material from concretions extracted from different levels of B_t2-C_k1 and C_k2 horizons of the OK Planosol profile.

As seen from Fig. 4.6.3, although all concretions are characterized by a magnetically soft component dominating the total IRM, a weak high coercivity component is present as well. The coercivity component analysis carried out, using the IRM-CLG1.0 software (Kruiver et al., 2001), quantifies the earlier observations and provides further insight into the magnetic mineralogy of the concretions from the OK Planosol (Table 4.6.5 and Fig. 4.6.3).

Table 4.6.5 Summary parameters from coercivity analysis by fitting IRM acquisition curves with cumulative Gaussian functions, using IRM-CLG1.0 software (Kruiver et al., 2001) for samples from concretions extracted from different depth intervals of the OK Planosol profile

	IRM component 1				IRM component 2			
Soil concr. depth (cm)	Intensity (10^{-6} Am^2/kg)	% from IRM_{total}	$B_{1/2}$ (mT)	DP	Intensity (10^{-6} Am^2/kg)	% from IRM_{total}	$B_{1/2}$ (mT)	DP
Profile OK concretions								
42	1842.10	90	36.30	0.41	203.95	10	1000.0	0.34
50	1616.07	87	35.50	0.38	232.10	13	891.30	0.34
60	1960.40	89	38.00	0.39	234.70	11	891.30	0.34
84	1960.80	90	34.70	0.36	225.50	10	602.60	0.40
92	2565.22	91	38.00	0.32	260.87	9	501.20	0.20
100	1246.03	72	28.20	0.27	476.19	28	354.80	0.35
110	1661.63	81	30.20	0.29	392.75	19	416.90	0.40
120	2148.15	84	29.50	0.30	414.81	16	316.20	0.32
130	1854.30	84	29.50	0.30	357.62	16	316.20	0.32

The intensity, relative contribution (% from IRM_{total}), median acquisition field ($B_{1/2}$), and dispersion parameter (DP) for each coercivity component are shown.

Similarly to the bulk soil samples, two coercivity components in the IRM are separated in the concretions (Table 4.6.5)—magnetically soft (IRM1) and magnetically hard (IRM2). The soft component is ascribed to maghemite, while the hard one is ascribed to hematite (Dunlop and Özdemir, 1997; Özdemir and Dunlop, 2014). A comparison between the coercivities of the two components in the bulk soil and in the concretions from the depth of 42 cm (the closest one to the soil sample analyzed, OK38—Table 4.4.1) shows that the IRM1 component is of higher coercivity in the bulk soil, while its intensity is much lower there and vice versa for the IRM2—it has higher coercivity in the concretions compared with the soil. Variations in the intensity and the coercivity (the $B_{1/2}$ parameter) of the two components according to the depth of the concretions are shown in Fig. 4.6.4.

As demonstrated in Fig. 4.6.4, a striking horizon-bounded regularity in the intensity and coercivities of the IRM components is present. The mean coercivity $B_{1/2}$ of the soft component in the concretions from the B_t2 horizon is higher compared with $B_{1/2}$ for concretions in the underlying C_k1 horizon. Simultaneously, the coercivity of the hard component decreases from the top of B_t2 toward Ck1, where, however, it has a higher intensity (Fig. 4.6.4). In order to explain the observed variations in the magnetic properties, it is necessary to consider the major factors determining the hydrodynamic regime in the soil related to the mechanisms of the globules' formation—the potential for Fe^{2+} release and the pH. As discussed in Chan et al. (2007), the soil pH plays an important role because a lower pH allows a larger amount of Fe^{2+} to be available in the solution and allows a higher amount of hematite to precipitate on the nucleation core. As shown in Section 4.3 of this chapter, the OK Planosol is characterized by a very steep increase in pH along the solum, and this will inevitably influence the iron segregation processes. Since at a lower pH (e.g., in the upper part of the Bt2 horizon) more Fe^{2+} will be

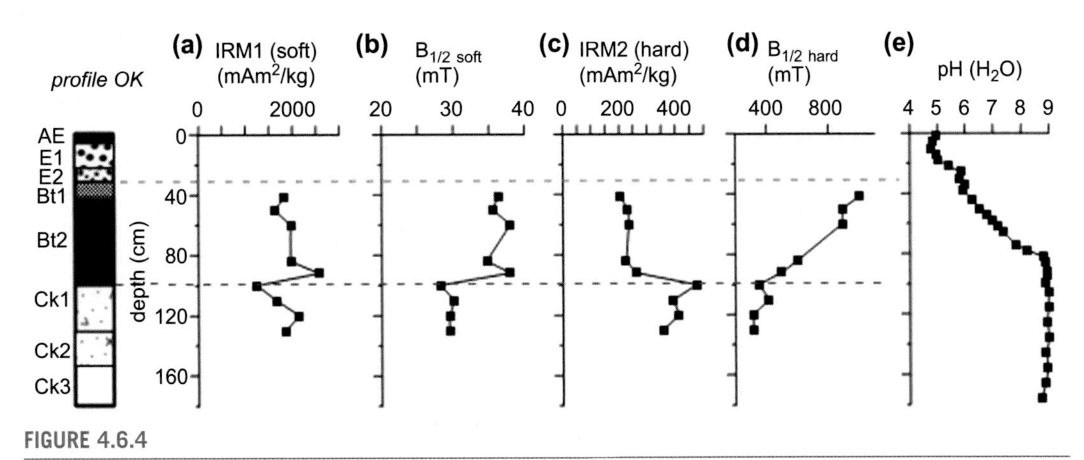

FIGURE 4.6.4

Depth variations in the intensity and coercivity of the magnetically soft [(a) IRM1 (soft) and (b) $B_{1/2}$ soft] and magnetically hard [(c) IRM2 (hard) and (d) $B_{1/2}$ (hard)] fractions, identified in the IRM acquisition curves. Depth change in the soil pH is shown in (e).

present in the solution for a longer time, hematite precipitation will last the longest there and, according to Chan et al. (2007), the hematite will be of higher crystallinity. The latter is presumably related to higher coercivity, as observed in Fig. 4.6.4. The soil reaction pH sharply increases with increasing depth and thus the hematite's crystallinity would decrease, leading to a low $B_{1/2}$. There is no unanimous view on the mode of occurrence of maghemite in concretions—some authors suppose that it is inherited from the parent rock (Mitsuchi, 1976), or it is a result of pedogenic precipitation (Singh and Gilkes, 1996), while others consider it to be as a result of the thermal alteration of the nonmagnetic nodules located at the top of the profile during severe bushfires (Löhr et al., 2013). Since the nodules are extracted from the bottom part of the profile, the last hypothesis can be neglected. A lithogenic origin of the maghemite is also not very probable because the shape of the thermomagnetic curve (Fig. 4.6.2) and strongly horizon-related changes in the parameters of the maghemite component suggest a leading role for the pedogenic processes. As discussed in Singh and Gilkes (1996) referring to Coventry et al. (1983), a fluctuating water table leads to an increased Fe^{2+} concentration in the soil pores and then, depending on the rate of oxidation, either hematite or maghemite precipitate, or a mixture of these. In the case of slower oxidation, mixed Fe^{3+} and Fe^{2+} species coexist in the pores, allowing for maghemite precipitation. Thus, it can be anticipated that in the more clayey B_t2 horizon, this situation is met and a higher amount of maghemite precipitates, while down in the C_k1 horizon, there is a larger amount of hematite but it is of a lower crystallinity. Therefore, studies of the magnetic properties of concretions could yield very sensitive information about the dynamics of redox processes and iron segregation in concretions.

4.7 PODZOLS: MAIN CHARACTERISTICS, FORMATION PROCESSES, AND DISTRIBUTION

The name of the "Podzol" soil type has its roots in the Russian words "pod", which means "under" and "zola," meaning "ash" (Ponomareva, 1964). Therefore, the term "Podzol" relates to the soil morphology characterized by an ash-gray eluvial horizon, which overlies the illuvial horizon enriched in organic matter, Al, Si, and Fe.

Podzols have been extensively studied in soil science since the 1960s, and a wealth of literature can be found relating to their genesis and main characteristics (Buurman, 1984; Lundström et al., 2000; Buurman and Jongmans, 2005; Sauer et al., 2007; Schaetzl and Anderson, 2009). Podzols commonly occur on highly permeable sandy parent materials that are poor in base cations and iron and develop in cool humid climates under forest or heath vegetation (e.g., Lundström et al., 2000). Such conditions favor the development of a thick organic (mor) layer, overlying a weathered eluvial (E) horizon. The deeper illuvial horizon is dark in color (reddish brown or dark brown to black) and enriched in Al, Fe, and organics (Buurman, 1984; Lundström et al., 2000; Buurman and Jongmans, 2005). Despite the extensive studies on Podzol formation, several different theories exist that support the following three major pathways of their genesis (Schaetzl and Anderson, 2009; references therein; Sauer et al., 2007): (1) the chelate-complex theory ("organic acids theory"). This is the oldest and the most widely accepted theory that is based on the common finding that ~80% of soluble Al and Fe in the eluvial horizons of the Podzols can be bound to the organic matter. The theory suggests that Fe, Al, and Si form metal–organic complexes that are highly mobile and freely move down the profile along with the percolating water. They further precipitate in the B horizon due to the saturation of the organic molecules through metal complexation (Buurman and Jongmans, 2005). (2) (Proto)imogolite theory ("inorganic theory") is based on the finding that Al and Fe in humus-poor Podzols can exist in the form of inorganic compounds—imogolite and allophane. The theory suggests that soils containing (proto)imogolite/allophane are transported in the B horizon, where subsequently mobile humus can be absorbed (Anderson et al., 1982). (3) Two-stage Podzol development ("hybrid theory"), consisting of (i) the in situ formation of imogolite/allophane in the illuvial horizon by weathering, triggered by carbonic acid and (ii) the precipitation of fulvic acid on these Al-rich precipitates in the B_S horizon (Wang et al., 1986). Despite the ongoing discussions on the appropriateness of these theories, the mobilization-transport-precipitation model dictates the formation process in all of them (Buurman and Jongmans, 2005).

Podzols develop mainly in cool and wet climates, especially in the boreal zone (Buurman and Jongmans, 2005; Sauer et al., 2007 and references therein), and they are often referred to as "boreal Podzols." Usually, they are characterized by the accumulation of more sesquioxides and less organic matter in the B horizon compared with "nonboreal Podzols" developed in the high-mountain regions of lower latitudes. These "nonboreal Podzols" exhibit dark-gray to black colors at the top of the B horizon, indicating an accumulation of a significant amount of translocated organic matter. Podzols have also been found in the subarctic tundra and polar desert (Sauer et al., 2007). The third major occurrence of Podzols is in the humid to perhumid regions in the tropics (the "tropical Podzols"), where in some alpine or subalpine areas, climatic parameters provide suitable environments for podzolization (Do Nascimento et al., 2004; Sauer et al., 2007; Jien et al., 2010a). When the aquic moisture regime is established in highly porous clay-depleted laterites, the downward translocation of the organometallic complexes formed is possible and a Podzol develops.

The parent material is of crucial importance for the formation of Podzols. They are found only in coarse-textured and base-poor parent materials—most often sands, sandy tills, and Precambrian Shield granites/gneisses (Lundström et al., 2000; Schaetzl and Anderson, 2009). The major favorable feature of such parent material is its low surface area, allowing fast and effective leaching and the dissolution of the initially low amount of base minerals and strong acidification. The coarse texture also supports the rapid infiltration, translocation, and leaching of the weathering products (Schaetzl and Anderson, 2009).

Podzol development is mainly favored by vegetation producing slowly degradable and nutrient-poor litter such as coniferous forest and ericaceous shrubs. However, Podzols are also identified under deciduous forests (Lundström et al., 2000). When the vegetation changes, the podzolization process also changes. Thus, nonpodzolic soils like Leptosols, Regosols, Cambisols, Luvisols, and Acrisols may turn to Podzols when, under climate change or anthropogenic activities, the dominant vegetation change causes the inhibition of litter decomposition and the enhanced formation of water-soluble organic acids (Lundström et al., 2000; Sauer et al., 2007).

The rates of Podzol formation are extensively studied in different chronosequences. It is established that the initial signs of podzolization are already visible after ~40 years in poor sands or the extremely cool and wet climate in Scandinavia, while in milder conditions a longer period of ~2300–3800 years is necessary (Sauer et al., 2007 and references therein; Lundström et al., 2000). It has been established that indices related to the degree of podzolization show a linear dependence on the logarithm of time (Lundström et al., 2000).

Podzols cover ~485 million ha worldwide, mainly in the temperate and boreal regions of the Northern Hemisphere (IUSS Working Group WRB, 2014) (Fig. 4.7.1). They are most widespread in Scandinavia, the northwest of the Russian Federation, and Canada. Podzols are also common in high-mountain regions such as the Rocky Mountains and the Appalachian Mountains (USA), the West Sierra Madre (Mexico), the Bolivian Highlands, the European Alps, the Himalayas, and the Alps of New Zealand (Sauer et al., 2007). The exact distribution of tropical Podzols is disputable, but they are found along the Rio Negro and in French Guiana, Guyana and Suriname in South America, in

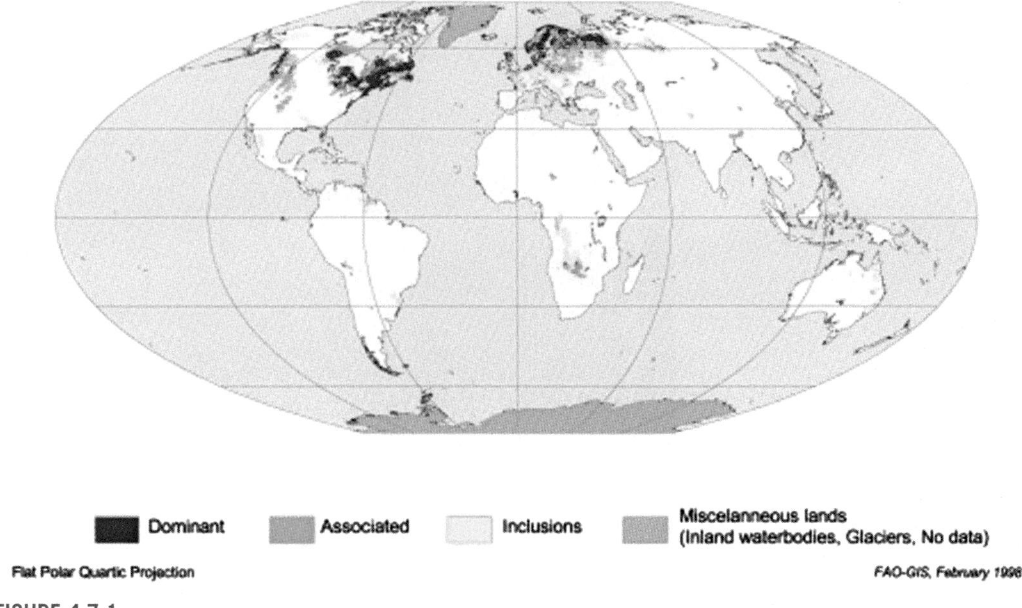

FIGURE 4.7.1

Spatial distribution of Podzols worldwide (FAO, 2001; Lecture notes on the major soils of the world: http://www.fao.org/docrep/003/y1899e/y1899e12.htm#P353_47671).

Southeast Asia, in Papua New Guinea, and in northern and eastern Australia (IUSS Working Group WRB, 2014).

Two profiles of typical boreal Podzols were sampled in South Finland (near Helsinki), thanks to the logistical support from the colleagues of the Geophysics Department at the University of Helsinki and thanks personally to Dr. Tomas Kohout as well as Dr. Mike Starr from the Department of Forest Sciences at the University of Helsinki.

DESCRIPTION OF THE PROFILES STUDIED

4.7.1 PODZOLIC SOIL, PROFILE DESIGNATION: FPZ1

Situated at Ruotsinkylä Experimental Forest, Finnish Forest Research Institute, Helsinki (Finland)

Location: N 60°20′58.4″; E 25°00′51.5″, h = 60 m a.s.l.

Present-day climate conditions: MAT = 5.0°C; MAP = 680 mm, vegetation cover: 79-year-old Scots pine (*Pinus sylvestris*); relief: flat to slightly sloping site

Profile description: general view and details from the horizons are shown in Figs. 4.7.2–4.7.5.

0–2 cm: L horizon, raw organics, nondecayed needles layer
2–6 cm: O horizon, organics, black
6–7 cm: E horizon, ash-gray, bleached, irregular depth, at some places extends up to 10 cm

FIGURE 4.7.2

View of the profile of Podzol (profile FPZ1).

FIGURE 4.7.3

Podzol FPZ1. L-O-E-B_{hs} horizons, detail.

FIGURE 4.7.4

Podzol FPZ1. B_{hs} horizon, detail.

FIGURE 4.7.5

Podzols in the Ruotsinkylä Experimental Forest, Finnish Forest Research Institute, Helsinki (Finland). View of the location sampled.

7–15 cm: B_{hs} horizon, dark brown with organics
15–27 cm: B_s1 horizon, light brown, sandy
27–38 cm: B_s2 horizon, dark brown, more Fe oxides
38–50 cm: BC transitional horizon
50–70 cm: C horizon, gray-brown sandy deposits
Parent material: medium- to coarse-grained sorted glaciofluvial sand deposit

4.7.2 PODZOLIC SOIL, PROFILE DESIGNATION: FPZ2

Situated at Lukki area, NW of Helsinki (Finland)

Location: N 60°19′15.8″; E 24°41′43.4″, h = 73 m a.s.l.

Present-day climate conditions: MAT = 5.0°C; MAP = 655 mm, vegetation cover: Scots pine (*Pinus sylvestris*); relief: slightly sloping site.

Profile description: general view and details from the horizons are shown in Figs. 4.7.6–4.7.9

0–2 cm: L horizon, brown, vegetation roots
2–6 cm: O horizon, black, vegetation roots also present
6–8 cm: OE horizon, bleached, ash-gray with signs of organics translocated
8–20 cm: B_h horizon, organics at the upper boundary, brown-black on top, light-brown at the bottom
20–60 cm: B_s horizon, red-brown, sandy, coarse pebbles also present
60–75 cm: C horizon, brown, sandy with pebbles
Parent material: glaciofluvial sand deposit

FIGURE 4.7.6

View of the profile of Podzol FPZ2.

FIGURE 4.7.7

Podzol FPZ2. B_h horizon—detail.

FIGURE 4.7.8

Podzol FPZ2. B_s horizon—detail.

FIGURE 4.7.9

Podzol FPZ2. C horizon—detail.

4.8 SOIL REACTION, ORGANIC CARBON, AND MAGNETIC PROPERTIES OF PODZOLS

The soil reaction (pH) along the depth of the Podzols studied is shown in Fig. 4.8.1. It is very acidic in the upper part, comprising the organic-rich and eluvial horizons, in accordance with the data for typical boreal Podzols (e.g., Buurman, 1984; Lundström et al., 2000). In both profiles studied, the pH is ~4 at the surface and drops down to pH ~3–3.3 in the thin eluvial horizon, followed by a gradual increase in the underlying illuvial horizons (Fig. 4.8.1a and d). The maximum pH values in the profiles do not reach 5.9, a criterion set as a diagnostic feature for the definition of Podzols (Lundström et al., 2000 and references therein). The organic carbon content $_{(Corg)}$ shows different results for the two profiles. In the FPZ1 Podzol, the uppermost L and O horizons contain a high amount of organic matter, reaching 16% and decreasing down to <1% (Fig. 4.8.1b). In the second profile (FPZ2), the organic-rich layers

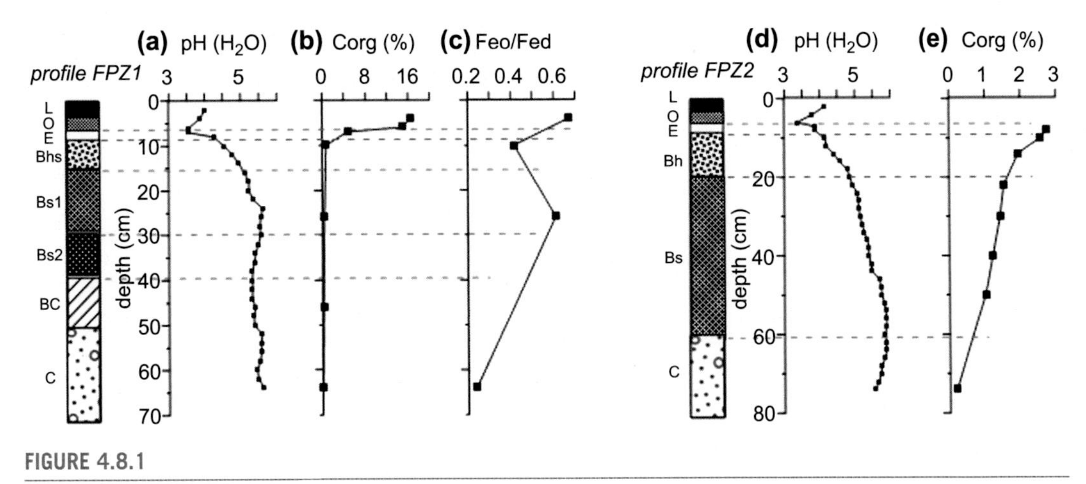

FIGURE 4.8.1

Soil reaction (a), organic carbon content (b), and the ratio Fe_o/Fe_d for selected samples from the FPZ1 Podzol. Soil reaction (d) and organic carbon (e) for the FPZ2 Podzol determined in selected samples.

contain only ~3% Corg, but downward its content decreases slowly and falls below 1% in only the C horizon (Fig. 4.8.1e). Thus, according to the distribution of Corg, the two Podzol profiles from Finland show typical characteristics for boreal Podzols.

Several samples from the FPZ1 profile were analyzed for the content of dithionite- and oxalate-extractible iron, and the ratio Fe_o/Fe_d is displayed in Fig. 4.8.1c. High values of Fe_o/Fe_d of ~0.4–0.7 are obtained in the upper part of the profile, suggesting a very important share of amorphous iron oxides in the "free" iron pool in the FPZ1 Podzol. This feature is also characteristic for Podzols and is related to the intense destruction processes and transport of dissolved iron compounds, which, under such extremely acid conditions, precipitate mainly in noncrystalline minerals (Cornell and Schwertmann, 2003).

The magnetic mineralogy in the two Podzol profiles is investigated through a stepwise thermal demagnetization of composite IRM and ARM. Representative examples are shown in Fig. 4.8.2. The thermal demagnetization of ARM (Fig. 4.8.2a–d) reveals stable magnetite and/or maghemite with an unblocking temperature of ~600°C as the main phase in both profiles. It should be noted, however, that in the FPZ1 profile and especially the uppermost levels, the unblocking temperature spectra of ARM is much wider, as suggested by the concave shape of the curve (Fig. 4.8.2a). For the sample FPZ1-64 from the bottom part of the profile, a kink on the ARM(T) is also visible at ~250°C (Fig. 4.8.2b), which could be attributed to the presence of unstable maghemite and its transformation temperature (Dunlop and Özdemir, 1997).

The samples from the FPZ2 profile are characterized by the convex shape of the ARM(T) curve (Fig. 4.8.2c and d) related to a narrower grain-size distribution of the magnetically stable particles. The thermal demagnetization of the composite IRM provides further information about the magnetic mineralogy on the basis of the coercivity of the different fractions. All the samples show the presence of hematite's T_{ub} on the intermediate (0.2–0.6 T) and on the hard (0.6–5.0 T) coercivity IRM components (Fig. 4.8.2e–h), which are demagnetized at 700°C. A fast decrease in all the three IRM components at ~280–300°C for the FPZ2–4 sample (the topmost organic layer) (Fig. 4.8.2g) could

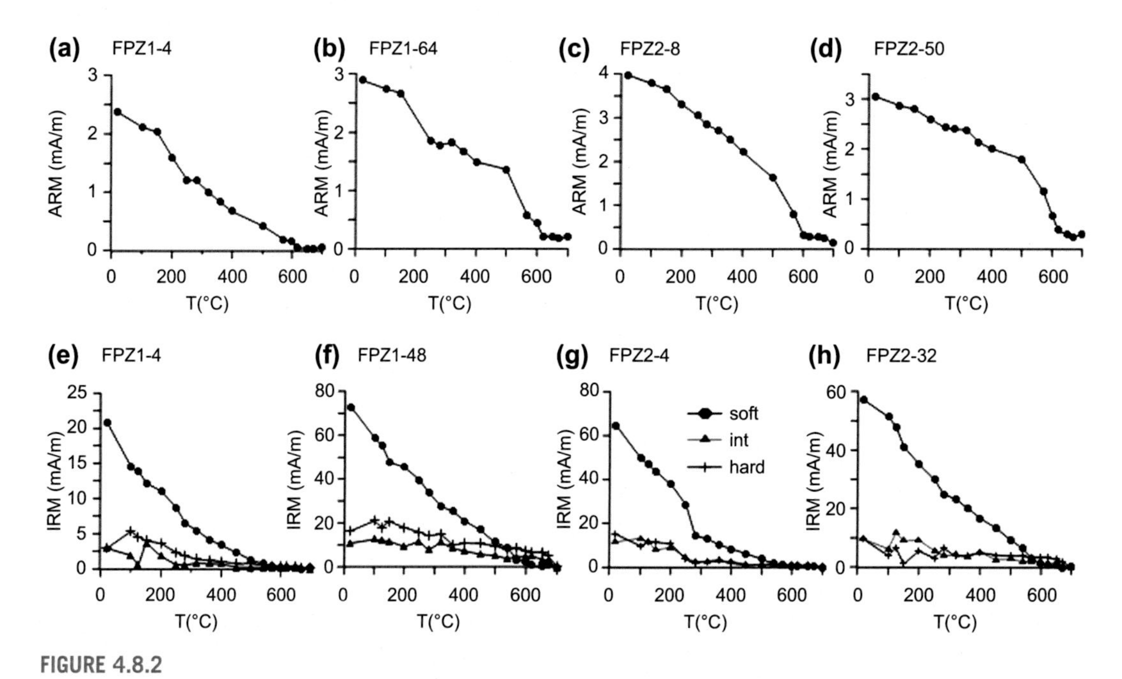

FIGURE 4.8.2

Stepwise thermal demagnetization of anhysteretic remanence (ARM) for selected samples from the FPZ1 Podzol (a and b) and FPZ2 Podzol (c and d). Stepwise thermal demagnetization of composite IRM for two samples from FPZ1 (e and f) and two samples from FPZ2 Podzol (g and h). Saturating fields: soft component (0–300 mT); intermediate component (300–600 mT); hard component (600 mT–5.0 T).

be due either to maghemite's transformation or to the effect of weight loss as a result of the destruction of organic matter during heating (Schnitzer et al., 1964; Plante et al., 2009). The presence of hematite in the intermediate- and hard-coercivity fractions suggests that it is of wide grain-size distribution or has highly varying Al substitutions, both factors strongly influencing its coercivity (Dunlop and Özdemir, 1997; Özdemir and Dunlop, 2014).

The above conclusion is further supported by the results from the stepwise acquisition of IRM carried out up to fields of 5 T and coercivity analysis (Kruiver et al., 2001). Fig. 4.8.3 shows the IRM acquisition curves for selected samples from the FPZ1 and FPZ2 profiles. It can be seen that most of the samples do not saturate in fields of ~1 T, but the magnetization continues increasing up to 5 T. An analysis of the coercivity distributions shows the presence of two components (Table 4.8.1).

The magnetically soft component is the strongest in all the samples, accounting for 70–90% of the total IRM, but the non-negligible contribution of the hard component, reaching 20–40% of the total IRM, is also revealed (Table 4.8.1). According to the obtained coercivities, the soft component (IRM1) could be ascribed to magnetite/maghemite and the hard component to hematite of varying grain size and/or Al substitutions. The samples from the FPZ1 profile display horizon-related regularity in the coercivity of the two IRM components: in the uppermost O + E horizons (samples from depths of 4 cm and 10 cm), the soft component has $B_{1/2}$ of 47 mT and high values for the $B_{1/2}$ of the hard component (~800–870 mT). Immediately below the E horizon, in the B_{hs}, the coercivity of the soft

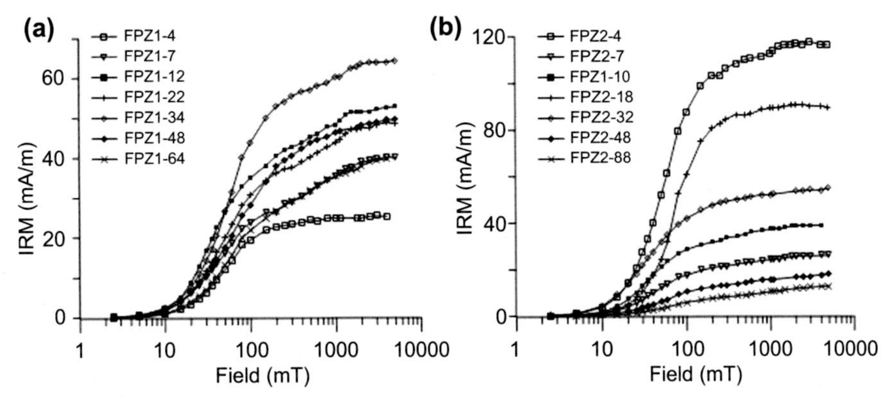

FIGURE 4.8.3

Stepwise IRM acquisition curves up to a field of 5 T for selected samples from the FPZ1 Podzol (a) and FPZ2 Podzol (b).

Table 4.8.1 Summary parameters from coercivity analysis using IRM-CLG1.0 software (Kruiver et al., 2001) for selected samples from the Podzol profiles FPZ1 and FPZ2

	IRM component 1				IRM component 2			
Sample depth (cm)	Intensity (10^{-3} Am^2/kg)	% from IRM_{total}	$B_{1/2}$ (mT)	DP	Intensity (10^{-3} Am^2/kg)	% from IRM_{total}	$B_{1/2}$ (mT)	DP
Profile FPZ1								
4	1.15	89	46.8	0.33	0.14	11.0	794.3	0.45
7	1.5	76	47.9	0.42	0.47	9.4	871.0	0.24
12	2.0	77	38.0	0.36	0.6	23.0	631.0	0.40
22	1.75	73	43.7	0.36	0.65	27.0	691.8	0.40
34	2.65	83	50.1	0.36	0.55	17.0	707.9	0.41
48	1.95	78	56.2	0.42	0.55	22.0	631.0	0.45
64	1.5	76	58.9	0.42	0.48	24.0	891.3	0.31
Profile FPZ2								
4	5.15	87	49.0	0.32	0.75	13.0	794.3	0.38
7	1.0	77	39.8	0.32	0.31	23.0	691.8	0.43
10	1.5	76	33.9	0.30	0.47	24.0	501.2	0.40
18	4.1	90	66.1	0.30	0.45	10.0	398.1	0.35
32	2.25	83	31.6	0.36	0.45	17.0	380.2	0.46
48	0.6	69	49.0	0.37	0.28	31.0	631.0	0.46
68	0.38	60	56.2	0.38	0.25	40.0	794.3	0.42

The intensity, relative contribution to the total IRM (% from IRM_{total}), median acquisition field ($B_{1/2}$), and dispersion parameter (DP) for each coercivity component are shown.

component is low ($B_{1/2}$ of 38 mT) and progressively increases in depth, reaching 58 mT in the C horizon (Table 4.8.1). At the same time, the coercivity of the hard component remains relatively constant in the illuvial horizons ($B_{1/2}$ varying between 630–708 mT) and is again high in the C horizon. As pointed out earlier, additional analyses are needed to judge whether the coercivity changes are due to grain-size changes in the hematite's grains or the degree of Al substitutions. Similar regularities in the coercivities of the two IRM components are obtained for the FPZ2 profile (Table 4.8.1), except for the depth of 18 cm, where the $B_{1/2}$ of the soft component is unusually high compared with the other samples. As in the FPZ1 profile, the samples from O and E horizon again show a higher coercivity of both IRM components, while starting from a lower value in the upper part of the illuvial horizon downwards, the coercivities increase (Table 4.8.1).

The depth variations of the main magnetic characteristics for the two Podzols are shown in Figs. 4.8.4 and 4.8.5. The magnetic susceptibility of the FPZ1 profile is very low, in the order of $\sim 20 \times 10^{-8}$ m^3/kg (Fig. 4.8.4), and almost constant along the depth with the exception of a sharp maxima at a 4-cm depth, followed by a sharp minimum in the eluvial horizon. The uppermost peak χ value is observed in the organic-rich layer, while the minimum is linked to the bleached eluvial zone with the most extreme acidic soil reaction (Fig. 4.8.1a). The ferrimagnetic susceptibility ($\chi - \chi_{hf}$) variations are similar to the bulk χ, revealing the important role of the ferrimagnetic component, though in a very small amount. The remanent magnetizations IRM_{2T} and ARM (Fig. 4.8.4b and c) show, in general, a pattern similar to the χ pattern but with relatively stronger enhancement in the illuvial horizons. The SD grain-size proxy ratio $\chi_{ARM}/IRM_{0.1T}$ (Fig. 4.8.4d) shows a maximum in the thin eluvial horizon, similar to the other soil types with a well-expressed eluvial horizon (e.g., Luvisols, Vertisols), but reaching a value of only $\sim$0.3, far below the estimated SD magnetite or maghemite threshold value (e.g., Geiss et al., 2008).

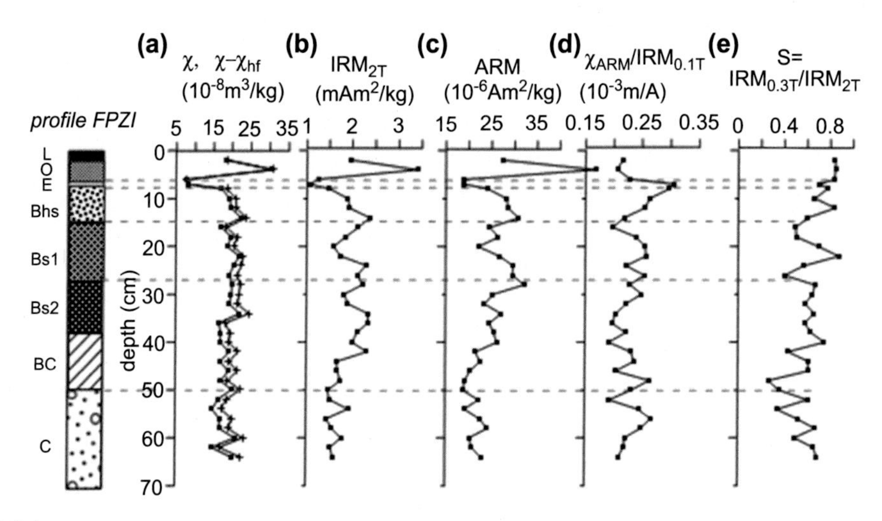

FIGURE 4.8.4

Depth variations of selected magnetic parameters and ratios for FPZ1 Podzol: (a) magnetic susceptibility (χ) (*full dots*) and ferrimagnetic susceptibility ($\chi - \chi_{hf}$) (*crosses*); (b) isothermal remanence acquired in a field of 2 T (IRM_{2T}); (c) Anhysteretic remanence (ARM), (d) ratio between anhysteretic susceptibility (χ_{ARM}) and IRM acquired in 0.1 T field ($IRM_{0.1T}$); (e) S ratio.

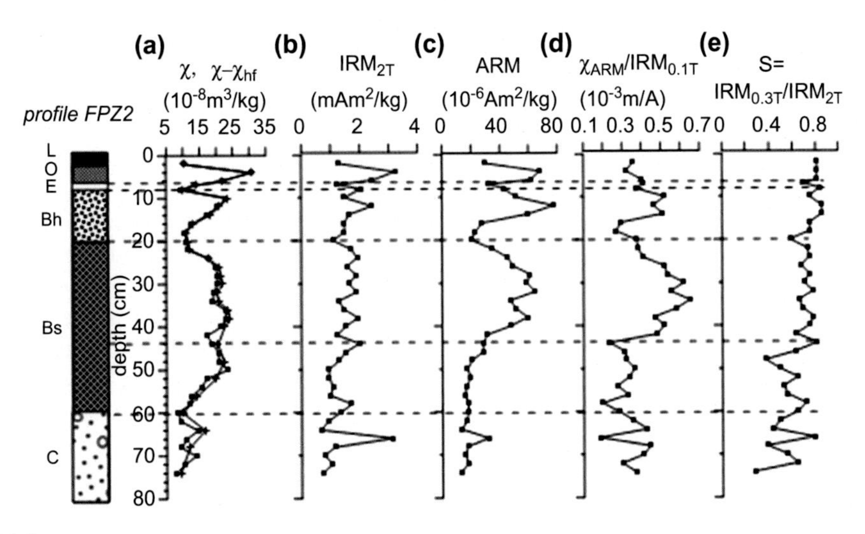

FIGURE 4.8.5

Depth variations of selected magnetic parameters and ratios for FPZ2 Podzol: (a) magnetic susceptibility (χ) (*full dots*) and ferrimagnetic susceptibility ($\chi-\chi_{hf}$) (*crosses*); (b) isothermal remanence acquired in a field of 2 T (IRM_{2T}); (c) anhysteretic remanence (ARM), (d) ratio between anhysteretic susceptibility (χ_{ARM}) and IRM acquired in 0.1 T field ($IRM_{0.1T}$); (e) S ratio.

As judged from the mineral magnetic diagnostics earlier, hematite makes an important contribution to the magnetic properties of the Podzols studied. This is emphasized in the depth variations of the S ratio (Fig. 4.8.4e) as well, showing a noisy pattern but with an unambiguous decay trend from ~0.8 at the surface to ~0.3–0.4 at the bottom of the profile.

The second Podzol profile FPZ2 shows more constrained variations in the magnetic characteristics along the depth. The magnetic susceptibility once again shows a maximum in the organic-rich O horizon at a 4-cm depth, followed by a minimum in the E horizon (Fig. 4.8.5a), similar to the FPZ1 profile. In contrast to the first profile, however, downward χ variations are well bound to the described genetic soil horizons—distinct susceptibility maxima are obtained in the B_h and B_s horizons. The IRM does not display such pronounced regularities (Fig. 4.8.5b), implying an important role for the grain size of the strongly magnetic fraction on the magnetic properties of this Podzol. Variations in the ARM (Fig. 4.8.5c) clearly underline the effect of the stable SD-like ferrimagnetic grains, pointing to an enhancement of the upper-middle parts of the E, B_h, and B_s horizons. These sections contain not only a larger amount but also finer stable grains, as inferred from the behavior of the SD grain-size proxy ratio $\chi_{ARM}/IRM_{0.1T}$ (Fig. 4.8.5d). The increasing contribution of high-coercivity hematite with increasing depth is revealed by the S ratio variations (Fig. 4.8.5e), reaching values of ~0.2 at the bottom of the profile.

4.9 PEDOGENESIS OF IRON OXIDES IN PODZOLS, AS REFLECTED IN SOIL MAGNETISM

The classic concept of podzolization assumes the organic complexation of Al and Fe as a major factor in their mobilization and translocation into the spodic horizons (Lundström et al., 2000). Mineral weathering at a low pH and high concentrations of low molecular organic acids promote the rapid loss of easily weathered minerals and their transfer down-profile by “rock-eating mycorrhiza” (van Breemen et al., 2000). On the other hand, different studies report the presence of large amounts of iron and aluminum in the O horizon (Giesler et al., 2000 and references therein), which is explained to be a result of the combination of several processes: mixing mineral soil into the O horizon through mechanical disturbances; upward transport by soil water; and root uptake and redistribution via litterfall or via roots or mycorrhizal hyphae. In addition, Giesler et al. (2000) conclude that the largest Al and Fe pool in the humus layer of the Podzols is the mineral part, mixed into the humus. In accordance with these observations are the maximum values of magnetic susceptibility obtained in the O horizons of the two Podzols (Figs. 4.8.4 and 4.8.5) at a ~4-cm depth. The magnetic data further suggest that, in this thin subsurface depth interval in the organic-rich layer, a strongly magnetic iron oxide mineral is formed. More detailed investigations are needed to infer the mechanism of this enhancement. A similar strong magnetic enhancement of the organic-rich Podzol horizons is also reported in other studies, such as the podzolic soils from Poland (Sandgren and Thompson, 1990). The authors attribute the obtained magnetic enhancement to the local influence of forest wildfires and thermal transformations of ferrihydrite/lepidocrocite to maghemite (Le Borgne, 1955). It is, however, unlikely that in all the different locations where Podzols occur, fire has always played a role in the formation of maghemite. It is more reasonable to suppose the creation of locally favorable micro-environmental conditions leading to ferrihydrite's transformation into maghemite. Furthermore, in a study of podzolic soils from the former USSR (Vodyanitskiy et al., 1983), the authors also report strong magnetic enhancement in the organic horizon of the Podzols and attribute this to the presence of ferrimagnetic Mn-containing oxides.

The cool humid conditions characteristic for Podzol environmental settings encode the major forms of iron oxides in their genetic horizons. High moisture and low temperatures in combination with abundant organics and a low pH lead to a predominance of poorly crystalline ferrihydrite formation (Schwertmann, 1988; Cornell and Schwertmann, 2003; Vodyanitskii, 2010). Due to extensive ferrihydrite–organic matter complexation and a high amount of dissolved silica, the secondary transformation of ferrihydrite to more-crystalline Fe minerals is significantly inhibited (Cornell and Schwertmann, 2003) and, when occurring, its transformation to goethite is favored by the low temperatures (Schwertmann, 1988). The presence of a high amount of amorphous Fe minerals in the upper Podzol horizons is evidenced by the high values of the Fe_o/Fe_d ratio obtained for the FPZ1 profile (Fig. 4.8.1). On the other hand, as reported earlier, the two boreal Podzols from Finland show the presence of hematite with an increasing concentration toward the C horizons based on the IRM coercivity analyses and the unblocking temperatures of magnetically hard IRM (Fig. 4.8.2). Therefore, one possible explanation for its occurrence is the lithogenic origin (e.g., inheritance from the parent rock). Likewise, Sandgren and Thompson (1990) have also found an increasing contribution by high coercivity antiferromagnetic minerals from the surface toward the C horizon of podzolic soils from Poland. The room temperature magnetic analyses carried out on the two Finnish Podzols do not reveal the presence of goethite, similarly to a Stagnopodsol from the UK (Maher, 1986). However, as also suggested by Maher (1986), the slight magnetic enhancement observed (in the magnetic susceptibility and IRM) in the illuvial horizon of the Stagnopodsol could be attributed to the neoformation of superparamagnetic goethite. As shown in Figs. 4.8.4 and 4.8.5, the illuvial horizons of the Podzols FPZ1 and FPZ2 are also magnetically enhanced, as reflected in χ, IRM, and ARM variations. Thus, we cannot exclude SP goethite as a possible iron oxyhydroxide existing in these soil horizons. Vodyanitskii (2010) also states that podzolic soils from the Russian plain frequently contain goethite.

As revealed by the thermal demagnetization of laboratory remanences (Fig. 4.8.2), a magnetically soft Fe-containing mineral is detected in Podzols as well. This finding is in agreement with the results reported by Vodyanitskiy et al. (1983) for Podzols from the former USSR, where a magnetically soft mineral saturated in fields of ~150–200 mT was found, as well as a high-coercivity phase attributed to hydrogoethite. A similar conclusion is drawn by Sandgren and Thompson (1990), who pointed out that local conditions in different parts of the Podzol profile lead to the pedogenic formation of goethite, magnetite, maghemite, and hematite. In contrast, a classic geochemical study for establishing the "normative mineralogy" of soils from Fennoscandia and northwestern Russia (Salminen et al., 2008) accepts that magnetite enrichment of the upper horizons of the Podzols is due to their relative enhancement as a result of the intense weathering of the other major lithogenic minerals (feldspars, biotite, chlorite, etc), while magnetite as a resistant mineral remains in place, similarly to zircon and quartz. The authors also accept that Fe released through weathering from the easily weathered minerals is translocated to the illuvial horizon of the Podzols, where it precipitates as goethite. Very fine-grained goethite has also been identified through Möessbauer studies in Scottish podzolic soils (Goodman and Berrow, 1976). The authors further suggest a close association between goethite and the organic matter and relate this association to the properties of the ferritin.

Extensive geochemical and physical studies of tropical Podzols from the Amazon basin (e.g., Fritsch et al., 2011 and references therein) show the presence of hematite and goethite in the soil profiles. An upward decreasing proportion and diminishing size of hematite particles along with an increasing content of structural Al in fine-grained goethite is reported for these tropical Podzols. Since podzolization in the tropical environments is regarded as a secondary process, occurring after the laterite development, it is supposed that highly substituted goethite is a product of podzolization. In contrast, studies of Spodosol-like soils in alpine forest soils in the subtropics and tropics (Chiang et al., 1999) show that crystalline Fe oxides are enriched in the B horizons, but only Al-substituted lepidocrocite and goethite were identified, while no hematite is inferred.

A limited number of environmental magnetic studies of Podzols do not allow for a detailed and well-grounded hypothesis to be set for the pedogenic expression of the magnetic signature obtained. Further detailed geochemical studies, combined with low-temperature magnetic measurements, could provide more data for obtaining additional information about the genesis and transformations of magnetic minerals in Podzols.

4.10 ANDOSOLS: MAIN CHARACTERISTICS, FORMATION PROCESSES, AND DISTRIBUTION

Andosols are a specific soil order that includes soils developed on different volcanic materials. The characteristics of the parent material depend on the type of lava that is spread out of the volcanic vent—when it is of a low viscosity, it hardens to basalt. The explosive eruptions are most often associated with silicic lava, which is more viscous. Other types of volcanic parent material are the various ash-fall pyroclastic materials (dust, ash, lapilli, scoria, pumice) (Schaetzl and Anderson, 2009). Pyroclastic materials settled down by wind are known as tephra. They are dominated by glass and amorphous glass-like materials of low crystallinity, predetermined by the rapid cooling of the molten ejecta. Since these volcanic materials contain such a large amount of low-crystallinity minerals, even minimal weathering causes the dominance of amorphous materials and short-range order minerals dominated by Al and Si, such as allophane, imogolite, and ferrihydrite. Thus, in Andosols, secondary Al–Fe and Si–humus complexes are common. Because of the particular type of parent material and its weathering characteristics, Andosols have a thick dark A horizon with abundant humic acids (Schaetzl and Anderson, 2009). They develop rapidly under humid climates in volcanically active areas, where the weathering of the amorphous volcanic materials proceeds very quickly. In dry climates, volcanic materials weather more slowly and the Andosol development takes longer. When the primary cations Al, Si, and Fe are released into the soil solution via weathering, they become complexed by the organic acids and form stable organo-metallic complexes at low pH (Parfit and Saigusa, 1985; Schaetzl and Anderson, 2009). The formation of immobile Al—organic matter complexes leads to the accumulation of a high amount of organic matter in the near surface soil horizons and a dark soil color. The major characteristics of Andosols are high porosity, low bulk density, commonly low pH (but not as a prerequisite), a high amount of exchangeable Al, and high amounts of phosphorous (Schaetzl and Anderson, 2009). Unlike podzolization described in the previous section, organo-metallic complexes in the humic horizon are immobile, probably because the immobilization of organic matter is caused by a very high amount of exchangeable Al, so that the humic compounds become saturated in the A horizon and are immobilized (Aran et al., 2001). Another possible reason is sought in the type of organic acids—while in the Podzols, aggressive fulvic acids dominate and form chelate complexes with Al and Fe that become mobile, in Andosols, organic acids are predominantly of a humic type, which are not mobile to such an extent.

One of the most important diagnostic features for the classification of a soil as an Andosol is the criterion that the $Al_o + ½Fe_o$ content (extractions in ammonium oxalate) in the fine earth (0—2 mm) fraction equal 2.0% or more (IUSS Working Group WRB, 2014). This criterion is set down on the premises that the acid—ammonium oxalate extractable Fe (Fe_o) is a reliable proxy of the ferrihydrite content and the Fe substituting Al in allophane (e.g., Schwertmann, 1988). The "½" in the formula is taking into consideration the fact that Fe has approximately twice the atomic weight of Al and thus, by using one-half the weight of Fe, about the same amount of Fe and Al is required (Algoe et al., 2012 and references therein). This substantial decisive factor, however, has been recently questioned by Algoe et al. (2012). The authors show in a detailed study of dolerite sample that extraction with oxalate took out not only amorphous or short-range-order (SRO) Fe minerals, typical of andic properties, but is influenced by the dissolution of Fe in (titano)magnetite and maghemite as well. The Tiron (catechol disulfonic acid disodium salt, $C_6H_4Na_2O_8S_2$) extraction method is recommended as a better alternative to the oxalate method as a measure of amorphous/SRO Fe constituents in soils containing magnetic minerals, and especially in order to identify andic properties (Algoe et al., 2012).

Based on the mineralogical and colloidal properties, three subtypes of andic horizons are distinguished: (1) vitr-andic – dominated by volcanic glass and other primary minerals – the following relation must also be met: $Al_o + ½\ Fe_o = 0.4–2\%$; volcanic glass and other primary minerals > 60%; (2) sil-andic – dominated by allophane and similar poorly crystalline secondary aluminosilicate materials; and (3) alu-andic – dominated by aluminum complexed by organic acids (WRB, 2006). Usually sil-andic Andosols are found on recent pyroclastics, while alu-andic soils develop on older volcanic bedrocks (Quantin, 2004). The aluminum—humus complexes form preferentially in pedogenic environments that are rich in organic matter and have a pH ~5 (Shoji et al., 1996). The allophane and imogolite form mostly when the pH is 5—7 and there is a low amount of complexing organic compounds. The criteria set for the discrimination of these subtypes of Andosols are based mainly on the estimated content of the amorphous Al- and Fe-bearing minerals through ammonium-oxalate and pyrophosphate extractions (Shoji et al., 1996). Commonly, the soil's pH is also used in this classification, being set to $pH > 4.5$ if $C_{org} > 5\%$ or $pH > 5.0$ if $C_{org} > 6\%$ for sil-andic properties and to $pH < 4.5$ if $C_{org} > 5\%$ or $pH < 5.0$ if $C_{org} < 5\%$ for alu-andic properties. It is shown by Shoji et al. (1996), however, that the pH range for Andosols with alu-andic or sil-andic properties largely overlap.

Andosols are typically found in humid, volcanically active mountain ranges, such as Iceland, Alaska, Japan, Indonesia, the Andes Mountains (South America), and the Kamchatka peninsula (Schaetzl and Anderson, 2009). Recently, however, non-volcanic Andisols have also been reported to form on parent material rich in acidic silicate minerals and in warm, moist climates promoting fast weathering. Andosols in non-volcanic mountain ecosystems have been identified in different parts of the world (Nepal, India, Austria, the Alps, etc.) (Mileti et al., 2013 and references therein, Colombo et al., 2014; Vingiani et al., 2014). Fig. 4.10.1 shows the global distribution of Andosols according to the Food and Agriculture Organization.

As is obvious, Andosols occur all over the world in the volcanic regions—South and Central America, the United States, the Philippine Archipelago, and on many islands in the Pacific (IUSS Working Group WRB, 2014). In Europe, large continuous areas with Andosols are found in the Massif Central of France, in the northeastern Carpathians in Romania, and in the coastal volcanic areas of Sardinia and continental Italy (Tóth et al., 2008).

In the Bulgarian classification system, Andosols are not considered, but some recent investigations on soils developed on zeolitic materials in the Rhodope mountain suggest that they possess some features typical of Andosols (Ninov, 2002). A soil profile close to the Beli Plast open pit zeolite mine near the town of Kurdzhali consisting of zeolitized tuffs deposits has been studied.

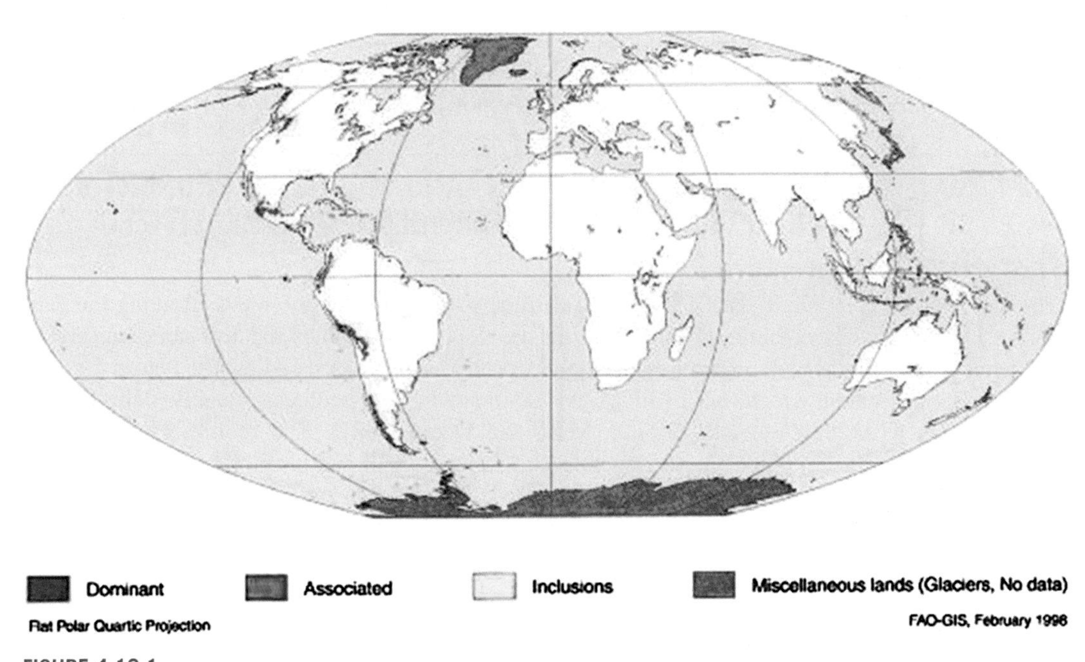

FIGURE 4.10.1

Spatial distribution of Andosols worldwide (FAO, 2001; Lecture notes on the major soils of the world: http://www.fao.org/docrep/003/y1899e/y1899e12.htm#P353_47671).

4.10.1 DESCRIPTION OF THE PROFILE STUDIED

Andic soil (shallow brown forest soil), profile designation: BP

Situated close to the village of Beli Plast, Kurdzhali district (South Bulgaria)

Location: N 41°46′49.5″; E 25°25′41.3″, h = 456 m a.s.l.

Present-day climate conditions: MAT = 12.5°C; MAP = 595 mm, vegetation cover: grassland; relief: hilly

Profile description:

0–8 cm: A_h1 horizon, light-brown, sandy-loamy with tuffs
8–40 cm: A_h2 horizon, humic, dark-brown, clayey, friable
40–50 cm: illuvial B horizon, brown, clayey, friable, glossy
50–73 cm: transitional BC horizon, light brown
73–82 cm: C horizon, weathered tuffs, light-yellow, sandy-clayey
Parent rock: Paleogene tuffs composed mainly of zeolites (clinoptilolite)

The tectonic processes during the Paleogene caused the microplates of the African continent to collide with the southern edge of the Eurasian plate, triggering significant collision-related magmatism and the formation of extensive volcanic areas, which extended from the Alps to northwest Turkey (Yanev et al., 2006 and references therein). The Eastern Rhodopes is such an area, where the studied profile was sampled. The acid volcanic glass deposited in the marine environment was transformed into zeolites, mainly clinoptilolite, less mordenite and analcime, as well as clay minerals, opal-CT [a silica phase mineral (e.g., Jones and Segnit, 1971)], and adularia (Yanev et al., 2006). Zeolitized tuffs, representing the parent rock of the profile BP, are a product of the First Early Oligocene acid phase volcanism. At the Beli Plast deposit, located to the north of the area of pyroclastic flow distributions, vitroclastic and pumice fall-out tuffs are exposed (Raynov et al., 1997).

4.11 SOIL REACTION, ORGANIC CARBON, AND MAGNETIC PROPERTIES OF THE ANDIC SOIL (PROFILE BP), DEVELOPED ON ZEOLITIZED TUFFS

The soil reaction (pH) along the BP profile is alkaline with an average of ~8.0 all along the depth (Fig. 4.11.1a), which is an unusual feature for soils developed on volcanics and does not correspond to the major criteria for the alu-andic and even sil-andic properties discussed earlier. The reason for this is the particular parent material, which is dominated by zeolites, mainly clinoptilolite $[(Na,K)_{6-2x}Ca_x]\cdot(Al_6Si_{30}O_{72})\cdot 24H_2O$ (e.g., Ming and Dixon, 1987). The zeolites are crystalline, microporous, hydrated aluminosilicate minerals that contain alkali and alkaline earth metals, which cause the establishment of a strongly alkaline pH, since obviously the weathering did not lead to the zeolite's breakdown. The organic carbon content is near 2% in the A_h1 horizon and ~1% in the A_h2 (Fig. 4.11.1b).

The clinoptilolitic tuffs from Beli Plast contain ~80% clinoptilolite and are widely used in different studies on zeolite applications in various fields (e.g., Bogdanov et al., 2009; Lihareva et al., 2009; Popov et al., 2012). The zeolitized tuffs from the Beli Plast deposit are Ca-K-Na dominant and, according to Lihareva et al. (2009) and Popov et al. (2012), the chemical composition is dominated by $SiO = 68.05\%$, $Al_2O_3 = 12.63\%$, $K_2O = 3.02\%$, $Fe_2O_3 = 0.83\%$, $Na_2O = 0.51\%$, $CaO = 3.53\%$, and

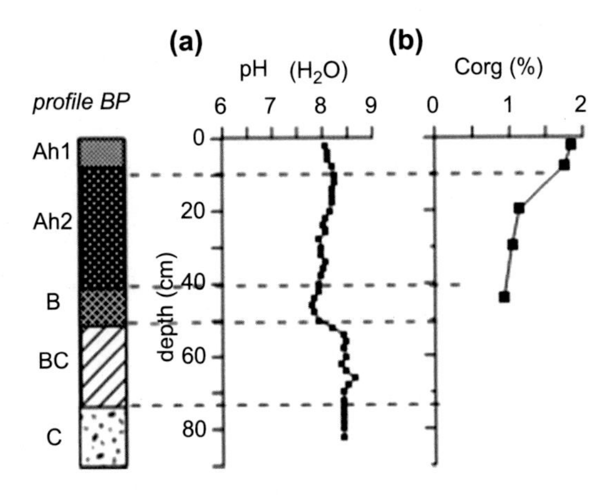

FIGURE 4.11.1

Depth variations of soil reaction (pH) (a) and organic carbon content (b) for the profile of andic soil BP.

$H_2O = 11.24\%$. The ratio $Si/Al = 5.39$. The major physical properties as reported in Popov et al. (2012) include density: 1.16 g/m^3, CEC = 110.61 mEq/100 g, pH 6.9–7.0, pore volume = 0.1 cm^3/g.

The magnetic mineralogy of the BP profile was studied through a stepwise thermal demagnetization of the composite IRM (Fig. 4.11.2). The examples shown point out that the magnetically soft

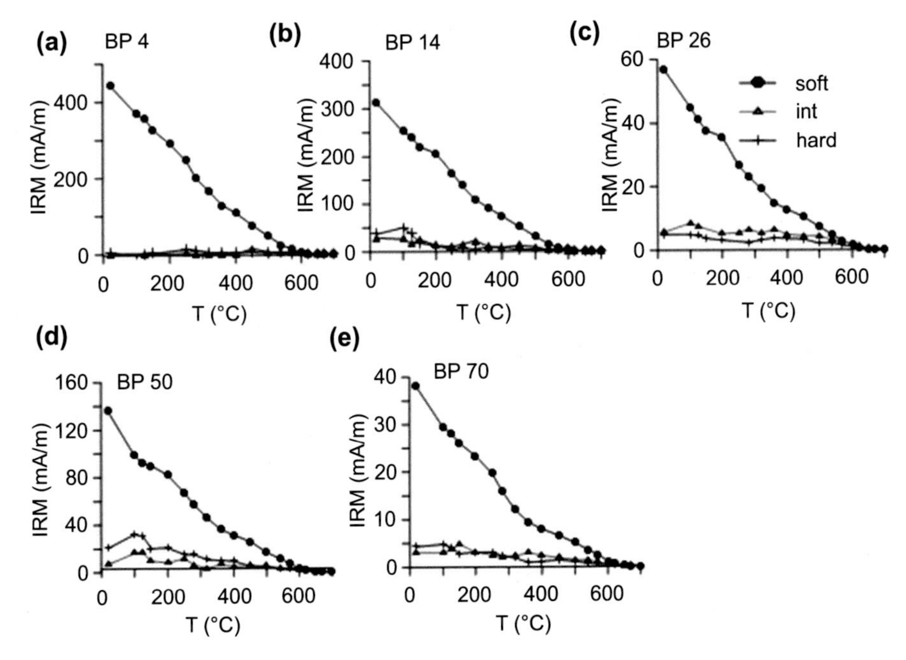

FIGURE 4.11.2

Examples of stepwise thermal demagnetization of composite IRM carried out for soil samples from different depths of the BP profile. The number in the sample name indicates the sampling depth. (a) sample from the A_h1 horizon, (b and c) samples from the A_h2 horizon, (d) sample from B horizon, and (e) sample from BC horizon. Saturating fields: soft component (0–300 mT); intermediate component (300–600 mT); hard component (600 mT–5.0 T).

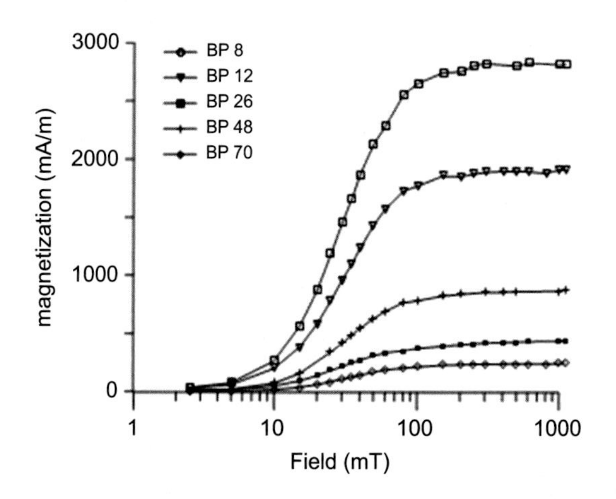

FIGURE 4.11.3

Stepwise IRM acquisition curves up to a field of 1 T for selected samples from the BP profile.

IRM component acquired in a field of 300 mT dominates the total magnetization. In the uppermost sample from the A_h1 horizon (Fig. 4.11.2a), this soft fraction is practically the only one, since the other two components are negligible. The unblocking temperature spectrum of the soft component is rather wide, as seen from the concave shape of the demagnetization curves. The final unblocking occurs at the Curie temperature of (oxidized) magnetite—580−600°C. The samples from deeper horizons are characterized by the presence of relatively weak high-coercivity components, represented by the intermediate (300−600 mT) and hard (600 mT−5 T) IRM fractions (Fig. 4.11.2b−e). Both of them are demagnetized at ~700°C, suggesting hematite as a possible mineral carrier.

The above results are supplemented by the data from the IRM acquisition in fields up to 1.1 T (Fig. 4.11.3). The magnetization curves suggest wide differentiation in the concentration of the strongly magnetic fraction along the profile since an equal quantity of soil material was used from all depths. The prevailing influence of the magnetically soft fraction is evidenced by the almost saturated acquisition curves in fields of 100−200 mT (Fig. 4.11.3). The coercivity analysis through IRM curve deconvolution by Gaussian functions (Kruiver et al., 2001), however, reveals, except for this dominant soft fraction, the presence of a high-coercivity component as well (Table 4.11.1).

Only the IRM acquisition curve for the uppermost sample (BP 8) from the A_h1 horizon is approximated by a single component of low coercivity (Table 4.11.1), related most probably to fine-grained magnetite/maghemite or coarse magnetite grains (Egli, 2004). Samples from the deeper horizons are dominated by a well-defined magnetically soft component (IRM1), accounting for ~90−95% of the total IRM and possessing the same coercivity characteristics as in the A_h1 horizon. In addition, a second component of moderate coercivity (IRM2 with $B_{1/2}$ varying between 150−250 mT and relatively narrow grain size, as suggested by the small DP values, is separated. This component could be ascribed to hematite of a different grain size and/or Al substitutions in the lattice (Özdemir and Dunlop, 2014). The samples from the A_h2 horizon (BP 26 and 48 in Table 4.11.1) display an IRM2 component of higher coercivity ($B_{1/2}$ of 250 mT) compared with the other samples. Thus, the magnetic signal in the studied soil

Table 4.11.1 Summary parameters from coercivity analysis using IRM-CLG1.0 software (Kruiver et al., 2001) for selected samples from the andic soil profile BP

	IRM component 1				IRM component 2			
Soil depth (cm)	Intensity (10^{-3} Am^2/kg)	% from IRM_{total}	$B_{1/2}$ (mT)	DP	Intensity (10^{-3} Am^2/kg)	% from IRM_{total}	$B_{1/2}$ (mT)	DP
Profile BP								
8	14.0	100	30.2	0.34	—	—	—	—
12	9.1	95	29.5	0.32	0.5	5	158.5	0.20
26	1.95	90	28.2	0.32	0.225	10	251.2	0.25
48	4.08	92	30.2	0.32	0.35	8	251.2	0.25
70	1.13	93	31.6	0.32	0.09	7	199.5	0.28

The intensity, relative contribution to the total IRM (% from IRM_{total}), median acquisition field ($B_{1/2}$), and dispersion parameter (DP) for each coercivity component are shown.

developed on zeolitized tuffs is governed by highly magnetic magnetite and a small amount of relatively low-coercivity hematite.

The depth variations of a set of major magnetic parameters are summarized in Fig. 4.11.4. The concentration-dependent characteristics (χ, χ_{fd}, IRM_{2T}, ARM) demonstrate strong magnetic enhancement in the upper 20 cm of the profile, spanning the A_h1 and the upper part of A_h2. A well-expressed maximum in all of the mentioned parameters suppose an enrichment with magnetite/maghemite of a broad grain-size distribution, spanning SP (χ_{fd}), SD (ARM) and coarse PSD (IRM_{2T})

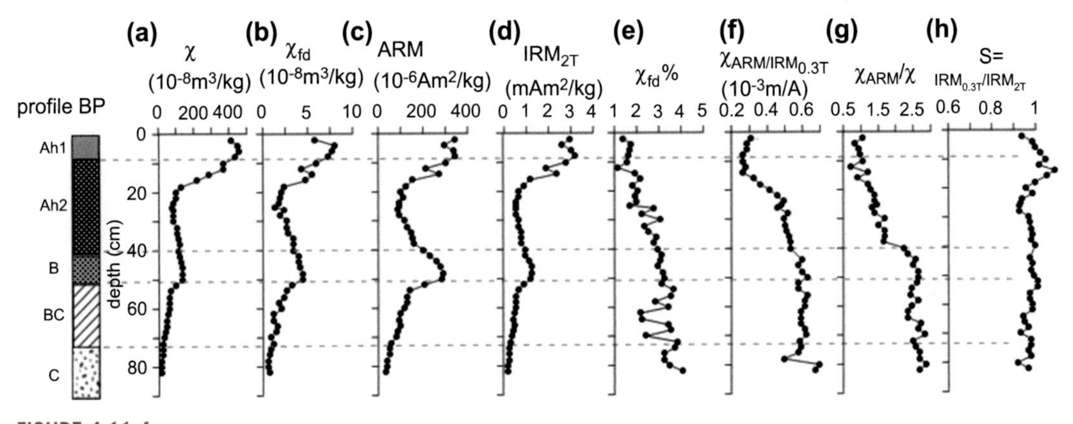

FIGURE 4.11.4

Depth variations of: (a) magnetic susceptibility (χ); (b) frequency-dependent magnetic susceptibility (χ_{fd}); (c) anhysteretic remanence (ARM); (d) IRM acquired in 2 T field (IRM_{2T}); (e) percent frequency-dependent magnetic susceptibility (χ_{fd}%); (f) ratio between the anhysteretic susceptibility (χ_{ARM}) and the IRM acquired in a field of 0.3 T ($IRM_{0.3T}$); (g) ratio between the anhysteretic susceptibility (χ_{ARM}) and magnetic susceptibility (χ); (h) S ratio.

domain ranges. The A_h2 horizon is relatively magnetically depleted, while a smaller maximum is seen in the B horizon (Fig. 4.11.4a–d).

In contrast to the A_h1 horizon, the enhancement of the B horizon is the most expressed in the SP proxy parameter (χ_{fd}) and in the SD grain-size proxy (ARM). The C horizon is depleted in magnetic minerals (Fig. 4.11.4), thus suggesting a pedogenic origin for the magnetic fraction in the studied soil. The grain-size related magnetic proxies (Fig. 4.11.4e–g) consistently suggest a refinement of the magnetic grain size from the top toward the bottom of the profile. This tendency is expressed in the increasing trend in $\chi_{fd}\%$, $\chi_{ARM}/IRM_{0.3T}$ and χ_{ARM}/χ in depth. In particular, the A_h1 horizon is characterized by a minima in these three ratios, implying the presence of the coarsest magnetic grains there, while downward, the finer fractions increase in proportion. The relatively sharp kink observed in the χ_{ARM}/χ variations at the boundary between the A_h2-B horizons is most probably mineralogy-determined, rather than grain-size related, since, as became clear from the IRM acquisition curves analyses (Table 4.11.1), the hematite's coercivity is higher in A_h2 and decreases downprofile. The S ratio (Fig. 4.11.4h) confirms the prevailing role of the magnetically soft fraction.

The obtained magnetic signature in the BP soil profile developed on zeolitized tuffs suggests an intense pedogenic formation of strongly magnetic magnetite/maghemite restricted to the upper soil horizons. Its formation is probably favored by the characteristic for the Andosol's stable complexation of Fe and the organic matter (Schaetzl and Anderson, 2009). The alkaline soil reaction in the BP profile may thus promote magnetite nucleation (Cornell and Schwertmann, 2003) in the well-aerated solum. The dominating ferrimagnetic mineral phase does not allow us to infer a potential contribution of amorphous short-range Fe-containing minerals like ferrihydrite. At the present stage of magnetic experimental research on this soil profile, and without chemical extraction data to check the criteria for andic properties, it is not possible to better constrain its genesis and classification.

A summary of the research on zeolitic soils from the Deccan basalt areas in India (Pal et al., 2013) shows that, in humid-tropical climates, acidic Mollisols, Alfisols and Vertisols form, rather than Andosols. As pointed out by Egli et al. (2008) and Dubroeucq et al. (1998), the weathering products from the volcanic ashes in drier climates are quite different from those in humid climates and soil formation in such an environment is not yet entirely understood. In a study of Mexican volcanic ash–derived soils, Dubroeucq et al. (1998) demonstrate that weathering and pedogenesis under semi-arid conditions on acid pyroclastics result in the interplay of several sub-processes: the breakdown of coarse minerals into small fragments; the diagenesis of non-crystalline minerals at the contact with the parent minerals; the transformation of non-crystalline minerals into halloysite in compact soil microstructures and the preservation of the amorphous minerals in the topsoil; and the desiccation and condensation of crystalline and noncrystalline minerals into microaggregates in the topsoil. Nevertheless, the magnetic signature obtained in the BP profile, characterized effectively by two maxima in the concentration of the strongly magnetic fraction—in A_h1 and in A_h2/B, would suggest an expression of two depositional events in the tuffs accumulation, separated by a pedogenic development. It is well documented that pyroclastic deposits may exhibit well-defined layering corresponding to separate phases in volcanic activity (Schaetzl and Anderson, 2009; Colombo et al., 2014). The systematic differences in the grain size–related magnetic characteristics in the two A_h horizons also support such a hypothesis. The identified hematite phase in the deeper (e.g., older) soil layer (A_h2-B) may be a result of the magnetite's oxidation over time, while the younger (upper) layer (A_h1) contains only a magnetite-like mineral.

The magnetic data for the BP profile may be compared with the data reported by Vingiani et al. (2014) in a study of andic soils in Calabria in a non-volcanic region, located between major volcanic areas and thus affected by pyroclastic deposition. The authors have found a several-fold increase in magnetic susceptibility in soils compared with the parent material. In addition, a decrease in the absolute χ values with diminishing andic properties is suggested as a possible criterion for a magnetic proxy for distinguishing Andic soils in the area.

4.12 MAGNETIC STUDIES OF ANDOSOLS DEVELOPED ON BASALTIC LAVAS

Massif Central in France is one of the most famous areas where typical Andosols have been found and extensively studied (Quantin, 2004). In the study by Soubrand-Colin et al. (2009), two soil profiles from this region developed in basalts have been investigated through magnetic methods—an alu-andic Andosol and an Andic Cambisol. The authors show that the primary lithogenic titanomagnetites control the magnetic properties of the soils, though the soil's magnetic minerals have been transformed in the course of pedogenesis. Their magnetic data evidence an evolution of Curie temperatures from the parent rock to the soil horizons, which is attributed to a maghemitization of the iron titanium oxides. The primary magnetites and titanomagnetites of the basaltic rocks are supposed to be converted in soils into maghemites and titanomaghemites, respectively. Recently, Grison et al. (2015, 2016) published detailed magnetic and geochemical studies on six soil profiles with andic properties, sampled in the eastern part of the French Massif Central. The sampled locations have been chosen in such a way as to represent the basaltic parent material of different ages. The results from the classic geochemical analyses reported in these two publications confirm the well-expressed andic properties of the profiles studied, but they will not be discussed here. According to Grison et al. (2015), the profiles of sil-andic Andosols showed similar magnetic phases in the soil and the corresponding bedrock, while the magnetic minerals in the alu-andic soils changed along the depth. The most typical feature in the depth variations of the magnetic susceptibility in Andosols developed on basaltic lavas was the increasing susceptibility (χ) with depth (Fig. 4.12.1).

This behavior underlines the overriding role of the lithogenic iron oxide–rich minerals inherited from the basaltic lava. Weathering and pedogenesis, therefore, are expressed in these profiles through the decreased magnetic signal in the upper soil horizons due to the acid dissolution of the coarse magnetic grains. Nevertheless, as Grison et al. (2016) suppose, pedogenic transformations lead to a well-expressed (χ_{fd}% between 4% and 9%) enhancement with SP grains of all the genetic horizons. In addition, a correlation between the χ_{fd} and SOM (soil organic matter) has been found, suggesting a pedogenic origin for the ultrafine SP grains. However, the relatively high χ_{fd}% obtained even in the C horizons of the profiles needs further consideration and analysis. The authors also found a significant correlation between the magnetic S ratio and (Fe_o−Fe_p) (oxalate-extractable and pyrophosphate-extractable Fe) in alu-andic samples, suggesting that well-crystallized Fe oxides (approximated by the S ratio) are associated with Fe organic complexes. On the basis of this finding, Grison et al. (2016) suggest using the S ratio as an indicator for the presence of alu-andic properties in soils. Thus, the authors claim that magnetic susceptibility measurements of soil samples can be used as a reliable indicator for detecting andic properties, while additional thermomagnetic analyses could help in the discrimination between alu-andic and sil-andic subtypes.

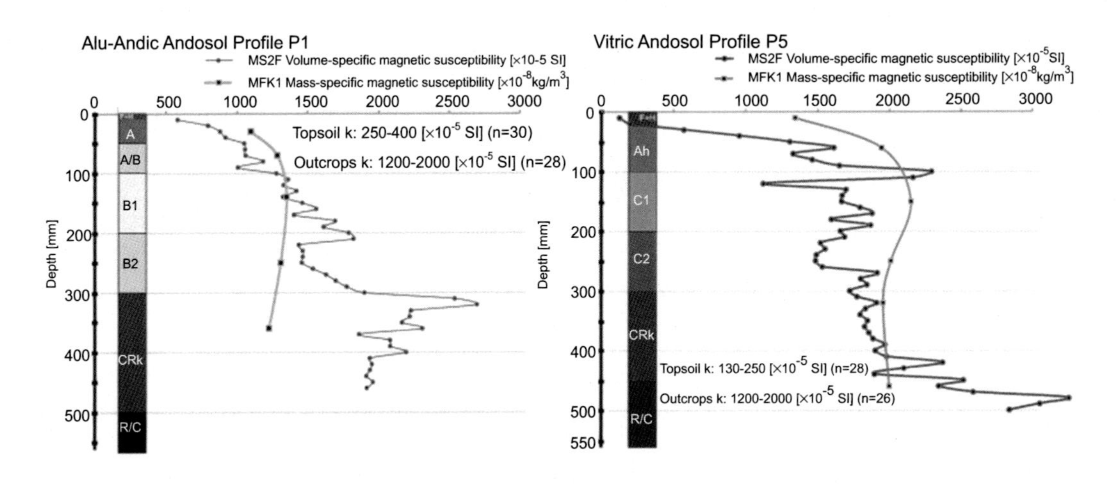

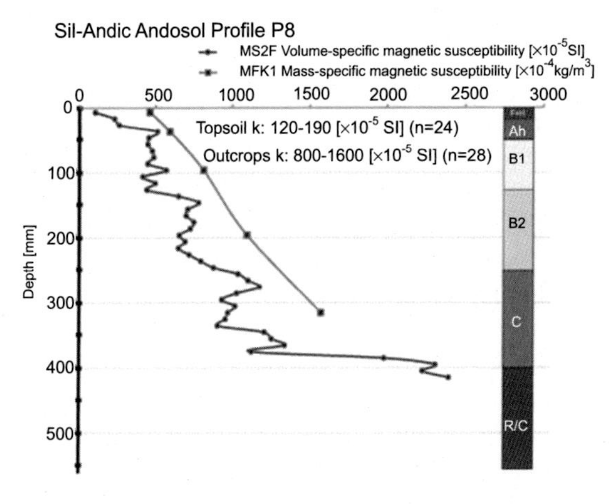

FIGURE 4.12.1

Magnetic susceptibility (volume and mass specific) variations along the depth of Andosols developed on basalts from Massif Central (France), according to Grison et al. (2016). Measured values of the topsoil κ (after removing the leaves), κ of outcropping rocks and soil column description are shown.

Reprinted with permission from Elsevier.

Another relevant case study involving soils with andic properties is reported by Thompson et al. (2011). The authors provide an excellent extensive study on the fate and transformations of the initial iron oxyhydroxide mineralogy in response to climate-induced redox fluctuations across a natural rainfall gradient imposed on tropical soils, developed from a basaltic lava flow. Thompson et al. (2011) studied a set of Hawaiian soils where rainfall increased from 2200 to 4200 mm/yr. These soils developed in mixed lava and tephra series that deposited $\sim$400 ka ago. Different techniques were used to characterize the soils, including Moessbauer spectroscopy, XRD, and selective chemical extractions, as well as magnetic susceptibility measurements. The classification criteria for the andic properties are not examined for the soils studied, but, according to the data provided (Thompson et al., 2011) soils demonstrate a large share of oxalate-extractible Fe (e.g., Fe_o/Fe_d in the interval 0.4−1.0) and significant organic matter content, accompanied by the specified evidence for the presence of an increasing amount of noncrystalline SRO (short-range-ordered oxides). This points to a high probability that the soils possess andic properties. As pointed out in Section 4.10, volcanic glass weathering in climates with high precipitation proceeds very quickly, releasing large amounts of Fe, Al, and Si, which favors the precipitation of nanocrystalline or SRO phases (allophane, ferrihydrite, and opaline silica) over crystalline phases. Further transformations of these secondary phases into crystalline minerals depend strongly on the local environmental conditions and aging time. Thompson et al. (2011) assume that most of the primary ferrous iron was oxidized during early weathering stages and the large decrease obtained in the total Fe across the soil sequence studied is a result of the pedogenic transformations of these secondary nanocrystalline minerals. The major iron oxyhydroxides identified in the basalt-derived soils are nanocrystalline goethite and/or ferrihydrite, hematite, and maghemite. The systematic decrease in the total Fe with increasing rainfall was observed for the soils developed in different precipitation regimes, accompanied by a related decrease in the magnetic susceptibility of the uppermost soil layers (Fig. 4.12.2).

As a resu-lt, the upper part of the solum is magnetically depleted in contrast to deeper soil horizons, as also observed by Grison et al. (2015, 2016) for Andosols from the Massif Central. The magnetic

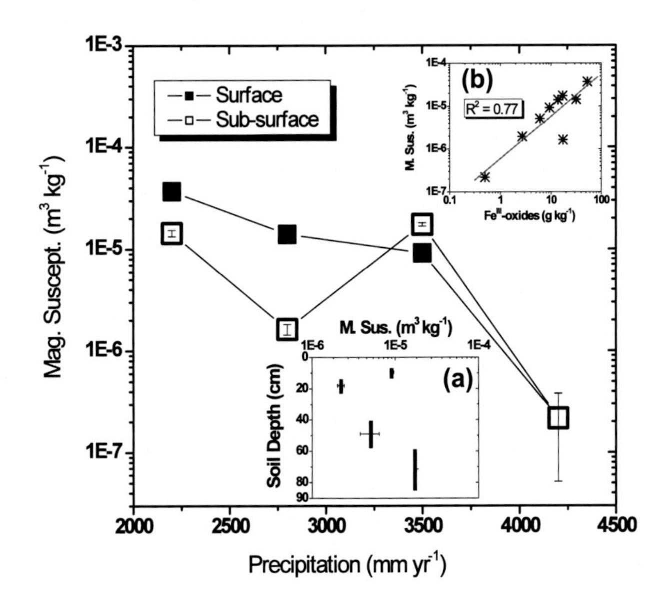

FIGURE 4.12.2

Magnetic susceptibility as a function of rainfall for the surface and sub-surface transects of tropical soils developed on basalts from the Hainan Island. The *inset (a)* shows magnetic susceptibility along the depth at the 3500 mm/yr site. *Inset (b)* displays log–log plot of magnetic susceptibility vs Fe(III)-oxide concentration calculated from Moessbauer spectra for all soil samples.

Reprinted from Thompson, A., Rancourt, D., Chadwick, O., Chorover, J., 2011. Iron solid-phase differentiation along a redox gradient in basaltic soils. Geochim. Cosmochim. Acta 75, 119–133, with permission from Elsevier.

susceptibility measurements of the soil profile developed under the highest rainfall (~3500 mm/yr) show a relatively low χ (~919×10^{-8} m^3/kg) in the uppermost organic-rich levels and an almost two-fold increase in χ with depth (Thompson et al., 2011). The authors conclude that the crystallinity of the Fe-(oxy)hydroxide phases decreases with increasing rainfall, which they ascribe to organic matter coprecipitation and cation substitutions retarding mineral ripening and the development of more crystalline phases. The events of periodic water saturation in such high rainfall tropical forests cause intense Fe reduction, accompanied by the mobilization of substantial amounts of dissolved and colloidal carbon (Buettner et al., 2014).

4.13 FERRALSOLS: MAIN CHARACTERISTICS, FORMATION PROCESSES, AND DISTRIBUTION. MAGNETIC STUDIES OF FERRALSOLS

The Ferralsols are highly weathered soils found in the humid tropics. The pedogenic processes in tropical regions are influenced by the occurrence of abundant moisture (which could be seasonal as well), which reinforces chemical weathering and plant growth. High temperatures assist in inorganic chemical reactions and biological activity, while old landscapes without active tectonics provide the necessary long time-frame for all the reactions to go on to completion. An additional strong pedogenic factor in the tropics is the mechanical pedoturbation caused by termites and ants (Schaetzl and Anderson, 2009). Under the combined influence of these factors, the rapid mineralization of organic matter and intense, nearly complete weathering of primary minerals occurs. Since Al and Fe are relatively insoluble, they remain in the soil, while the other weathered products (including even Si because of the high rainfall) are translocated out of the profile. The residual Al and Fe form large amounts of stable oxides (hematite, goethite) and clays, which cause the cementation of the solum (Schaetzl and Anderson, 2009 and references therein). The accumulation of Fe in tropical soils is considered to result from two processes—residual accumulation (latosolization) and processes involving transfers of iron (laterization), although the precise routes causing Fe accretion in the soil profile are disputable.

One widely accepted theory for the genesis of tropical soils is that proposed by Duchaufour (1982)—the three-phase weathering/pedogenesis sequence. Phase 1 is labeled "fersiallitization" and includes soils that retain a high amount of Fe, Al, and Si. Such are the red Mediterranean soils developed in hot semi-arid conditions mainly on calcareous parent material (see also Chapter 3). These soils are not very strongly weathered but contain a significant amount of iron oxides and the illuvial horizon contains 2:1 clay minerals. Phase 2 is termed "ferrugination" and is a transitional stage, when weathering increases but some primary minerals can still be preserved. At this stage, desilication increases and kaolinite may neoform. Soils showing such signs are typical of subtropical savanna areas (Schaetzl and Anderson, 2009). The third phase is termed "ferralitization" and is indicative of strongly weathered acid soils where kaolinite and oxide clays prevail. Only quartz remains from the primary minerals in the profile, the soil is enriched in Fe and Al oxides, silica has been lost, and gibbsite and oxide clays may form.

Ferralsols are deeply weathered red or yellow soils. They do not exhibit clear horizon boundaries and they have a high content of iron oxides, and the clay complex is dominated by kaolinite (IUSS Working Group WRB, 2014). Ferralsols are known as Oxisols (USA), Latosols (Brazil), Sols ferralitiques (France), and Ferralitic soils (Russia). They develop in the flat lands of the Pleistocene age or

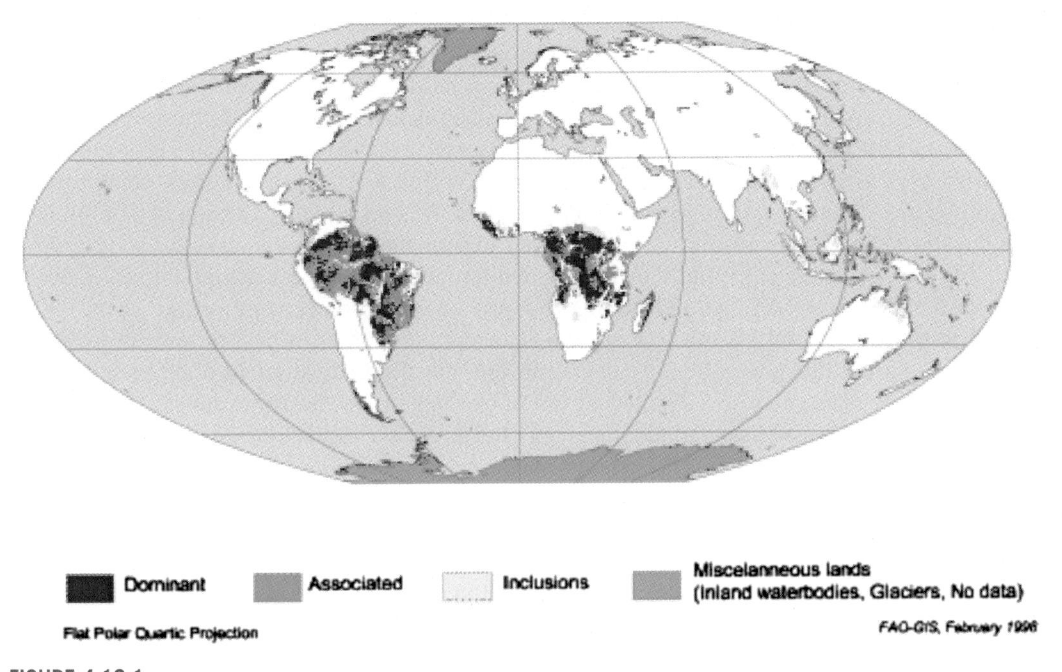

FIGURE 4.13.1

Spatial distribution of Ferralsols worldwide (FAO, 2001). Lecture notes on the major soils of the world: http://www.fao.org/docrep/003/y1899e/y1899e12.htm#P353_47671.

older, commonly in perhumid and humid tropics. When such soils occur elsewhere (in different climate conditions), they are considered to be relict soils. Ferralsols are widespread in the humid tropics of South America (especially Brazil), Africa, and less so in Southeast Asia (Fig. 4.13.1).

4.13.1 MAGNETIC STUDIES OF FERRALSOLS

Detailed environmental magnetic studies on tropical Ferralsols have been undertaken just recently, while numerous investigations on soils developed in mid-latitude temperate areas and in the Chinese Loess Plateau have been accumulating for a long time (e.g., Maher, 1998; Maher et al., 2003; Liu et al., 2007). On the other hand, Ferralsols provide an excellent archive of iron oxide transformations, neoformation, and growth due to their extensive pedogenic development in the nearly unchanging climatic conditions. The tropical climate is characterized by excessive annual rainfall, combined with high temperatures and evaporation. Most of the Ferralsols studied in different tropical regions develop on basalts—material providing a rich Fe source that is released during the intense weathering. The inherited strongly magnetic lithogenic iron oxides also contribute to the very high ferrimagnetic concentration commonly observed in basalt-derived soils. Van Dam et al. (2008) have studied three soil profiles from the Hawaii island Kaho'olawe Island, where the mean annual rainfall varies between ~200 and 1500 mm/y. The magnetic susceptibility along depth showed a large enhancement in the old soils, compared with the signal from the fresh basalt sample, thus indicating an enhanced pedogenic formation of strongly magnetic minerals. A particular feature in the depth variations of χ is the lower values obtained in the topsoil part, while a maximum χ is linked to the B horizons (van Dam et al., 2008). According to the authors, the range of iron oxide forms detected—goethite, hematite, magnetite/maghemite—is related to the degree of weathering and exposure to soil erosion. A specific characteristic that, however, is not discussed by van Dam et al. (2008) is the obtained constant or even increasing $\chi_{fd}\%$ from the top toward the bottom of the soil profiles. The same $\chi_{fd}\%$ depth behavior along a Ferralsol from a subtropical area in Parana (Brazil) is reported by Cervi et al. (2014), along with the corresponding χ variations, again showing lower values at the top and increasing χ downward. The rock-magnetic properties are also reported for lateritic soil profiles from tropical India in a monsoonal climate (MAP ~ 3500 mm and the annual temperature range 16°C–37°C) (Ananthapadmanabha et al., 2014). The authors studied three profiles of lateritic soils developed on charnockite/charnockitic gneiss from the Archaean age. The detailed variations of different magnetic parameters along the depth give a detailed picture of changes in concentration and grain size–related magnetic proxies. The obtained variations are consistently similar among the profiles and Fig. 4.13.2 depicts the results for one of them. The relative magnetic enhancement of the upper part of the profile could be related to enrichment with a strongly magnetic fraction with a wide grain-size distribution, as reflected in the χ_{fd}, IRM_{20mT}, SIRM, and ARM.

The presence of coarser PSD-MD grains is supposed in the topmost part of the profile, while SP and SD fractions (contributing to χ_{fd} and χ_{ARM}) are linked to the middle part (Fig. 4.13.2). Similar to the earlier-cited works on Ferralsols from Hawaii and Brazil, the $\chi_{fd}\%$ effectively increases in depth, reaching ~15% despite the obtained decrease in χ_{fd}. The authors report a significant influence from the lithogenic iron oxide fraction, consisting of magnetite and hematite, thus giving the overall high "background" signal in the parent material horizon. The pedogenic magnetic fraction containing SP and SD maghemite is suggested by Ananthapadmanabha et al. (2014) to prevail in the middle and

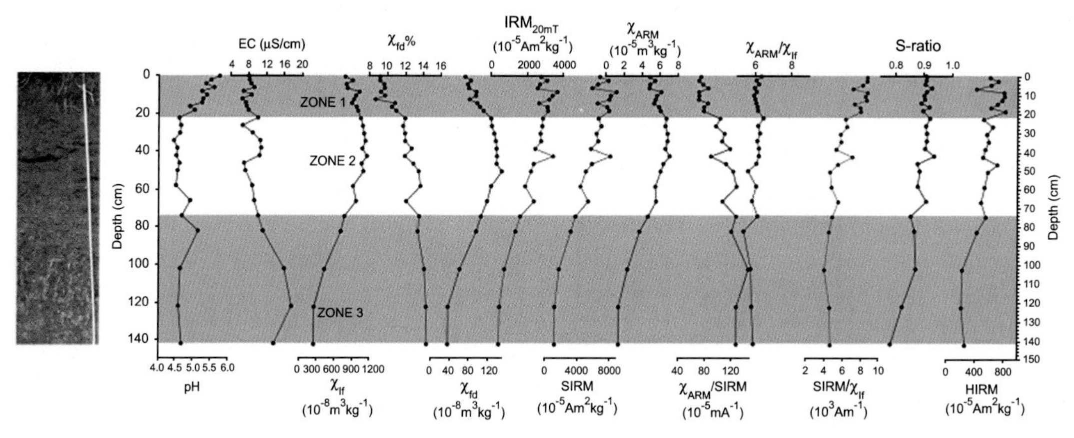

FIGURE 4.13.2

pH, electrical conductivity (EC), rock magnetic parameters, and interparametric ratios for Ferralsol (Uliyathadka soil profile) from the Kasaragod District of Kerala State (India), according to Ananthapadmanabha et al. (2014).

bottom parts of the solum, based on the obtained $\chi_{fd}\%$ pattern, while the topmost layers are supposed to contain more lithogenic grains. This conclusion, however, contradicts "natural" top-down pedogenesis and must be considered with caution.

The most extensive magnetic studies on Ferralsols are reported for soils developed on basalts from Hainan Island, south China (Li et al., 2011; Long et al., 2011, 2015; Ouyang et al., 2015), because of the high rainfall gradient across the island (1440–2020 mm/yr), insignificant temperature variation (23–24°C), and extreme weathering (Long et al., 2011). The latter authors have found that the hematite content and magnetic susceptibility decrease with increasing rainfall, whereas goethite concentration increases. At the same time, no systematic trend in the total amount of iron oxides related to chemical weathering intensity has been observed along the transect. The ferrimagnetic minerals enrichment in the uppermost soil part, based on the observed synchronous change of χ_{fd} and χ, Long et al. (2011) assign to the accumulation of fine-grained ferrimagnets. In a detailed consideration of the intrinsic links and relations among the magnetic data, selective chemical extractions data, and DRS determinations of hematite/goethite abundance, Long et al. (2011) suggest that excessive rainfall above the inflection point of ~1440 mm/y in the relation between the magnetic enhancement and amount of rainfall can block the neoformation of hematite and maghemite and favors the neoformation of goethite. Moreover, the authors propose a modified conceptual model of magnetic change in response to the climate parameters, which takes into account the effect of both temperature and rainfall. The model elucidates the reason as to why a rainfall infection point exists and the manner in which it changes with temperature in aerobic soils. In a further study of a set of three Ferralsol profiles from the same area, Long et al. (2015) investigate the mechanism responsible for the magnetic enhancement in these soils and the processes responsible for this. The profiles studied are characterized by similar total concentrations of iron oxides but different Hm/(Hm + Gt) ratios, varying between 0.14 and 0.74. It was found that χ was often 3–20 times higher than that of the unweathered rock, indicating the pedogenic character of the magnetic enhancement. Based on a detailed consideration of the different bi-parametric plots of the magnetic characteristics of the three soil profiles, Long et al. (2015) conclude that the accumulation of SP and SD particles was correlated to the formation of hematite. In addition, the rates of the SP and SD fractions' build-up decrease with the increasing formation of hematite, estimated by Hm/(Hm + Gt). The authors attribute this decreasing rate to the growth of fine pedogenic magnetic particles and their transformation into Hm when a large amount of SP plus SD magnetic particles accumulate and they grow above a certain size. Thus, this study contributes to the widely discussed theories of the mechanism of the magnetic enhancement of aerated soils (e.g., Chapter 1) and its relation to climate parameters, supporting the model proposed by Barrón and Torrent (2002). The observed depth variations of magnetic susceptibility χ and the $\chi_{fd}\%$ demonstrate features already discussed earlier—in two of the profiles characterized by a higher Hm/(Hm + Gt) ratio, the percent frequency-dependent magnetic susceptibility maintains very high values (~16% in HNA profile and ~20% in HNK profile from Long et al., 2011) along all the profiles despite the decrease in χ_{fd} and χ at the bottom. This peculiarity is not discussed by the authors, but it needs further elucidation because (1) it is illogical to accept that an increasing relative share of the SP fraction in the magnetic assemblage (e.g., increase of $\chi_{fd}\%$) could be possible when the absolute amount of this fraction strongly decreases (e.g., decrease of χ_{fd}) along with the coarser fractions, and (2) the realistic upper limit of $\chi_{fd}\%$ is shown to be ~15% (Worm, 1998) since, for obtaining $\chi_{fd}\% = 22\%$, a very narrow grain-size distribution of the SP fraction must be assumed, which rarely occurs in nature.

Tropical Ferralsols from south China (including profiles from Hainan Island) have also been studied by Li et al. (2011), who compared the magnetic properties of Ferralsols with those of other tropical soils – Primosols and Ferrosols. The authors found the highest χ and $\chi_{fd}\%$ values in the Ferralsol profiles ($\chi_{fd}\% > 10\%$; $\chi > 7000 \times 10^{-8}$ m^3/kg) and, again, $\chi_{fd}\%$ kept this high value along the whole profile. Li et al. (2011) propose using the magnetic susceptibility and $\chi_{fd}\%$ as magnetic proxies for the identification and location of Ferralsols in the tropics. In a study of lateritic soils from Indonesia, Safiuddin et al. (2011) observe typical high magnetic susceptibilities in the upper part of the soil profiles ($\chi \sim 1500–8000 \times 10^{-8}$ m^3/kg), but in this case study, only in three of six profiles does $\chi_{fd}\%$ sustain its high values all along the depth.

Thus, an unresolved question in the studies of Ferralsols appears to be the reason for the very high $\chi_{fd}\%$ observed up to the bottom of the soil profiles. This question will be addressed in a later chapter of this book.

4.14 GLEYSOLS: MAIN CHARACTERISTICS, FORMATION PROCESSES, AND DISTRIBUTION

When soil is experiencing permanent or prolonged periods of water saturation, reducing conditions are created and thus gleying occurs. Gleying advances when an ample amount of organic matter is present. It contains abundant microorganisms that need oxygen for their metabolism. Since, upon permanent soil water-logging, the main oxygen source—the air—is exhausted, the microorganisms first use the oxygen dissolved in the water. After that, their metabolism proceeds through an electron exchange with the most redox-sensitive elements in the soil—Mn^{3+}, Mn^{4+}, and Fe^{3+} (van Breemen, 1988a). Gleying could be considered both as a pedogenic and a geochemical process. The reductive conditions provoke geochemical reactions, which lead to the formation of new minerals, thus changing the soil's color toward hues of gray, green, and blue (Schaetzl and Anderson, 2009). Thus, in Gleysols, conditions favoring ferrihydrite formation exist (Cornell and Schwertmann, 2003). Brown hues typical of the initial soil matrix could be observed in the Gleysol profile despite the permanent water saturation, when microorganisms are missing or are present in a small amount. Usually, Gleysols are regarded as groundwater-influenced soils; that is, they have a permanently water-logged (non-mottled, grayish) subsoil and a seasonally wet (brown and gray mottled) shallower horizon (Van Breemen and Buurman, 2003). The gleying can be regarded as a several-stage process: (1) solid Fe(III) oxides are reduced and dissolved when organic matter is present, the soil is saturated with water so the diffusion of oxygen is slowed down, and microorganisms capable of reducing Fe(III) are abundant in the soil; (2) The dissolved Fe^{2+} moves down-profile via diffusion or mass flow. The diffusion takes place along a concentration gradient, which is usually caused by a spatial variation in the rates of production or the removal of Fe^{2+}. (3) Dissolved Fe^{2+} is removed from the solution via precipitation of solid Fe(II) compounds, through the absorption on clay or Fe(III) oxides, or by the oxidation of dissolved, adsorbed, or solid Fe(II) by free Fe^{2+} to Fe(III) oxide. The transport of Fe^{2+} can occur over distances varying from millimeters to kilometers. Groundwater gley soils in valleys may exhibit an accumulation of iron transported from the surrounding areas (van Breemen and Buurman, 2003). The iron (oxy)hydroxide forms detected in Gleysols usually include ferrihydrite, nanocrystalline goethite, and siderite (van Breemenn, 1988a; Mansfeld et al., 2012). The study by Mansfeld et al. (2012) evidenced that, when Fe^{2+} accelerates the transformation of ferrihydrite to more-crystalline Fe oxides, then the Fe(III) pool in the soils with reducing conditions is rather dynamic. Usually the permanently saturated soil horizons are Fe poor, while illuvial horizons can be very rich in iron and, as in the gley soil studied by Mansfeld et al. (2012), goethite can accumulate in large amounts. Nevertheless, as discussed earlier in this chapter, the reductive dissolution of ferrihydrite may result in siderite and magnetite precipitation if the Fe^{2+} concentration in the soil solution is high (e.g., Fredrickson et al., 1998). In soils containing a greater amount of sulfur, part or most of the reduced forms of iron is in the form of FeS (mackinawite) or cubic FeS_2 (pyrite) (van Breemen, 1988b). These soils are acid sulfate soils that have a pH < 4. They often occur in recent coastal plains, where pyritic tidal marsh sediments have been drained and thus exposed to the air. A more-detailed discussion on acid sulfate soils can be found in van Breemen (1988b).

Gleysols are frequently found in lowlands with a high groundwater table, tidal areas, shallow lakes, and sea-shores (IUSS Working Group FAO, 2014). They occur all over the world in different climates, from perhumid to arid (Fig. 4.14.1). Abundant Gleysols are found in sub-arctic areas in the north of the Russian Federation, Canada, Alaska, as well as in humid temperate and sub-tropical areas in China, Bangladesh, in the Amazon region, equatorial Africa, and the coastal swamps of Southeast Asia.

Meadow-gley soils in Bulgaria develop under the influence of meadow vegetation when the groundwater level is high (0.5–1.0 m) and can even reach the surface (Penkov, 1983). Gleysols most

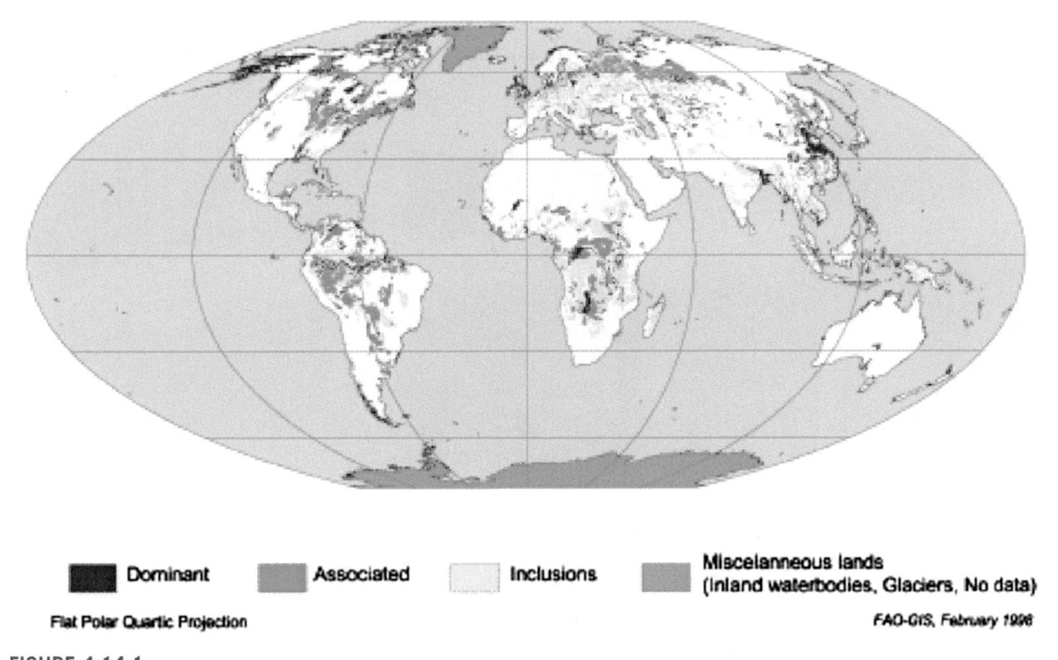

FIGURE 4.14.1

Spatial distribution of Gleysols worldwide (FAO, 2001) Lecture notes on the major soils of the world: http://www.fao.org/docrep/003/y1899e/y1899e12.htm#P353_47671.

frequently occur in the lowlands, as well as in the peripheral parts of the bogs. They are characterized by a heavier mechanical composition because their development is related to the finer river sedimentation. The meadow-gley soils display a direct genetic relationship to the meadow soils and a number of transitional varieties have been observed—alluvial-meadow, meadow-boggy, and meadow-gley (Shishkov and Kolev, 2014). Their development is characterized by water-logging and the accumulation of organic matter from the meadow and bog vegetation. The major genetic horizon is the gley horizon (B_g) with permanent anerobic conditions.

4.14.1 DESCRIPTION OF GLEYSOL (MEADOW GLEY SOIL) PROFILE STUDIED, PROFILE DESIGNATION: GL

Situated close to the town of Sadovo, Plovdiv district (South Bulgaria)

Location: N 42°08′22.4″; E 24°59′29.1″, h = 150 m a.s.l.

Present day climate conditions: MAT = 12.1°C; MAP = 552 mm, vegetation cover: boggy vegetation; relief: flat

Profile description: general view and details from the horizons are shown in Figs. 4.14.2–4.14.6.

0–30 cm: humic (A) horizon, dark brown–black, humus rich, crumby-grainy structure, sandy-loamy
30–47 cm: transitional AC_{gk} horizon, mottled orange-brown, gray mottled gley spots, clayey
47–57 cm: $C_{kg}1$ horizon, gray-brown color, bluish gley spots around the vegetation roots, heavy clayey
57–73 cm: $C_{kg}2$ horizon, yellow-brownish color, sandy, crumbly, structureless

FIGURE 4.14.2

View of the profile of Gleysol (GL).

FIGURE 4.14.3

Gleysol profile GL. A_h horizon—detail.

FIGURE 4.14.4

Gleysol profile GL. AC_{kg} horizon—detail.

73–87 cm: $C_{kg}3$ horizon, gray-brown color, orange redox spots, heavy clayey
87–133 cm: $C_{kg}4$ horizon, brown-black color, clayey, bluish gley spots around the vegetation roots
Parent material: alluvial deposits

FIGURE 4.14.5

Gleysol profile GL. $C_{kg}3$ horizon—detail.

FIGURE 4.14.6

Gleysol profile GL. $C_{kg}4$ horizon—detail.

4.15 TEXTURE, GEOCHEMICAL CHARACTERISTICS, AND MAGNETIC PROPERTIES OF GLEYSOLS

Variations in the three texture classes—sand, silt, and clay—along the depth of the GL profile are shown in Fig. 4.15.1. The obtained results relate closely to the description of the profile's horizons and suggest that we are dealing with a complex of two successive gley soils. Most probably, these two soils have been formed during two stages of ceased sediment accumulation. As seen from Fig. 4.15.1, the uppermost soil has a more clayey texture in the A horizon, which gradually transforms into a sand-dominated horizon. The buried soil is revealed at an ~80-cm depth and is dominated by the silt fraction (Fig. 4.15.1). The soil's reaction is alkaline, but in the uppermost soil it increases with depth until $C_{kg}1$, while in the buried soil (depth 80–130 cm), pH decreases. The obtained pH values are in agreement with the data for similar meadow-gley soil from the Sadovo district, published in the *Soil Atlas for Bulgaria* (Koinov et al., 1998).

Depth variations in the content of the major chemical elements are shown in Fig. 4.15.2 and clearly indicate the presence of a batch of two soils separated by sandy deposits. Both soils display high Mg, K, Al, and Fe contents, while the $C_{kg}1$ horizon is depleted. In contrast, it is enriched in Ca, Si, and Ti (Fig. 4.15.2).

The profile's parts that are enriched in organics contain a higher amount of sulfur (S) and chlorine (Cl) (not shown), more enhanced in the recent gley soil (Fig. 4.15.2). Iron content along the depth varies strongly, being the highest in the humic (A_h) horizon and in the buried gley soil. The presence of amorphous Fe and Al phases in the gley soil is demonstrated by acid oxalate extraction (Table 4.15.1). Oxalate extractable iron (Fe_o) content reaches 4–5% of the total iron for samples from the uppermost level (2 cm depth) and at an 82-cm depth. The calculated ratio ($Al_o + ½Fe_o$) used for the estimation of the amount of SRO phases in soils (e.g., Algoe et al., 2012) indeed indicates a significant share of such a fraction in both soil depth intervals, as mentioned earlier (Table 4.15.1).

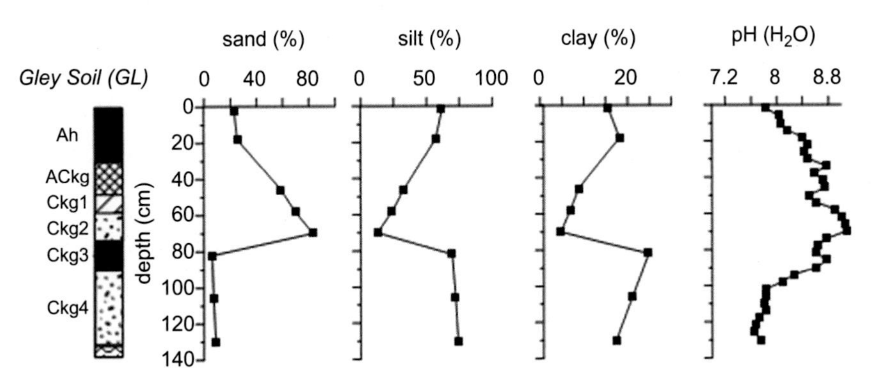

FIGURE 4.15.1

Mechanical fractions: sand, silt, clay, and soil reaction pH along the GL profile.

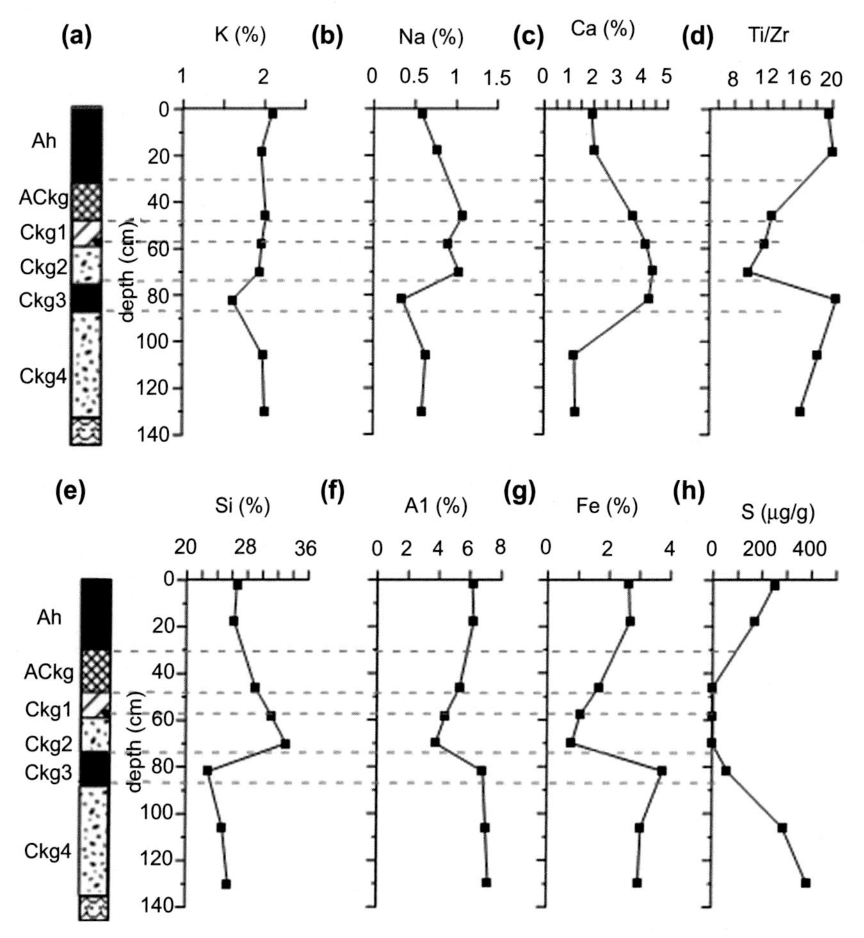

FIGURE 4.15.2

Content of some major and trace elements (K, Na, Ca) (a–c), (Si, Al, Fe,and S) (e–h), and the ratio Ti/Zr (d) measured for selected samples from the GL profile.

Table 4.15.1 Chemical extraction data for selected samples from the GL Gleysol profile: oxalate-extractable Aluminum (Al_o), Iron (Fe_o), and Manganese (Mn_o), total Fe content (Fe_{tot}) and the ratios Fe_o/Fe_{tot} and (Al_o + ½Fe_o)

Sample depth (cm)	Al_o (g/kg)	Fe_o (g/kg)	Mn_o (g/kg)	Fe_{tot} (g/kg)	Fe_o/Fe_{tot}	Al_o + 1/2Fe_o
2	0.924	1.128	0.348	26.28	0.043	1.49
46	0.419	0.591	0.027	16.65	0.036	0.71
70	0.069	0.000	0.000	7.935	0.00	0.07
82	1.103	2.036	0.082	37.57	0.054	2.12
130	0.874	0.784	0.000	29.73	0.026	1.27

4.15.1 MAGNETIC PROPERTIES OF THE STUDIED GLEYSOL

The high-temperature behavior of magnetic susceptibility for the selected samples from the GL profile is shown in Fig. 4.15.3a–c. The uppermost surface sample, GL2, exhibits strong mineralogical transformations during heating, resulting in a many-fold increase in χ on cooling. Two transformation peaks on the heating curve are present—at 300°C and at 500°C (Fig. 4.15.3a). The first peak is most probably due to the presence of maghemite that inverts to hematite at ~300°C (Dunlop and Özdemir, 1997). The second maximum at ~500°C is a Hopkinson peak of the magnetite, newly created during heating and possibly originally present, registered through the T_c ~580°C. The other two samples studied (Fig. 4.15.3b and c) do not show a strong enhancement of magnetic susceptibility on cooling. Despite the noisy curves obtained because of the very weak signal, the magnetite's T_c is registered for these two soil depths. In addition, a weakly expressed maximum at ~300°C is again seen on the heating curves, probably related to maghemite presence. Experiments on stepwise thermal demagnetization of ARM (Fig. 4.15.3d–g) confirm the presence of stable SD-like magnetite in all the samples studied.

The main unblocking temperature in the uppermost sample (GL7) is 600°C (Fig. 4.15.3d), while in the samples from deeper horizons, a small share of ARM is demagnetized at 640°C, except for the major unblocking at 600°C. All demagnetization curves are characterized by a convex shape suggesting rather narrow unblocking temperature spectra and, therefore, a narrow grain size of the magnetite's fraction. Stepwise acquisition of IRM provides further information about the magnetic characteristics of the magnetic mineral phases in the gley soil studied. The IRM acquisition curves are shown in Fig. 4.15.4, and the results from the coercivity analysis according to Kruiver et al. (2001) are

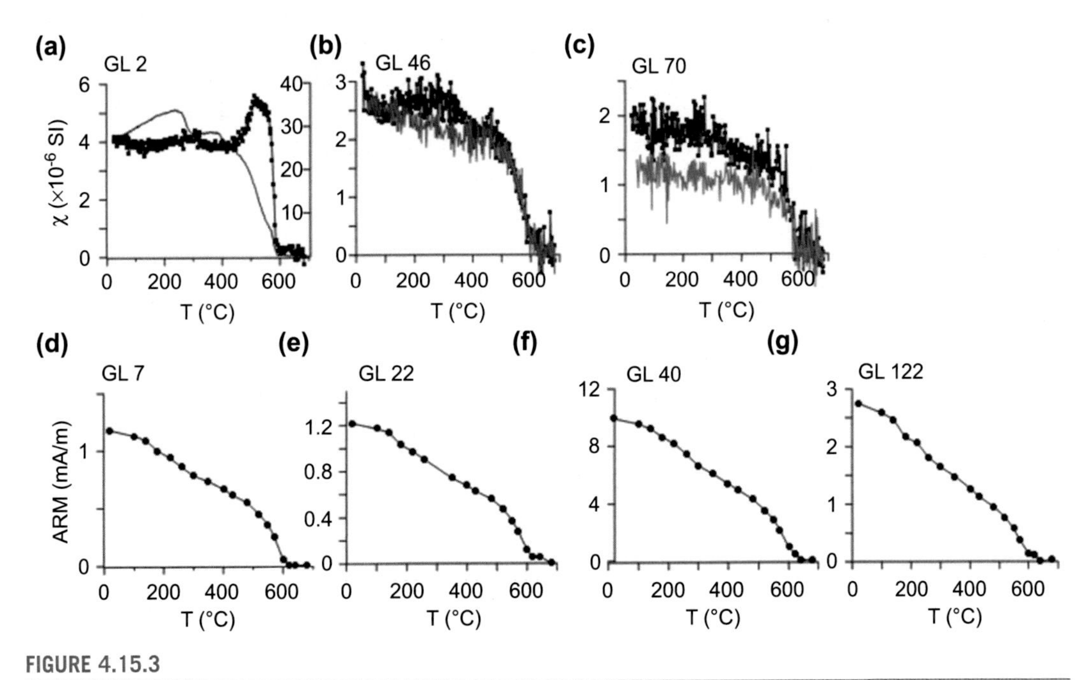

FIGURE 4.15.3

Thermomagnetic analysis of magnetic susceptibility for selected samples from the GL Gleysol (a–c). Stepwise thermal demagnetization of ARM for selected samples from the GL profile (d–g).

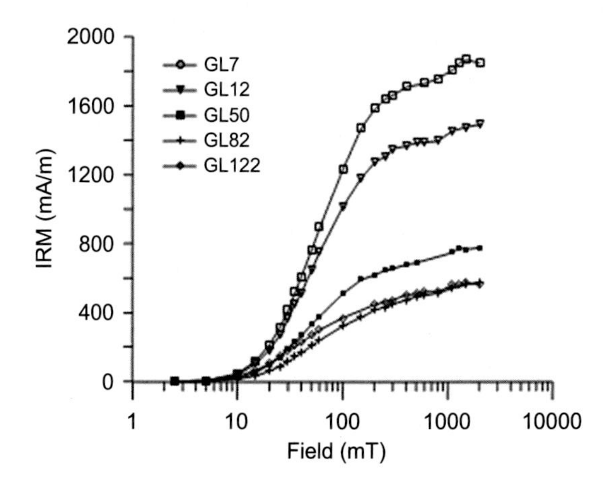

FIGURE 4.15.4

Stepwise IRM acquisition curves up to a field of 2 T for selected samples from the GL Gleysol.

summarized in Table 4.15.2. As is seen from Fig. 4.15.4, the intensity of IRM at the maximum applied field of 2 T regularly decreases with increasing depth, together with a change in the shape of the curve. For the uppermost three samples (depths of 7, 12, and 50 cm), an initial fast increase of IRM up to 300 mT suggests a significant contribution of the magnetically soft fraction. In higher fields, the IRM still grows, but more slowly (Fig. 4.15.4).

In contrast, samples from depths of 82 cm and 122 cm are magnetized more slowly without a clear separation of the magnetically soft contribution. The coercivity analysis (Kruiver et al., 2001) permits an elucidation of the observed behavior and allows us to link the obtained components to certain magnetic minerals. The first magnetically soft component (IRM1) (Table 4.15.2) makes the biggest contribution in the total IRM, but its share diminishes with increasing depth. Simultaneously, IRM1 coercivity (see the $B_{1/2}$ parameter) decreases significantly from ~(50–55) mT in the upper levels to

Table 4.15.2 Summary parameters from coercivity analysis using IRM-CLG1.0 software (Kruiver et al., 2001) for selected samples from the gleysol profile GL

	IRM component 1				IRM component 2			
Soil depth (cm)	Intensity (10^{-6} Am^2/kg)	% from IRM_{total}	$B_{1/2}$ (mT)	DP	Intensity (10^{-6} Am^2/kg)	% from IRM_{total}	$B_{1/2}$ (mT)	DP
Profile GL								
7	1402.64	92	53.70	0.39	123.76	8	616.6	0.40
12	1158.94	92	55.00	0.41	99.34	8	1000.0	0.55
50	691.43	84	51.30	0.41	134.16	16	1000.0	0.55
82	427.97	69	50.10	0.41	187.89	31	489.8	0.38
122	471.91	71	37.20	0.38	196.62	29	446.7	0.43

The intensity, relative contribution to the total IRM (% from IRM_{total}), median acquisition field ($B_{1/2}$), and dispersion parameter (DP) for each coercivity component are shown.

~37 mT in the deepest ones. Thus, it can be supposed that the soft component (IRM1) is carried by progressively coarser particles with increasing depth. This supposition is also in agreement with the decrease in the intensity of IRM1 (Table 4.15.2). The magnetically harder component (IRM2) shows the opposite behavior—its share in the total IRM increases in depth. This component, however, is probably carried by different mineral phases because its coercivity, expressed through the $B_{1/2}$ parameter, differes significantly in the two soils, building up the sampled profile. While in the recent gleysol, the $B_{1/2}$ of IRM2 varies between 600 and 1000 mT, in the buried soil (samples GL82, GL122) it is rather low (440–490 mT). While coercivities of the order 400–600 mT are typical of hematite (Dunlop and Özdemir, 1997; Özdemir and Dunlop, 2014), the $B_{1/2}$ of ~1000 mT may indicate the presence of goethite as well (Evans and Heller, 2003). On the other hand, the thermomagnetic analyses (Fig. 4.15.3) do not indicate goethite's presence, which could be due to its very weak magnetic susceptibility.

Therefore, magnetic diagnostic analyses suggest that the Gleysol studied contains mainly magnetite/maghemite of a decreasing amount in depth, in combination with hematite of varying coercivity but increasing concentration toward the bottom of the profile.

The magnetic susceptibility along the depth of the Gleysol is quite low, varying between 10 and 20×10^{-8} m^3/kg (Fig. 4.15.5a), which is a typical property of reduced soils where reductive dissolution of the iron oxides proceeds in anaerobic conditions (Maher, 1998; de Jong, 2002; Owliaie et al., 2006; Lu et al., 2012). The low concentration of the strongly magnetic fraction is also underlined by the ferrimagnetic susceptibility ($\chi - \chi_{hf}$), which is notably lower than the bulk χ. Thus, the influence of the nonmagnetic clay minerals and/or the weakly magnetic antiferromagnetic minerals in the profile is non-negligible. Despite the low χ values, the upper soil shows a higher magnetic susceptibility in the A_h horizon, which decreases in the AC_{kg}. In contrast, the buried soil beneath has a permanently lower χ (Fig. 4.15.5a).

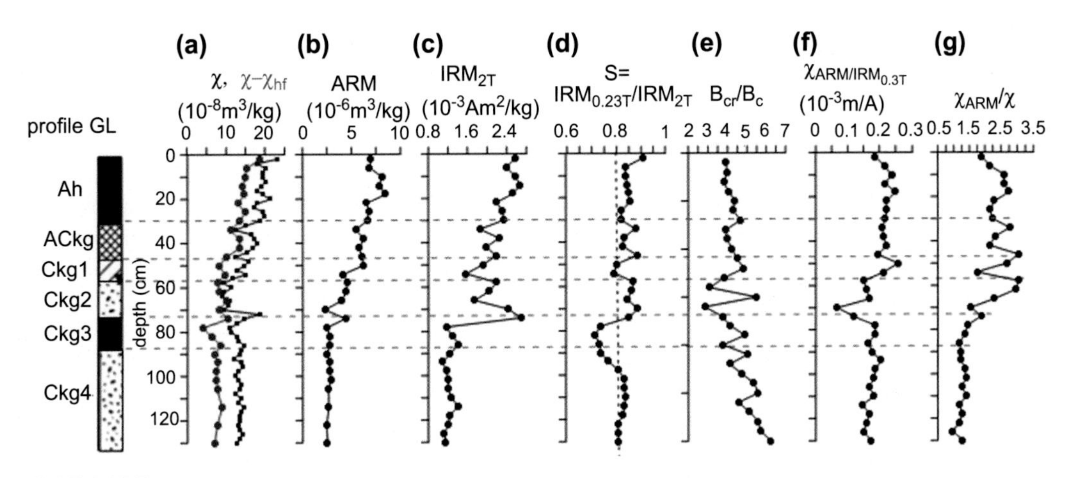

FIGURE 4.15.5

Depth variations of: (a) magnetic susceptibility (χ) (*full dots*) and ferrimagnetic susceptibility ($\chi - \chi_{hf}$) (*red dots*); (b) anysteretic remanence (ARM); (c) IRM acquired in 2 T field (IRM); (d) S ratio; (e) ratio between coercivity of remanence (B_{cr}) and coercive force (B_c); (f) ratio between the anhysteretic susceptibility (χ_{ARM}) and the IRM acquired in a field of 0.3 T ($IRM_{0.3T}$); (g) ratio between the anhysteretic susceptibility (χ_{ARM}) and magnetic susceptibility (χ).

The remanent magnetizations ARM and IRM_{2T} also underline the stronger magnetic enhancement of the upper solum with stable (SD, PSD) as well as coarser magnetic grains, compared with the deepest part of the GL profile. The changes in the proportion of magnetically soft/hard minerals is clearly reflected in the S ratio (Fig. 4.15.5d), confirming the conclusions from the coercivity analysis that the upper recent Gleysol contains a higher amount of the soft fraction (S is varying ~0.85), while the buried soil is enriched with high-coercivity minerals (S ~0.7 in the $C_{kg}3$). The Bcr/Bc ratio, which increases with depth, probably reflects an enhanced content of antiferromagnetic minerals rather than a coarsening of the magnetite's grains, since the M_{rs}/M_s ratio (not shown) does not show any systematic variations along the profile. The grain-size proxies for the magnetite particles—the $\chi_{ARM}/IRM_{0.23T}$ as well as χ_{ARM}/χ (Fig. 4.15.5f and g) —corroborate with the conclusions from the coercivity analysis of IRM acquisition curves (Table 4.15.2) that the upper part of the profile is enriched with finer grains, while downward in the buried soil the magnetite's particles are coarser.

The pedogenic processes in the Gleysols are strongly influenced by the large amounts of organic matter accumulated from the swamp vegetation. The important role of microorganisms in the biogeochemical cycling of iron is well known (Lovley, 1991). One of the most widely spread bacteria in soils and sediments is dissimilatory iron-reducing bacteria (DIRB), which links the oxidation of organic matter or hydrogen to a reduction of different Fe(III) iron oxides for sustaining their energy for growth and functioning (Lovley, 1991; Hansel et al., 2003). The microbial reduction of Fe at a neutral pH leads to Fe(III)—mineral dissolution and the formation of soluble Fe^{2+} or secondary mineral phases precipitate (Hansel et al., 2003, 2011; O'Loughlin et al., 2010). As was discussed already in this chapter, a high concentration of Fe^{2+} induces magnetite and goethite precipitation and, over time, the goethite's concentration diminishes while magnetite continues to form (Hansel et al., 2003). When the initial Fe^{2+} concentration in the soil solution is low, only goethite forms. As was shown earlier, gley horizons in the GL profile are enriched with stable SD-PSD magnetite particles. Since lithogenic iron oxide minerals are typically coarse grained (e.g., Cornell and Schwertmann, 2003), it can be supposed that the stable fine-grained magnetite fraction is a product of the reductive dissolution of ferrihydrite. Clearly, the concentration of this fraction is very low, as evidenced by the low magnetization and susceptibility along the GL profile (Fig. 4.15.5). A similar weak magnetic enhancement of gley soils is reported by other authors (Grimley and Arruda, 2007; Wang et al., 2008; Lu et al., 2012). An environmental magnetic study of Gley soils from Germany (Jordanova et al., 2012) also reveals depleted magnetic susceptibility, showing maximum values in the interval ($20-30 \times 10^{-8}$ m^3/kg) in the upper near-surface (up to ~30-cm depth) parts of the soil cores, while at the bottom, χ reaches near-zero values. A limiting factor for the secondary magnetite or goethite precipitation during the reductive dissolution of ferrihydrite is the Al substitution or absorption into it (Hansel et al., 2011). Increasing the Al substitutions leads to a linear decrease in the magnetite's concentration. One possible speculation about the shift in the magnetite's grain size toward coarser grains at the bottom of the GL profile is if we suppose that, in the presence of more organics in the recent Gleysol, the biotic process of ferrihydrite reductive dissolution produces smaller grains, while a lower amount of organics or its exhaustion with time leads to an inorganic pathway when coarser grains are produced (Hansel et al., 2003).

The depleted magnetic properties of Gleysols are widely reported in the relevant literature, and this feature is applied as a measurement for delineating hydric soils in the landscape through field measurements of volume magnetic susceptibility. Grimley and Vepraskas (2000) and Grimley et al. (2004) suggest a threshold susceptibility value for hydric soil delineation between 22 and 33×10^{-5} SI in silty loess or alluvial soils in Illinois, but as high as 61×10^{-5} SI at a site with fine sandy soil (Fig. 4.15.6).

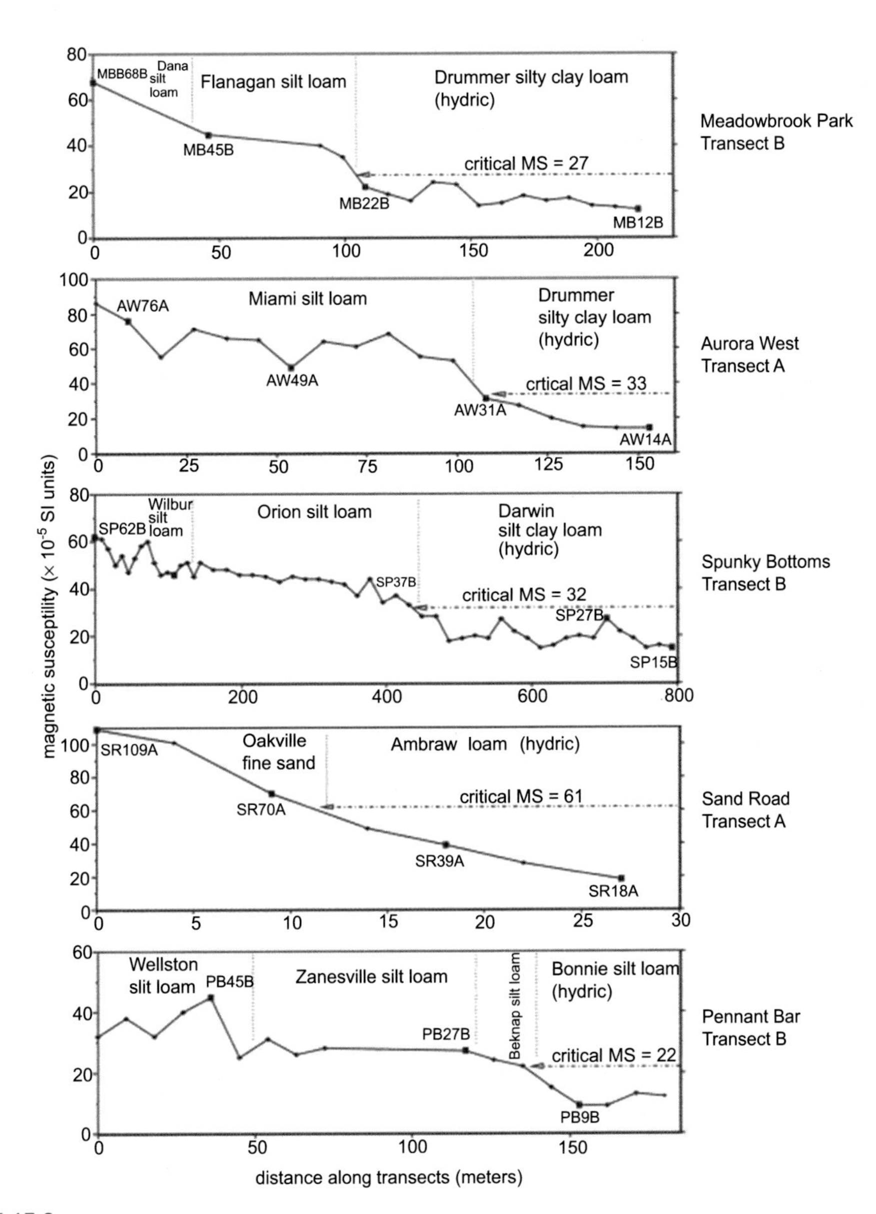

FIGURE 4.15.6

Volume magnetic susceptibility of surface soils of diverse hydric regimes along five transects in geographically different areas, representing different lithologies, climate and vegetation. "Critical" magnetic susceptibility values, suggested for delineation of hydric soil of the corresponding lithology, is marked for each transect. According to Grimley et al. (2004).

Reprinted with permission from Elsevier.

The lowest critical values suggested came from locations where the soil pH was <5. The authors underline that the critical values should be determined on a site-specific basis; that is, there is no unique value applicable universally because of the differences in the lithogenic iron oxide content, as well as the pedogenically formed fine grained magnetite/maghemite in different environments. Furthermore, Grimley et al. (2008) extend their studies by also considering, together with the soil moisture regime, the tree species associations in different landscapes where soils with gley properties occur. The authors conclude that soil susceptibility mapping can provide a simple, rapid, and quantitative means for precise guidance in reforestation. This conclusion is based on the established adherence of specific tree species to well-drained soils and others to poorly drained soils. In a similar study, Valaee et al. (2016) evaluate the efficiency of magnetic measurements for discriminating between four soil moisture regimes in northern Iran. Based on the statistical evaluation of a set of magnetic parameters obtained for 25 soil profiles, the authors conclude that a combination of magnetic measurements could successfully discriminate between four different moisture regimes (aridic, xeric, udic, and aquic with a mean annual precipitation of 700–900 mm, 300–700 mm, 200–450 mm, and <300 mm, respectively). The biggest contrast between magnetic susceptibility was found between soils developed under udic and aquic regimes.

REFERENCES

Algoe, C., Stoops, G., Vandenberghe, R., Van Ranst, E., 2012. Selective dissolution of Fe–Ti oxides — extractable iron as a criterion for andic properties revisited. Catena 92, 49–54.

Ananthapadmanabha, A., Shankar, R., Sandeep, K., 2014. Rock magnetic properties of lateritic soil profiles from southern India: evidence for pedogenic processes. J. Appl. Geophys. 111, 203–210.

Anderson, H., Berrow, M., Farmer, V., Hepburn, A., Russell, J., Walker, A., 1982. A reassessment of podzol formation process. J. Soil Sci. 33, 125–136.

Aran, D., Gury, M., Jeanroy, E., 2001. Organo-metallic complexes in an Andisol: a comparative study with a Cambisol and a Podzol. Geoderma 99, 65–79.

Baize, D., 1989. Planosols in the "Champagne Humide" region, France. A multi-approach study. Pedologie XXXIX (2), 119–151.

Barrón, V., Torrent, J., 2002. Evidence for a simple pathway to maghemite in Earth and Mars soils. Geochim. Cosmochim. Acta 66 (15), 2801–2806.

Bogdanov, B., Georgiev, D., Angelova, K., Yaneva, K., 2009. Natural zeolites: clinoptilolite. Review. In: Proceedings of International Science Conference "Economics and Society Development on the Base of Knowledge", 4th – 5th June 2009, Stara Zagora, Bulgaria, Volume IV 11, Natural & Mathematical Science, pp. 6–11.

van Breemen, N., 1988a. Long-term chemical, mineralogical and morphological effects of iron-redox processes in periodically flooded soils. In: Stucki, J., Goodman, B., Schwertmann, U. (Eds.), Iron in Soils and Clay Minerals. NATO ASI Series. D. Reidel Publishing company.

van Breemen, N., 1988b. Redox processes of iron and sulfur involved in the formation of acid sulfate soils. In: Stucki, J.W., Goodman, B.A., Schwertmann, U. (Eds.), Iron in Soils and Clay Minerals, NATO ASI Series C: Mathematical and Physical Sciences, vol. 217. D. Reidel Publishing Company, pp. 825–841.

Brinkman, R., 1970. Ferrolysis, a hydromorphic soil forming process. Geoderma 3, 199–206.

Buettner, S., Kramer, M., Chadwick, O., Thompson, A., 2014. Mobilization of colloidal carbon during iron reduction in basaltic soils. Geoderma 221–222, 139–145.

Buurman, P., Jongmans, A., 2005. Podzolization and soil organic matter dynamics. Geoderma 125, 71–83.

Buurman, P., 1984. In: Buurman, P. (Ed.), Podzols. Van Nostrand Reinhold Company Inc, ISBN 0-442-21129-5, p. 450.

Le Borgne, E., 1955. Susceptibilité magnétique anormale du sol superfi ciel. Ann. de Geophysique 11, 399–419.

van Breemen, N., Buurman, P., 2003. Soil Formation, second ed. Kluwer Academic Publishers, New York, Boston, Dordrecht, London, Moscow, ISBN 1-4020-0718-3.

van Breemen, N., Lundström, U.S., Jongmans, A.G., 2000. Do plants drive podzolization via rock-eating mycorrhizal fungi? Geoderma 94, 163–171.

Cervi, E., da Costa, A., de Souza Jr., I., 2014. Magnetic susceptibility and the spatial variability of heavy metals in soils developed on basalt. J. Appl. Geophys. 111, 377–383.

Chan, M., Ormö, J., Park, A., Stich, M., Souza-Egipsy, V., Komatsu, G., 2007. Models of iron oxide concretion formation: field, numerical, and laboratory comparisons. Geofluids 7, 1–13.

Chiang, H.C., Wang, M.K., Houng, K.H., White, N., Dixon, J., 1999. Minerology of B horizons in alpine forest soils of Taiwan. Soil Sci. 164, 111–122.

Colombo, C., Sellitto, V., Palumbo, G., Di Iorio, E., Terribile, F., Schulze, D., 2014. Clay formation and pedogenetic processes in tephra-derived soils and buried soils from Central-Southern Apennines (Italy). Geoderma 213, 346–356.

Cornell, R., Schwertmann, U., 2003. The Iron Oxides. Structure, Properties, Reactions, Occurrence and Uses (Weinheim, New York).

Cornu, S., Deschatrettes, V., Salvador-Blanes, S., Clozel, B., Hardy, M., Branchut, S., Le Forestier, L., 2005. Trace element accumulation in Mn-Fe-oxide nodules of a planosolic horizon. Geoderma 125, 11–24.

Coventry, R., Taylor, R., Fitzpatrick, R., 1983. Pedological significance of the gravels in some red and gray earths of central North Queensland. Aust. J. Soil Res. 21, 219–240.

Dubroeucq, D., Geissert, D., Quantin, P., 1998. Weathering and soil forming processes under semi-arid conditions in two Mexican volcanic ash soils. Geoderma 86, 99–122.

Duchaufour, P., 1982. Pedology: Pedogenesis and Classification. George Allen and Unwin, London, UK.

Dumon, M., Tolossa, A.R., Capon, B., Detavernier, C., Van Ranst, E., 2014. Quantitative clay mineralogy of a Vertic Planosol in southwestern Ethiopia: impact on soil formation hypotheses. Geoderma 214–215, 184–196.

Dunlop, D., Argyle, K., 1997. Thermoremanence, anhysteretic remanence and susceptibility of submicron magnetites: nonlinear field dependence and variation with grain size. J. Geophys. Res. 102 (B9), 20199–20210.

Dunlop, D., Özdemir, Ö., 1997. Rock magnetism. Fundamentals and frontiers. In: Edwards, D. (Ed.), Cambridge Studies in Magnetism. Cambridge University Press.

Van Dam, R., Harrison, J., Hirschfeld, D., Meglich, T., Li, Y., North, R., 2008. Mineralogy and magnetic properties of basaltic Substrate soils: Kaho'olawe and big island. Hawaii Soil Sci. Soc. Am. J. 72, 244–257.

Egli, M., Nater, M., Mirabella, A., Raimondi, S., Plötze, M., Alioth, L., 2008. Clay minerals, oxyhydroxide formation, element leaching and humus development in volcanic soils. Geoderma 143, 101–114.

Egli, R., 2004. Characterization of individual rock magnetic components by analysis of remanence curves. 3. Bacterial magnetite and natural processes in lakes. Phys. Chem. Earth 29, 869–884.

Evans, M., Heller, F., 2003. Environmental Magnetism: Principles and Applications of Enviromagnetics. Academic Press, San Diego, CA.

FAO World Soil Resources Reports, 2001. Lecture Notes on the Major Soils of the World. ISSN:0532–0488. http://www.fao.org/DOCREP/003/Y1899E/Y1899E00.HTM.

Favre, F., Tessier, D., Abdelmoula, M., Genin, J., Gates, W.P., Boivin, P., 2002. Iron reduction and changes in cation exchange capacity in intermittently waterlogged soil. Eur. J. Soil Sci. 53, 175–184.

Fine, P., Singer, M., La Ven, R., Verosub, K., Southard, R., 1989. Role of pedogenesis in distribution of magnetic susceptibility in two California chronosequences. Geoderma 44, 287–306.

Fredrickson, J., Zachara, J., Kennedy, D., Dong, H., Onstott, T., Hinman, N., Li, S.-M., 1998. Biogenic iron mineralization accompanying the dissimilatory reduction of hydrous ferric oxide by a groundwater bacterium. Geochim. Cosmochim. Acta 62 (19/20), 3239–3257.

Fritsch, E., Balan, E., Do Nascimento, N., Allard, T., Bardy, M., Bueno, G., Derenne, S., Melfi, A., Calas, G., 2011. Deciphering the weathering processes using environmental mineralogy and geochemistry: towards an integrated model of laterite and podzol genesis in the Upper Amazon Basin. C. R. Geosci. 343, 188–198.

Gasparatos, D., Tarenidis, D., Haidouti, C., Oikonomou, G., 2005. Microscopic structure of soil Fe-Mn nodules: environmental implication. Environ. Chem. Lett. 2, 175–178.

Gasparatos, D., 2013. Sequestration of heavy metals from soil with Fe-Mn concretions and nodules. Environ. Chem. Lett. 11 (1), 1–9.

Geisler, R., Ilvesniemi, H., Nyberg, L., van Hees, P., Starr, M., Bishop, K., Kareinen, T., Lundström, U., 2000. Distribution and mobilization of Al, Fe and Si in three podzolic soil profiles in relation to the humus layer. Geoderma 94, 249–263.

Geiss, C.E., Egli, R., Zanner, C.W., 2008. Direct estimates of pedogenic magnetite as a tool to reconstruct past climates from buried soils. J. Geophys. Res. 113, B11102. http://dx.doi.org/10.1029/2008JB005669.

Goodman, B., Berrow, M., 1976. The characterization by Mössbauer spectroscopy of the secondary iron in pans formed in Scottish podzolic soils. J. Phys. Colloq. 37 (C6). C6–C849–C6–855.

Grimley, D., Arruda, N., 2007. Observations of magnetite dissolution in poorly drained soils. Soil Sci. 172 (12), 968–982.

Grimley, D., Vepraskas, M., 2000. Magnetic susceptibility for use in delineating hydric soils. SSSAJ 64 (6), 2174–2180.

Grimley, D., Arruda, N., Bramstedt, M., 2004. Using magnetic susceptibility to facilitate more rapid, reproducible and precise delineation of hydric soils in the midwestern USA. Catena 58, 183–213.

Grimley, D., Wang, J.-S., Liebert, D., Dawson, J., 2008. Soil magnetic susceptibility: a quantitative proxy of soil drainage for use in ecological restoration. Restor. Ecol. 16 (4), 657–667.

Grison, H., Petrovský, E., Stejskalova, S., Kapička, A., 2015. Magnetic and geochemical characterization of Andosols developed on basalts in the Massif Central, France. Geochem. Geophys. Geosyst. 16, 1348–1363.

Grison, H., Petrovský, E., Kapička, A., Stejskalova, S., 2016. Magnetic and chemical parameters of andic soils and their relation to selected pedogenesis factors. Catena 139, 179–190.

Hanesch, M., Stanjek, H., Petersen, N., 2006. Thermomagnetic measurements of soil iron minerals: the role of organic carbon. Geophys. J. Int. 165, 53–61.

Hansel, C., Benner, S., Netss, J., Dohnalkova, A., Kukkadapu, R., Fendorf, S., 2003. Second mineralization pathways induced by dissimilatory iron reduction of ferrihydrite under advective flow. Geochim. Cosmochim. Acta 67 (16), 2977–2992.

Hansel, C., Benner, S., Fendorf, S., 2005. Competing Fe(II)-induced mineralization pathways of ferrihydrite. Environ. Sci. Technol. 39, 7147–7153.

Hansel, C., Learman, D., Lentini, C., Ekstrom, E., 2011. Effect of adsorbed and substituted Al on Fe(II)-induced mineralization pathways of ferrihydrite. Geochim. Cosmochim. Acta 75, 4653–4666.

Henry, B., Naydenov, K., Dimov, D., Jordanova, D., Jordanova, N., 2012. Relations between the emplacement and fabric-forming conditions of the Kapitan-Dimitrievo pluton and the Maritsa shear zone (Central Bulgaria): magnetic and visible fabrics analysis. Int. J. Earth Sci. 101 (3), 747–759.

IUSS Working Group WRB, 2014. World Reference Base for Soil Resources 2014. International Soil Classification System for Naming Soils and Creating Legends for Soil Maps. World Soil Resources Reports No. 106. FAO, Rome.

Jien, S.-H., Wu, S.-P., Chen, Z.-S., Chen, T.-H., Chiu, C.-Y., 2010a. Characteristics and pedogenesis of podzolic forest soils along a toposequence near a subalpine lake in northern Taiwan. Bot. Stud. 51, 223–236.

Jien, S.-H., Hseu, Z.-Y., Chen, Z.-S., 2010b. Hydropedological implications of ferromanganiferous nodules in rice-growing plinthitic Ultisols under different moisture regimens. Soil Sci. Soc. Am. J. 74, 880–891.

Jokova, M., Boyadjiev, T., 1993. Forms of Fe, Al and Mn compounds in some soils from Bulgaria. Forset Sci. 2, 62–71 (in Bulgarian).

Jokova, M., 1999. Extractable compounds of Fe, Al and Mn as participants and indicators of soil processes. Soil Sci. Agro-Chem. Ecol. XXXIV (4–5), 95–102.

Jones, J.B., Segnit, E.R., 1971. The nature of opal. 1. Nomenclature and constituent phases. J. Geol. Soc. Australia 18 (1), 57–68.

de Jong, E., 2002. Magnetic susceptibility of Gleysolic and Chernozemic soils in Saskatchewan. Can. J. Soil Sci 82, 191–199.
Jordanova, N., Jordanova, D., Petrov, P., 2011. Magnetic imprints of pedogenesis in Planosols and stagnic Alisol from Bulgaria. Geoderma 160, 477–489.
Jordanova, D., Jordanova, N., Werban, U., 2012. Environmental significance of magnetic properties of Gley soils near Rosslau (Germany). Environ. Earth Sci. 69 (5), 1719–1732.
King, J., Channell J., 1991. Sedimentary magnetism, environmental magnetism, and magnetostratigraphy. In: U.S. National Report to the International Union of Geodesy and Geophysics, Rev. Geophys. Suppl., vol. 29, 358–370.
Koinov, V., Kabakchiev, I., Boneva, K., 1998. Soil Atlas of Bulgaria. Zemizdat, Sofia, p. 320.
Kruiver, P., Dekkers, M., Heslop, D., 2001. Quanti¢cation of magnetic coercivity components by the analysis of acquisition curves of isothermal remanent magnetization. Earth Planet. Sci. Lett. 189, 269–276.
Laveuf, C., Cornu, S., Guilherme, R.L., Guerin, A., Juillot, F., 2012. The impact of redox conditions on the rare earth element signature of redoximorphic features in a soil sequence developed from limestone. Geoderma 170, 25–38.
Li, Y., Guo, J., Velde, B., Zhang, G., Hu, F., Zhao, M., 2011. Evolution and significance of soil magnetism of basalt-derived chronosequence soils in tropical southern China. Agric. Sci. 2, 536–543.
Lihareva, N., Dimova, L., Petrov, O., Tzvetanova, Y., 2009. Investigation of Zn sorption by natural clinoptilolite and mordenite. Bulgarian Chem. Commun. 41 (3), 266–271.
Liu, F., Colombo, C., Adamo, P., He, J., Violante, A., 2002. Trace elements in manganese-iron nodules from Chinese Alfisol. Soil Sci. Soc. Am. J. 66, 661–671.
Liu, Q., Deng, Ch, Torrent, J., Zhu, R., 2007. Review of recent developments in mineral magnetism of the Chinese loess. Quat. Sci. Rev. 26 (3–4), 368–385.
Löhr, S., Grigorescu, M., Cox, M., 2013. Iron nodules in ferric soils of the Fraser Coast, Australia: relicts of laterisation or features of contemporary weathering and pedogenesis? Soil Res. 51, 77–93.
Long, X., Ji, J., Balsam, W., 2011. Rainfall-dependent transformations of iron oxides in a tropical saprolite transect of Hainan Island, South China: spectral and magnetic measurements. J. Geophys. Res. 116, F03015. http://dx.doi.org/10.1029/2010JF001712.
Long, X., Ji, J., Balsam, W., Barrón, V., Torrent, J., 2015. Grain growth and transformation of pedogenic magnetic particles in red Ferralsols. Geophys. Res. Lett. 42, 5762–5770. http://dx.doi.org/10.1002/2015GL064678.
Lovley, D.R., 1991. Magnetite formation during microbial dissimilatory iron reduction. In: Frankel, R.B., Blakemore, R.P. (Eds.), Iron Biominerals. Plenum, New York, pp. 151–166.
Lowrie, W., 1990. Identification of ferromagnetic minerals in a rock by coercivity and unblocking temperature properties. Geophys. Res. Lett. 17 (2), 159–162.
Lu, S.-G., Zhu, L., Yu, J.-Y., 2012. Mineral magnetic properties of Chinese paddy soils and its pedogenic implications. Catena 93, 9–17.
Lundström, U., van Breemen, N., Bain, D., 2000. The podzolization process. A review. Geoderma 94, 91–107.
Maher, B., Yu, H.-M., Roberts, H., Wintle, A., 2003. Holocene loess accumulation and soil development at the western edge of the Chinese Loess Plateau: implications for magnetic proxies of palaeorainfall. Quat. Sci. Rev. 22, 445–451.
Maher, B., 1986. Characterization of soils by mineral magnetic measurements. Phys. Earth Planet. Inter 42, 76–92.
Maher, B., 1998. Magnetic properties of modern soils and quaternary loessic paleosols: paleoclimatic implications. Palaeogeogr. Palaeoclimat. Palaeoecol. 137, 25–54.
Mansfeldt, T., Schuth, S., Häusler, W., Wagner, F., Kaufhold, S., Overesch, M., 2012. Iron oxide mineralogy and stable iron isotope composition in a Gleysol with petrogleyic properties. J. Soils Sediments 12, 97–114.
Mileti, F., Langella, G., Prins, M., Vingiani, S., Terribile, F., 2013. The hidden nature of parent material in soils of Italian mountain ecosystems. Geoderma 207–208, 291–309.
Ming, D., Dixon, J., 1987. Quantitative determination of clinoptilolite in soils by a cation-exchange capacity method. Clays Clay Miner. 35 (6), 463–468.

Mitsuchi, M., 1976. Characteristics and genesis of nodules and concretions occurring in soils of the R. Chinit area, Kompong Thom Province, Cambodia. Soil Sci. Plant Nutr. 22 (4), 409–421.
Montagne, D., Cornu, S., Le Forestier, L., Hardy, M., Josiére, O., Caner, L., Cousin, I., 2008. Impact of drainage on soil-forming mechanisms in a French albeluvisol: input of mineralogical data in mass-balance modelling. Geoderma 145, 426–438.
Do Nascimento, N., Bueno, G., Fritsch, E., Herbillon, A., Allard, T., Melfi, A., Astolfo, R., Boucher, H., Li, Y., 2004. Podzolisation as a deferralitization process. A study of an Acrisol-Podzol sequence derived from Paleozoic sandstones in the northern upper Amazon Basin. Eur. J. Soil Sci. 55, 523–538.
Ninov, N., 2002. Soils. In: Kopralev, I., Yordanova, M., Mladenov, C. (Eds.), Geography of Bulgaria. Institute of Geography – Bulg. Acad. Sci. ForCom, Sofia, pp. 277–317.
O'Loughlin, E., Gorski, C., Scherer, M., Boyanov, M., Kemner, K., 2010. Effects of oxyanions, natural organic matter, and bacterial cell numbers on the bioreduction of lepidocrocite (gamma-FeOOH) and the formation of secondary mineralization products. Environ. Sci. Technol 44, 4570–4576.
Ouyang, T., Tang, Z., Zhao, X., Tian, C., Ma, J., Wei, G., Huang, N., Li, M., Bian, Y., 2015. Magnetic mineralogy of a weathered tropical basalt, Hainan Island, South China. Phys. Earth Planet. Inter. 240, 105–113.
Owliaie, H., Heck, R., Abtahi, A., 2006. The magnetic susceptibility of soils in Kohgilouye, Iran. Can. J. Soil Sci 86, 97–107.
Özdemir, Ö., Dunlop, D., 2014. Hysteresis and coercivity of hematite. J. Geophys. Res. Solid Earth 119, 2582–2594. http://dx.doi.org/10.1002/2013JB010739.
Pal, D., Wani, S., Sahrawat, K., 2013. Zeolitic soils of the Deccan basalt areas in India: their pedology and edaphology. Curr. Sci. 105 (3), 309–318.
Palumbo, B., Bellanca, A., Neri, R., Roe, M., 2001. Trace metal partitioning in Fe-Mn nodules from Sicilian soils, Italy. Chem. Geol. 173, 257–269.
Parfit, R., Saigusa, M., 1985. Allophane and humus-alluminium in Spodosols and Andepts formed from the same volcanic ash beds in New Zealand. Soil Sci. 139, 149–155.
Penkov, M., 1983. Soils in Bulgaria – Protection and Preservation. Nauka i Izkustvo, Sofia, p. 223 (in Bulgarian).
Plante, A., Fernández, J., Leifeld, J., 2009. Application of thermal analysis techniques in soil science. Geoderma 153, 1–10.
Ponomareva, V.V., 1964. Theory of podzolization. Isr. Progr. Sci. Transl., Jerusalem 1969, 309 p.
Popov, N., Popova, T., Rubio, J., Taffarel, S., 2012. Use of natural and modified zeolites from Bulgarian and Chilian deposits to improve adsorption of heavy metals from aqueous solutions. Geochemistry. Mineral. Petrol. 49, 83–93.
Quantin, P., 2004. Volcanic soils of France. Catena 56, 95–109.
Raynov, N., Popov, N., Yanev, Y., Petrova, P., Popova, T., Hristova, V., Atanasova, R., Zankarska, R., 1997. Geological, mineralogical and technological characteristics of zeolitized (clinoptilolitized) tuffs deposits in the Eastern Rhodopes, Bulgaria. In: Kirov, G., Filizova, L., Petrov, O. (Eds.), Natural Zeolites, Sofia '95. Pensoft, Sofla-Moscow, pp. 263–275.
Van Ranst, E., De Coninck, F., 2002. Evaluation of ferrolysis in soil formation. Eur. J. Soil Sci. 53, 513–519.
Van Ranst, E., Dumon, M., Tolossa, A., Cornelis, J.-T., Stoops, G., Vandenberghe, R., Deckers, J., 2011. Revisiting ferrolysis processes in the formation of Planosols for rationalizing the soils with stagnic properties in WRB. Geoderma 163, 265–274.
Safiuddin, L., Haris, V., Wirman, R., Bijaksana, S., 2011. A preliminary study of the magnetic properties on laterite soils as indicators of pedogenic processes. Latinmag Lett 1 (1), 1–15.
Salminen, R., Gregorauskiene, V., Tarvainen, T., 2008. The normative mineralogy of 10 soil profiles in Fenno-scandia and north-western Russia. Appl. Geochem. 23, 3651–3665.
Sandgren, P., Thompson, R., 1990. Mineral magnetic characteristics of podzolic soils developed on sand dunes in the Lake Gosciaz catchment, central Poland. Phys. Earth Planet. Inter. 60, 297–313.

Sauer, D., Sponagel, H., Sommer, M., Giani, L., Jahn, R., Stahr, K., 2007. Podzol: soil of the Year 2007. A review on its genesis, occurrence, and functions. J. Plant Nutr. Soil Sci. 2007 (170), 581–597.

Schaetzl, R., Anderson, A., 2009. Soils. Genesis and Geomorphology. Cambridge Univ. Press, UK, ISBN 978-0-521-81201-6.

Schnitzer, M., Turner, R., Hoffmann, I., 1964. A thermogravimetric study of organic matter of representative Canadian podzol soils. Can. J. Soil Sci. 44, 7–13.

Schwertmann, U., Friedl, J., Stanjek, H., Schulze, D., 2000. The effect of clay minerals on the formation of goethite and hematite from ferrihydrite after 16 years' ageing at 25oC and pH 4-7. Clay Miner. 35, 613–623.

Schwertmann, U., 1988. Occurrence and formation of iron oxides in various pedoenvironments. In: Stucki, J., Goodman, B., Schwertmann, U. (Eds.), Iron in Soils and Clay Minerals. NATO ASI Series: Series C: Mathematical and Physical Sciences, 217. Reidel Publ. Company, pp. 267–308.

Shishkov, T., Kolev, N., 2014. The soils of Bulgaria. In: Hartemink, A.E. (Ed.), World Soils Book Series. Series. Springer Science+Business Media Dordrecht, ISBN 978-94-007-7784-2 (eBook).

Shoji, S., Nanzyo, M., Dahlyen, R., Quantin, P., 1996. Evaluation and proposed revisions of criteria for andosols in the World Reference Base for soil resources. Soil Sci. 161 (9), 604–615.

Singh, B., Gilkes, R., 1996. Nature and properties of iron rich glaebules and mottles from some south-west Australian soils. Geoderma 71, 95–120.

Soubrand-Colin, M., Horen, H., Courtin-Nomade, A., 2009. Mineralogical and magnetic characterisation of iron titanium oxides in soils developed on two various basaltic rocks under temperate climate. Geoderma 149, 27–32.

Stiles, C., Mora, C., Driese, S., 2003. Pedogenic processes and domain boundaries in a Vertisol climosequence: evidence from titanium and zirconium distribution and morphology. Geoderma 116, 279–299.

Thompson, A., Chadwick, O., Rancourt, D., Chorover, J., 2006. Iron oxide crystallinity increases during soil redox oscillations. Geochim. Cosmochim. Acta 70, 1710–1727.

Thompson, A., Rancourt, D., Chadwick, O., Chorover, J., 2011. Iron solid-phase differentiation along a redox gradient in basaltic soils. Geochim. Cosmochim. Acta 75, 119–133.

Timofeeva, Y., Karabtsov, A., Semal¢, V., Burdukovskii, M., Bondarchuk, N., 2014. Iron–manganese nodules in Udepts: the dependence of the accumulation of trace elements on nodule size. Soil Sci. Soc. Am. J. 78, 767–778.

Torrent, J., Liu, Q.S., Barrón, V., 2010. Magnetic susceptibility changes in relation to pedogenesis in a Xeralf chronosequence in northwestern Spain. Eur. J. Soil Sci. 61, 161–173.

Tóth, G., Montanarella, L., Máté, S.F., Bódis, K., Jones, A., Panagos, P., Van Liedekerke, M., 2008. Soils of the European Union.

Tufano, K., Benner, S., Mayer, K., Marcus, M., Nico, P., Fendorf, S., 2009. Aggregate-scale heterogeneity in iron (Hydr)oxide reductive transformations. Vadose Zone J. 8 (4), 1004–1012.

Usman, M., Abdelmoula, M., Hanna, K., Gregoire, B., Faure, P., Ruby, C., 2012. Fe(II) induced mineralogical transformations of ferric oxyhydroxides into magnetite of variable stoichiometry and morphology. J. Solid State Chem. 194, 328–335.

Valaee, M., Ayoubi, S., Khormali, F., Lu, S.-G., Karimzadeh, H., 2016. Using magnetic susceptibility to discriminate between soil moisture regimens in selected loess and loess-like soils in northern Iran. J. Appl. Geophys. 127, 23–30.

Vingiani, S., Scarciglia, F., Mileti, F., Donato, P., Terribile, F., 2014. Occurrence and origin of soils with andic properties in Calabria (southern Italy). Geoderma 232–234, 500–516.

Vodyanitskiy, Yu, Bagin, V., Mymrin, V., 1983. Ferromagnetic minerals distribution in the podzolic soil profile. Pochvovedenie 3, 104–111 (in Russian).

Vodyanitskii, Y., 2010. Iron hydroxides in soils: a review of publications. Eurasian Soil Sci. 43 (11), 1244–1254.

Wang, C., McKeague, J., Kodama, H., 1986. Pedogenic imogolite and soil environments: case study of Spodosols in Quebec, Canada. Soil Sci. Soc. Am. J. 50, 711–718.

Wang, J.-S., Grimley, D., Xu, C., Dawson, O., 2008. Soil magnetic susceptibility reflects soil moisture regimes and the adaptability of tree species to these regimes. Forest Ecology and Management 255 (5–6), 1664–1673.

Wells, M., Fitzpatrick, R., Gilkes, R., Dobson, J., 1999. Magnetic properties of metalsubstituted haematite. Geophys. J. Int. 138, 571–580.

World Reference Base for Soil Resources, 2006. World Soil Resources Report No. 103. FAO, Rome. Available at. http://www.fao.org/ag/ag1.

Worm, H.-U., 1998. On the superparamagnetic-stable single domain transition for magnetite, and frequency dependence of susceptibility. Geophys. J. 133 (1), 201–206.

Yanev, Y., Cocheme, J., Ivanova, R., Grauby, O., Burlet, E., Pravchanska, R., 2006. Zeolites and zeolitization of acid pyroclastic rocks from paroxysmal Paleogene volcanism, Eastern Rhodopes, Bulgaria. N. Jb. Miner. Abh. 182/3, 265–283.

Yu, X.-L., Fu, Y.-N., Brookes, P., Lu, S.-G., 2015. Insights into the formation process and environmental fingerprints of iron–manganese nodules in subtropical soils of China. Soil Sci. Soc. Am. J. 79, 1101–1114.

Zachara, J., Kukkadapu, R., Fredrickson, J., Gorby, Y., Smith, S., 2002. Biomineralization of poorly crystalline Fe(III) oxides by dissimilatory metal reducing bacteria (DMRB). Geomicrobiol. J. 19 (2), 179–207.

Zachara, J., Kukkadapu, R., Peretyazhko, T., Bowden, M., Wang, C., Kennedy, D., Moore, D., Arey, B., 2011. The mineralogic transformation of ferrihydrite induced by heterogeneous reaction with bioreduced anthraquinone disulfonate (AQDS) and the role of phosphate. Geochim. Cosmochim. Acta 75, 6330–6349.

Zhang, M., Karathanasis, A., 1997. Characterization of iron-manganese concretions in Kentucky Alfisols with perched water tables. Clays Clay Miner. 45, 428–439.

Zhang, G.-Y., He, J.-Z., Liu, F., Zhang, L.-M., 2014. Iron-manganese nodules harbor lower bacterial diversity and greater proportions of Proteobacteria compared with bulk soils in four locations spanning from north to south China. Geomicrobiol. J. 31, 562–577.

CHAPTER

MAGNETISM OF SOILS WITH LIMITATIONS TO ROOT GROWTH: VERTISOLS, SOLONETZ, SOLONCHAKS, AND LEPTOSOLS

5

The specific soil factors that can limit plant and root growth are dispersion, compaction, water-logging, salinity, soil pH, and physical barrier to root growth. Soils included in this WRB Reference Soil Group are those affected by the factors above.

5.1 VERTISOLS: MAIN CHARACTERISTICS, FORMATION PROCESSES, AND DISTRIBUTION

The most characteristic peculiarity of the Vertisols is their deep black color and seasonal severe cracking on drying and swelling on being water-logged (IUSS Working Group WRB, 2014). Vertisols develop when a combination of several processes/factors take place—enough rain that leads to the weathering of primary rock-forming minerals but without their outwash outside the profile; weathering leads to the crystallization of new clay minerals; impeded drainage, hindering the leaching of weathering products; and high temperatures, which speed up the weathering process (Schaetzl and Anderson, 2009). The combination of these factors leads to the formation of smectite clays in the presence of basic cations (Ca^{2+}, Mg^{2+}). Vertisols are characterized by high clay content, represented mainly by the group of smectite phyllosilicate minerals. The major clay mineral is commonly montmorillonite [$Na_{0.33}(Al_{1.67}Mg_{0.33})Si_4O_{10}(OH)_2$] (Schaetzl and Anderson, 2009). Vertisols develop on different parent materials including igneous rocks, metamorphic-sedimentary rocks, and alluvial and colluvial materials from the weathering of rocks enriched in basic cations (Driese et al., 2003; Schaetzl and Anderson, 2009; Pal et al., 2012). Due to their ability to adsorb a large amount of water in the interlayers of the clay minerals, smectites can significantly change their volume. Vertisols commonly occur in low and flat relief forms, like lowlands, valleys, forefront of mountain ranges, former lake bottoms, lower river terraces, etc., where often conditions favoring the accumulation and stagnation of water exist. During dry periods, wide cracks appear on the surface and penetrate deep in the solum. They become filled with soil material due to landslides, aeolian input, heavy rains, etc. Deeper in the soil profile, where this process is not expressed, slickensides appear as a result of the shear stresses produced (Schaetzl and Anderson, 2009). Slickensides usually appear at between 25- and 125-cm depth and are best expressed in soils developing under seasonal interchange between extremely dry and extremely wet periods. As a result of the interaction between the processes of pedoturbation and shrinking/swelling, a specific microrelief of lows and highs (*gilgai*) appear on the surface.

Soil Magnetism. http://dx.doi.org/10.1016/B978-0-12-809239-2.00005-X

Iron (Fe) and manganese (Mn) are among the most redox-sensitive elements in soil (Cornell and Schwertmann, 2003), and their behavior in Vertisols is regarded as a major factor in establishing different models for the genesis and functions of Vertisols (Nordt and Driese, 2009). The occurrence of Vertisols in lowlands with impeded drainage determines the establishment of reducing conditions in the soil during wet seasons. The major conditions required for a soil to become reduced are that the soil should be water saturated; it must contain organic tissues that can be decomposed; the microbial population must be present in order to use organics as a respirable medium; and the water should be stagnant or moving very slowly (Vepraskas and Faulkner, 2001). A respiring microbial population is fundamental for the formation of reduced soils. Recent studies suggest that bacteria are widespread and abundant in many soils and adapted to function in different climates (Bazylinski, 1996; Konhauser, 1998; Fredrickson et al., 1998; Zachara et al., 2002). Seasonal changes in the oxidation–reduction conditions lead to the formation of Fe-Mn concretions and nodules, gley spots, etc., accompanied by reductive processes in iron oxyhydroxides (van Breemen, 1988). Periodic alternation of oxidative and reducing conditions in soils invokes increased Fe mobility as a result of the transfer of electrons from decomposing organic matter to Fe^{3+}. When reduced soils with a high Fe^{2+} concentration become oxidized, mixed-valency Fe oxides [mostly magnetite (Fe_3O_4) and hydrated oxides] often form (Brennan and Lindsay, 1998). Another process related to the cyclic reduction/oxidation of iron in an open system is so-called ferrolysis (Brinkman, 1970; van Breemen, 1988). The ferrolysis process is put forward to explain the formation of more sandy, strongly acid surface soil horizons, which lay above more clayey and not so acid horizons in soils. Such are soils developed in lowlands with seasonal surface water-logging. It has been proved that, under certain conditions, ferrolysis also can take place in Vertisols (Barbiero et al., 2010), where amorphous silica and $CaCO_3$ concretions form as a result. However, most recent studies demonstrate that ferrolysis as a process in soils subjected to seasonal water-logging is overestimated (Van Ranst et al., 2011). Instead, processes of clay translocation (Montagne et al., 2008) or geogenic processes (Barbiero et al., 2010) are proposed to explain the contrasting soil characteristics observed.

Most of the Vertisols are considered as young soils formed during the Quaternary, because their parent materials are often Quaternary alluvial deposits (Singh et al., 1998; Pal et al., 2006). Recent studies using different dating techniques (i.e., the isotopic dating of ^{14}C from pedogenic and non-pedogenic carbonates), show that a Vertisol can form within 500 years and most of the Vertisols from tropical, subtropical, and more arid areas are of Holocene age (Kovda et al., 2006; Pal et al., 2012). Ancient, palaeo Vertisols, formed during the Pliocene, however, are also frequent (Nordt et al., 2004; Achyuthan et al., 2010).

Vertisols develop in various climates—they are often found in tropical and subtropical regions with high precipitation, as well as in subarid areas (Pal et al., 2012; IUSS Working Group WRB, 2014). Vertisols developed in the arid and semiarid areas are characterized by the presence of a carbonate horizon containing primary and/or secondary carbonates in the form of diffuse masses, concretions, and crystals (Kovda et al., 2006; Nordt and Driese, 2009; Pal et al., 2012). It has been demonstrated (Nordt and Driese, 2010a) that when the mean annual precipitation (MAP) is <900 mm, the main part of CaO is in the form of $CaCO_3$, while at MAP >900 mm, the calcium is in the form of exchangeable cations (Ca^{2+}) and the Vertisols are decalcified. The precipitation of pedogenic carbonates leads to an increase in both the pH and the relative content of Na^{2+} in the soil and in the soil solution. These sodium cations, in turn, cause a dispersion of smectite particles, which can move along the depth. This is the reason why, despite the presence of carbonates, clay particles in Vertisols are illuviated (Pal et al., 2012).

Vertisols and soils having vertic properties are identified and described in many countries, but the largest share of Vertisols worldwide is concentrated in India (25%), Australia (22%), the United States (6%), Africa (5%), and China (4%) (Soil Survey Staff, 2003). The worldwide distribution of Vertisols is shown in Fig. 5.1.1 (FAO, 2001). In Europe, Vertisols are characteristic for Mediterranean and Balkan countries (Italy, Cyprus, Bulgaria, Hungary, Romania, Spain) (Tóth et al., 2008).

In Bulgaria, Vertisols are widespread in the lowlands and the valley plains in central and south Bulgaria (Koinov et al., 1998; Shishkov and Kolev, 2014). Similar to other regions, the local name for Vertisols is indicative of the soil color—"Smolnitza" ("tar black"). Bulgaria is among the European countries where Vertisols are found as a dominant soil type (Tóth et al., 2008). Vertisol distribution in Bulgaria is related to the former Pliocene plateaus and old Quaternary terraces (Koinov et al., 1998). During the late Miocene, Pliocene lakes dried out and marshland conditions became dominant (Koinov et al., 1998; Shishkov and Kolev, 2014). This favored the development of gley soils, which further transformed into Vertisols (Ninov, 2002). Later on, Vertisols were not buried to form palaeosol, but remained on the surface during the Quaternary. The age of the Vertisols found in the different areas in Bulgaria is uncertain (Shishkov and Kolev, 2014). It is usually accepted that the age of the soil is determined by the age of the rock-forming parent material. Several varieties of Vertisols are recognized in Bulgaria—carbonaceous, typical, leached, and degraded (Shishkov and Kolev, 2014). The most widespread is the leached variety.

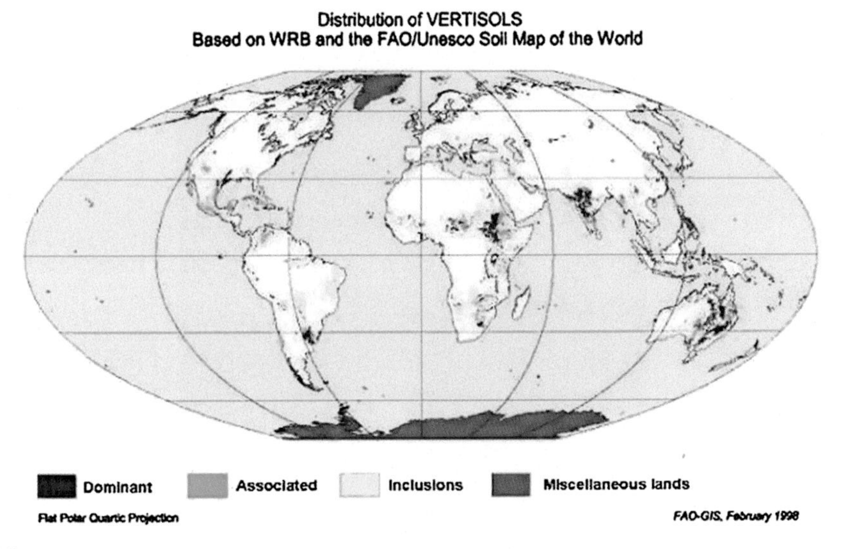

FIGURE 5.1.1

Spatial distribution of Vertisols worldwide (FAO, 2001). Lecture notes on the major soils of the world). http://www.fao.org/docrep/003/y1899e/y1899e00.htm#toc.

5.2 DESCRIPTION OF THE PROFILES STUDIED

Four profiles of Vertisols from Bulgaria were studied: a profile of typical Vertisol (Epicalcic Mollic Vertisol), two profiles of leached Vertisol (Mollic Vertisol), and degraded (pseudo-podzolic) Vertisol (Stagnic Endogleyic Vertisol).

5.2.1 EPICALCIC MOLLIC VERTISOL (TYPICAL SMOLNITZA), PROFILE DESIGNATION: VR

Location: N 42°29′00.9″; E 27°20′37.0″, h = 70 m a.s.l.

Present-day climate conditions: MAT = 12.5°C; MAP = 543 mm, vegetation cover: broadleaf forest (hornbeam); relief: flat

Profile description: general view and details from the horizons are shown in Figs. 5.2.1–5.2.4.

0–25 cm: humic (A_{k1}) horizon, tarry black, heavy sandy-clayey, carbonates, vegetation roots

25–50 cm: humic (A_{k2}) horizon, sandy-clayey, slickensides present

50–88 cm: humic-illuvial horizon (AB_k), brown-black, clayey, dense, slickensides present, carbonate concretions, and orange-yellow spots

88–130 cm: C_{kg}, horizon, yellow-brown, carbonate concretions present

Parent rock: Pliocene carbonate-rich clays

FIGURE 5.2.1

View of the profile of Epicalcic Mollic Vertisol—VR profile.

FIGURE 5.2.2

Epicalcic Mollic Vertisol VR. A_{k1}-A_{k2} horizons—detail.

FIGURE 5.2.3

Epicalcic Mollic Vertisol VR. AB_k horizon—detail.

FIGURE 5.2.4

Epicalcic Mollic Vertisol VR. C_k horizon—detail.

5.2.2 MOLLIC VERTISOL (LEACHED SMOLNITZA), PROFILE DESIGNATION: JAS

Location: N 42°16′58.8″; E 25°39′18″, h = 160 m a.s.l.

Present-day climate conditions: MAT = 12.9°C; MAP = 576 mm, vegetation cover: grassland/pasture; relief: flat

Profile description: general view is shown in Fig. 5.2.5:

0–6 cm: humic (A1) horizon, dark gray, sandy-clayey, no carbonates

6–40 cm: humic (A2) horizon, black-graygray, heavy clayey, dense, slickensides present

40–100 cm: humic (A3) horizon, tarry black, no carbonates, heavy clayey, slickensides present

100–140 cm: transitional AB horizon, dark brown, dense, carbonate and Fe-Mn concretions

140–155 cm: illuvial (B_k) horizon, graygray-brown, clayey, carbonate concretions, Fe-Mn concretions present

155–183 cm: transitional (BC_k) horizon, light brown, carbonates and carbonate concretions present

183–240 cm: C_k horizon, light brown-beige, sandy

Parent rock: Pliocene clays

FIGURE 5.2.5

View of the Mollic Vertisol—profile JAS.

5.2.3 STAGNIC ENDOGLEYIC VERTISOL (PSEUDO-PODZOLIZED SMOLNITZA), PROFILE DESIGNATION: SM

Location: N 41°56′06.3″; E 25°01′14.4″, h = 358 m a.s.l.

Present-day climate conditions: MAT = 12.6°C; MAP = 768 mm, vegetation cover: grassland/pasture; relief: flat

Profile description: general view and details from the horizons are shown in Figs. 5.2.6–5.2.10.

0–20 cm: humic-eluvial (AE) horizon, light beige-brown, bleached, sandy

20–40 cm: A_g horizon, dark brown-black, sandy-clayey, Fe-Mn concretions, and slickensides present

40–83 cm: illuvial (B_{tg}) horizon, no carbonates, clayey, Fe-Mn concretions, and slickensides present

83–116 cm: BC_{kg} horizon, light beige, carbonate coatings and films, abundant Fe-Mn concretions, carbonate concretions, small yellow-orange redox spots

116–140 cm: C_{kg} horizon, beige, carbonate coatings, abundant yellow-orange redox spots; Fe-Mn concretions present

Parent material: alluvial-delluvial deposits

FIGURE 5.2.6

View of the profile of Stagnic Endogleyic Vertisol—profile SM.

FIGURE 5.2.7

Stagnic Endogleyic Vertisol SM—A horizon—detail.

FIGURE 5.2.8

Stagnic Endogleyic Vertisol SM—$B_{tg}1$ horizon—detail.

FIGURE 5.2.9

Stagnic Endogleyic Vertisol SM—Fe-Mn concretions in the BC_{kg} horizon—detail.

FIGURE 5.2.10

Stagnic Endogleyic Vertisol SM—C_{kg} horizon—detail.

5.2.4 MOLLIC VERTISOL (LEACHED SMOLNITZA), PROFILE DESIGNATION: SF

Location: N 42°40′34.0″; E 23°21′40.0″, h = 565 m a.s.l.

Present-day climate conditions: MAT = 10.6°C; MAP = 582 mm, vegetation cover: grassland, relief: flat

Profile description:

0—72 cm: humic (A1) horizon, dark gray, clayey, no carbonates

72—170 cm: A2 horizon, black, heavy clayey, dense, slickensides present

170—330 cm: BC_k horizon, gray-brown, clayey, carbonate diffuse masses and carbonate concretions, Fe-Mn concretions present

330—386 cm: C_k horizon, yellowish Pliocene clays

5.3 TEXTURE, SOIL REACTION, AND MAJOR GEOCHEMISTRY OF VERTISOLS

A detailed magnetic and geochemical study of the three profiles (VR, JAS, and SM) is reported in Jordanova and Jordanova (2016) and several magnetic parameters of SF profile are given in Jordanova and Jordanova (1999).

Data obtained for the soil reaction (pH) and texture (clay/silt ratio) of the three profiles are shown in Fig. 5.3.1. The profiles of Mollic Vertisols (VR and JAS) have an alkaline reaction, although the uppermost A horizon shows relatively lower pH, compared with the bottom part of the profiles

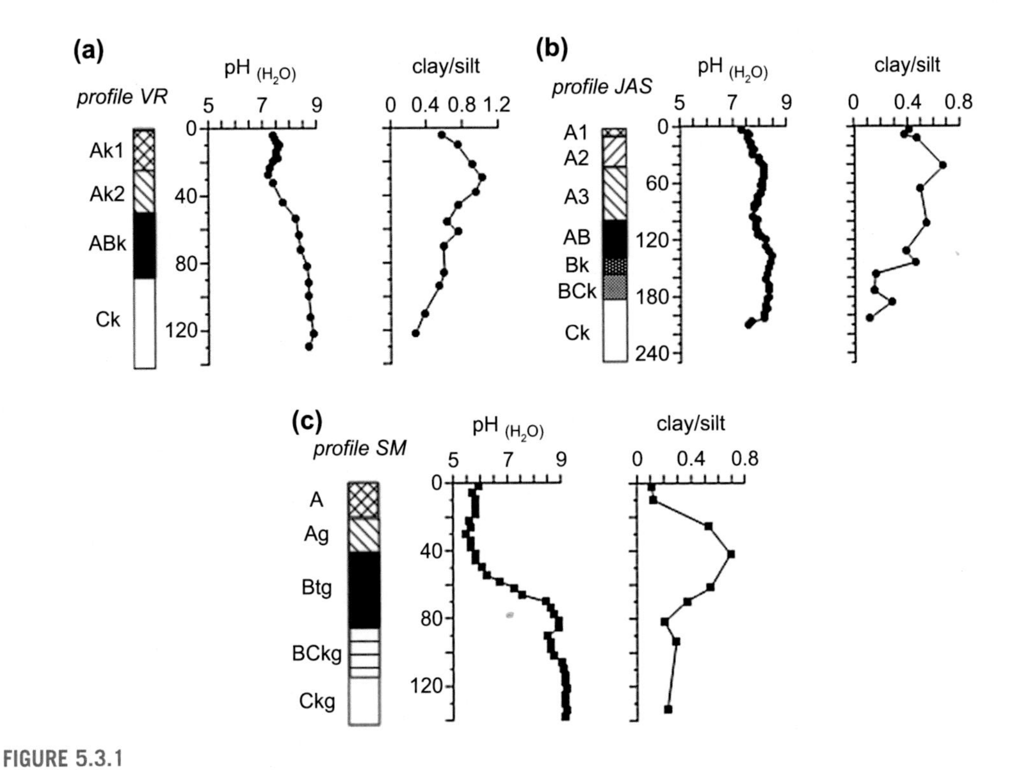

FIGURE 5.3.1

Depth variations in soil pH and the ratio clay/silt for (a) Epicalcic Mollic Vertisol (VR), (b) Mollic Vertisol (JAS), and (c) Stagnic Endogleyic Vertisol (SM).

(Fig. 5.3.1a,b). In contrast, the Stagnic Vertisol (pseudo-podzolized soil) shows an abrupt change in pH at the middle of the B_{tg} horizon (Fig. 5.3.1c). The mechanical composition is dominated by the silt and clay fractions, while sand comprises usually 2–4%. The clay/silt ratio (Fig. 5.3.1) points to an enhancement of the middle part of the soil profiles with clay. The strongest textural contrast is observed in the profile of Stagnic Vertisol (SM) (Fig. 5.3.1c). Most of the studies of Vertisols worldwide report the dominant role of the clay fraction in the soil texture (Soil Survey Staff, 2003; Pal et al., 2012), while Vertisols from Bulgaria and other neighboring Balkan countries (like Serbia) are characterized by a lower clay content (Koinov et al., 1998; Jelić et al., 2011). This is reflected in the clay/silt ratio (Fig. 5.3.1), which is the highest in profile VR and lower in the other two profiles. The clay fraction varies between 30% and 50% (Jordanova and Jordanova, 2016).

Variations along the depth of some of the major chemical elements in the Vertisols studied are presented in Fig. 5.3.2. The content of Silica (Si) is relatively constant in depth (Fig. 5.3.2) and in general, the lowest content is found in the VR profile (~18–20%), while the content in the JAS profile and the SM profile is higher (~25%). The calcium content is enriched at the bottom parts of the profiles (Fig. 5.3.2), achieving the highest values in the gleyed profile (SM). The redox-sensitive elements Fe and Mn, as well as Al, are enriched in the upper horizons where carbonates are missing. An exception is the SM profile, which has a well-developed eluvial horizon (AE). There, the

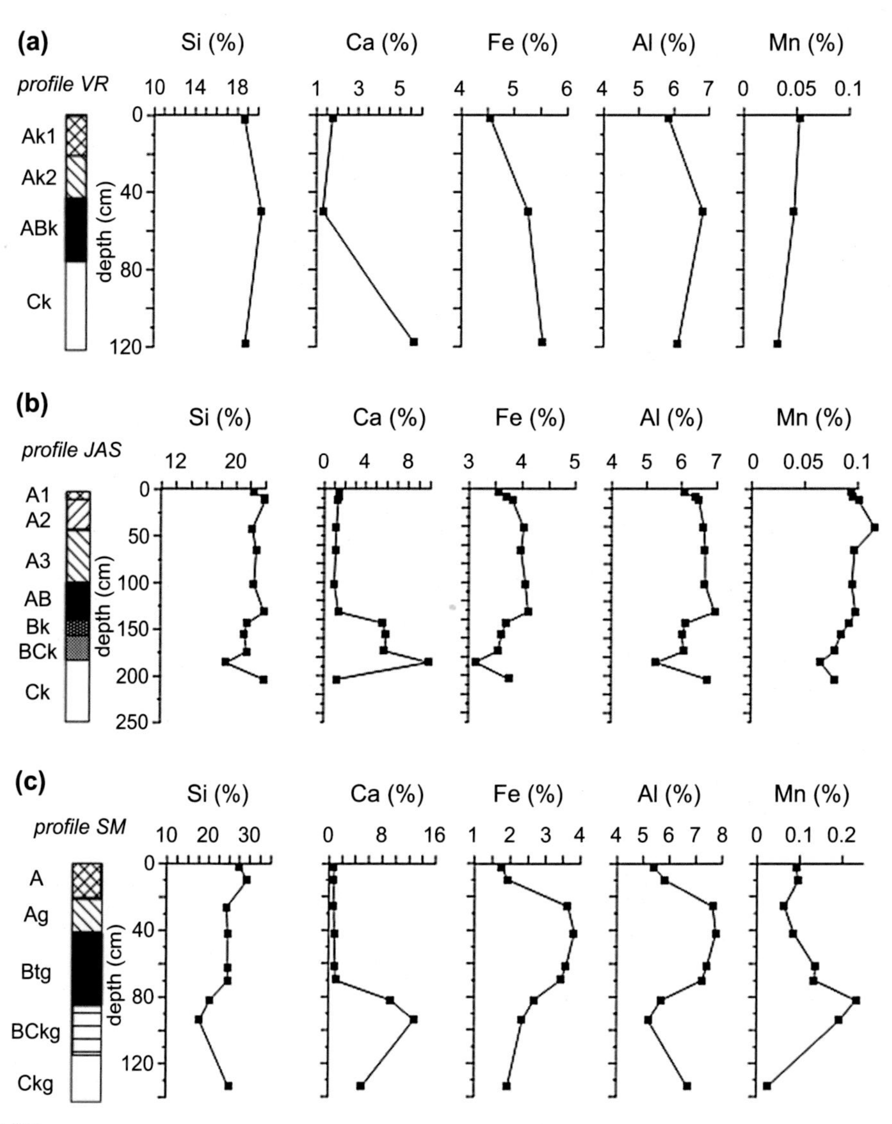

FIGURE 5.3.2

Content of some major and trace elements (Si, Ca, Fe, Al, Mn) for selected samples along the depth of profiles of (a) VR, (b) JAS, and (c) SM Vertisols.

Fe and Al contents are depleted (Fig. 5.3.2c). The manganese (Mn) content in the VR profile is the lowest, while it increases in the JAS profile and further in the SM profile. Based on the iron extraction data (oxalate and dithionate extractions), the ratio Feo/Fed is used to infer the content of poorly crystalline iron oxyhydroxides in the "free iron" pool in the soil (Fig. 5.3.3.)

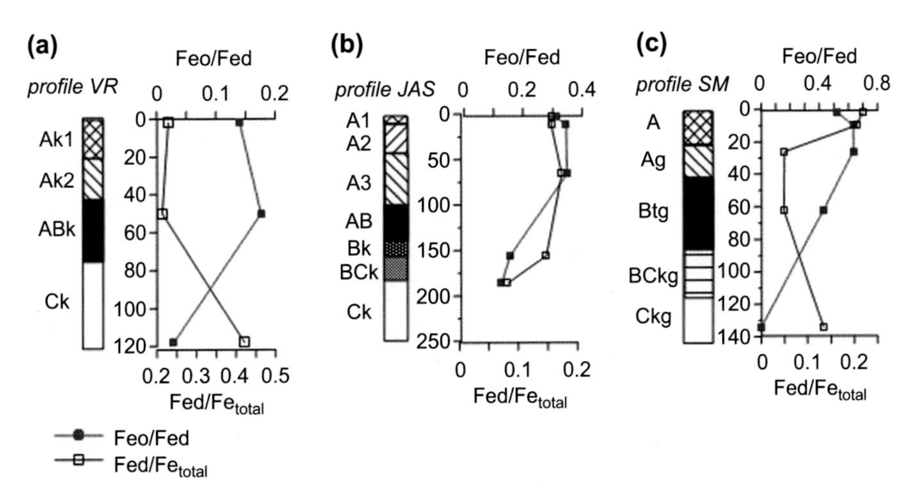

FIGURE 5.3.3

Chemical extraction data for Vertisols—variations in ratios Fe_o/Fe_d (in red) and Fe_d/Fe_{total} (in black) for (a) VR profile, (b) JAS profile, and (c) SM profile.

The three Vertisols studied are characterized by relatively high Feo/Fed values in the A and B horizons of the profiles, also typically observed for other Vertisols (Fisher et al., 2008; Moustakis, 2012). Moreover, a comparison between the profiles reveals a systematic increase in Feo/Fed values obtained for samples from the A and B horizons in the following order: Epicalcic Vertisol (VR profile) through Mollic Vertisol (JAS) toward the Endogleyc Vertisol (SM). In the latter, the Feo/Fed ratio reaches values of 0.6 in the A horizon (Fig. 5.3.3c). The Fe_d/Fe_{total} ratio (dithionite-extractable Fe to the total Fe) is widely used as an indication of the degree of soil maturity (Cornell and Schwertmann, 2003) and is also shown in Fig. 5.3.3. This ratio is higher in the upper humic horizons of the Vertisols, reaching a value of ~0.2. It is notable, however, that in the SM profile such a higher value is retained only in the uppermost AE horizon, while downward it drops sharply. This is most likely an indication that, in the A_g horizon and below, iron is mostly involved in nonoxide forms (e.g., in clay minerals), which is confirmed by the other magnetic parameter variations—the high-field magnetic susceptibility (which will be discussed later), as well as the sharp enhancement obtained in the clay/silt ratio in the lower parts of the SM profile (Fig. 5.3.1c). The increase in Fe_d/Fe_{total} in the C_k horizon of the VR profile (Fig. 5.3.3a) most likely relates to the parent rock mineralogy.

5.4 MAGNETIC PROPERTIES OF VERTISOLS

The magnetic mineralogy of the Vertisol profiles VR, JAS, and SM is studied through thermomagnetic analysis of magnetic susceptibility as well as stepwise thermal demagnetization of laboratory-induced isothermal remanent magnetization (IRM) and anhysteretic remanent magnetization (ARM). Selected examples of these analyses are presented in Fig. 5.4.1. The heating curves for all the samples from the uppermost levels of the three soils show a distinct behavior characterized by a small peak at ~300°C, followed by a decrease and a further stronger increase with a maximum

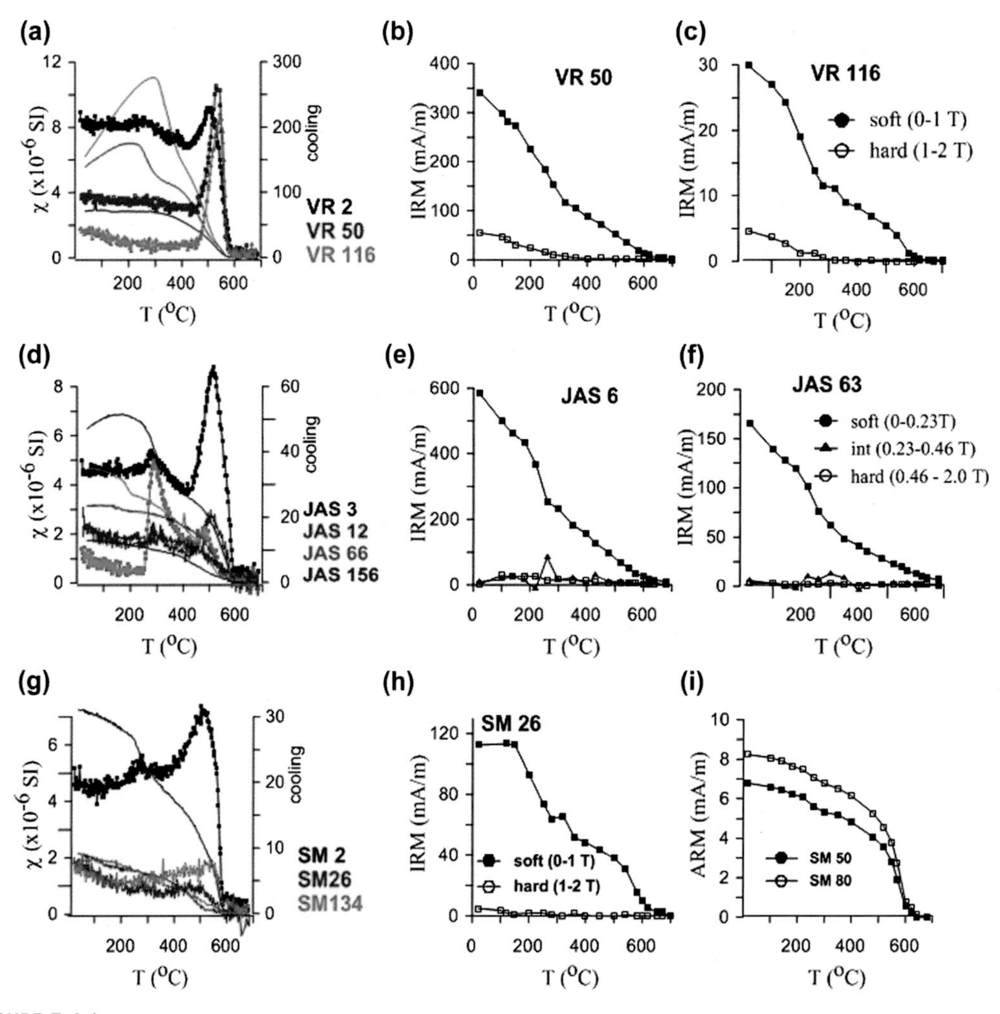

FIGURE 5.4.1

Magnetic mineral identification data for Vertisols: (a) thermomagnetic analysis of magnetic susceptibility for selected samples from VR profile; (b) and (c) stepwise thermal demagnetization of composite IRM for two samples from the VR profile; (d) thermomagnetic analysis of magnetic susceptibility for selected samples from JAS profile; (e) and (f) stepwise thermal demagnetization of composite IRM for two samples from the JAS profile; (g) thermomagnetic analysis of magnetic susceptibility for selected samples from SM profile; (h) stepwise thermal demagnetization of composite IRM for a sample from the A_g horizon of the SM profile; (i) stepwise thermal demagnetization of ARM for two samples from the $B_{tg}1$ horizon of the SM profile.

peak at 500°C. The final drop in χ at 580°C is indicative of magnetite (Dunlop and Özdemir, 1997). As was discussed in the previous chapters, as well as illustrated in other works (Liu et al., 2005, 2010; Jordanova et al., 2010; Jelenska et al., 2010), this behavior could be ascribed to the presence of fine-grained pedogenic maghemite, formed under well-aerated condition in the surface levels.

Thermomagnetic analyses of the soil samples from the deeper horizons of the Vertisols are characterized by strongly expressed mineral transformations, occurring during heating (Fig. 5.4.1a,d,g), which are envisaged as sharp peaks in the susceptibility signal. This is an indication of the presence of (amorphous) thermally unstable Fe-containing minerals, which are often present in seasonally water-logged soils (Schwertmann, 1988; van Breemen, 1988; Cornell and Schwertmann, 2003). However, as demonstrated by Hanesch et al. (2006), different iron oxyhydroxides show different magnetic behavior during heating, which could be also used for a mineral's identification. Phase transformations occurring in the range (200–300°C) could be due to the presence of two Fe oxyhydroxides—ferrihydrite and lepidocrocite (Gehring and Hofmeister, 1994; Mitov et al., 2002; Hanesch et al., 2006), while transformations beginning at ~400°C are more indicative of goethite in combination with organic matter (Hanesch et al., 2006). Therefore, the observed transformation behavior in the VR profile (Fig. 5.4.1a) indicates goethite and organic matter; in the JAS profile (Fig. 5.4.1d) it likely suggests the presence of ferrihydrite and organic matter, considering the model laboratory heating of Hanesch et al. (2006), while in the SM profile (Fig. 5.4.1g), mostly magnetite is expressed.

The identification of coarser, remanence-carrying magnetic minerals at different soil depths is carried out using stepwise thermal demagnetization of composite IRM and ARM. Several results for the three Vertisols are shown in Fig. 5.4.1b,c,e,f,h,i. In all of the samples, the magnetically soft component saturated in a (0–1) Tesla field (samples from VR and SM profiles) or in a lower (0–0.23) Tesla field (JAS profile) dominates the isothermal remanence. Two unblocking temperatures are visible for all the samples, at 300°C and 580°C (Fig. 5.4.1), which could be ascribed to maghemite and magnetite, respectively. The hard component is significant only in the samples from the VR profile (Fig. 5.4.1b,c), having unblocking temperatures in the range (200–250°C), likely related to goethite's dehydration (Dekkers, 1988). Stepwise thermal demagnetization of anhysteretic remanence (ARM) for two samples from the illuvial horizon of the SM profile (Fig. 5.4.1i) shows a specific and nearly rectangular demagnetization curve with a final unblocking at 580°C, which is characteristic for the presence of a stable fraction of single-domain (SD) magnetite (Hartstra, 1983; Dunlop and Özdemir, 1997). A small fraction of thermally stable maghemite (oxidized magnetite) is unblocked at higher temperatures of ~640°C.

The components′ deconvolution of IRM acquisition curves (Kruiver et al., 2001) provides highly useful information regarding the magnetic carriers in soil, approximating the coercivity distribution with cumulative Gaussian functions. Selected samples from the three Vertisol profiles were stepwise magnetized up to a field of 5 T (Fig. 5.4.2).

All the samples contain predominantly magnetically soft minerals, since only for several soil depths is there a clear high-coercivity component visible on the IRM-acquisition curves. These are the samples mostly from the VR profile. A summary of the characteristics (median coercivity $B_{1/2}$ and intensity) of the discriminated coercivity components (Kruiver et al., 2001) is shown in Fig. 5.4.3. The highest is the intensity of the soft component in the uppermost levels in all the three profiles. Corroborating this information with the data from the thermomagnetic analyses (Fig. 5.4.1a,d,f), it could be supposed that the soft magnetic component is carried by pedogenic maghemite, which is formed in the surface levels under locally existing well-aerated conditions. The coercivity of the soft component (expressed through the $B_{1/2}$ parameter) in these samples is similar, varying between 45 and 50 mT (Fig. 5.4.3a, black bars). Down-profiles, the soft component is also the dominant one. A well-observed systematic increase in the $B_{1/2}$ parameter from ~35 to 40 mT in the VR profile, through ~42–50 mT in the JAS profile, up to ~60–66 mT in the SM profile needs further examination and will be discussed in the following sections. The second, high-coercivity component is identified only in

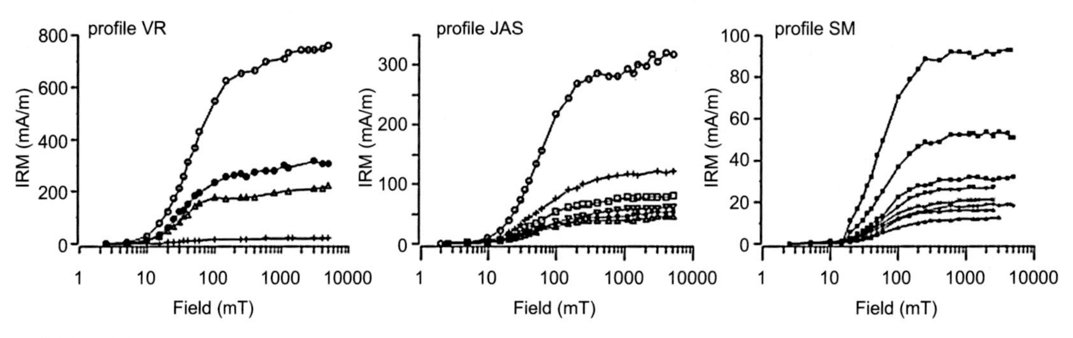

FIGURE 5.4.2

Stepwise acquisition of isothermal remanence (IRM) up to field of 5 T for selected samples from the Vertisol profiles.

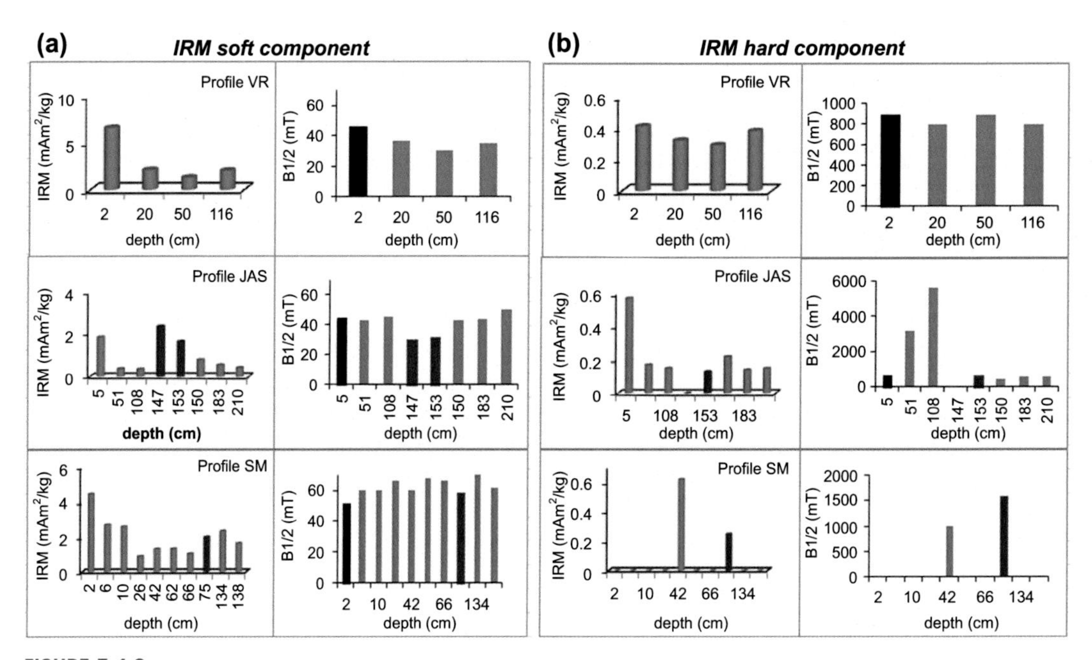

FIGURE 5.4.3

Intensity (IRM) and coercivity (B½) of the **soft** IRM component (a), extracted through coercivity analysis of IRM acquisition curves using IRM-CLG1.0 software (Kruiver et al., 2001) for selected samples from the VR Vertisol, JAS Vertisol and SM Vertisol. Intensity (IRM) and coercivity (B½) of the **hard** IRM component (b) for the three Vertisols. Data bars in red relate to IRM coercivity analysis of soil concretions from the respective depths; Coercivity parameter $B_{½}$ for the uppermost (0–2 cm) interval of the three profiles is shown in black in order to underline its uniform value among the three Vertisols.

samples from the VR and JAS profiles, while in the SM profile such a component has been separated only in one depth level and in a soil concretion (Fig. 5.4.3b). The high-coercivity component in the VR profile has $B_{1/2}$ between (790–890) mT, while in JAS it is softer at the bottom of the profile (Fig. 5.4.3b). The coercivities of the hard IRM component isolated in the VR and JAS profiles are ascribed to hematite (Jordanova and Jordanova, 2016), which saturates in different fields, depending on the grain size and shape of the particles, foreign substitutions in the crystal lattice, and formation pathway (Dunlop and Özdemir, 1997; de Boer and Dekkers, 1998; Jiang et al., 2012; Hu et al., 2013). The hard components with very high coercivity (Fig. 5.4.3b) could be ascribed to goethite.

The magnetic signature in Vertisols is further studied through tracking the depth variations in different concentration- and grain-size–dependent parameters and ratios. A detailed discussion can be found in Jordanova and Jordanova (2016, 1999). A summary of the most relevant parameters is shown in Fig. 5.4.4 for the VR profile, Fig. 5.4.5 for the JAS profile, in Fig. 5.4.6 for the SF profile, and in Fig. 5.4.7 for the SM profile. As expected for soils influenced by seasonal water-logging and resulting reducing conditions in the soil, the magnetic characteristics demonstrate the presence of a low concentration of strongly magnetic iron oxide minerals. A typical magnetic susceptibility value in the upper humic horizons is $\chi \sim 20 \times 10^{-8}$ m^3/kg (Figs. 5.4.5–5.4.7), except for the VR profile, where a higher χ of the order of $\sim 80 \times 10^{-8}$ m^3/kg is observed in the A_{k1} horizon (Fig. 5.4.4). The importance of clay minerals containing structural Fe is confirmed by the relatively high values of the χ_{hf} obtained in the middle parts of the profiles, compared with the bulk χ. This feature is especially pronounced in the most texturally differentiated profile SM (Fig. 5.4.7), where the important difference between the variations of χ and isothermal remanence (IRM) underlines a significant share of the paramagnetic fraction in the magnetic susceptibility as well.

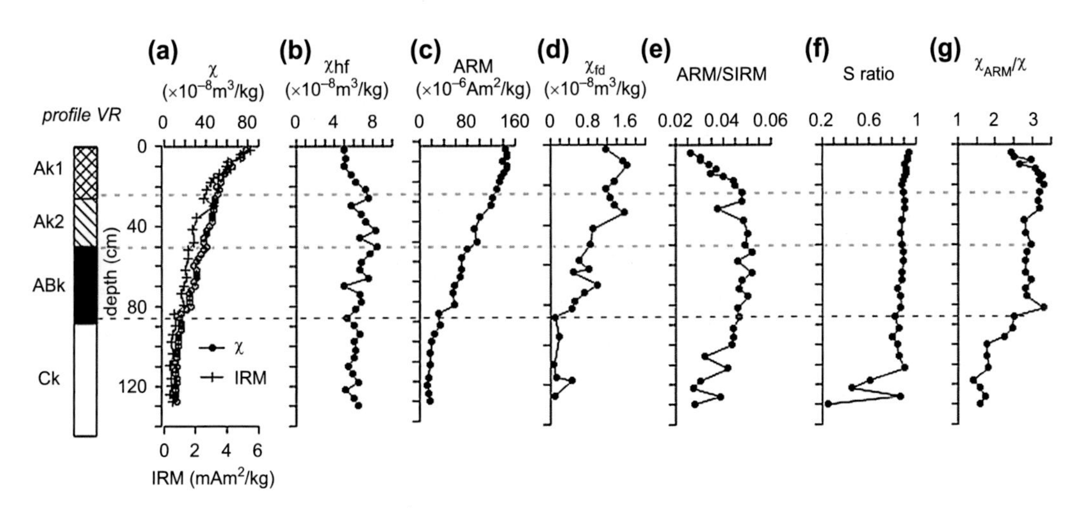

FIGURE 5.4.4

Magnetic characteristics of Epicalcic Mollic Vertisol (VR): depth variations of (a) magnetic susceptibility (χ) (full dots) and IRM acquired in 2 T field (IRM) (crosses); (b) high-field magnetic susceptibility (χ_{hf}); (c) anhysteretic remanence (ARM); (d) frequency-dependent magnetic susceptibility (χ_{fd}); (e) ratio between the ARM and saturation remanent magnetization, acquired in a 2 T field (SIRM); (f) S ratio; (g) ratio between the anhysteretic susceptibility (χ_{ARM}) and magnetic susceptibility (χ).

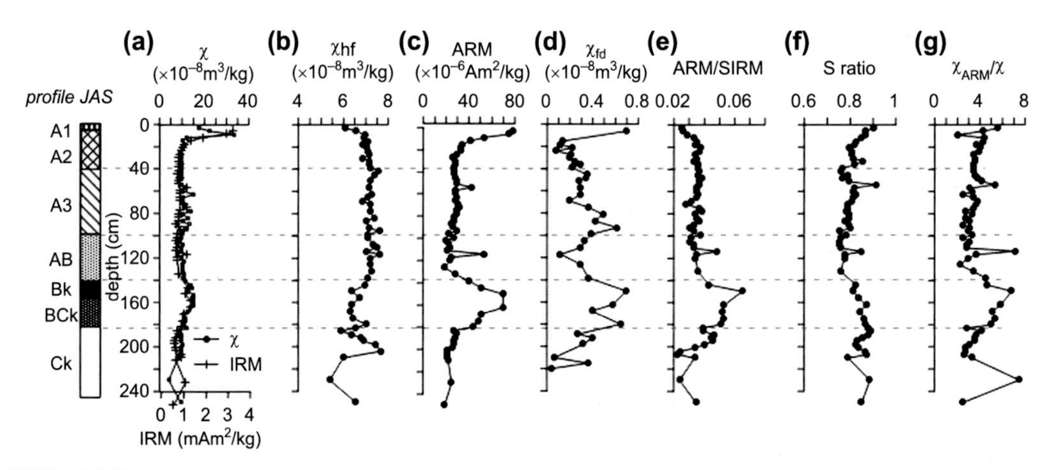

FIGURE 5.4.5

Magnetic characteristics of Mollic Vertisol (JAS): depth variations of (a) magnetic susceptibility (χ) (full dots) and IRM acquired in 2 T field (IRM) (crosses); (b) high-field magnetic susceptibility (χ_{hf}); (c) anhysteretic remanence (ARM); (d) frequency-dependent magnetic susceptibility (χ_{fd}); (e) ratio between the ARM and saturation remanent magnetization, acquired in a 2 T field (SIRM); (f) S ratio; (g) ratio between the anhysteretic susceptibility (χ_{ARM}) and magnetic susceptibility (χ).

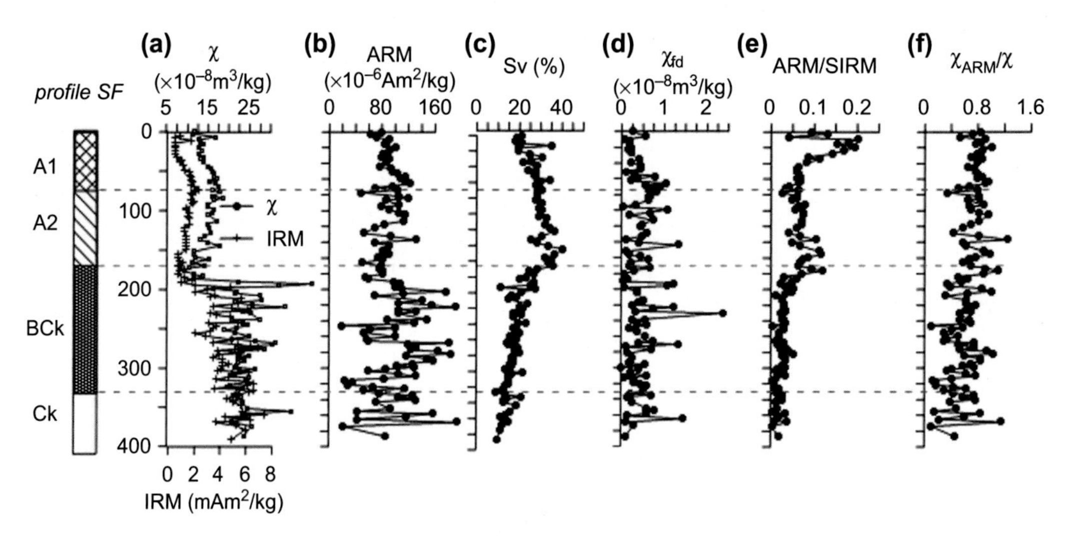

FIGURE 5.4.6

Magnetic characteristics of Mollic Vertisol (SF): depth variations of (a) magnetic susceptibility (χ) (full dots) and IRM acquired in 2 T field (IRM) (crosses); (b) anhysteretic remanence (ARM); (c) viscosity coefficient (S_v %); (d) frequency-dependent magnetic susceptibility (χ_{fd}); (e) ratio between the ARM and saturation remanent magnetization, acquired in a 2 T field (SIRM); (f) ratio between the anhysteretic susceptibility (χ_{ARM}) and magnetic susceptibility (χ).

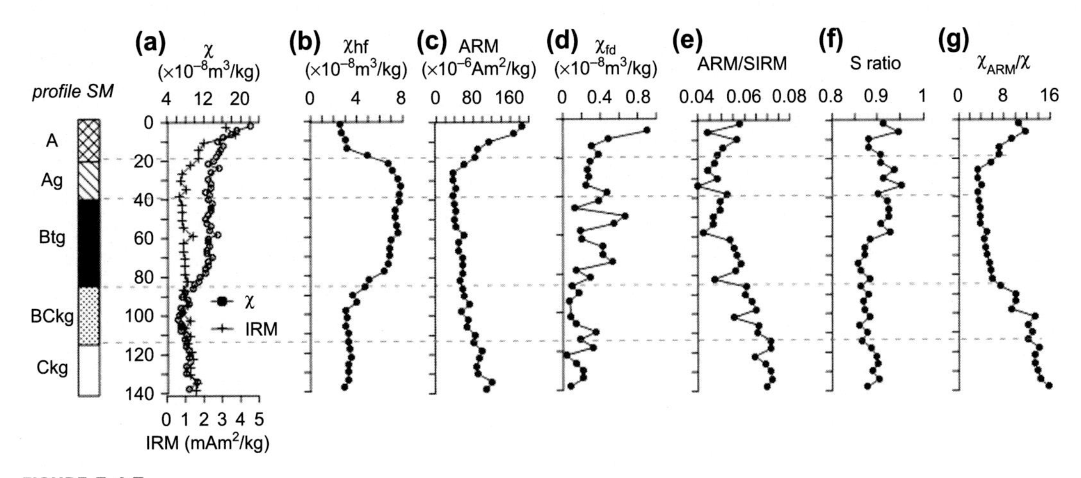

FIGURE 5.4.7

Magnetic characteristics of Stagnic Endogleyic Vertisol (SM): depth variations of (a) magnetic susceptibility (χ) (full dots) and IRM acquired in 2 T field (IRM) (crosses); (b) high-field magnetic susceptibility (χ_{hf}); (c) anhysteretic remanence (ARM); (d) frequency-dependent magnetic susceptibility (χ_{fd}); (e) ratio between the ARM and saturation remanent magnetization, acquired in a 2 T field (SIRM); (f) S ratio; (g) ratio between the anhysteretic susceptibility (χ_{ARM}) and magnetic susceptibility (χ).

Magnetic grain-size variations are studied through the behavior of several parameters—χ_{ARM} as a proxy for the concentration of SD grains, χ_{fd} as a proxy for the amount of ultrafine superparamagnetic particles, and ARM/SIRM and χ_{ARM}/χ for the relative change in the amount of SD-like (submicrometer)—sized grains. For the SF profile, the "viscosity coefficient" (S_v) reflecting the presence of near-SP grains that carry very unstable remanence ("viscous remanence") (Dunlop, 1981; Dunlop and Özdemir, 1997) is obtained in addition. Despite the fact that the four profiles are developed on different parent material with a specific mineralogy, in all of them the magnetic proxy parameters related to the variations in the SD fraction of magnetic grains are the best expressed. In the VR profile, a gradual decrease in χ_{ARM} from its maxima at the surface toward the bottom of the AB_k horizon is accompanied by a clearly expressed trend in ARM/SIRM and $\chi_{ARM/\chi}$, suggesting the finest SD grains concentrated in A_{k2} and AB_k soil horizons and coarser (pseudo-SD [PSD]) grains – in the uppermost A_{k1}. Therefore, the concentration of coarser stable grains in the uppermost A_{k1} horizon is higher, while downward the magnetic grain size decreases simultaneously with a decrease in concentration of the SD-like fraction (Fig. 5.4.4). A certain small amount of superparamagnetic fraction is, however, also present in the solum, as seen from the χ_{fd} parameter. Its concentration diminishes in depth toward the bottom of AB_k. In the Mollic Vertisol (JAS profile), the concentration of the SD-like fraction deduced by the χ_{ARM} shows maxima of similar intensity in the uppermost A1 and deeper B_k plus BC_k horizons (Fig. 5.4.5). However, the grain-size proxy ARM/SIRM suggests the finest SD grains are enhanced in the middle part of the profile (B_k plus BC_k), while in A1 the ratio has a minimum. Therefore, in A_{k1}, a higher amount of coarser SD-PSD grains is present, while in the illuvial horizon, a high amount of fine SD particles is detected. A very weak χ_{fd} maximum is also bound to the illuvial horizon, as well as to the bottom of the A3 horizon (Fig. 5.4.5). In the other Mollic

Vertisol—the SF profile (Fig. 5.4.6)—the χ_{ARM} variations along the depth are not as clear as in the other profiles, most likely because of the obvious considerable change in magnetic mineralogy between the A1 plus A2 horizons and BC_k plus C_k, as seen in the magnetic susceptibility and IRM changes. Nevertheless, looking at the grain-size proxy ARM/SIRM (Fig. 5.4.6e), it could be concluded that, in A1, the finest SD grains are concentrated, while in the deeper part (A2), these stable magnetic grains are of a coarser size. Such a conclusion is also supported by the observed relative minimum in χ in the A1 (due to the fact that the "true" SD grains have minimum magnetic susceptibility; e.g., Dunlop and Özdemir, 1997). Simultaneously, the fraction of SP/SD grains (the "viscous" fraction expressed by the Sv coefficient) increases progressively in depth, reaching a maximum at the bottom of A2 (Fig. 5.4.6c). Thus, an apparent decrease in the grain size of the stable SD-like fraction is seen. However, if we consider the nature of the viscous magnetization, it could also be supposed that, because of the increase in the concentration of magnetically "viscous" grains, the ARM intensity decreases and, thus, the ARM/SIRM decreases as well. But, in fact, there is an effective decrease in the bulk grain size from the surface toward the A2 horizon, since the amount of the SP/SD grains increases in depth.

The profile of the Endogleyc Vertisol (the SM profile) showing the greatest textural differentiation is characterized by a specific systematic behavior in the variations in ARM, χ_{ARM}/χ, and ARM/SIRM (Fig. 5.4.7c,e,g). Considering these parameters together, it could be supposed that the A1 horizon is enriched in coarse SD grains, while going deeper into the solum, the grain size decreases toward a "true" SD threshold, judging from the continuous ARM/SIRM increase. A very weak frequency-dependent magnetic susceptibility (χ_{fd}) is observed at the surface and in the gleyed horizon (B_{tg}).

In summary, taking into account the depth variations of the magnetic parameters in the Vertisols considered above, the near-surface horizons in all of them are enriched with coarser magnetically stable magnetite of PSD grain size (in all profiles, except the SF profile, where SD characteristics are obtained). Deeper Vertisols′ horizons contain a decreasing amount of smaller near-SD grains of magnetite, as well as traces of SP grains.

5.5 PEDOGENESIS OF IRON OXIDES IN VERTISOLS, AS REFLECTED IN SOIL MAGNETISM

Vertisols develop under repetitive cycles of drying and pronounced water-logging during the wet seasons. This inevitably encompasses an iron oxide reduction and thus influences the soil's magnetic properties. Usually, studies on the magnetism of soils influenced by permanent or seasonal water-logging report significant magnetic depletion in the soil profile (Dearing et al., 1995; Maher, 1998; de Jong, 2002; Grimley and Arruda, 2007; Lu et al., 2012; Chen et al., 2015), due to the destruction of iron oxide minerals under the influence of an oxygen deficit. Bulgarian Vertisols are also characterized by very low magnetic susceptibility (Figs. 5.4.4–5.4.7) and the presence of a significant amount of amorphous iron oxides, as revealed by the iron oxalate-extraction data (Fig. 5.3.3.). On the other hand, these soils are considered as relict in Bulgarian soil science (Shishkov and Kolev, 2014), and an important question arising in their pedological characterization is how to decipher their initial pedogenic signature from the changes imposed during the Quaternary. We try to address this problem by comparing the magnetic signature obtained in Bulgarian Vertisols with those of typical contemporary Vertisols. Unfortunately, there are very few magnetic studies of Vertisols and they are limited to a tropical Vertisol from Mali (Gehring et al., 1997; Fisher et al., 2007, 2008) and a Vertisol from Texas (USA) (Lindquist et al., 2011). Considering the magnetic characteristics obtained for the soil from

Texas (Lindquist et al., 2011), it seems that the soil studied is more likely a Fluvisol, rather than a Vertisol, because a very strong magnetic enhancement ($\chi \sim 200 \times 10^{-8}$ m^3/kg) and high χ_{fd}% (χ_{fd}% ~10–12%) in the upper soil horizons are reported by the authors. Thus, we will concentrate on the comparison between the Bulgarian Vertisols and the Vertisol from Mali, studied by Gehring et al. (1997) and Fisher et al. (2007, 2008).

A detailed discussion on the comparison between these profiles is provided in Jordanova and Jordanova (2016). In general, the magnetic characteristics of the tropical Vertisol have been explained by Gehring et al. (1997) and Fisher et al. (2007, 2008) by the strong effect of reductive dissolution of lithogenic iron oxides, which results in surface maghemitization of the large (e.g., multidomain [MD]) lithogenic magnetite particles, which is effectively registered as a decrease in the magnetic grain size (e.g., van Velzen and Zijderveld, 1995; Dunlop and Özdemir, 1997). These conclusions are supported by scanning electron microscopy coupled with X-ray analysis (SEM-EDX analyses) as well as electron paramagnetic resonance (EPR) analyses on large lithogenic grains from magnetic separates (Fisher et al., 2007, 2008). The authors suggest that in the upper A_p horizon of their Vertisol, magnetic grains are predominantly in an MD state. In addition, large Feo/Fed values in the soil horizons most affected by water-logging supported the existence of a high amount of amorphous iron oxyhydroxides (Gehring et al., 1997). The data for the Bulgarian Vertisols, however, support the presence of an SD/PSD magnetite fraction, based on the data from ARM measurements and different grain-size ratios, indicating SD presence (Figs. 5.4.4–5.4.7). Based on the magnetic parameters shown, we cannot support the idea of MD grains present in the surface soil horizons, because of the obtained evidence for a higher concentration of ARM-carrying grains in the upper horizons and finer grains down-profile. A brief summary of the main magnetic characteristics obtained from Fisher et al. (2008) as well as the same parameters for Bulgarian Vertisols, are shown in Table 5.5.1.

As is demonstrated by the data, the VR profile shows very similar values of χ down-profile, compared with the tropical Vertisol. However, the coercivity parameters (B_c and B_{cr}) reveal magnetically softer minerals in the VR profile. On the other hand, coercivity parameters in the different horizons of the Endogleyic Vertisol (SM) are very similar to those reported for the tropical Vertisol (Table 5.5.1). It should be kept in mind, however, that hysteresis parameters obtained for the bulk soil samples reflect, in fact, the effective magnetic grain size, which could be a superposition of a mixture of different grain sizes or different mineral phases (Dunlop and Özdemir, 1997; Dunlop, 2002a,b). As was shown by the magnetic mineral analyses (Figs. 5.4.1–5.4.3), various amounts of a high-coercivity fraction (mainly hematite) were identified in the VR profile, as well as in the JAS profile. Thus, in order to evaluate the grain size of the magnetite component, we have estimated the ratio χ_{ARM}/IRM_{soft}—the ratio of the anhysteretic susceptibility to the intensity of the magnetically soft fraction, extracted through the IRM coercivity component analysis (Kruiver et al., 2001, Fig. 5.4.3). Thus, the ratio will characterize the grain size of the magnetically soft fraction, eliminating the effects of the concentration of coarser ferrimagnetic phases and high-coercivity mineral components. As is shown in a number of studies (Maher, 1986; Geiss et al., 2008; Liu et al., 2012; Hatfield, 2014), this interparametric ratio is very sensitive toward the content of stable SD (pedogenic) particles in the mineral assembly in soils/sediments. Using the numerical estimates reported by Geiss et al. (2008) for the χ_{ARM}/IRM characteristic for pedogenic magnetite (χ_{ARM}/IRM $\sim 1.2 \times 10^{-3}$ m/A) and maghemite (χ_{ARM}/IRM $\sim 1.4 \times 10^{-3}$ m/A) and comparing with the obtained parameters for the studied Vertisols (Table 5.5.1), we can make the following conclusion: High values of the ratio χ_{ARM}/IRM $\sim 0.8–1.0 \times 10^{-3}$ m/A are obtained in the Vertisols' horizons, which are most influenced by

Table 5.5.1 Selected magnetic parameters of Vertisols according to literature data and the present study

Site, location, climate (MAT, MAP)	Soil horizon	$\chi_{average}$ $(10^{-8}\ m^3/kg)$	B_{cr} (mT) range	B_c (mT) range	Fe_o/Fe_d	χ_{ARM}/IRM_{soft} $(10^{-3}\ m/A)$	Reference
Mali	A_p	55	29.7–32.1	8.9–10.7	0.38		Fisher et al., 2008
N10°55′0″	A2	42	29.6–35.6	9.2–11.1	0.44		
W5°40′0″	A3	16	53.8–56.2	21.4–27.8	0.65		
MAT = 27°C	Ag	24	41.9–58.3	15.4–22.2	0.96		
MAP = 1186 mm							
VR profile, Burgas	A_{k1}	60.4	22.7–25.8	5.0–6.8	0.14	0.27	Jordanova and
N42°29′0.9″	A_{k2}	44.4	20.7–22.6	4.7–7.0	0.18	**0.86**	Jordanova, 2016
E27°20′37.0″	AB_k	28.3	21.5–23.9	4.0–6.9			and this study
MAT = 12.5°C	C_{kg}	13.0	20.0–25.6	3.0–13.0	0.03	0.09	
MAP = 543 mm							
JAS profile, Jastrebovo	A1	20.0	30.0–30.9	12.0–13.7	0.31	0.50	Jordanova and Jordanova, 2016
N42°16′58.8″	A2	12.3	28.4–35.4	9.7–12.7	0.35		and this study
E25°39′18.0″	A3	11.0	17.4–38.6	7.2–13.2	0.35	**0.96**	
MAT = 12.9°C	AB	10.2	27.1–44.3	10.4–13.3		**0.80**	
MAP = 576 mm	B_k	13.9	24.0–34.8	9.5–11.2		**0.88**	
	BC_k	12.9	24.0–33.0	9.5–11.2	0.14	0.34	
	C_k	9.3	29.8–37.8	9.0–12.3		0.28	
SM profile, Topolovo	A	16.7	34.5–48.8	11.1–15.0	0.58	0.65	Jordanova and Jordanova, 2016
N41°56′6.3″	A_g	13.4	44.4–48.5	11.0–16.2	0.63	0.58	and this study
E25°01′14.4″	B_{tg}	12.9	42.5–56.8	12.6–22.0	0.43	0.64	
MAT = 12.6°C	BC_{kg}	7.8	36.7–52.4	17.6–22.5		**0.80**	
MAP = 768 mm	C_{kg}	8.7	34.8–54.9	17.6–22.6	0.0	**0.87**	
SF profile, Sofia	A1	14.6				**1.27**	Jordanova and
N42°40′34″	A2	15.1				0.64	Jordanova, 1999
E23°21′40″	BC_k	23.0				0.43	and this study
MAT = 10.6°C	C_k	24.2				0.22	
MAP = 582 mm							

The ratio χ_{ARM}/IRM_{soft} used as an indication of the pedogenic magnetite/maghemite grain size (e.g., Geiss et al., 2008) is also shown and the data in bold indicate values characteristic for fine-grained SD magnetite. The range of variations in magnetic susceptibility, coercivity of remanence (B_{cr}), and coercive force (B_c) are shown according to the genetic soil horizons.

prolonged water-logging and multiple reduction–oxidation cycles. These are the A_k2 horizon in the VR profile; the A3 plus AB plus B_k horizons in the JAS profile, the BC_{kg} plus C_{kg} horizons in the SM profile, and the A1 horizon in the SF profile. Relatively high values of the ratio χ_{ARM}/IRM corroborate well with the $\chi_{ARM}/IRM \sim 1.0 \times 10^{-3}$ m/A obtained for the stable SD pedogenic fraction in loess-derived Chernozem-like soils (Geiss et al., 2008). Therefore, we assume that a certain amount of stable SD magnetite exists in the middle-bottom parts of the Vertisol profiles. This conclusion contradicts the hypothesis of Fisher et al. (2007, 2008) that the magnetic signature in the tropical Vertisol is due to the presence of coarse lithogenic magnetite grains that experienced surface reductive dissolution during active seasonal water-logging. As is discussed in detail in Jordanova and Jordanova (2016), the observed contradiction required seeking another possible mechanism for the formation of the stable SD-magnetite like fraction in the Bulgarian Vertisols. A hypothesis put forward for explaining the origin of the stable magnetite fraction is related to the widely studied process of ferrihydrite reductive dissolution in the presence of Fe^{2+} in the soil solution, which leads to the formation of magnetite when the Fe^{2+} concentration in the soil solution is high enough (Hansel et al., 2005, 2011; Piepenbrock et al., 2011; Usman et al., 2012). In addition, the presence of an advective flow drives the mineral precipitation kinetics toward a complete reductive dissolution of ferrihydrite (Hansel et al., 2003). Vertisols are subjected to seasonal multiple reduction–oxidation cycles and match the above kinetic conditions. Also, Thompson et al. (2006) demonstrate that, as a result of periodic oscillations in the reductive conditions, the crystallinity of the precipitated secondary minerals increases with increasing the number of cycles. Considering the latter experimental finding, it could be supposed that the Endogleyic Stagnic Vertisol (profile SM) experienced the most frequent reduction–oxidation cycles among the profiles studied. Thus, the resulting grain size of the secondary magnetite is the finest, as evidenced by the coercivity analysis (Fig. 5.4.3).

An important question to be considered in the study of relict Vertisols is whether the magnetic mineral signature obtained is a product of recent soil development or a sign of its initial formation. Jordanova and Jordanova (2016) suggest that the stable fraction of magnetite detected in the middle/bottom part of the Vertisols (VR, JAS, and SM profiles) is formed in a recent Quaternary climate, since ferrihydrite is a thermodynamically unstable mineral and its preservation in the Vertisols since the Pliocene is very unlikely. Nevertheless, future work should be focused on the elucidation of this question by seeking direct [e.g., SEM, transmission electron microscopy (TEM), high-resolution TEM (HRTEM)] evidence for the presence and characterization of the fine-grained magnetite phase in Vertisols.

As far as Vertisols are found as both surface soils and buried palaeosols, an extensive amount of research is devoted to the establishment of a pedotransfer function, relating Vertisols' geochemical characteristics to the climate (mainly precipitation) (Stiles et al., 2003; Nordt et al., 2004; Driese et al., 2005; Nordt and Driesse, 2010a,b), in order to reconstruct palaeoclimate conditions during palaeosol formation. Different weathering indexes are usually employed in rainfall estimates (Sheldon et al., 2002), but Nordt and Driese (2010b) showed that, especially for Vertisols, it is more appropriate to use a new weathering index, which they label CALMAG. According to its definition, CALMAG $= Al_2O_3/(Al_2O_3 + CaO + MgO) \times 100$ (Nordt and Driese, 2010b). The authors demonstrate that the regression between the new weathering index and the mean annual precipitation (MAP) for a suite of Vertisols developed in a wide climosequence is defined by the relation: MAP = 22.69*CALMAG-435.8. Additional research carried out by Adams et al. (2011) also shows that the CALMAG index provides a more robust MAP estimate for Vertisols, compared with the widely used CIA-K index (Sheldon et al., 2002). In the study of Bulgarian Vertisols, Jordanova and Jordanova (2016) estimate the weathering index

CALMAG for the three Vertisols described in this chapter (VR, JAS, and SM profiles). The estimated CALMAG was used for MAP calculation. The estimated MAP for the three Vertisols are 1079 ± 58 mm (VR profile), 1108.0 ± 79 mm (JAS profile), and 1125 ± 68 mm (SM profile). The obtained MAP estimates are significantly higher than the contemporary precipitation (see the profile description, MAP = 543 mm for VR, MAP = 576 mm for JAS, and MAP = 768 mm for SM; Koinov et al., 1998). Therefore, the weathering state in the studied relict Vertisols reflects the ancient climate conditions during the Pliocene, when these soils were initially formed in a more humid and warmer climate. In addition, Jordanova and Jordanova (2016) study Fe-Mn nodule from the SM profile through SEM with EDS analysis in different parts of the nodule. Based on the element composition obtained, and applying the correlation proposed by Stiles et al. (2001) between the total iron content in nodules from Vertisols and the MAP, an approximate MAP value of ~1000 mm is estimated for the SM profile. As can be seen, it correlates relatively well with the MAP estimate from the CALMAG index. Thus, another important conclusion regarding the Vertisols' formation is that the nodules found in the profiles have a relict character (i.e., they are formed during the initial Vertisol development).

5.6 SALINE AND SODIC SOILS (SOLONCHAK, SOLONETZ, SOLOD): MAIN CHARACTERISTICS, FORMATION PROCESSES, AND DISTRIBUTION

Salts in the soils have different origins, depending on the major processes acting in each specific locality. One of the soluble salt sources in soils are eolian processes—sea aerosols transported over the mainland, the presence of sea salts as coatings onto the clastic particles, or salts from pyroclastic materials (Berger and Cooke, 1997). The second most important source are secondary products of the chemical weathering of the rocks (mainly evaporites of calcium and magnesium chlorides, nitrates, etc.) in a hyperarid climate and low mean annual temperatures. These weathering products contain a large amount of soluble salts—sodium chlorite, sulfates, and bicarbonates (Schaetzl and Anderson, 2009). The third foremost mechanism of soil salinization is the influence of a fluctuating level of strongly mineralized groundwater in the frames of the soil profile (Morgan and Jankowski, 2004). The salts formed as a result of some of these mechanisms are accumulated in soils with a weak permeability, where they cannot be washed out of the soil profile, or in the conditions of a dry climate, when leaching is negligible and, at the same time, there is a constant source of salts. The salts are most intensively formed and accumulated in soils developed under a dry climate with clearly separated dry and wet seasons. During the wet season, the salts weather from the minerals and they are transported together with the soil solution toward the soil horizons or relief forms, where they accumulate. The most widespread are the sodium salts, while magnesium salts are less common. Special attention is paid to the sodium cations (Na^+) and its salts, because the presence of even a small amount of such cations causes significant changes in soil morphology. Because of its small size, similar to that of the hydrogen (H^+), the Na^+ is able to disperse the clay colloids, which become highly susceptible to translocations. The accumulation of such colloids with absorbed Na^+ cations in the illuvial soil horizon leads to the creation of sticky mold masses, while, on drying, a strongly impermeable clayey layer forms—the "nitric" horizon. In fact, the Na^+ plays the role of a cementing agent in saline soils. Other salts found in soils are the so-called neutral salts, comprising sodium chloride (NaCl), gypsum, sodium sulfate (Na_2SO_4), as well as other sulfates, bicarbonates, and chlorides (Schaetzl and Anderson, 2009). Sodium ions are easily soluble and they are usually washed out of the profiles of well-aerated soils. Therefore, salt accumulation occurs under the influence of the factors, which (1) impede the Na^+ export out of the soil profile (like impermeable parent rock or buried clayey palaeosol below the surface soil) or (2) when the newly formed Na^+ cations are more abundant than the amount of the cations exported out of the profile. Frequently, in lowland relief forms, salinization results from the presence of a high level of Na-mineralized groundwater, while at the same time, the soils developed on the slopes are Na-deficient. This is why saline soils do not form a continuous soil cover in a given region but instead are spread in patches.

Saline and sodic soils comprise three main soil types—Solonchak, Solonetz, Solod—which represent different stages in the evolution of the salinization/solodization processes. Solonchaks develop as a result of salinization (accumulation of neutral Na and Mg salts) and are characterized by the presence of >1% of water-soluble salts along the soil profile. Solonetz forms as a result of the desalinization process—the transport/leaching of salts from the upper parts of the soil profile as a result of different factors—climate change, artificial irrigation, or changes in the surface water or groundwater hydrology. As a result, an illuvial horizon develops and the salts are exported deeper in the soil profile, but the exchangeable sodium cations (Na^+) remain in place. This causes soil alkalization and the pH frequently becomes >8.5 in certain parts of the profile (Schaetzl and Anderson, 2009). Solods represent the final stage in the evolution of the salt-affected soils, when the process of solodization (dealkalization) proceeds. The sodium is gradually washed out of the profile, because the hydrolysis becomes dominant and Na^+ are replaced by H^+, leading to the acidification of the upper soil horizons.

Saline and sodic soils cover ~2.6% of the land worldwide. They are most abundant in Russia, China, Argentina, Iran, India, and Paraguay (Abrol et al., 1988). Solonchaks are most extensive in the arid and semiarid parts of Africa, the Middle East, Russia, central Asia, Australia, and the Americas (FAO, 2001) (Fig. 5.6.1a). Solonetz soils are known internationally as sodic soils or alkali soils and they are strongly alkaline ($pH > 8.5$). Solonetz soils occur mainly in areas of a steppe climate (dry summers and low annual precipitation). They occupy the largest areas in Ukraine, Russia, Kazakhstan, Hungary, Romania, China, the United States, Canada, South Africa, and Australia (Fig. 5.6.1b) (FAO, 2001).

In Bulgaria, saline and sodic soils are represented by Solonetz, Solonchaks, and Solonetz-Solonchak. They occupy relatively restricted areas, mainly in South Bulgaria (in the Thracian lowland along the Tundzha river, Stara Zagora district), in the coastal area near the town of Burgas, as well as in small areas in north Bulgaria (some lowlands near the Danube River) and valleys in southeast Bulgaria (Koinov et al., 1998; Ninov, 2002; Shishkov and Kolev, 2014). Saline and sodic soils are usually developed in lowlands where the highly mineralized groundwater discharges. It is accepted in Bulgarian pedology that the main reason for the formation of saline and sodic soils in Bulgaria is the active tectonic movements that caused submergence of parts of the Thracian lowland and the Fore-Balkan plains. These processes triggered a rise in the groundwater level and increased its mineralization (Koinov et al., 1998; Ninov, 2002). Visually, saline and sodic soils are distinguished by the presence of light-gray areas on the surface without vegetation and are frequently covered by "salt flowers" of white soluble salts. Solonchaks in Bulgaria are considered as secondary-formed soils, a result of the salinization of marshland soils, meadow-bog soils, and alluvial-Chernozem−like soils or Vertisols (Shishkov and Kolev, 2014). Salinization is mostly chloride-sulfate and sodic-sulfate. Solonetz soils have a more restricted distribution than Solonchaks. They are characterized by a well-distinguished illuvial horizon with a prismatic structure, strongly bleached and structureless humic horizons, as well as a strongly alkaline soil reaction (Shishkov and Kolev, 2014).

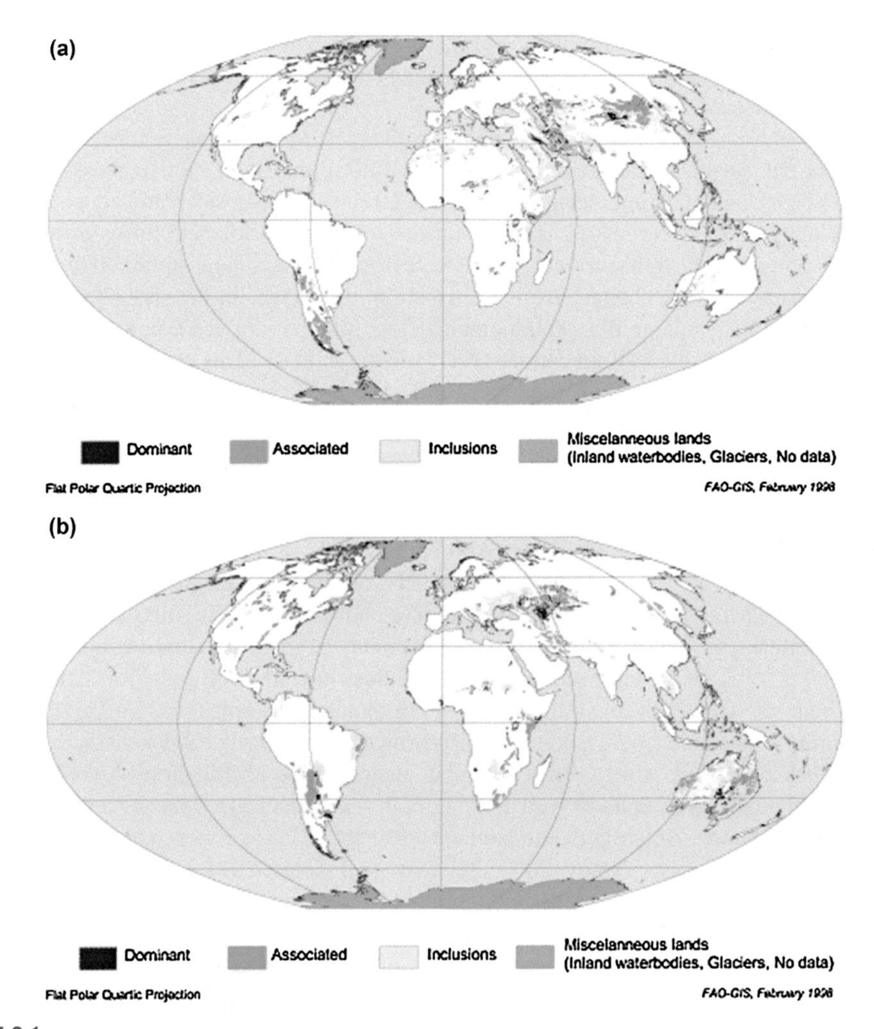

FIGURE 5.6.1

(a) Spatial distribution of Solonchaks worldwide (FAO, 2001; Lecture notes on the major soils of the world). http://www.fao.org/docrep/003/y1899e/y1899e12.htm#P353_47671. (b) Spatial distribution of Solonetz soils worldwide (FAO, 2001; Lecture notes on the major soils of the world). http://www.fao.org/docrep/003/y1899e/y1899e12.htm#P353_47671.

5.7 DESCRIPTION OF THE PROFILES STUDIED

5.7.1 SOLONCHAK-SOLONETZ, PROFILE DESIGNATION: S

Location: N 42°12′29.9″; E 25°03′52.2″, h = 145 m a.s.l.

Situated close to Sadovo town, Plovdiv district

Present-day climate conditions: MAT = 12.0°C; MAP = 561 mm, vegetation cover: sparse dry shrubs (Fig. 5.7.6); relief: flat

Profile description: general view, details from the horizons and terrain view are shown in Figs. 5.7.1–5.7.6.

0–23 cm: humic horizon (A_{zn}), gray-beige, grown salt crystals on the ped surfaces, crumby structure

23–70 cm: illuvial (B_z) horizon, brown, clayey, small carbonate concretions

70–118 cm: transitional (B_zC) horizon, brown-orange, sandy, Fe-Mn concretions present, grown salt crystals

118–135 cm: C1 horizon, yellowish-brown, grown salt crystals, grayish gleyed spots present

135–150 cm: C2 horizon, dark brown, sandy.

Groundwater level is at 140-cm depth at the time of sampling (25.08.2006).

Parent material: alluvial deposits

FIGURE 5.7.1

View of the profile of salt-affected soil—S profile.

FIGURE 5.7.2

Salt-affected soil (S). A_{zn} horizon—detail.

FIGURE 5.7.3

Salt-affected soil (S). B_z horizon—detail.

FIGURE 5.7.4

Salt-affected soil (S). B_zC horizon—detail.

FIGURE 5.7.5

Salt-affected soil (S). C1 horizon—detail.

FIGURE 5.7.6

Salt-affected soils in the area of Belozem (central south Bulgaria) —terrain view.

5.8 TEXTURE, SOIL REACTION, AND MAJOR GEOCHEMISTRY OF SOLONETZ-SOLONCHAK

Results from the texture analyses carried out on selected samples from the S profile are shown in Fig. 5.8.1a–c and demonstrate that the sand and silt fractions dominate the mechanical composition. These two fractions, however, have an opposite depth trend—while the sand content increases in depth, the silt content is maximum in the surface horizon and decreases steadily toward the C horizon. The clay fraction is relatively enriched in the upper A_{zn} and B_z horizons (Fig. 5.8.1c) and sharply decreases in the transitional and the C horizons, pointing to the pedogenic character of the clay within the S profile. The soil reaction (pH) shows very high values (pH between 9.5 and 10) in the A_{zn}, B_z, and B_zC horizons and a gradual decrease down to ~8.5 at the bottom of the profile (Fig. 5.8.1d). Such a very high pH is characteristic only for Solonetz soil, like the profile near Belozem described in Koinov et al. (1998), while for Solonchak and Solonetz-Solonchak the highest pH reported is ~9.2. The strongly alkaline pH obtained, as in the S profile, is comparable to some typic Haplaquepts and Aeric Haplaquepts in the United States, where pH reaches ~10.3, as well as sodic soils in Russia (Bockheim and Hartemink, 2013).

Variations in some of the major chemical elements in selected samples from the profile are shown in Fig. 5.8.1e–l. The concentrations of sodium (Na) and magnesium (Mg) are relatively high (Fig. 5.8.1f) compared with other soil types presented in the book. This suggests an increased content of Na and Mg salts, though no analyses for the content of water-soluble salts and/or exchangeable Na^+, Mg^{2+}, K^+, and Ca^{2+} are available here. The sulfur (S) content (Fig. 5.8.1l) is very high, most likely related to sodium sulfate accumulation in the soil. Iron (Fe) and titanium (Ti) contents vary synchronously in depth, showing an enhancement in the upper A_{zn} and B_z horizons, while Al is relatively

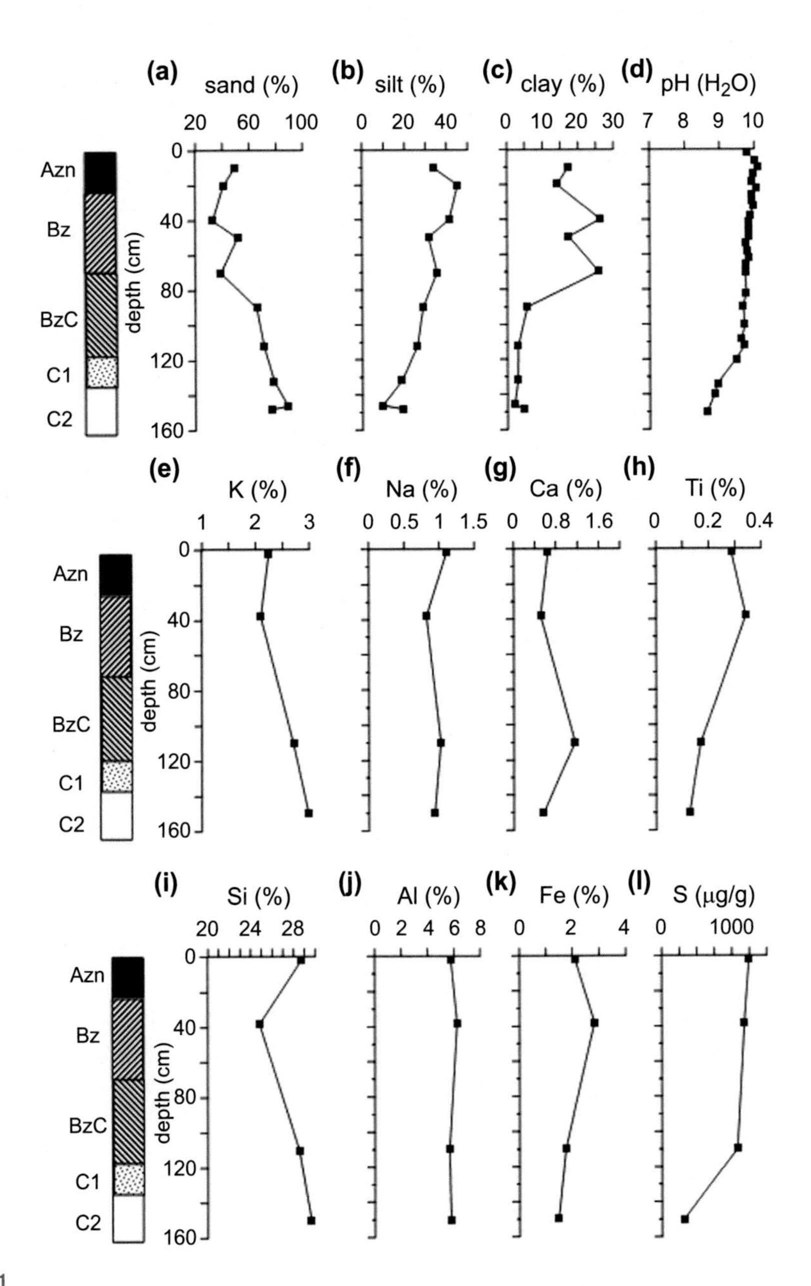

FIGURE 5.8.1

Texture and major geochemistry of salt-affected soil (S): (a) sand content; (b) silt content; (c) clay content; (d) soil pH; (e–l) selected major and trace elements variations in depth (K, Na, Ca, Ti, Si, Al, Fe, S).

Table 5.8.1 Chemical extraction data for selected samples from salt-affected soil (S)

Sample depth (cm)	Al_o (g/kg)	Fe_o (g/kg)	Mn_o (g/kg)	Fe_d (g/kg)	Fe_{tot}(g/kg)	Fe_o/Fe_d	Fe_d/Fe_{tot}
2	1.050	1.106	0.468	2.990	21.10	0.37	0.14
38	1.315	0.968	0.180	3.545	28.34	0.27	0.13
110	0.658	0.465	0.294	2.170	17.54	0.21	0.12
150	0.490	0.440	0.178	1.455	14.73	0.30	0.10

Oxalate-extractable Al, Fe, and Mn (Al_o, Fe_o, Mn_o), dithionite-extractable Fe (Fe_d), total Fe content (Fe_{tot}), and the ratios Fe_o/Fe_d and Fed/Fe_{tot}

constant in all soil horizons (Fig. 5.8.1h,j,k). The iron content is relatively low (~2% in average), compared with other soil types. Oxalate and dithionate extraction data for the selected samples is presented in Table 5.8.1. As can be seen, oxalate-extractable Fe is relatively high, but the highest is in the A_{zn} (sample from depth 2 cm) and B_z horizons (sample from depth 38 cm) (Table 5.8.1). The Feo/Fed ratio is high in all the soil horizons (0.37–0.2), pointing to an important share of pedogenic amorphous iron oxyhydroxides formed under high alkalinity and water-logging. Simultaneously, the Fed/Fe_{total} ratio is low, confirming the above conclusion.

5.9 MAGNETIC PROPERTIES OF SALT-AFFECTED SOIL

The magnetic signal in the salt-affected soil (the S profile) is very weak, which is the reason for the larger amount of noise on the thermomagnetic curves shown in Fig. 5.9.1. Despite this, the heating runs for all the samples analyzed clearly show the presence of magnetite's Tc (~580°C). The uppermost sample displays a weak peak at ~300°C (Fig. 5.9.1a), likely indicating a mineral transformation in the Fe-containing minerals initially present. Considering the model thermomagnetic curves for the different iron oxyhydroxides with or without organics (Hanesch et al., 2006), it could be supposed that the observed behavior for the S2 sample indicates the transformation of ferrihydrite into maghemite. The other samples show relatively similar heating–cooling curves without strongly expressed changes in the magnetic mineralogy after heating to 700°C (Fig. 5.9.1b–d). Here, the magnetite phase is more clearly expressed, especially in the deeper levels (samples S110 and S150) (Fig. 5.9.1c,d).

The magnetite's occurrence in the S profile is further evidenced by the stepwise thermal demagnetization of ARM (Fig. 5.9.1e–h) carried out for selected depth levels. The thermal decay curves all show a clearly distinguished unblocking temperature of magnetite (T_{ub} ~ 580°C) and convex-shaped curves. Thus, it could be supposed that the magnetite is rather well defined with a constrained grain-size distribution (Hartstra, 1983; Dunlop and Özdemir, 1997). A weak inflection point can be observed on the demagnetization curves at ~250°C for the samples belonging to the B_z horizon (S28 and S38 [curve not shown]) and the B_zC horizon (S108). It could be related to the presence of oxidized magnetite (maghemite), which inverts to hematite in this temperature range (Dunlop and Özdemir, 1997).

Magnetic mineralogy in the salt-affected soil is also evaluated through a stepwise acquisition of isothermal remanence (IRM) (Fig. 5.9.2) performed for the selected samples. It is obvious that all samples, except S112, do not reach saturation in fields up to 5 T. This suggests the presence of a high-coercivity component in the soil studied. The magnetization intensity in the outlier sample (S112) is

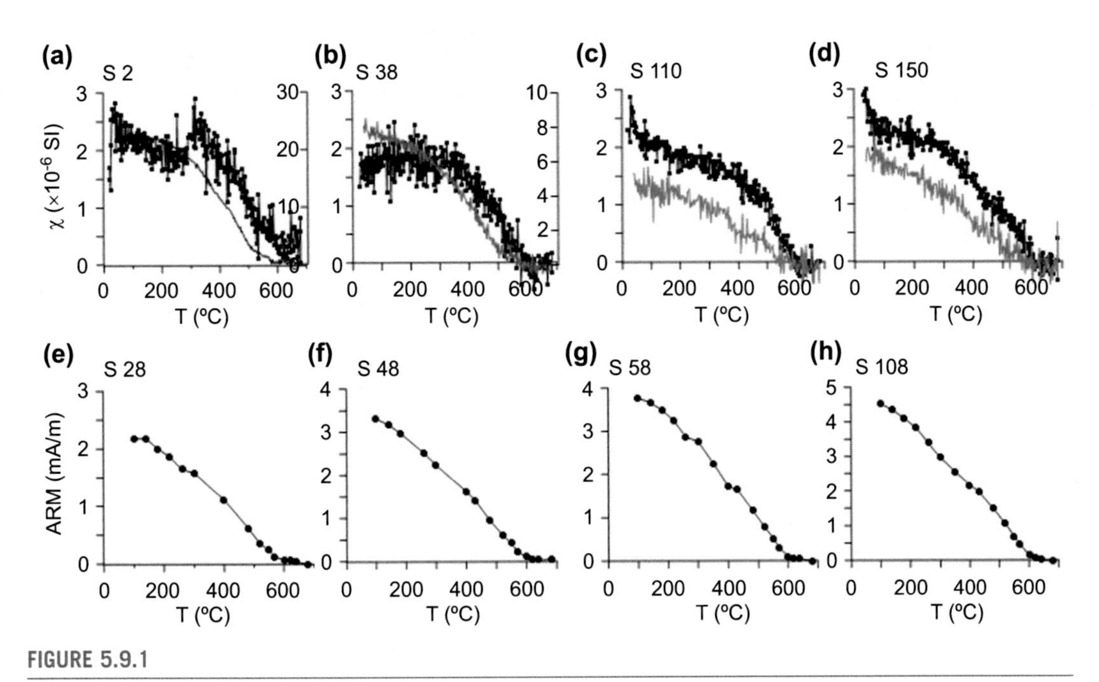

FIGURE 5.9.1

Thermomagnetic analysis of magnetic susceptibility for selected samples from the salt-affected soil (S profile) (a–d). Heating curves—thick black line, cooling curves—thin gray lines. Heating in air, fast heating rate (11°/min). Stepwise thermal demagnetization of ARM (anhysteretic remanence) for selected samples from the S profile (e–h).

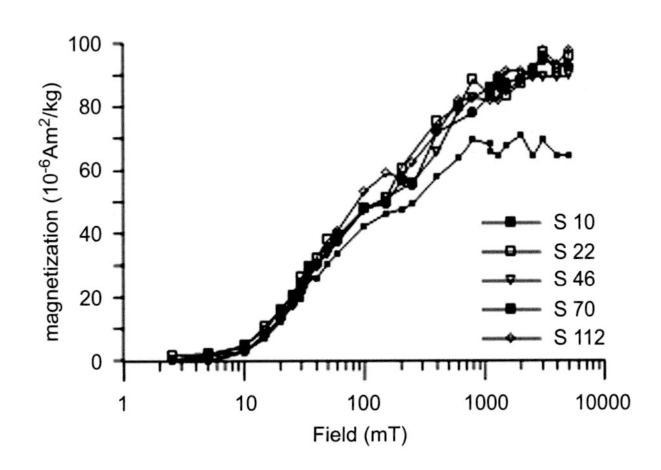

FIGURE 5.9.2

Stepwise acquisition of isothermal remanence (IRM) up to field of 5 T for selected samples from the profile of salt-affected soil (S).

Table 5.9.1 Summary parameters from coercivity analysis by fitting IRM acquisition curves with cumulative Gaussian functions, using IRM-CLG1.0 software (Kruiver et al., 2001) for selected samples from the salt-affected soil S

Sample depth (cm)	IRM component 1				IRM component 2			
	Intensity ($10^{-6}Am^2/kg$)	% from IRM_{total}	$B_{1/2}$ (mT)	DP	Intensity ($10^{-6}Am^2/kg$)	% from IRM_{total}	$B_{1/2}$ (mT)	DP
Profile S salt-affected soil								
10	34.78	58	39.80	0.40	22.58	35	562.30	0.37
22	30.74	56	34.70	0.40	20.58	35	478.60	0.37
46	32.09	56	34.70	0.40	21.49	35	478.60	0.28
70	33.47	51	36.30	0.40	14.04	17	446.70	0.28
112	36.50	60	38.90	0.37	24.08	37	631.00	0.70

The intensity, relative contribution (% from IRM_{total}), median acquisition field ($B_{1/2}$), and dispersion parameter (DP) for each coercivity component are shown.

very weak and noisy and does not show a high-coercivity component. The coercivity component analysis carried out (Kruiver et al., 2001) revealed the presence of two IRM components (Table 5.9.1).

The low-coercivity component (IRM1 component in Table 5.9.1) is characterized by uniform intensity along the profile ($30–36.5 \times 10^{-6}Am^2/kg$), as well as an equal share in the total IRM (Table 5.9.1). Therefore, it could be supposed that the IRM1 is carried by the same Fe oxide in all soil depths. The coercivity of the IRM1 varies between 35 and 40 mT, suggesting that it could be related to magnetite/maghemite (Egli, 2004). The second high coercivity component has a weaker intensity and the coercivity ($B_{1/2}$) varies in the interval 446–631 mT (Table 5.9.1). These coercivities are more characteristic for hematite, rather than for goethite. As shown in a number of works (Stanjek and Schwertmann, 1992; Wells et al., 1999; Jiang et al., 2012), hematite's coercivity strongly depends on the structural substitutions (and especially of Al) in its crystal lattice, being higher when Al substitutions are higher. As is also shown by Wells et al. (1999), the coercivity of remanence (B_{cr}) of hematite varies in a wide range between 200 and 600 mT, depending on the degree of Al substitution and the grain size. Thus, the IRM acquisition curves and their analysis suggest that, in the salt-affected soil, the main magnetic minerals are magnetite and hematite.

The concentration-dependent magnetic characteristics (Fig. 5.9.3) show the presence of a low amount of strongly magnetic minerals along the profile, as seen by the low values of M_s, IRM_{5T}, and χ, compared with other soil types (see the other chapters in this book). Nevertheless, the magnetic enhancement phenomena in the A_{zn}, B_z, and B_zC horizons in relation to the parent material (C1 and C2 horizons) is evident. Magnetic susceptibility (χ) is very low, reaching a maximum value of only 10×10^{-8} m^3/kg (Fig. 5.9.3d), and a significant part of this is due to paramagnetic Fe-containing minerals and/or hematite, as evidenced by the high-field magnetic susceptibility (χ_{hf}) (Fig. 5.9.3e). The concentration of the magnetic minerals, as reflected by the M_s (Fig. 5.9.3a), suggests a maximum occurrence of highly magnetic minerals (magnetite/maghemite) at the bottom of B_zC, while in the upper part (A_{zn} and B_z), M_s displays a concave shape. A comparison with the behavior of χ and χ_{hf} allows us to assume that, in the upper soil horizons, the weakly magnetic minerals make a greater

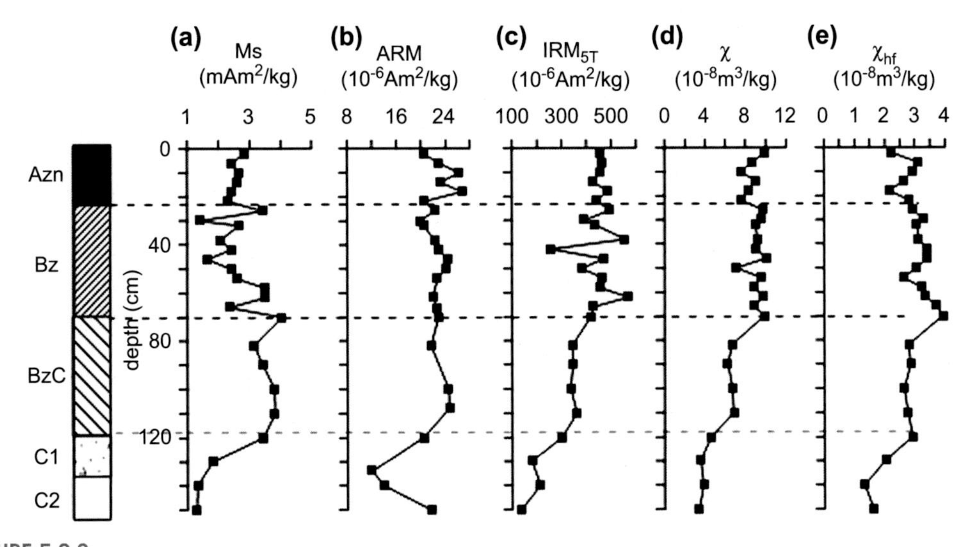

FIGURE 5.9.3

Depth variations in the concentration-dependent magnetic characteristics: (a) saturation magnetization (M_S); (b) anhysteretic remanence (ARM); (c) saturation isothermal remanence, acquired in a field of 5 T (IRM_{5T}); (d) magnetic susceptibility (χ); (e) high-field magnetic susceptibility (χ_{hf}).

contribution to the magnetic assembly, while in the B_ZC, the highly magnetic magnetite-type minerals prevail. The restriction of the magnetic enhancement to the bottom of the BzC also suggests that the magnetic minerals in the soil are of pedogenic origin. Similarly, the variations in the ARM support the presence of SD-like magnetite grains in the solum with maximum concentration in the A_{zn} and B_ZC and a slightly lower but still high content in B_Z (Fig. 5.9.3b).

A set of magnetic grain-size proxies and hematite/goethite proxies is shown in Fig. 5.9.4. As is evidenced in Fig. 5.9.4a,b, the classic grain-size–related parameters M_{rs}/M_S and B_{cr}/B_c (Day et al., 1977)

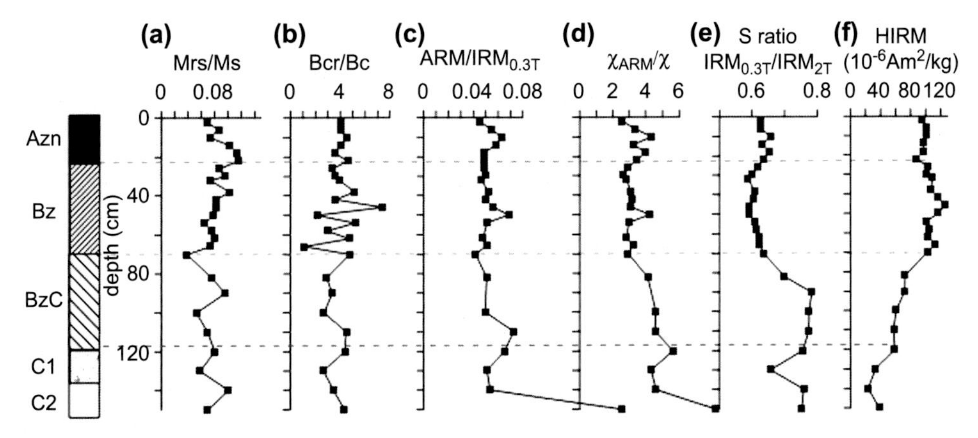

FIGURE 5.9.4

Depth variations in grain-size anhysteretic–sensitive magnetic biparametric ratios: (a) M_{rs}/M_S; (b) B_{cr}/B_c; (c) $ARM/IRM_{0.3T}$; (d) χ_{ARM}/χ; (e) S ratio; (f) "hard" isothermal remanence (HIRM) (e.g., Q. Liu et al., 2002).

do not display systematic changes along the depth, but only oscillations around M_{rs}/M_s ~0.08 and B_{cr}/B_c ~4, the values characteristic for large PSD magnetite grains (Dunlop and Özdemir, 1997; Dunlop, 2002b). However, as deduced by the above-presented rock magnetic data, as well as by the S ratio and the HIRM parameters (Fig. 5.9.4e,f), a significant hematite contribution to the magnetic properties is demonstrated. Thus, the M_{rs}/M_s and B_{cr}/B_c ratios cannot be used as grain-size proxies, because they are influenced by the differences in magnetic mineralogy. Low values of the S ratio (~0.6) and high HIRM in the upper part of the soil (A_{zn} and B_z horizons) strongly suggest an important contribution of hematite (Q. Liu et al., 2002; Liu et al., 2007), which decreases toward the C horizons. The magnetic grain size of the magnetically soft mineral component is deduced through the ratios ARM/$IRM_{0.3T}$ and χ_{ARM}/χ (Fig. 5.9.4c,d). In the B_zC horizon, the magnetic susceptibility is notably influenced by the SD fraction, since the magnetic susceptibility has a relative minimum there, while the ARM is at a maximum (Fig. 5.9.3b,d). Therefore, the χ_{ARM}/χ ratio will be biased there. Thus, a more robust grain-size proxy is the ARM/$IRM_{0.3T}$ ratio. Its variations (Fig. 5.9.4c) suggest a relatively uniform magnetite grain size in the upper A_{zn} and B_z horizons and an increased grain size down to the C1. The maximum in the C2 horizon is most likely due to differences in mineralogy. Because of the very weak magnetic susceptibility of samples from the S profile, the frequency-dependent susceptibility measurements do not give a reliable estimate of the SP fraction through the χ_{fd} parameter and this magnetic proxy is not discussed here.

5.10 STUDIES OF CONCRETIONS AND NODULES FROM THE SALT-AFFECTED SOIL PROFILE S

As mentioned in the profile's description, abundant Fe-Mn concretions/nodules were present in the profile. This fact implies the existence of cyclic reduction–oxidation changes (van Breemen, 1988). The presence of Fe-Mn concretions is also reported in salt-affected soil from India (Ram et al., 2001), as well as for gleyed Solonetz developed in the alluvial deposits of the Missouri River (USA) (Aide, 2005). Three concretions extracted from a depth of 70–72 cm in the S profile (B_z/B_zC transition) have been studied through SEM with EDX analysis. SEM images of polished cross-sections as well as details of the three concretions are shown in Figs. 5.10.1–5.10.3. A comparison of the SEM images shows that the concretions have a different shape and different internal structure. The S-1 concretion (Fig. 5.10.1) is ~1.6 mm in diameter and has an irregular shape. Its internal structure has no differentiated layers, but rather it is built up of soil particles incorporated into the nodule's matrix. The elemental analyses probed into two areas of the nodule (denoted by squares in Fig. 5.10.1a) show that the bright particles in the BSE (back-scattered electrons) mode are enriched in Fe and Ba (spectra Sp1-1, Sp2-2, Sp1-4, Sp1-5 in Table 5.10.1), while the analyses taken into the soil particles from the nodule (spectra Sp1-3, Sp2-3, Sp2-4) logically show an enhancement in Si and Al. Therefore, a strong inhomogeneity in the chemical elements' distribution into different parts of the nodule is observed. Such an inhomogeneity is a characteristic feature of most of the soil nodules (F. Liu et al., 2002; Cornu et al., 2005; Gasparatos, 2013).

The S-2 soil concretion shown in Fig. 5.10.2 has a spherical shape with a ~1.3-mm diameter, a well-established concentric structure, and characteristic outer shell. Among the different layers in the shell, soil particles are caught up as well. The elemental analyses (Table 5.10.2) show that the S-2 concretion is enriched in Fe, Mn, Al, and Si together with some 1.5 wt% of Na, Mg, and K. The outer

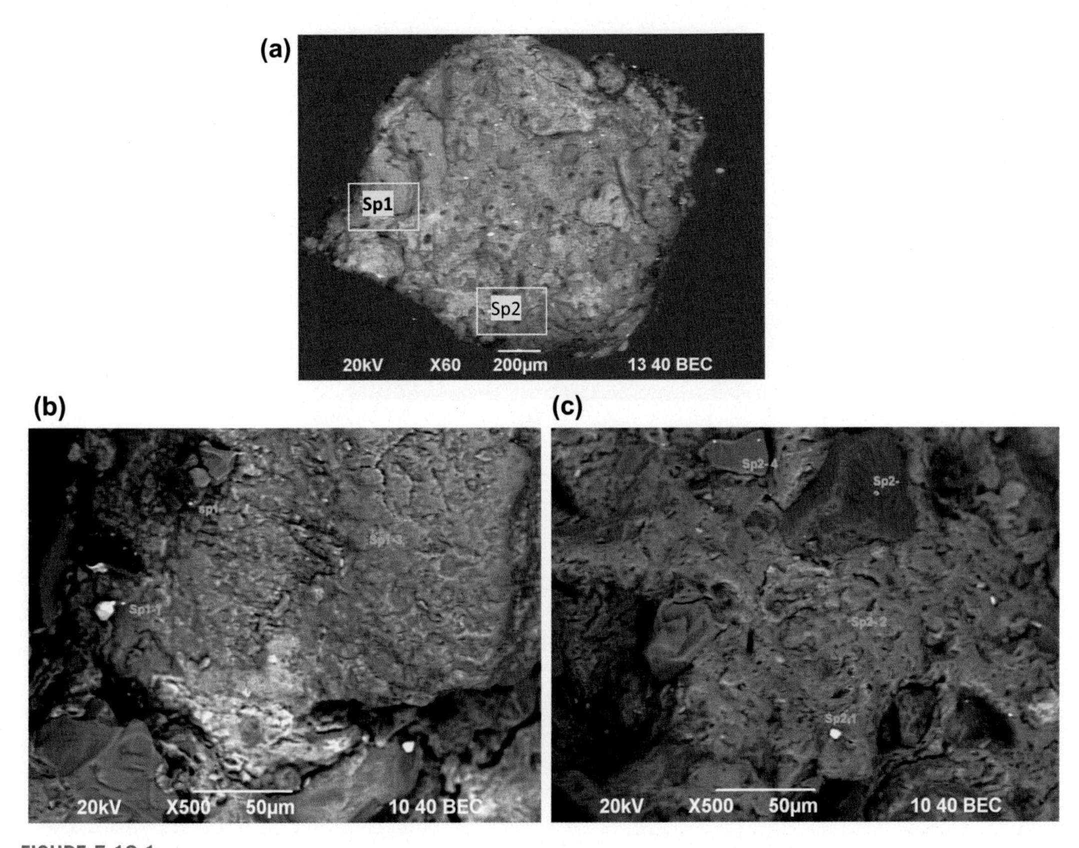

FIGURE 5.10.1

(a) Scanning electron microscopy (SEM) images (in a back-scattered electrons mode) of a nodule S-1 extracted from the salt-affected soil S. EDS spectra were taken at the two spots indicated by squares and labels "Sp1" and "Sp2." More detailed images of these two areas are shown in (b) and (c) together with the spots of the EDS analyses taken.

shell is inhomogeneous, and in some parts it is enriched in Fe, Mn, and Ba (spectra Sp2-2, Sp2-4 in Table 5.10.2). The occurrence of concentric layers in the spherical soil concretions is usually related to their formation during separate stages of alternating dry and moist periods (Gasparatos et al., 2005; Cornu et al., 2005; Gasparatos, 2013). A number of detailed elemental and mineralogical studies on soil concretions show a regular concentric layering in the Fe and Mn distribution as well (Zhang and Karathanasis, 1997; Palumbo et al., 2001; Gasparatos et al., 2005; Cornu et al., 2005), resulting from the difference in the reduction potentials necessary for the oxidation of Fe and Mn.

The third concretion studied (S-3, Fig. 5.10.3) is characterized by a very different shape and structure compared with the previous two. It is very large (~2.5 mm length and 1.5 mm wide) and irregular in shape (Fig. 5.10.3a). The internal structure is disordered, very dense, and without pores and cracks, and numerous soil particles are incorporated into the nodule's matrix (Fig. 5.10.3b,c). This nodule also has an outer shell built up of fine lamella of pedogenic layers and incorporated needle-shaped soil particles

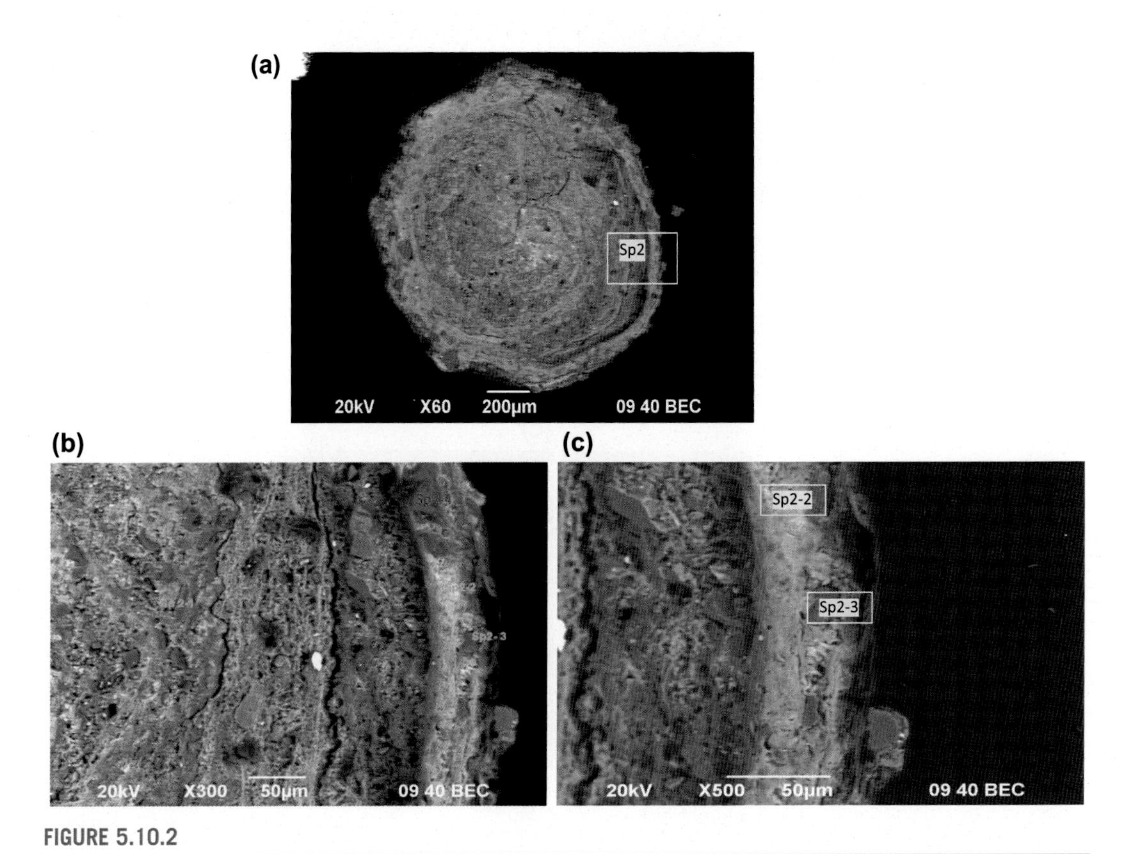

FIGURE 5.10.2

(a) Scanning electron microscopy (SEM) images (in a back-scattered electrons mode) of a nodule S-2 extracted from the salt-affected soil S. More-detailed images of the areas where EDS spectra were taken, are shown in (b) and (c).

(Fig. 5.10.3d). The elemental composition of the nodule determined in six points, as well as the "integral" spectrum (Table 5.10.3), suggests the presence of a lower Fe content and a relatively enhanced content of Al, Ca, and Na compared with S-1 and S-2 concretions.

The observed differences among the three concretions indicate the existence of different environmental settings during their formation and, therefore, their multistage genesis. Studies on soil concretions from soils with impeded drainage show that usually the spherical concretions with a concentric structure form in moderately clayey soil horizons experiencing periodic seasonal water-logging (Zhang and Karathanasis, 1997). On the other hand, in the heavy clayey impermeable soil horizons, the periods of water saturation are significantly longer and the cycles of water-logging more frequent, which provokes more intense reductive dissolution of Fe and Mn and formation of dense concretions (Zhang and Karathanasis, 1997). Considering these findings in relation to the concretions from the S profile, the following hypothesis can be stated: The spherical concentric concretions in the salt-affected soil (like the concretion S-2) and the irregularly shaped concretions (e.g., concretion S-1) were formed in the soil before its salinization by the mineralized ground water, when the soil most likely existed as a stagnic endogleyic Vertisol (similar to the profile SM, described in this chapter). Such a view

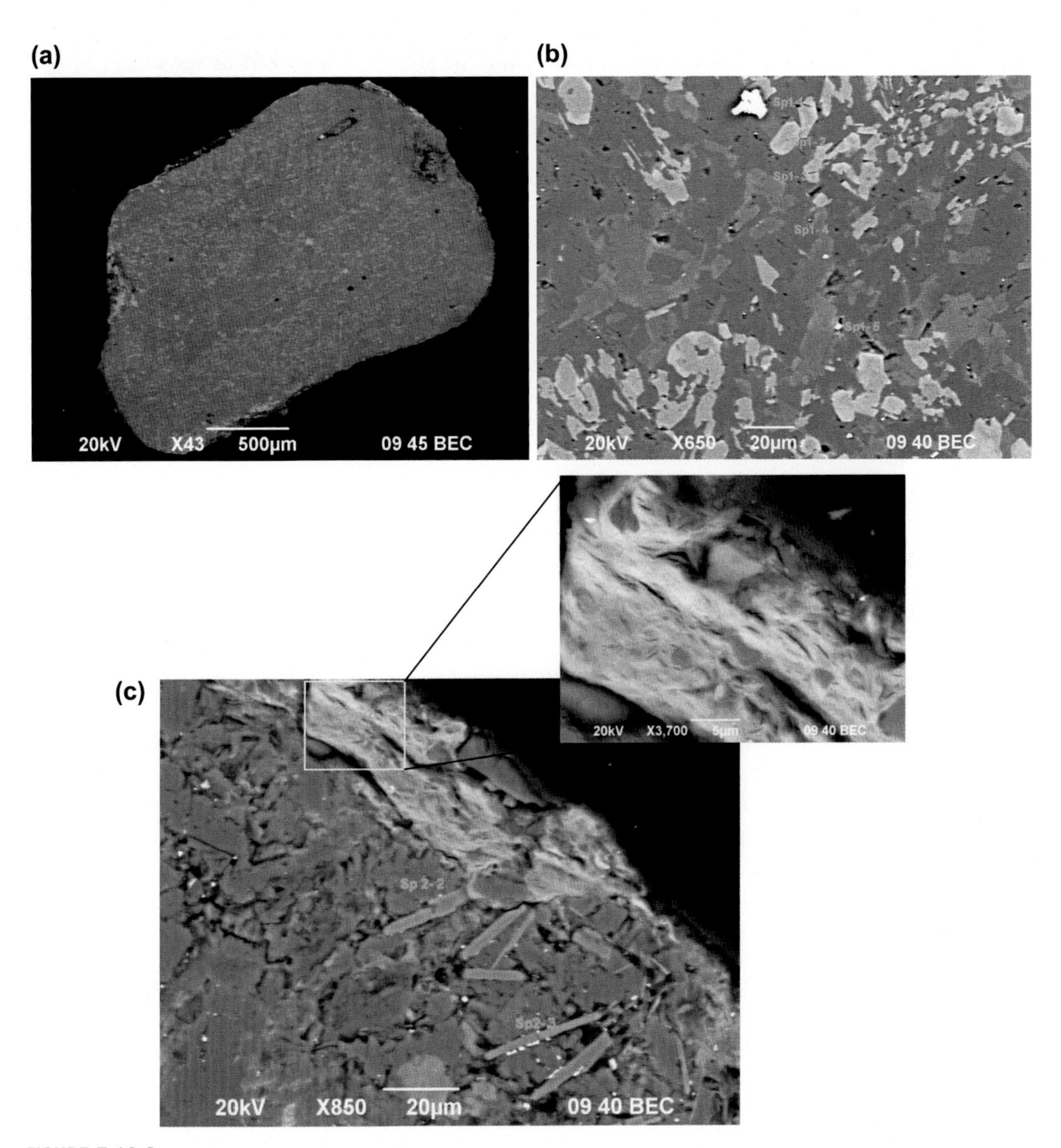

FIGURE 5.10.3

(a) Scanning electron microscopy (SEM) image (in a back-scattered electrons mode) of a cross-section of a dense nodule S-3 extracted from the salt-affected soil S. More detailed images in the center (b) and periphery parts (c) of the nodule with spots of the EDS analyses denoted are also shown.

Table 5.10.1 EDS spectra in two areas of concretion S-1 shown in the SEM image in Fig. 5.10.1. (S profile)

	O	Na	Mg	Al	Si	K	Mn	Fe	Ba	S
Sp1-1	42.99	—	—	—	—	—	—	57.01	—	—
Sp1-2	52.2	—	—	—	—	—	1.13	46.22	—	—
Sp1-3	52.09	0.92	—	8.58	24.96	12.46	—	0.99	—	—
Sp1-4	60.83	—	—	—	—	—	3.11	15.01	21.05	—
Sp1-5	51.57	—	—	—	—		8.81	14.39	25.23	—
Sp2-1	42.37	—	—	—	—	—	1.07	1.0	43.41	12.14
Sp2-2	60.84	0.59	1.63	6.79	15.87	1.92	6.93	4.51	—	—
Sp2-3	67.89	—	—	—	32.11	—	—	—	—	—
Sp2-4	54.49	—	—	45.51	—	—	—	—	—	—

Elements shown in weight%.

is supported by the accepted hypotheses in Bulgarian pedology concerning the genesis of salt-affected soils in Bulgaria (Shishkov and Kolev, 2014). The dense, irregularly shaped S-3 concretion (Fig. 5.10.3) from the S profile is likely related to a more recent stage of soil development. Salt accumulation and increased colloidal capacity likely lead to the creation of the above-described conditions for the formation of dense concretions with a nondifferentiated structure. In support of such a view is the higher Na content obtained in the S-3 concretion (Table 5.10.3), while in the other two concretions (Tables 5.10.1 and 5.10.2) sodium is lacking from the composition or is a much lower amount.

Concretions extracted from several depth levels from the S profile are used for magnetic measurements. A stepwise acquisition of IRM up to a field of 5 T is carried out for concretions extracted from three depths—70 cm (B_z horizon), 80 cm (B_zC horizon), and 120 cm (C1 horizon). The obtained acquisition curves are shown in Fig. 5.10.4, and the results from the coercivity component analysis carried out (Kruiver et al., 2001) are summarized in Table 5.10.4.

Two coercivity components are separated—a magnetically soft one (IRM comp. 1 in Table 5.10.4), which dominates the IRM, contributing 81–92% to the total IRM, and a high coercivity component accounting for a small part of IRM. The first component shows a low coercivity—the $B_{1/2}$ parameter varies between 35 and 43 mT, while the high-coercivity component is characterized by $B_{1/2} = 280$ mT (Table 5.10.4). A comparison between the component analysis for soil samples (Table 5.9.1) and the concretions evidences a much higher share of the high-coercivity component in the soil compared with concretions. The coercivities of the soft components in soil samples and in the concretions are comparable, suggesting the presence of magnetite/maghemite. At the same time, the intensity of the soft component in the concretions is about five times higher than in the soils (compare Tables 5.10.4 and 5.9.1). The high coercivity component may be related to hematite and shows higher $B_{1/2}$ values for the soils than for concretions. This could indicate smaller hematite grains in the soil and low crystalline hematite in concretions. Studies on the mineralogy of soil concretions most often use spectroscopy methods and prove the presence of goethite and ferrihydrite (F. Liu et al., 2002; Cornu et al., 2005) but also hematite and maghemite (Mitsuchi, 1976; Zhang and Karathanasis, 1997; Singh and Gilkes, 1996). The advantage of the magnetic methods to detect even a trace amount of strongly magnetic minerals (e.g., Oldfield, 1999; Evans and Heller, 2003) allows us to confirm the presence of magnetite-type mineral in the concretions from the salt-affected soil S.

Table 5.10.2 EDS spectra in selected spots of concretion S-2 shown in the SEM image in Fig. 5.10.2 and "integral spectrum" (Int. Sp.) obtained for the whole diametric surface of the concretion

	O	Na	Mg	Al	Si	S	K	Ca	Ti	Mn	Fe	Ba	Cr
Int. Sp.	49.1	1.54	1.55	6.65	14.02	0.36	1.83	0.37	0.63	0.30	17.4	—	—
Sp2-1	56.71	1.35	0.95	7.6	15.47	0.22	1.91	0.42	0.90	8.8	5.27	—	0.4
Sp2-2	13.31	1.14	1.10	4.02	10.13	0.7	2.62	—	—	43.54	21.63	—	1.81
Sp2-3	51.20	7.77	—	10.2	28.99	—	—	0.43	—	1.42	—	—	—
Sp2-4	12.77	—	—	—	—	4.36	—	—	—	3.43	—	79.45	—

Elements shown in weight%.

Table 5.10.3 EDS spectra in two selected spots (Sp1 and Sp2) of nodule S-3 shown in the SEM image in Fig. 5.10.3 and "integral spectrum" (Int. Sp.) obtained for the whole diametric surface of the concretion

	O	Na	Mg	Al	Si	S	K	Cl	Ca	Ti	Mn	Fe	Ni	Cr
Int. Sp.	55.11	5.09	—	12.3	20.87	—	1.15		4.06	—	0.54	0.70	—	0.23
Sp1-2	40.90	2.15	—	4.84	5.73	0.44	1.44	1.71	0.49	—	0.60	0.46	3.51	6.52
Sp1-3	54.47	1.46	0.7	15.5	19.46	—	6.94	—	—	—	—	1.45	—	—
Sp1-4	54.51	7.69	—	10.5	25.31	—	0.17	—	1.50	—	—	0.34	—	—
Sp1-6	55.04	2.97	—	12.5	17.28	—	—	—	9.84	—	—	2.38	—	—
Sp2-1	51.8	1.73	3.4	6.64	11.0	0.47	1.26	—	0.53	0.28	18.27	3.87	—	0.71
Sp2-2	52.28	—	—	15.5	15.41	—	—	—	14.6	—	0.75	1.41	—	—

Elements shown in weight%.

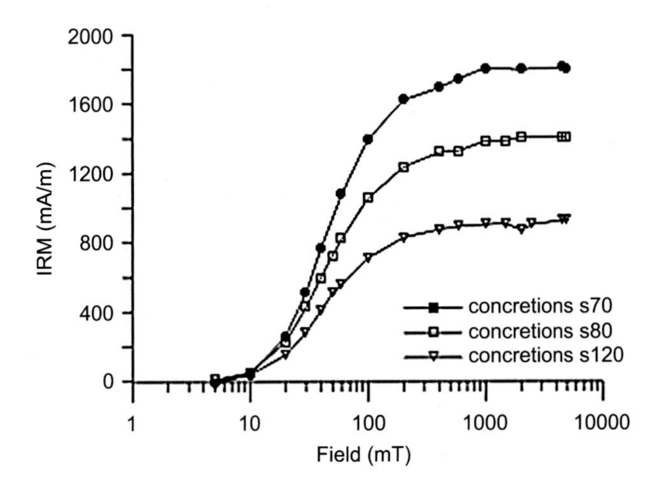

FIGURE 5.10.4

Stepwise acquisition of isothermal remanence (IRM) up to field of 5 T for samples from concretions, extracted from three profile depths from the salt-affected soil (S).

Table 5.10.4 Summary parameters from coercivity analysis of IRM acquisition curves, using IRM-CLG1.0 software (Kruiver et al., 2001) for samples from concretions extracted from different depth intervals of the salt-affected Soil S

Sample depth (cm)	IRM component 1				IRM component 2			
	Intensity ($10^{-6}Am^2/kg$)	% from IRM_{total}	$B_{1/2}$ (mT)	DP	Intensity ($10^{-6}Am^2/kg$)	% from IRM_{total}	$B_{1/2}$ (mT)	DP
Profile S concretions salt-affected soil								
70	1535.27	86	39.80	0.33	249.00	6	281.80	0.45
80	1277.78	92	43.70	0.35	111.11	8	354.80	0.50
120	751.14	81	35.50	0.33	173.40	19	281.80	0.45

The intensity, relative contribution (% from IRM_{total}), median acquisition field ($B_{1/2}$), and dispersion parameter (DP) for each coercivity component are shown.

5.11 PEDOGENESIS OF IRON OXIDES IN THE SALT-AFFECTED SOILS, AS REFLECTED IN SOIL MAGNETISM

Salt-affected soils are characterized by a strongly alkaline soil reaction (Schaetzl and Anderson, 2009). The soils containing sodic salts have an enhanced content of exchangeable Na^+ and pH > 8.5. The soil profile S shows a pH between 9.5 and 10.0 in the upper soil horizons (Fig. 5.8.1d), highlighting that it likely is of a meadow Solonetz type with an occurrence of soda (Na_2CO_3) (Shishkov and Kolev, 2014). According to Koinov et al. (1998), Shishkov and Kolev (2014) and references therein, meadow

Solonetz are secondarily salt-affected soils of initially meadow Vertisols, meadow Cinnamonic, and meadow gley soils. As was discussed in part 5.9 of this chapter, the salt-affected profile S is characterized by a very low enhancement of soil horizons with magnetite and more SD-like magnetite grains are concentrated in the A_z and B_zC horizons. It was also shown that Vertisols are distinguished by weak magnetism, but simultaneously, the presence of a small fraction SD-like magnetite. The magnetic susceptibility of salt-affected soil S is even lower than that of the Vertisols, though the characters of the variations along depth are similar (Figs. 5.4.4–5.4.7 and Fig. 5.9.3). Thus, it can be assumed that, initially, the soil S was formed as a Vertisol. In support of this assumption is the observed similarity in the morphology and the elemental content of concretions from the Vertisol SM (Jordanova and Jordanova, 2016) and the S-1 and S-2 concretions from the salt-affected profile S. Therefore, a question to be answered arises—at what stage of the soil development were the identified iron oxides (magnetite and hematite) formed in the salt-affected soil? As was discussed in the studies of Vertisols, the main processes influencing the iron oxyhydroxide forms in different soil horizons are the reductive dissolution of the lithogenic iron oxide minerals (Grimley and Arruda, 2007) and the suggested subsequent formation of a small amount of magnetite of fine grain size (Hansel et al., 2011). Thus, it is important to evaluate the characteristics of the magnetite phase found in the salt-affected soil versus magnetite, identified in the Vertisols. This comparison could be done through the results of the coercivity component analysis obtained for these soil profiles. Considering together Fig. 5.4.3 and Table 5.10.4, it is obvious that the coercivity parameter $B_{1/2}$ of the "magnetite" component in the Vertisols is significantly higher than the $B_{1/2}$ for the S profile. However, the coercivity of magnetite also depends on the degree of the isomorphous substitutions in the lattice, especially Al, which is the most common Fe-substituting ion in soils (Cornell and Schwertmann, 2003). Such substitutions cause an increase in the coercivity of magnetite. This could be a possible reason for the observed differences in the magnetic grain size of magnetite in the Vertisol SM and the salt-affected soil S.

Many laboratory studies provide evidence that, in alkaline solutions (with $pH > 8$), the Fe^{2+} oxidation proceeds through $Fe(OH)_2$ dehydration and results in magnetite formation (Cornell and Schwertmann, 2003), after which the pH becomes neutral (Usman et al., 2012). In these studies, the possible effects of salts in the solution are not discussed. Model experiments on the influence of highly alkaline and saline solutions on soils and sediments (Qafoku et al., 2007) show that, as a result of the reductive dissolution, Fe ions are detached from the soil minerals (biotite, smectite, chlorite) into the soil solution and precipitate afterward as hematite and goethite. The strongest effect is observed at a high pH and lack of Al ions. Another study (Das et al., 2011) demonstrates that at a high pH ($pH = 10$), ferrihydrite transforms to hematite as well. As was shown in the magnetic properties of the salt-affected soil, in addition to magnetite, hematite with a relatively high coercivity was also detected in the profile (Table 5.10.4). Taking into account that no hematite was identified in the study of Vertisols, one possible explanation of its presence in the S profile is its formation during the stage of salinization by the groundwater. If such a hypothesis is accepted, then it means that the most intensive processes occurred in the A_{zn} and B_z horizons, where the S ratio is the lowest and the HIRM is maximum (Fig. 5.9.4). The magnetite phase identified in the soil horizons, therefore, is formed at an earlier stage in the soil development—when it was formed as a Vertisol. The relatively high values of the Feo/Fed ratio (Table 5.8.1) obtained also suggest the presence of amorphous iron oxides, mainly ferrihydrite (Cornell and Schwertmann, 2003). The intense processes of weathering and reductive dissolution in the salt-affected soil studied are thus expressed in the exclusively weak magnetic properties related to the concentration of the iron oxides.

Very few studies present the magnetic properties of salt-affected soils and they come from Solonetz and Solonchak soils from Russia (Alekseeva et al., 2010; Tatyanchenko et al., 2013; Vodyanitskii et al., 2014). The studies, however, provide in general only the magnetic susceptibility of the different soil horizons, and only in Alekseeva et al. (2010) are a few other parameters shown. A summary of the available published magnetic data on salt-affected soils is shown in Table 5.11.1.

A comparison between the magnetic data obtained for the secondary salt-affected soil from Bulgaria (the S profile) and the saline/sodic soils from Russia show that the lowest magnetic susceptibility is obtained for the S profile. In the other Solonchak-Solonetz soils, magnetic susceptibility is relatively high in the uppermost A horizons, reaching 50–60 ($\times 10^{-8}$ m^3/kg) (Table 5.11.1), but it should be noted that most of these soils are identified as "chestnut soil" affected by salinization/sodification and no gleying is indicated, except for the profile reported by Vodyanitskii et al. (2014). But even in this gleyed Solonchak, the magnetic susceptibility is relatively high. The observed differences in the values of magnetic susceptibility could be due to the fact that the salt-affected soil S developed originally as a Vertisol and afterward was affected by mineralized groundwater causing salinity/sodicity. Moreover, as noted in the profile's description, the groundwater table was in the frames of the soil profile at the time of sampling. Thus, an active gleying most likely significantly influences the redox state of iron in the soil S. Despite the differences in magnetic susceptibility, in all salt-affected soils the main magnetic minerals identified are magnetite and hematite [only in Alekseeva et al. (2010) is goethite supposed, but on the basis of the $HIRM_{100}$ parameter, which is more indicative of hematite (e.g., Liu et al., 2012)]. Magnetic studies carried out on the Bulgarian profile (S) and the Russian Solonchakous Solonetz (Alekseeva et al., 2010) evidence the presence of stable SD-like magnetite in the upper soil horizons (A_{zn} and B_z in the S profile and in the B1 horizon of the Russian profiles). Magnetite's presence is also confirmed in the gley Solonchak (Vodyanitskii et al., 2014) by a Moessbauer analysis, but the authors argue that the magnetite is stoichiometric and well crystalline. Hematite in this study is claimed to be of lithogenic origin and mostly dissolved as a result of water-logging (Vodyanitskii et al., 2014), which is also reflected in very high Fe_o/Fe_d ratios (Table 5.11.1). In contrast, hematite in the S profile is supposed to be of pedogenic origin, a result of iron transformations in strongly alkaline media (Qafoku et al., 2007). The values of the CIA index (the chemical index of alteration) obtained for the S profile are comparable to the values reported for the Russian profiles (Table 5.11.1) and indicate relatively strong weathering.

Therefore, magnetism of salt-affected soils strongly depends on the source of salinization/sodification, as well as on the mechanism of the salt's influence—as a result of climate conditions (hyperarid climate, sea salt aerosols influence, groundwater-induced salinization, etc.). In the case of a secondary imposed salinization (as is the case of the S profile), the initial soil type and related forms of iron oxides play an important role in further iron oxide transformations.

5.12 LEPTOSOLS - MAIN CHARACTERISTICS, FORMATION PROCESSES, AND DISTRIBUTION

Leptosols are shallow genetically young soils in which pedogenic development is restricted to the formation of a thin humic (A) horizon, overlying an incipient illuvial (B) horizon or the parent rock directly (FAO, 2001). Leptosols correlate with "Lithic" subgroups in many other national classification systems. Commonly, Leptosols developed on hard limestone, marls, marbles, and other parent rocks are rich in carbonates and termed "Rendzinas." The main pedogenic process in Rendzina soils is the

Table 5.11.1 Summary of the major magnetic parameters of salt-affected soils according to published data and the present study

Location	Soil type	MAT (°C)	MAP (mm)	χ (10^{-8} m^3/kg)	χ_{ARM}/SIRM (10^{-3} m/A)	Magnetic minerals	Fe_o/Fe_d	CIA	Reference
Bulgaria	Solonetz-Solonchak (S profile)	12.0	561	A_z—9.0 B_z—6.1	A_z—0.52 B_z—0.57	A_z and B_z: SD/PSD magnetite hematite	A_z–0.37 B_z–0.27	61 67	This study
Russia lower Volga region, Abganerovo group of kurgans	Abganerovo site, D-451 pit Medium-deep Solonchakous Solonetz	7.0	380	A1—46 B1–63	A1—0.65 B1—0.74	A1–goethite/hematite B1—SD magnetite		75 79	Alekseeva et al. (2010)
Russia Lower Volga region Malyaevka group of kurgans	Pit D-482 deeply Solonchakous Solonetz	7.0	370	A1—40 B1 —43	A1—0.47 B1—0.59	A1–goethite/hematite B1—SD magnetite		77 82	Alekseeva et al. (2010)
Russia Lower Volga region Malyaevka group of kurgans	Malyaevka Pit D-484. Solonetzic and deeply Solonchakous light chestnut soil	7.0	370	A1—38 B1—44	A1—0.63 B1—0.84	A1–goethite/hematite B1—SD magnetite		78 83	Alekseeva et al. (2010)
Russia Novonukutskii District of Irkutsk oblast	Gley Solonchak (profile 4)			AU—56 S—31 C_{ca}—32 $D_{1g,ca,s}$—9.0		Magnetite hematite	AU—0.63 S—0.75 C_{ca}—0.69 $D_{1g,ca,s}$—0.8		Vodyanitskii et al. (2014).
Ergeni Upland (the Iki-Burul district of the Republic of Kalmykia)	Profile B-210, modern deeply Solonchakous light chestnut soil			A1–56 B1–66 B2–29 BC_{ca}–17 C–22				66 69 44 35 53	Tatyanchenko et al. (2013).
Ergeni Upland (the Iki-Burul district of the Republic of Kalmykia)	Profile K-4, residual Solonetzic deeply Solonchakous chestnut paleosol (600 BP)			A1–63 B1–55 $B2_{ca}$–24 BC_{ca}–16				67 69 45 32	Tatyanchenko et al. (2013).

Mean value per horizon of magnetic susceptibility (χ), ratio χARM/SIRM, and the main magnetic minerals identified are shown. Climate parameters (MAT, MAP) and the weathering index CIA are displayed as well.

dissolution of carbonates and their leaching out of the soil profile. In the Ca-rich soils, H_2O and H_2CO_3 are the main reagents at the initial phase of soil formation. As a result, Ca^{2+}, Mg^{2+}, K^+, and Na^+ cations are released from the soil and alkalinity increases due to HCO_3 (Egli et al., 2008). The atmosphere provides a pool for CO_2 and for oxidants needed in the weathering. The soil biota is involved in the weathering by providing organic ligands and acids and by delivering increased CO_2 in the soil. The formation of smectite clays during the pedogenesis causes the appearance of a blocky structure in some Leptosols developed on carbonate-rich parent rocks. Because of the presence of a significant amount of carbonates in the solum, the soil reaction (pH) in Leptosols is neutral-to alkaline, which prevents the clay minerals' translocation. As a result, the pedogenesis is expressed in the formation of a thin but homogeneous solum rich in humic substances (Schaetzl and Anderson, 2009). The impeded mobility of the clay and the thin soil profile that becomes available to biological activity throughout its depth condition the homogeneity of the Rendzina's composition. Humification and the following mineralization of the humic substances cause the creation of a thin humus coating on the mineral particles and on the surface of pedons, causing the observed brown-to-black soil color. Because of the restricted depth of the soil profile, rock fragments (limestone pieces) are usually abundant in the solum. Leptosols develop in different climatic conditions and on different lithology, but they are usually well-drained soils with good aeration. The physical, chemical, and biological properties of non-calcareous Leptosols are determined by the characteristics of the parent material and the climate. Rendzinas usually have better physical and chemical properties than do non-calcareous Leptosols. In mountainous areas, Leptosols are developed on acid rocks (e.g., granites) and they are referred to as "Rankers." Rankers have an acid soil reaction (usually pH varies between 4.5 and 5.9), and they form in a cold climates with high precipitation (Tóth et al., 2008).

Leptosols are the most extensive Reference soil group worldwide – they are spread in all climatic zones and altitudinal belts, especially in the mountainous regions (Fig. 5.12.1). They occupy large areas in Asia and South America, in the Saharan and Arabian deserts, the Ungava peninsula of northern Canada and in the Alaskan Mountains (FAO, 2001). Leptosols are also characteristic for areas largely influenced by erosion.

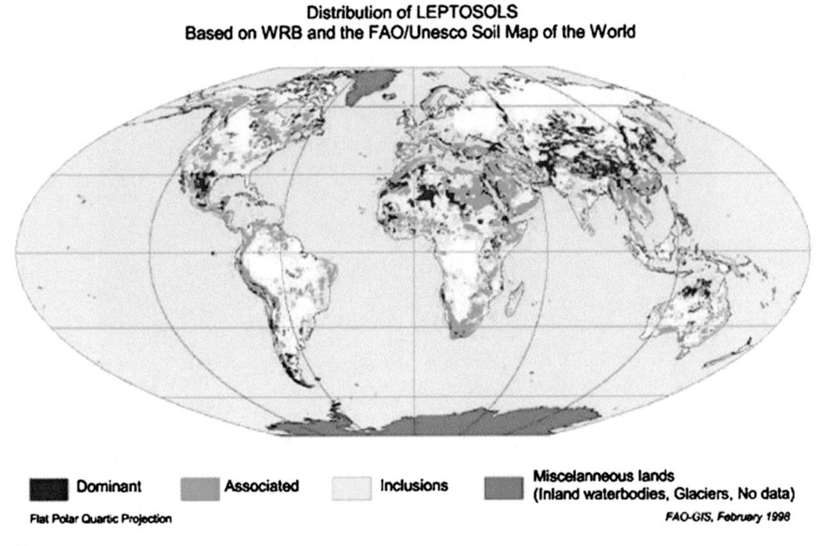

FIGURE 5.12.1

Spatial distribution of Leptosols worldwide (FAO, 2001). Lecture notes on the major soils of the world). http://www.fao.org/docrep/003/y1899e/y1899e12.htm#P353_47671.

In the European Union, Leptosols are also widely spread and occupy >10% of the area (Tóth et al., 2008). The largest regions covered by Leptosols are characteristic for the Mediterranean, and they are the main soil group in Cyprus, Greece, Spain, and France.

In Bulgaria, Rendzinas are identified in the karst hilly and mountain areas in the western Balkans area and the Fore-Balkan, Kraishte, and western Rhodopi mountain, Slavjanka mountain, and southeastern Bulgaria (Ninov, 2002; Shishkov and Kolev, 2014). Rendzina's soil cover is patchy, disrupted by bare limestones and karst forms. Shallow humus-carbonate soils are commonly 15–20 cm thick, while others could have a thickness of >50 cm (Koinov et al., 1998).

Rankers are developed on silicate rocks, and the profile is usually 10–50 cm thick. They are spread in mountainous areas on relatively steep slopes. Parent rocks are mostly granites, rhiolites, sandstones, granite-gneisses, crystalline shistst, andesites, etc. (Koinov et al., 1998). Rankers are found in the eastern Rhodopes, the Fore-Balkan, Sakar, Strandja, Osogovo, Ograjden, and Belasitza mountains (Teoharov, 1991; Ninov, 2002).

5.13 DESCRIPTION OF THE PROFILES STUDIED

5.13.1 RENDZIC LEPTOSOL (HUMUS-CARBONATE RENDZINA), PROFILE DESIGNATION: RZ

Situated close to Ignatievo village, Varna district (northeast Bulgaria)

Location: N 43°15′40.5″; E 27°45′59.7″, h = 194 m a.s.l.

Present-day climate conditions: MAT = 11.5°C; MAP = 522 mm, vegetation cover: grassland; relief: flat

Profile description: profile view is shown in Fig. 5.13.1:

FIGURE 5.13.1

View of the Rendzic Leptosol—profile Rz.

0–30 cm: humic (A_h) horizon, light-brown color, crumbly, carbonate inclusions

30–45 cm: transitional AC horizon, gray-brown color, abundant carbonates, and rock pieces

45–50 cm: R horizon, white-yellow color, weathered limestone

5.13.2 RENDZIC LEPTOSOL (CHESTNUT HUMUS-CARBONATE RENDZINA), PROFILE DESIGNATION: RZB

Situated close to Dobrich village, Dimitrovgrad district, south central Bulgaria

Location: N 41°59′49.3″; E 25°31′18.2″, h = 199 m a.s.l.

Present day climate conditions: MAT = 12.6°C; MAP = 620 mm, vegetation cover: grassland; relief: moderate slope

Profile description: profile details are shown in Figs. 5.13.2–5.13.4:

0–40 cm: humic AC horizon, brown color, crumbly, sandy, vegetation roots, and rock fragments

40–60 cm: R horizon, disintegrated rock pieces

Parent rock: Neogene sandstones/gravels

FIGURE 5.13.2

Rendzic Leptosol RzB. AC horizon—topmost part, detail.

FIGURE 5.13.3

Rendzic Leptosol RzB. AC horizon—bottom part, detail.

FIGURE 5.13.4

Rendzic Leptosol RzB—parent rock.

5.13.3 RANKER, PROFILE DESIGNATION: DRG

Situated close to Dragoman town, northwest Bulgaria

Location: N 42°54′41.0″; E 22°55′40.9″, h = 760 m a.s.l.

Present day climate conditions: MAT = 8.9°C; MAP = 535 mm, vegetation cover: grassland; relief: flat

Profile description:

0–10 cm: humic (A) horizon, light-brown color, crumbly, carbonates abundant

10–25 cm: transitional AC horizon, dark-brown color, carbonates, and rock pieces

25–30 cm: R horizon, beige-yellow color, weathered tephroidal flish sediments with andesitic composition

5.14 TEXTURE, SOIL REACTION, AND MAJOR GEOCHEMISTRY OF LEPTOSOLS

The mechanical composition is studied for selected samples from the profiles of Rendzinas Rz and RzB. The obtained results are shown in Fig. 5.14.1. Both profiles are dominated by the silt content, comprising ~60% in the Rz profile and even higher, ~75–80%, in the RzB profile (Fig. 5.14.1a–f). The clay fraction accounts for only 10–15% of the textural composition, while the sand fraction

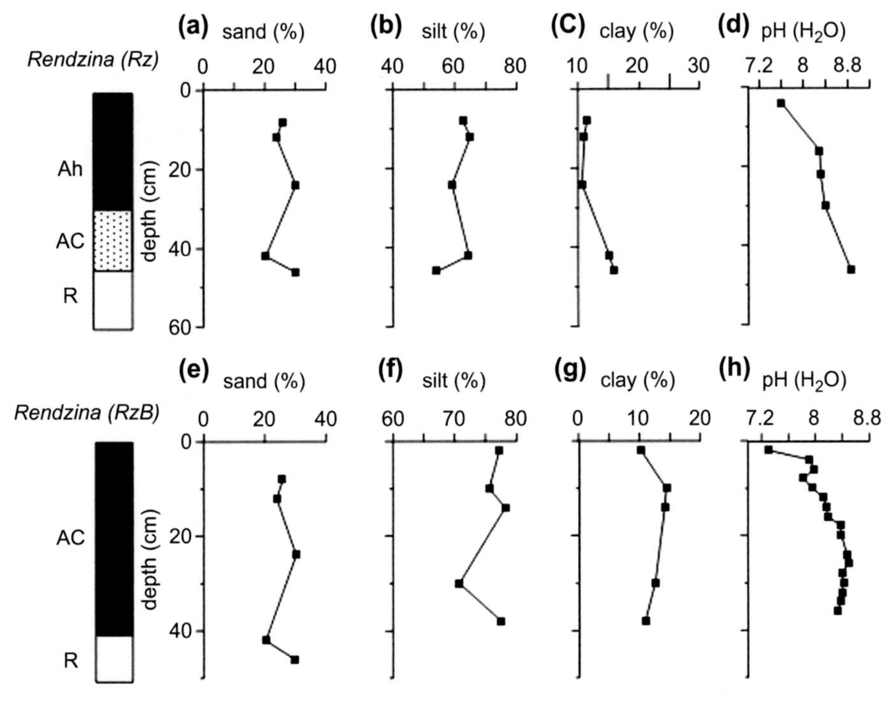

FIGURE 5.14.1

Texture (sand, silt, clay) and soil reaction (pH) for selected samples from Leptosols Rz (a–d) and RzB (e–h).

comprises ~25% in the Rz and ~15% in the RzB profiles. No systematic depth variations could be inferred, and the mechanical composition remains relatively homogeneous along the profiles. This result is in agreement with the view that Rendzic Leptosols are characterized by a homogeneous soil profile (Schaetzl and Anderson, 2009). The soil reaction is neutral to slightly alkaline in the lower part of the two profiles (Fig. 5.14.1d,h). A well-expressed trend of increasing pH in depth reflects the pedogenic formation of a soil driven by bioclimatic factors.

Variations in the major chemical elements along the Rendzic Leptosols Rz and RzB show a consistent behavior (Figs. 5.14.2 and 5.14.3), characterized by maximum concentration at the surface and a gradual decrease toward the R horizon. An exception is the Ca content (Figs. 5.14.2b and 5.14.3b), which increases due to the Ca-rich parent rocks. Both profiles (Rz and RzB) display similar variations in Fe, Ti, and Al, despite the fact that the concentration of these elements is enhanced in RzB, compared with the Rz profile.

The observed distribution of the major chemical elements reveals a gradual decay in the intensity of the pedogenic processes with increasing depth, so, as in the R horizon, the limestone (the parent rock) composition is uncovered. Carbonate dissolution is the key process, dictating the relative enhancement of the soil with the other chemical elements (Fe, Al, Si, etc.) (Egli et al., 2008). The obtained geochemical data on the elemental composition of the Rendzic Leptosols Rz and RzB agree well with the data for similar soils integrated into the systematic database of the soils in Bulgaria (Teoharov, 2009).

Dithionite and oxalate extraction data on selected samples from the profiles of Rendzic Leptosols are summarized in Table 5.14.1. Similar to the observed difference in the content of the major elements

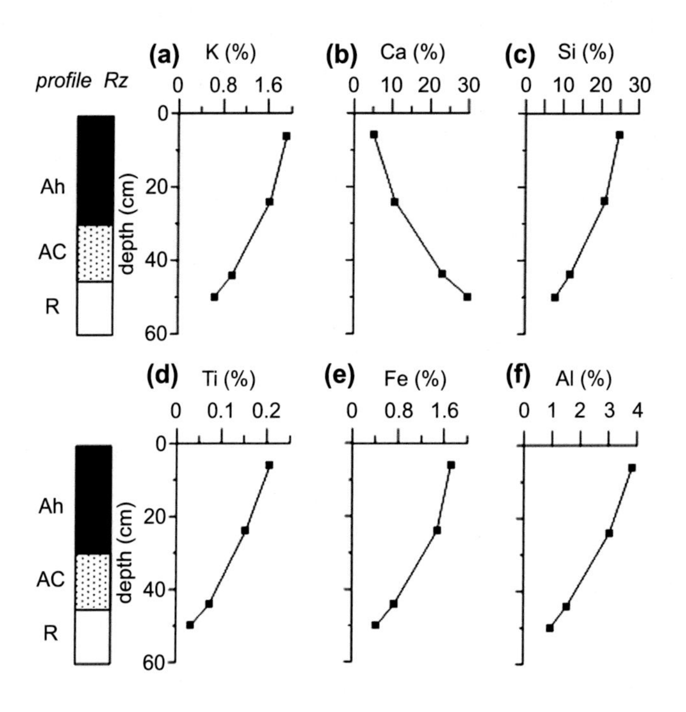

FIGURE 5.14.2

Content of some major and trace elements (K, Ca, Si, Ti, Fe, Al) for selected samples along the depth of Rendzic Leptosol Rz.

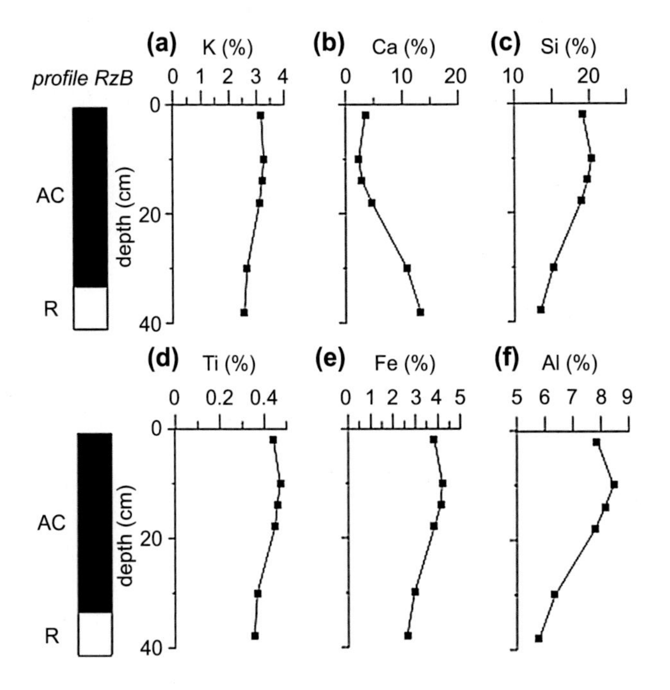

FIGURE 5.14.3

Content of some major and trace elements (K, Ca, Si, Ti, Fe, Al) for selected samples along the depth of Rendzic Leptosol RzB.

Table 5.14.1 Chemical extraction data for selected samples from Leptosols—Rendzic Leptosol (Rz Profile) and RzB Leptosol

Sample depth (cm)	Al_o (g/kg)	Fe_o (g/kg)	Mn_o (g/kg)	Fe_d (g/kg)	Fe_{tot} (g/kg)	Al_{tot} (g/kg)	Fe_o/Fe_d	Fe_d/Fe_{tot}
Profile Rz								
6	0.713	0.209	0.31	2.00	17.18	38.22	0.10	0.12
24	0.757	0.086	0.340	1.63	14.82	30.33	0.05	0.11
44	0.447	n.d.	0.115	0.765	7.383	15.25	—	0.10
50	0.143	n.d.	0.049	0.61	4.285	9.374	—	0.14
Profile RzB								
2	2.432	0.921	1.234	16.53	38.42	78.68	0.06	0.43
18	2.897	0.859	1.156	15.7	38.36	77.85	0.06	0.41
38	2.672	0.236	0.535	10.905	26.14	57.81	0.02	0.42

Oxalate-extractable Al, Fe, and Mn (Al_o, Fe_o, Mn_o), dithionite-extractable Fe (Fe_d), total Fe content (Fe_{tot}), total Al content (Al_{tot}), and the ratios Fe_o/Fe_d and Fe_d/Fe_{tot}.

in both profiles, the extracted Fe and Al in the RzB profile are higher compared with the amounts extracted from the Rz profile. Thus, the dominant role of the parent rock's composition on the amount of the pedogenic Fe-containing fraction is demonstrated. Except for the uppermost sample of the Rz profile, the Fe_o/Fe_d ratio in all other samples from both profiles is ~0.05 (Table 5.14.1), pointing to the presence of a relatively low amount of amorphous low-crystalline pedogenic iron oxides. On the other hand, in the Rz Rendzina, the relative Fe enhancement of soil compared with the R horizon is about a factor of 4 (Table 5.14.1), while the same relation in the RzB profile is less than 2. This fact implies a much stronger weathering of the limestone in the Rz profile compared with the processes in the RzB Rendzina. Simultaneously, however, the share of the dithionite-extractable Fe in the total Fe pool (expressed by the ratio Fed/Fe_{tot}) in soil is much higher in the RzB profile (Table 5.14.1) (~0.4), while in the Rz, Fed/Fe_{tot} ~0.11. Similar relations are also evident in the Al content.

5.15 MAGNETIC PROPERTIES OF LEPTOSOLS

The magnetic mineralogy in the Rendzic Leptosols is studied through thermomagnetic analysis of magnetic susceptibility and thermal stepwise demagnetization of laboratory-induced remanence. Thermomagnetic heating–cooling runs for pilot samples from the Rz and RzB profiles are shown in Fig. 5.15.1. At first glance, the similarity between the samples from the upper soil depths and the deeper counterparts is obvious, which again underlines the important influence of the parent rock mineralogy. The shape of the heating curves $\chi(T)$ for the soil samples from the upper soil depths (Fig. 5.15.1a,c) is

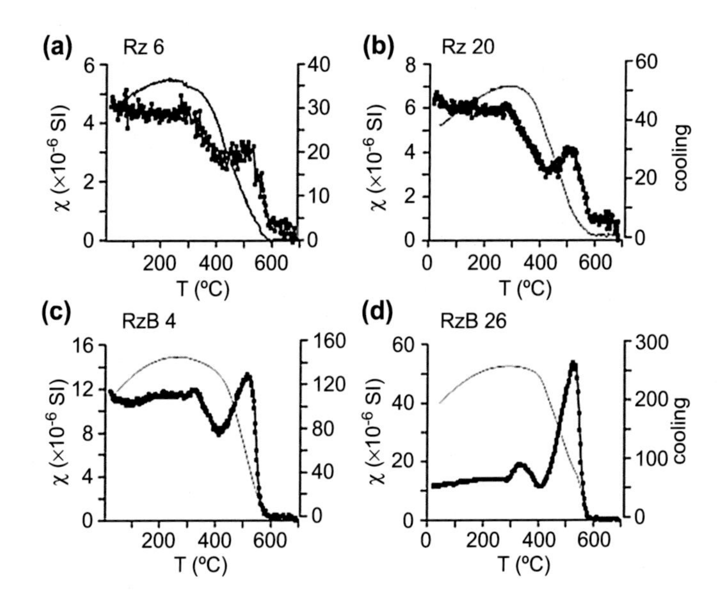

FIGURE 5.15.1

Thermomagnetic analysis of magnetic susceptibility for samples from different depths of Rz profile (a, b) and RzB profile (c, d). Heating curves are denoted by thick black lines, cooling curves relate to the secondary *y*-axis (cooling) and are denoted by thin gray lines. Heating in air. Fast heating rate applied.

typical for the behavior of other well-aerated soils (e.g., Chernozems, Phaeozems, Luvisols). The specific χ(T) heating run characterized by a weak maximum at ~300°C, followed by a decrease and a second sharp maximum at ~500°C, is related to the presence of pedogenic fine-grained maghemite and, possibly, initially existing magnetite. All the samples demonstrate strong thermal alteration on heating to 700°C, resulting in much stronger magnetic susceptibility on cooling, most likely provoked by the thermal transformations in the clay minerals during heating (e.g., Zhang et al., 2012).

Sample RzB26 (Fig. 5.15.1d) shows a very strong magnetic susceptibility peak at ~500°C compared with the other samples analyzed. Another specific feature in this heating curve is a well-distinguished peak with a maximum at ~320°C, which could be related to the presence of a well-defined maghemite fraction, which, after thermal inversion (e.g., creation of hematite) in the presence of organics, results in magnetite formation. This newly created magnetite phase is responsible for the strong maximum at 500°C and the observed Curie temperature of 590°C. Thus, the thermomagnetic analyses reveal similar pedogenic iron oxide mineralogy in the profiles of Rendzina soils, but present in a different relative concentration and grain size.

Additional information about the magnetically stable (coarser) iron oxide minerals in both profiles is obtained through stepwise thermal demagnetization of ARM and IRM (Fig. 5.15.2). Maghemite's occurrence is verified by the almost 40–50% share of a component in the ARM, unblocking at ~300°C. Another part of the ARM unblocks at 600°C in the samples from the Rz profile (Fig. 5.15.2a,b) and at 700°C in the samples from the RzB profile (Fig. 5.15.2e,f). Therefore, according to the obtained unblocking temperatures, maghemite and oxidized magnetite are the stable magnetic

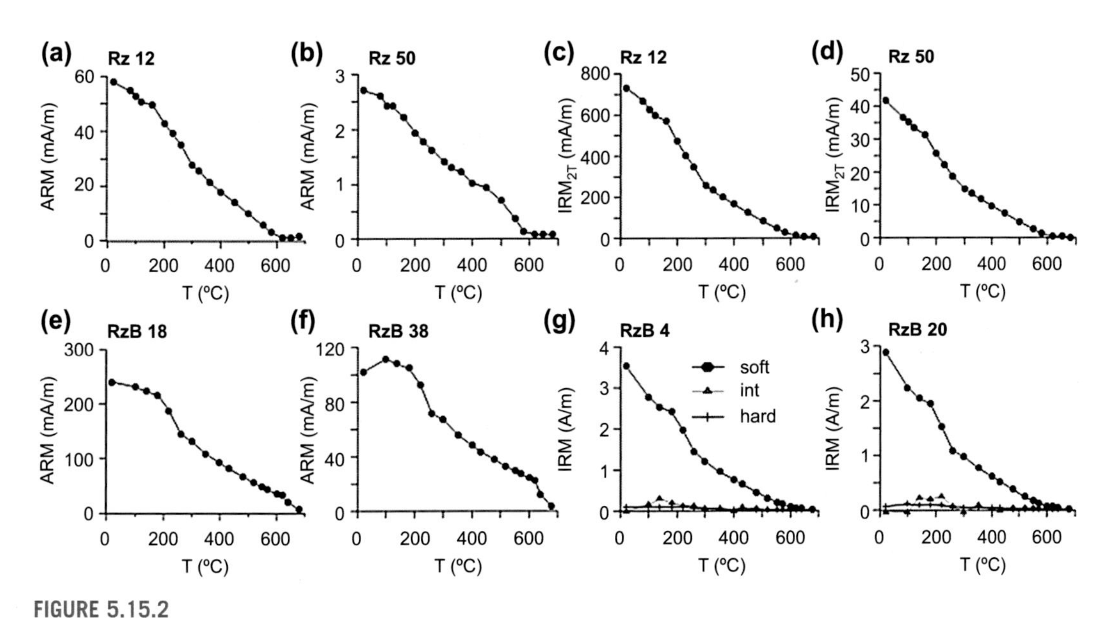

FIGURE 5.15.2

Stepwise thermal demagnetization of ARM (a, b) and IRM_{2T} (c, d) for selected samples from the Rendzic Leptosol Rz. Stepwise thermal demagnetization of ARM (e, f) and composite IRM (g, h) for selected samples from the RzB profile. Saturating fields for the composite IRM: soft component (0–0.3 T), intermediate component (0.3–0.6 T), hard component (0.6–2.0 T).

minerals in the Rz profile, while maghemite and hematite are the most stable in the RzB profile. It is remarkable that the hematite phase is involved in the ARM of the samples from the RzB profile, since usually magnetite particles acquire the most stable and intense ARM (Dunlop and Özdemir, 1997). Another possibility is to assign this unblocking temperature to thermally stable maghemite with extensive Al substitutions (Özdemir and Banerjee, 1984; Dunlop and Özdemir, 1997).

Thermal demagnetization of IRM reveals a wider unblocking temperature spectra compared with ARM (Fig. 5.15.2c,d,g,h) but essentially the same mineral phases. No high-coercivity phase indicating goethite/hematite is involved in the IRM components (Fig. 5.15.2g,h), therefore supporting the assumption that the ARM unblocking at ~700°C most likely indicates a thermally stable maghemite presence in the RzB profile.

Depth variations in magnetic concentration and grain-size-related proxies along the Rz profile are shown in Fig. 5.15.3. Magnetic susceptibility (χ), saturation magnetization (M_s), and remanences (ARM, IRM) indicate an intense magnetic enhancement through the profile, gradually decreasing from the surface toward the parent rock and a sharp drop at the bottom of the solum (Fig. 5.15.3a,b,c). An apparent internal boundary in the A_h horizon at ~12- to 15-cm depth is indicated through a systematic kink in all concentration-dependent parameters, which, in these uppermost depths of the profile, show a faster decrease in the magnetic enhancement, while downward it becomes slower. This feature is even more highlighted in the variations of the grain-size–related parameters – χ_{fd}, ARM/SIRM, χARM/χ, and χ_{fd}% (Fig. 5.15.3d–g). The superparamagnetic fraction, approximated by the χ_{fd}, is present all along the profile with a maximum concentration in the middle part of the solum, while the observed high values of χ_{fd}%, varying between 8% and 14%, suggest a prevailing influence of this ultrafine magnetic fraction in the magnetic susceptibility signal (Dearing et al., 1996).

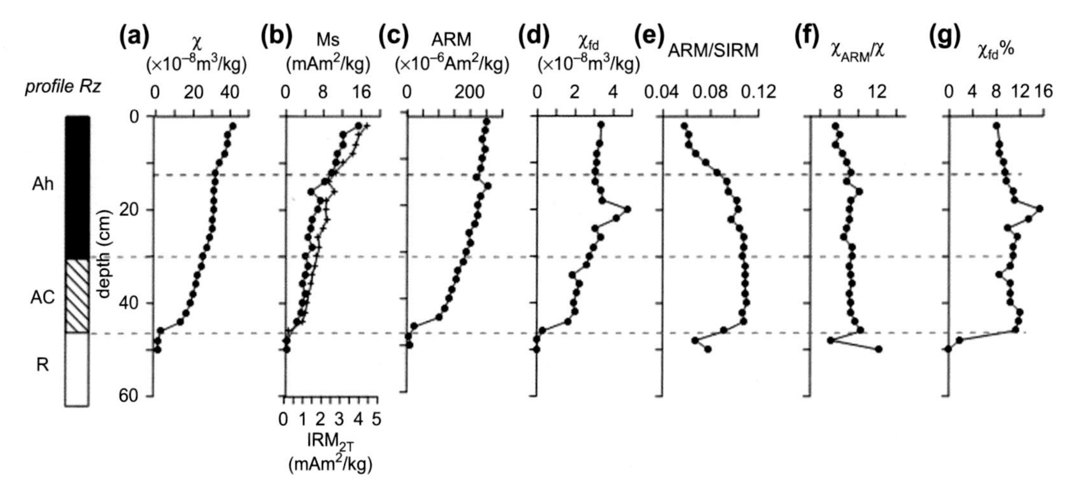

FIGURE 5.15.3

Depth variations of selected magnetic parameters and ratios for Rz Leptosol: (a) magnetic susceptibility (χ); (b) saturation magnetization (M_s) (full dots) and isothermal remanent magnetization acquired in a field of 2 T ($IRM_{2\,T}$) (crosses); (c) Anhysteretic remanence (ARM), (d) frequency-dependent magnetic susceptibility (χ_{fd}); (e) Ratio between the anhysteretic and isothermal remanences (ARM/IRM); (f) ratio between anhysteretic susceptibility (χ_{ARM}) and magnetic susceptibility (χ); (g) percent frequency-dependent magnetic susceptibility (χ_{fd}%).

A comparison between the χ_{fd} and $\chi_{fd}\%$ variations indicates that the absolute concentration of the SP grains decreases in depth, but its contribution to the total magnetic susceptibility is the highest in the bottom parts of the profile (Fig. 5.15.3d,g). SD magnetic grain-size proxies ARM/SIRM and χ_{ARM}/χ evidence coarser stable grains in the surface 0–15 cm of the profile, where the two ratios show lower values, and a well-outlined part with maximum ARM/SIRM (Fig. 5.15.3e), where the grain size is closer to the SD threshold. A sharp drop of the ratios in the R horizon indicates the pedogenic origin of the magnetic enhancement. Similar to other soil types, the ARM/SIRM ratio seems to be a more robust indicator of the stable SD magnetic grains in the Rendzic Leptosol Rz.

Depth variations of the magnetic parameters in the second Rendzic Leptosol(the RzB profile) are displayed in Fig. 5.15.4. Similarly to the Rz profile, the concentration of the magnetic minerals is maximum at the surface and decreases towards the parent rock horizon, as evidenced by χ, M_s, $IRM_{2\ T}$, ARM, as well as χ_{fd} variations (Fig. 5.15.4a–d). However, in the uppermost 0–15 cm, the magnetic parameters are almost constant, suggesting a uniform concentration of the magnetic minerals and a decline starting below. Simultaneously, grain-size magnetic proxies for the content of the SD and SP particles (Fig. 5.15.4e,f) do not show significant changes along depth, also suggesting an invariable grain size of these two magnetic fractions in the soil.

The S parameter (Fig. 5.15.4g) indicates the dominance of magnetically soft minerals in the upper ~20–25 cm of the profile and an increase in the higher coercivity fraction toward the bottom. As was discussed, such a coercivity increase could be related to the presence of highly (perhaps Al) substituted maghemite, which displays higher coercivity compared with magnetite (Dunlop and Özdemir, 1997).

The third Leptosol sampled in West Stara Planina represents a typical shallow mountainous soil developed on acid volcanoclastic sediments. Magnetic mineralogy is strongly influenced by the high content of the lithogenic iron oxides. This is evident from the thermomagentic analyses carried out, shown in Fig. 5.15.5. All heating–cooling curves from the three depth intervals are essentially the same,

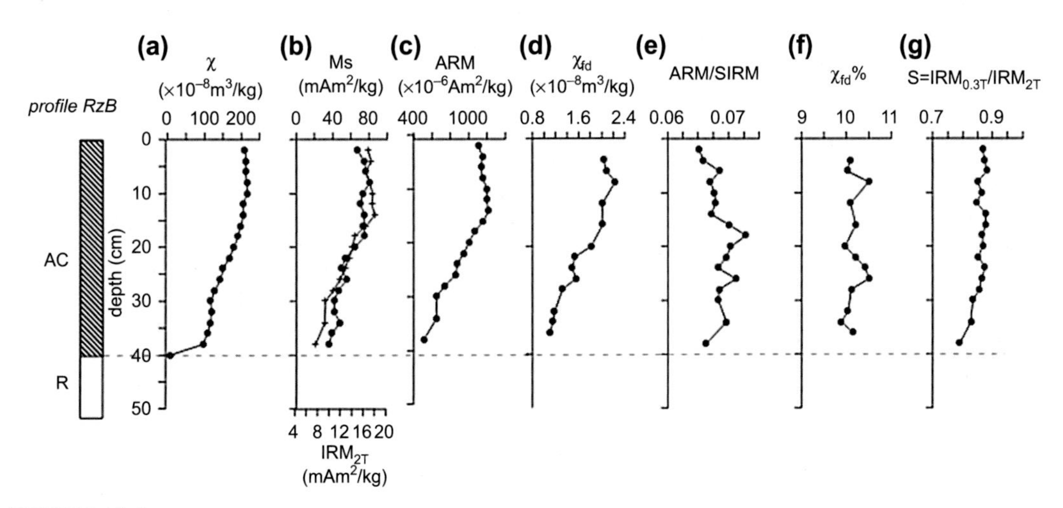

FIGURE 5.15.4

Depth variations of selected magnetic parameters and ratios for RzB Leptosol: (a) magnetic susceptibility (χ); (b) saturation magnetization (M_S) (full dots) and isothermal remanence acquired in a field of 2 T ($IRM_{2\ T}$) (crosses); (c) Anhysteretic remanence (ARM), (d) frequency-dependent magnetic susceptibility (χ_{fd}); (e) Ratio between the anhysteretic and isothermal remanences (ARM/IRM); (f) percent frequency-dependent magnetic susceptibility ($\chi_{fd}\%$); (g) S ratio.

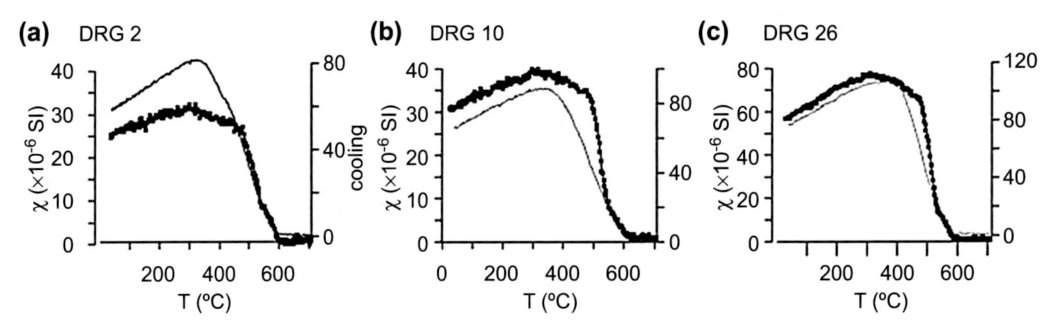

FIGURE 5.15.5

Thermomagnetic analysis of magnetic susceptibility for samples from different depths of DRG Leptosol (Ranker). Heating curves are denoted by thick black lines, cooling curves relate to the secondary *y*-axis (cooling) and are denoted by thin gray lines. Heating in air. Fast heating rate applied.

displaying a weak bump at ~300°C and two well-expressed Curie temperatures (T_c) at ~500°C and 580–600°C (Fig. 5.15.5). These T_c most likely indicate the occurrence of lithogenic fractions of titanomagnetite with a relatively high Ti content and (oxidized) magnetite (Dunlop and Özdemir, 1997). The cooling curves reveal much higher magnetic susceptibility, thus suggesting strong thermally induced transformations in the initially present mineral phases.

Variations of different magnetic characteristics along the DRG profile (Fig. 5.15.6) also show the dominant role of the lithogenic iron oxides in determining the soil's magnetic properties. In contrast to

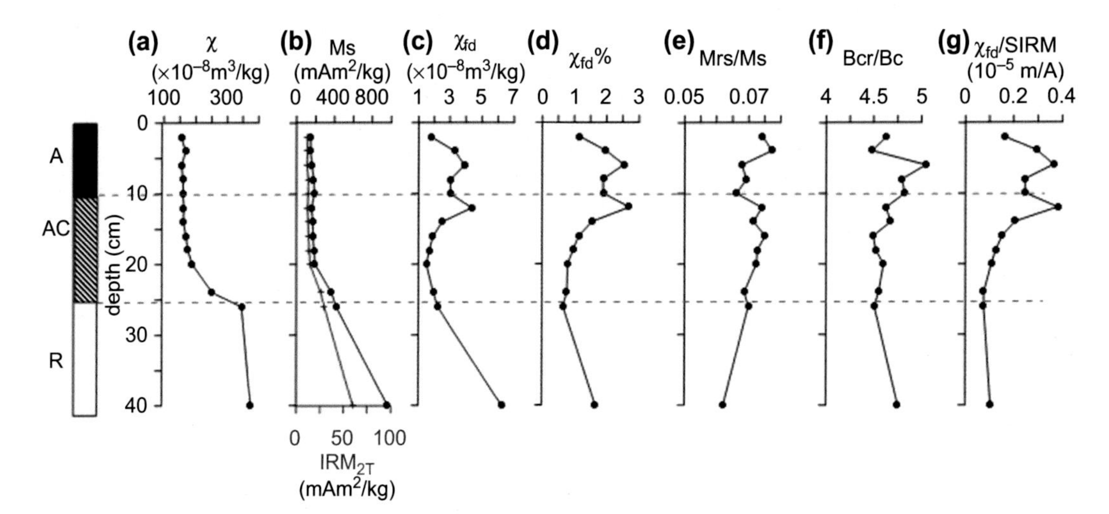

FIGURE 5.15.6

Depth variations of selected magnetic parameters and ratios for DRG Leptosol: (a) magnetic susceptibility (χ); (b) saturation magnetization (M_s) (full dots) and isothermal remanence acquired in a field of 2 T ($IRM_{2\ T}$) (crosses); (c) frequency-dependent magnetic susceptibility (χ_{fd}); (d) percent frequency-dependent magnetic susceptibility (χ_{fd}%); (e) M_{rs}/M_s; (f) B_{cr}/B_c; (g) ratio between frequency-dependent magnetic susceptibility (χ_{fd}) and saturation isothermal remanence (SIRM), acquired in 2 T field.

the Rendzic Leptosols developed on weakly magnetic calcareous rocks, the DRG soil profile is effectively magnetically depleted, as seen from the concentration-dependent characteristics χ, M_s, and IRM_{2T} (Fig. 5.15.5a,b). The concentration of iron oxides in the rock piece (the deepest sample at ~40 cm) is about four times as high as in the soil. However, the frequency-dependent magnetic susceptibility (χ_{fd}), despite its low absolute value as related to the bulk χ, clearly shows a maximum in the middle part of the soil profile (Fig. 5.15.5c,d). Thus, a certain amount of pedogenic ultrafine grained magnetite/maghemite is detected in the soil, otherwise dominated by coarse lithogenic magnetic grains. The relative share of this SP fraction in the total magnetic susceptibility is low, as evidenced by χ_{fd}% reaching only 2.5−3.0%. Nevertheless, this SP fraction biases the hysteresis parameters, lowering the B_c and M_{rs}, which is reflected in the M_{rs}/M_s and B_{cr}/B_c variations (Fig. 5.15.5e,f). Minute but clearly denoted pedogenic enhancement with SP grains is further expressed in the ratio χ_{fd}/SIRM (Fig. 5.15.5g) showing maximum values at a depth interval of 5−15 cm.

5.16 PEDOGENESIS OF IRON OXIDES IN LEPTOSOLS, AS REFLECTED IN SOIL MAGNETISM

Because of the small thickness of the Leptosols, their magnetic and geochemical characteristics are directly related to the parent rock and the specifics of its weathering (Schaetzl and Anderson, 2009; Žigova et al., 2014). Calcium-rich rocks on which the Rendzic Leptosols develop, the good aeration within the whole solum, the large amount of organics, and enhanced biological activity determine an active pedogenesis, resulting in the formation of fine-grained pedogenic iron oxides (Cornell and Schwertmann, 2003) and strong magnetic enhancement of the soil relative to the parent rock signal. Environmental factors, like the high pH, good drainage, available organics, wetting−drying interchange, etc., favor the synthesis of pedogenic magnetite/maghemite via an abiotic route or biologically induced by iron-reducing bacteria (Dearing et al., 1996; Maher, 1998; Maxbauer et al., 2016). The observed similarities in the thermomagnetic curves of magnetic susceptibility for different Rendzinas (Fig. 5.15.1) show that the pedogenic component (fine-grained magnetite/maghemite) is identical. Coarser, magnetically stable grains show maximum concentration in the surface layer and decrease with depth, while the SP fraction extends into the whole solum.

The pedogenic formation of iron oxides in the Leptosols developed on iron-rich lithology is impeded because of the radically different parent rock composition, rich in lithogenic coarse iron oxide particles, as in the case of the DRG profile. Due to the much slower weathering rate of the volcanogenic minerals, the magnetic properties of soils developed on such materials are ruled by the lithogenic minerals. Geochemical data available for the different iron forms in Leptosols from the Ograjden and Sakar mountains in Bulgaria (Teoharov, 2009) show that the crystalline Fe forms dominate all along the soil profile, accounting for >80% of the total Fe and the main rock-forming minerals are evenly distributed inside it (Teoharov, 1991). Magnetic studies fully comply with the pedologic studies but give in addition new specific information related to the finest pedogenic magnetic fraction (Fig. 5.15.5). The pedogenic SP fraction accounts for a very limited part of the magnetic signal in Leptosols developed on volcanoclastic sediments, as opposed to the Rendzinas (Figs. 5.15.3 and 5.15.4), but χ_{fd} very sensitively reflects the pedogenic processes in the soil profile. In contrast to the Rendzina soils, this SP fraction is related to the upper part of the profile.

REFERENCES

Abrol, I., Yadav, J., Massoud, F., 1988. Salt-affected Soils and Their Managment. FAO Soils Bulletin, 39, Rome.

Achyuthan, H., Flora, O., Braida, M., Shankar, N., Stenni, B., 2010. Radiocarbon ages of pedogenic carbonate nodules from Coimbatore region, Tamil Nadu. J. Geol. Soc. India 75, 791–798.

Adams, J., Kraus, M., Wing, S., 2011. Evaluating the use of weathering indices for determining mean annual precipitation in the ancient stratigraphic record. Palaeogeogr. Palaeoclimatol. Palaeoecol. 309, 358–366.

Aide, M., 2005. Elemental composition of soil nodules from two Alfisols on an alluvial terrace in Missouri. Soil Sci. 170 (12), 1022–1033.

Alekseeva, T., Alekseeva, A., Demkina, V., Alekseeva, V., Sokolowska, Z., Hajnos, M., Kalinin, P., 2010. Physicochemical and mineralogical diagnostic features of solonetzic process in soils of the Lower Volga region in the late Holocene. Eurasian Soil Sci. 43 (10), 1083–1101.

Barbiero, L., Mohan Kumard, M., Violette, A., Oliva, P., Braun, J., Kumar, C., Furian, S., Babic, M., Riotte, J., Valles, V., 2010. Ferrolysis induced soil transformation by natural drainage in Vertisols of sub-humid South India. Geoderma 156 (3–4), 173–188.

Bazylinski, D., 1996. Controlled biomineralization of magnetic minerals by magnetotactic bacteria. Chem. Geol. 132, 191–198.

Berger, I., Cooke, R., 1997. The origin and distribution of salts on alluvial fans in the Atacama Desert, Northern Chile. Earth Surf. Proc. Landf. 22, 581–600.

Bockheim, J., Hartemnik, A., 2013. Salic horizons in soils of the USA. Pedosphere 23 (5), 600–608.

Brennan, E., Lindsay, W., 1998. Reduction and oxidation effect on the solubility and transformation of iron oxides. Soil Sci. Soc. Am. J. 62 (4), 930–937.

Brinkman, R., 1970. Ferrolysis, a hydromorphic soil forming process. Geoderma 3, 199–206.

Chen, L., Zhang, G., Rossiter, D., Cao, Z., 2015. Magnetic depletion and enhancement in the evolution of paddy and non-paddy soil chronosequences. Eur. J. Soil Sci. 66, 886–897.

Cornell, R., Schwertmann, U., 2003. The Iron Oxides. Structure, Properties, Reactions, Occurrence and Uses (Weinheim, New York).

Cornu, S., Deschatrettes, V., Salvador-Blanes, S., Clozel, B., Hardy, M., Branchut, S., Le Forestier, L., 2005. Trace element accumulation in Mn—Fe—oxide nodules of a planosolic horizon. Geoderma 125, 11–24.

Das, S., Hendry, M., Essilfie-Dughan, J., 2011. Transformation of two-line ferrihydrite to goethite and hematite as a function of pH and temperature. Environ. Sci. Technol. 45, 268–275.

Day, R., Fuller, M., Schmidt, V.A., 1977. Hystereis properties of titanomagnetites: grain size and compositional dependence. Phys. Earth Planet. Inter. 13, 260–267.

de Boer, C., Dekkers, M., 1998. Thermomagnetic behavior of haematite and goethite as a function of grain size in various non-saturating magnetic fields. Geophys. J. Int. 133, 541–552.

de Jong, E., 2002. Magnetic susceptibility of Gleysolic and chernozemic soils in Saskatchewan. Can. J. Soil Sci. 82, 191–199.

Dearing, J.A., Hay, K.L., Baban, S.M.J., Huddleston, A.S., Wellington, E.M.H., Loveland, P.J., 1996. Magnetic susceptibility of soil: an evaluation of conflicting theories using a national data set. Geophys. J. Int. 127, 728–734.

Dearing, J., Lees, J., White, C., 1995. Mineral magnetic properties of acid gleyed soils under oak and Corsican Pine. Geoderma 68, 309–319.

Dekkers, M., 1988. Magnetic behavior of natural goethite during thermal demagnetization. Geophys. Res. Lett. 15 (5), 538–541.

Driese, S., Jacobs, J., Nordt, L., 2003. Comparison of modern and ancient Vertisols developed on limestone in terms of their geochemistry and parent material. Sediment. Geol. 157, 49–69.

Driese, S., Nordt, L., Lynn, W., Stiles, C., Mora, C., Wilding, L., 2005. Distinguishing climate in the soil record using chemical trends in a Vertisol climosequence from the Texas coast prairie, and application to interpreting Paleozoic paleosols in the Appalachian Basin, USA. J. Sediment. Res. 75, 339–349.

Dunlop, D., Özdemir, O., 1997. Rock magnetism. Fundamentals and frontiers. In: Edwards, D. (Ed.), Cambridge Studies in Magnetism. Cambridge University Press.

Dunlop, D., 1981. The rock magnetism of fine particles. Phys. Earth Planet. Inter. 26, 1–26.

Dunlop, D., 2002a. Theory and application of the day plot (M_{rs}/M_s versus H_{cr}/H_c) 1. Theoretical curves and tests using titanomagnetite data. J. Geophys.Res. 107 (B3), 2056. http://dx.doi.org/10.1029/2001JB000486.

Dunlop, D., 2002b. Theory and application of the day plot (M_{rs}/M_s versus H_{cr}/H_c) 2. Application to data for rocks, sediments and soils. J. Geophys.Res. 107 (B3), 2057. http://dx.doi.org/10.1029/2001JB000487.

Egli, R., 2004. Characterization of individual rock magnetic components by analysis of remanence curves. 3. Bacterial magnetite and natural processes in lakes. Physics and Chemistry of the Earth 29, 869–884.

Egli, M., Merkli, C., Sartori, G., Mirabella, A., Plötze, M., 2008. Weathering, mineralogical evolution and soil organic matter along a Holocene soil toposequence developed on carbonate-rich materials. Geomorphology 97, 675–696.

Evans, M., Heller, F., 2003. Environmental Magnetism: Principles and Applications of Enviromagnetics. Academic Press, San Diego, CA.

FAO World soil resources reports, 2001. Lecture Notes on the Major Soils of the World. http://www.fao.org/DOCREP/003/Y1899E/Y1899E00.HTM.

Fischer, H., Luster, J., Gehring, A.U., 2007. EPR evidence for maghemitization of magnetite in a tropical soil. Geophys. J. Int. 169 (3), 909–916.

Fisher, H., Luster, J., Gehring, A., 2008. Magnetite weathering in a Vertisol with seasonal redox-dynamics. Geoderma 143, 41–48.

Fredrickson, J., Zachara, J., Kennedy, D., Dong, H., Onstott, T., Hinman, N., Li, S., 1998. Biogenic iron mineralization accompanying the dissimilatory reduction of hydrous ferric oxide by a groundwater bacterium. Geochim. Cosmochim. Acta 62, 3239–3257.

Gasparatos, D., 2013. Sequestration of heavy metals from soil with Fe-Mn concretions and nodules. Environ. Chem. Lett. 11 (1), 1–9.

Gasparatos, D., Tarenidis, D., Haidouti, C., Oikonomou, G., 2005. Microscopic structure of soil Fe-Mn nodules: environmental implication. Environ. Chem. Lett. 2, 175–178.

Gehring, A., Guggenberger, G., Zech, W., Luster, J., 1997. Combined magnetic, spectroscopic, and analytical-chemical approach to infer genetic information for a Vertisol. Soil Sci. Soc. Am. J. 61 (1), 78–85.

Gehring, A., Hofmeister, A., 1994. The transformation of lepidocrocite during heating: a magnetic and spectroscopic study. Clays Clay Miner. 42 (4), 409–415.

Geiss, C.E., Egli, R., Zanner, C.W., 2008. Direct estimates of pedogenic magnetite as a tool to reconstruct past climates from buried soils. J. Geophys. Res. 113 (B11), 102. http://dx.doi.org/10.1029/2008JB005669.

Grimley, D., Arruda, N., 2007. Observations of magnetite dissolution in poorly drained soils. Soil Sci. 172 (12), 968–982.

Hanesch, M., Stanjek, H., Petersen, N., 2006. Thermomagnetic measurements of soil iron minerals: the role of organic carbon. Geophys. J. Int. 165, 53–61.

Hansel, C., Benner, S., Fendorf, S., 2005. Competing Fe(II)-induced mineralization pathways of ferrihydrite. Environ. Sci. Technol. 39, 7147–7153.

Hansel, C., Benner, S., Netss, J., Dohnalkova, A., Kukkadapu, R., Fendorf, S., 2003. Second mineralization pathways induced by dissimilatory iron reduction of ferrihydrite under advective flow. Geochim. Cosmochim. Acta 67 (16), 2977–2992.

Hansel, C., Learman, D., Lentini, C., Ekstrom, E., 2011. Effect of adsorbed and substituted Al on Fe(II)-induced mineralization pathways of ferrihydrite. Geochim. Cosmochim. Acta 75, 4653–4666.

Hartstra, R., 1983. TRM, ARM and Isr of two natural magnetites of MD and PSD grain size. Geophys. J. R. Astron. Soc. 73, 119–131.

Hatfield, R., 2014. Particle size-specific magnetic measurements as a tool for enhancing our understanding of the bulk magnetic properties of sediments. Minerals 4, 758–787.

Hu, P., Liu, Q., Torrent, J., Barron, V., Jin, C., 2013. Characterizing and quantifying iron oxides in Chinese loess/ paleosols: implications for pedogenesis. Earth Planet. Sci. Lett. 369-370, 271–283.

IUSS Working Group WRB, 2014. World Reference Base for Soil Resources 2014. International Soil Classification System for Naming Soils and Creating Legends for Soil Maps. World Soil Resources Reports No. 106. FAO, Rome.

Jelenska, M., Hasso-Agopsowicz, A., Kopcewicz, B., 2010. Thermally induced transformation of magnetic minerals in soil based on rock magnetic study and Mössbauer analysis. Phys. Earth Planet. Inter. 179, 164–177.

Jelić, M., Milivojević, J., Trifunović, S., Dalović, I., Milošev, D., Šeremešić, S., 2011. Distribution and forms of iron in the vertisols of Serbia. J. Serb. Chem. Soc. 76 (5), 781–794.

Jiang, Zh, Liu, Q., Barrón, V., Torrent, J., Yu, Y., 2012. Magnetic discrimination between Al-substituted hematites synthesized by hydrothermal and thermal dehydration methods and its geological significance. J. Geophys. Res. 117, B02102.

Jordanova, D., Jordanova, N., 1999. Magnetic characteristics of different soil types from Bulgaria. Stud. Geophys. Geod 43, 303–318.

Jordanova, D., Jordanova, N., 2016. Rock magnetic and geochemical characteristics of relict Vertisols – signs of past climate and recent pedogenic development. Geophys. J. Int. 205, 1437–1454.

Jordanova, D., Jordanova, N., Petrov, P., Tsacheva, T., 2010. Soil development of three Chernozem-like profiles from North Bulgaria revealed by magnetic studies. Catena 83, 158–169.

Koinov, V., Kabakchiev, I., Boneva, K., 1998. In: Koinov, V., Boneva, K. (Eds.), Atlas of Soils in Bulgaria. Zemizdat, Sofia.

Konhauser, K., 1998. Diversity of bacterial iron mineralization. Earth-Sci. Rev. 43, 91–121.

Kovda, I., Mora, C., Wilding, L., 2006. Stable isotope compositions of pedogenic carbonates and soil organic matter in a temperate climate Vertisol with gilgai, southern Russia. Geoderma 136, 423–435.

Kruiver, P., Dekkers, M., Heslop, D., 2001. Quantification of magnetic coercivity components by the analysis of acquisition curves of isothermal remanent magnetization. Earth Planet. Sci. Lett. 189, 269–276.

Lindquist, A., Feinberg, J., Waters, M., 2011. Rock magnetic properties of a soil developed on an alluvial deposit at Buttermilk Creek, Texas, USA. Geochem. Geophys. Geosyst. 12 (12) http://dx.doi.org/10.1029/ 2011GC003848. Q12Z36.

Liu, F., Colombo, C., Adamo, P., He, J., Violante, A., 2002. Trace elements in manganese-iron nodules from Chinese Alfisol. Soil Sci. Soc. Am. J. 66, 171–661.

Liu, Q., Banerjee, S., Jackson, M., Zhu, R., Pan, Y., 2002. A new method in mineral magnetism for the separation of weak antiferromagnetic signal from a strong ferrimagnetic background. Geophys. Res. Lett. 29 (12), 1565. http://dx.doi.org/10.1029/2002GL014699.

Liu, Q., Deng, Ch, Yu, Y., Torrent, J., Jackson, M., Banerjee, S., Zhu, R., 2005. Temperature dependence of magnetic susceptibilty in an argon environment: implications for pedogenesis of Chinese loess/palaeosols. Geophys. J. Int. 161, 102–112.

Liu, Q., Hu, P., Torrent, J., Barrón, V., Zhao, X., Jiang, Z., Su, Y., 2010. Environmental magnetic study of a Xeralf chronosequence in northwestern Spain: indications for pedogenesis. Palaeogeogr. Palaeoclimatol. Palaeoecol. 293, 144–156.

Liu, Q., Roberts, A., Larrasoaña, J., Banerjee, S., Guyodo, Y., Tauxe, L., Oldfield, F., 2012. Environmental magnetism: principles and applications. Rev. Geophys. 50. RG4002.

Liu, Q., Roberts, A., Torrent, J., Horng, Ch-Sh, Larrasoaña, J., 2007. What do the HIRM and *S*-ratio really measure in environmental magnetism? Geochem. Geophys. Geosyst. 8 http://dx.doi.org/10.1029/2007GC001717. Q09011.

Lu, S.-G., Zhu, L., Yu, J.-Y., 2012. Mineral magnetic properties of Chinese paddy soils and its pedogenic implications. Catena 93, 9–17.

Maher, B., 1986. Characterization of soils by mineral magnetic measurements. Phys. Earth Planet. Inter. 42, 76–92.

Maher, B.A., 1998. Magnetic properties of modern soils and Quaternary loessic paleosols: paleoclimatic implications. Palaeogeogr. Palaeoclimatol. Palaeoecol 137, 25–54.

Maxbauer, D., Feinberg, J., Fox, D., 2016. Magnetic mineral assemblages in soils and paleosols as the basis for paleoprecipitation proxies: a review of magnetic methods and challenges. Earth-Sci. Rev. 155, 28–48.

Mitov, I., Paneva, D., Kunev, B., 2002. Comparative study of the thermal decomposition of iron oxyhydroxides. Thermochim. Acta 386, 179–188.

Mitsuchi, M., 1976. Characteristics and genesis of nodules and concretions occurring in soils of the R. Chinit area, Kompong Thom Province, Cambodia. Soil Sci. Plant Nutr. 22 (4), 409–421.

Montagne, D., Cornu, S., Le Forestier, L., Hardy, M., Josiére, O., Caner, L., Cousin, I., 2008. Impact of drainage on soil-forming mechanisms in a French Albeluvisol: input of mineralogical data in mass-balance modelling. Geoderma 145, 426–438.

Morgan, K., Jankowski, W., 2004. Saline groundwater seepage zones and their impact on soil and water resources in the Spicers Creek catchment, central west, New South Wales, Australia. Environ. Geol. 46, 273–285.

Moustakis, N., 2012. A study of Vertisol genesis in North Eastern Greece. Catena 92, 208–215.

Ninov, N., 2002. Soils. In: Kopralev, I., Yordanova, M., Mladenov, C. (Eds.), Geography of Bulgaria. Institute of Geography – Bulg. Acad. Sci., ForCom, Sofia, pp. 277–317.

Nordt, L., Driese, S., 2009. Hydropedological model of vertisol formation along the Gulf Coast Prairie land resource area of Texas. Hydrol. Earth Syst. Sci. 13, 2039–2053.

Nordt, L., Driese, S., 2010a. A modern soil characterization approach to reconstructing physical and chemical properties of paleo-vertisols. Am. J. Sci. 310, 37–64.

Nordt, L., Driese, S., 2010b. New weathering index improves paleorainfall estimates from Vertisols. Geology 38 (5), 407–410. http://dx.doi.org/10.1130/G30689.1.

Nordt, L., Wilding, L., Lynn, W., Crawford, C., 2004. Vertisol genesis in a humid climate of the coastal plain of Texas, USA. Geoderma 122, 83–102.

Oldfield, F., 1999. The rock magnetic identification of magnetic mineral and magnetic grain size assemblages. In: Walden, J., Oldfield, F., Smith, J. (Eds.), Environmental Magnetism: A Practical Guide. Quaternary Research Association, Technical Guide No. 6, London, ISBN 9780907780427, pp. 98–112.

Özdemir, Ö., Banerjee, S., 1984. High temperature stability of maghemite (γ-Fe_2O_3). Geophys. Res. Lett. 11 (3), 161–164.

Pal, D., Bhattacharyya, T., Chandran, P., Ray, S., Srivastava, P., Durge, S., Bhuse, S., 2006. Significance of soil modifiers (Ca-zeolites and gypsum) in naturally degraded Vertisols in the peninsular India in redefining the sodic soils. Geoderma 136, 210–228.

Pal, D., Wani, S., Sahrawat, K., 2012. Vertisols of tropical Indian environments: pedology and edaphology. Geoderma 189-190, 28–49.

Palumbo, B., Bellanca, A., Neri, R., Roe, M., 2001. Trace metal partitioning in Fe–Mn nodules from Sicilian soils, Italy. Chem. Geol. 173, 257–269.

Piepenbrock, A., Dippon, U., Porsch, K., Appel, E., Kappler, A., 2011. Dependence of microbial magnetite formation on humic substance and ferrihydrite concentrations. Geochim. Cosmochim. Acta 75, 6844–6858.

Qafoku, N., Qafoku, O., Ainsworth, C., Dohnalkova, A., McKinley, S., 2007. Fe-solid phase transformations under highly basic conditions. Appl. Geochem. 22, 2054–2064.

Ram, H., Singh, R.P., Prasad, J., 2001. Chemical and mineralogical composition of Fe-Mn concretions and calcretes occurring in sodic soils of Eastern Uttar Pradesh, India. Aust. J. Soil Res. 39 (3), 641–648.

Schaetzl, R., Anderson, A., 2009. Soils. Genesis and Geomorphology. Cambridge University Press, UK, ISBN 978-0-521-81201-6.

Schwertmann, U., 1988. Occurrence and formation of iron oxides in various pedoenvironments. In: Stucki, J., Goodman, B., Schwertmann, U. (Eds.), Iron in Soils and Clay Minerals. NATO ASI Series: Series C: Mathematical and Physical Sciences, vol. 217. Reidel Publisher Company, pp. 267–308.

Sheldon, N., Retallack, G., Tanaka, S., 2002. Geochemical climofunctions from North American soils and application to paleosols across the Eocene-Oligocene boundary in Oregon. J. Geol. 110, 687–696.

Shishkov, T., Kolev, N., 2014. The Soils of Bulgaria. World Soils Book Series. In: Hartemink, A.E. (Ed.). Springer Science + Business Media, Dordrecht, ISBN 978-94-007-7784-2, p. 205 (eBook).

Singh, B., Gilkes, R., 1996. Nature and properties of iron rich glaebules and mottles from some south-west Australian soils. Geoderma 71, 95–120.

Singh, L., Parkash, B., Singhvi, A., 1998. Evolution of the Lower Gangetic Plain landforms and soils in West Bengal, India. Catena 33, 75–104.

Soil Survey Staff, 2003. Keys to Soil Taxonomy, ninth ed. United States Dept. Agriculture, Natural Resources Conservation Service, Washington, DC.

Stanjek, H., Schwertmann, U., 1992. The influence of aluminium on iron oxides, Part XVI: hydroxyl and aluminium substitution in synthetic hematites. Clays Clay Miner. 40, 347–354.

Stiles, C., Mora, C., Driese, C., 2001. Pedogenic iron-manganese nodules in Vertisols: a new proxy for paleoprecipitation? Geology 29, 943–946.

Stiles, C., Mora, C., Driese, S., 2003. Pedogenic processes and domain boundaries in a Vertisol climosequence: evidence from titanium and zirconium distribution and morphology. Geoderma 116, 279–299.

Tatyanchenko, T., Alekseeva, T., Kalinin, P., 2013. Mineralogical and chemical compositions of the paleosols of different ages buried under kurgans in the southern Ergeni region and their paleoclimatic interpretation. Eurasian Soil Sci. 46 (4), 341–354.

Teoharov, M., 1991. Influence of some ecological conditions on the bulk chemical composition of weakly developed soils (Rankers). Ecology 24, 20–25 (in Bulgarian).

Teoharov, M., 2009. Systematic Database for the Soils in Bulgaria. Agricultural Academy. Inst. of Soil Sci., Agrotechnologies and Plant Protection "N. Poushkarov", Sofia, ISBN 978-954-9467-26-0 (in Bulgarian).

Thompson, A., Chadwick, O., Rancourt, D., Chorover, J., 2006. Iron oxide crystallinity increases during soil redox oscillations. Geochim. Cosmochim. Acta 70, 1710–1727.

Tóth, G., Montanarella, L., Máté, S.F., Bódis, K., Jones, A., Panagos, P., Van Liedekerke, M., 2008. Soils of the European Union.

Usman, M., Abdelmoula, M., Hanna, K., Greágoire, B., Faure, P., Ruby, C., 2012. Fe^{II} induced mineralogical transformations of ferric oxyhydroxides into magnetite of variable stoichiometry and morphology. J. Solid State Chem. 194, 328–335.

van Breemen, N., 1988. Long-term chemical, mineralogical and morphological effects of iron-redox processes in periodically flooded soils. In: Stucki, J., Goodman, B., Schwertmann, U. (Eds.), Iron in Soils and Clay Minerals. NATO ASI Series, D. Reidel Publishing Company.

Van Ranst, E., Dumon, M., Tolossa, A., Cornelis, J.-T., Stoops, G., Vandenberghe, R., Deckers, J., 2011. Revisiting ferrolysis processes in the formation of Planosols for rationalizing the soils with stagnic properties in WRB. Geoderma 163, 265–274.

van Velzen, A., Zijderveld, J., 1995. Effects of weathering on single domain magnetite in Early Pliocene marine marls. Geophys. J. Int. 121, 267–278.

Vepraskas, M., Faulkner, S., 2001. Redox chemistry of hydric soils. In: Richardson, J.L., Vepraskas, M.J. (Eds.), Wetland Soils. Genesis, Hydrology, Landscapes, and Classification. CRC Press LLC, Lewis Publishers, Boca Raton, London, New York Washington, D.C., pp. 85–105

Vodyanitskii, Yu, Shoba, S., Lopatovskaya, O., 2014. Iron compounds in sulfate-carbonate soils on red-colored Cambrian rocks in the southern Angara region. Eurasian Soil Sci. 47 (5), 407–415.

Wells, M., Fitzpatrick, R., Gilkes, R., Dobson, J., 1999. Magnetic properties of metal-substituted haematite. Geophys. J. Int. 138 (2), 571–580.

Zachara, J., Kukkadapu, R., Fredrickson, J., Gorby, Y., Smith, S., 2002. Biomineralization of poorly crystalline Fe(III) oxides by dissimilatory metal reducing bacteria (DMRB). Geomicrobiol. J. 19 (2), 179–207.

Zhang, Ch, Paterson, G., Liu, Q., 2012. A new mechanism for the magnetic enhancement of hematite during heating: the role of clay minerals. Stud. Geophys. Geod. 56, 845–860.

Zhang, M., Karathanasis, A., 1997. Characterization of iron-manganese concretions in Kentucky Alfisols with perched water tables. Clay Clay Miner. 45, 428–439.

Žigova, A., Štastny, M., Hladil, J., 2014. Mineral composition of rendzic leptosols in protected areas of the Czech Republic. Acta Geodyn. Geomater 11 (1 173), 77–88.

CHAPTER

THE MAGNETISM OF SOILS WITH LITTLE OR NO PROFILE DIFFERENTIATION: SOILS FROM MOUNTAIN AREAS (CAMBISOLS, UMBRISOLS) AND FLOODPLAINS (FLUVISOLS)

6.1 CAMBISOLS AND UMBRISOLS: MAIN CHARACTERISTICS, FORMATION PROCESSES, AND DISTRIBUTION

Cambisols are the most widespread soil order on the world's soil map distribution. Their pedogenesis is related to recent soil development and weak pedogenic clay formation during the weathering of the primary rock minerals. These processes lead to the growth of the characteristic "cambic" soil horizon. Cambisols mostly occur in mountainous areas where the primary factor for pedogenic development is the climate, along with the respective changes in its parameters with altitudinal change (Schaetzl and Anderson, 2009). Decreasing air temperature with increasing altitude is accompanied by increased precipitation, while decreased evapotranspiration assists in enhanced humidity, even in the highest mountain areas. With increasing altitude, both climate and vegetation change—from broadleaf forests, through mixed forest, coniferous, to shrub vegetation (Fig. 6.1.1).

This leads to a systematic increase in the "effective humidity" and the content of organic matter in soil (Egli et al., 2008). At the same time, organic matter mineralization declines with altitude increase. Soils developed in the area above the subalpine line (the upper tree line limit) are often considered as a separate soil group—the so-called alpine soils (mountainous meadow soils). For alpine soils, the major soil formation factor is the topography that predetermines the snow cover thickness, vegetation type, and, thus, the main characteristics of the soil profile (Schaetzl and Anderson, 2009). In the high mountain areas, weathering—both physical and chemical—plays a major role in the soil's formation. Chemical weathering is primarily controlled by water availability (Egli et al., 2006). A well-known fact is that the weathering rate decreases when the weathering is more prolonged (Egli et al., 2001, 2014), and the presence of nonweathered fresh mineral surfaces provided by the physical erosion determines the rate of chemical weathering. In areas with massive parent rocks, the surface of the primary minerals available for weathering increases to a certain maximum and after that declines (Egli et al., 2014). With an increasing alteration of the parent rock, the weathering intensity in soil becomes restricted by the dissolved Al ions or by the pedogenically synthesized iron (oxy)hydroxides bounded on the mineral surfaces. This provokes a decrease in the

Soil Magnetism. http://dx.doi.org/10.1016/B978-0-12-809239-2.00006-1

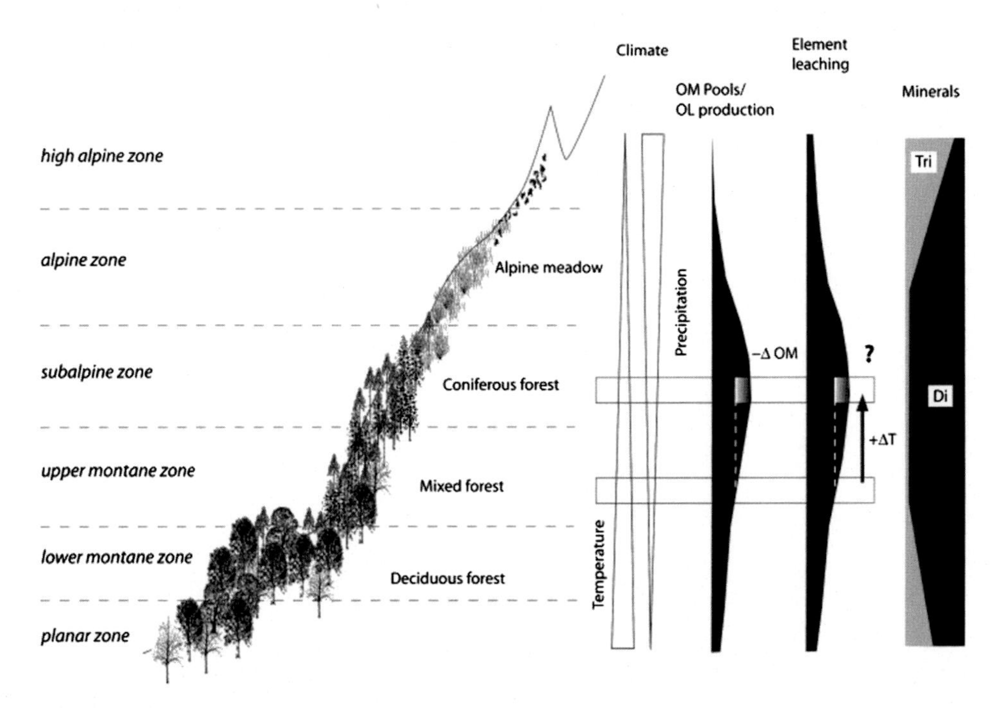

FIGURE 6.1.1

Summarized, typical processes along an alpine toposequence, according to Egli et al. (2008). The stored amount of organic matter (OM) in the soils, organic ligand production (OL), and element leaching intensities are represented by the width of the black graphs. The relative distribution of di- and trioctahedral minerals (Di, Tri) in the clay mineral fraction of surface soils is schematically given.

Reproduced with the permission from Elsevier.

chemical weathering because of the exhaustion of the mineral particles available for the process. Investigations carried out by Egli et al. (2008) demonstrate that chemical weathering, clay neoformation, and climate (and vegetation) are tightly connected in mountain-type soil pedogenesis. High weathering rates give rise to the preferential transformation of three-octahedral phyllosilicates (like mica and chlorite) into dioctahedral (smectite) (Fig. 6.1.1). Egli et al. (2008) provide evidence that the maximum weathering rate is established close to the upper altitudinal boundary of the spread of coniferous forests, which is caused by the most intense formation of organic acids in the soil solution there. As suggested by Egli et al. (2008), the link between the climate and the leaching of the chemical elements from the primary minerals is nonlinear due to the strongly expressed podzolization processes below and close to the altitudinal threshold of the coniferous vegetation (~1400–1900 m a.s.l). With an increase in altitude, the rainfall amount also increases, but evaporation decreases due to the lower mean annual temperatures. This results in a larger amount of water circulating into the soil. The effect is enhanced during winter when the snow cover allows intense water flows to circulate during the snowmelt. All of these processes favor podzolization (Egli et al., 2008). In addition to the amount of temperature and precipitation, another element of vital importance is the biotic factor. Vegetation and microbes in the soil influence the soil's reaction (pH) and lead to the creation of organic acids, carbon dioxide (CO_2), and thus govern the direction of the pedogenesis (Zanelli et al., 2007). Pedogenic processes in mountainous areas also depend on the mountain slope's aspect— southerly or north. A number of studies demonstrate that, on the slopes with a northerly aspect, the weathering and podzolization processes are significantly enhanced in respect to south-facing slopes (Egli et al., 2006, 2010).

A major pedogenic process in Cambisols is the so-called brunification. It consists of the detachment of Fe ions from the primary rock-forming minerals and the subsequent formation of goethite and other Fe oxides, which give the Cambisols a characteristic brown and/or brown-red color (Schaetzl and Anderson, 2009). A very important factor in this process is the presence of abundant organics, which are rapidly complexed with the Fe (hydr)oxides formed. Pedogenic Fe (hydr)oxides, clay minerals, and humus are bound together, creating immobile complexes. When, as a result of biological degradation, iron is released from these clay—humic complexes, it rapidly precipitates in a brown-colored goethite, which remains stable in most soils (Schaetzl and Anderson, 2009). In relatively young soil with weak element leaching, the organics turnover is accelerated and it is converted into forms that impede the podzolization. An additional factor hindering podzolization is the weakly acid to acid soil reaction, due to the presence of more basic cations in the surface organic-rich horizon. In such conditions, Al is mobile, while Fe is not and forms stable Fe—clay—humic complexes.

Cambisols are widespread in the temperate and boreal regions influenced by the Pleistocene glaciations. They are less common in the tropics and subtropics (Fig. 6.1.2).

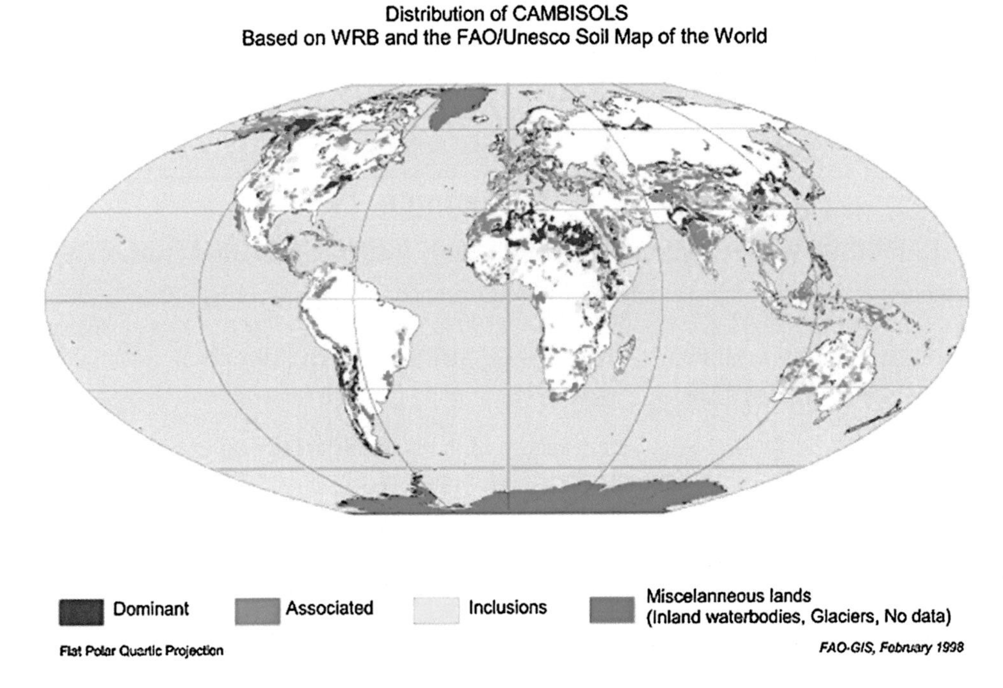

FIGURE 6.1.2

Spatial distribution of Cambisols worldwide (FAO, 2001). Lecture notes on the major soils of the world. http://www.fao.org/docrep/003/y1899e/y1899e00.htm#toc.

Umbrisols have not previously been recognized as a major taxonomic unit in soil classification and were classified as Humic Cambisols, Umbric Regosols (Food and Drug Organization [FAO]), and Umbrepts/Humitropepts (USA Soil Taxonomy). These soils are characterized by the presence of an umbric horizon with a large accumulation of organic material and with a low base saturation in the near surface layer (FAO, 2001). Umbrisols develop in a cold humid climate in mountainous areas with low or no moisture deficit but also in tropical and subtropical mountains. Umbric horizons are found in young, weakly developed soils that lack any other diagnostic horizon or have only a weak cambic horizon. Thus, their genesis is strongly influenced by the accumulation of organic material at the soil's surface. Umbrisols are defined as having a humus content within the interval 1–30% and an organic carbon content of >6%. Umbrisols are often found above the tree line in the Andean, Himalayan, and central Asian mountain ranges or at lower altitudes in northern and western Europe.

In the Bulgarian soil classification system, Cambisols are correlated to the brown forest soils (Ninov, 2002; Shishkov and Kolev, 2014). They occur in the mountainous part of the country under broadleaf or coniferous forests. They are commonly acid soils and develop in a mountain—forest climate with high humidity and precipitation. Parent rocks are most frequently noncarbonaceous—granites, granodiorites, shists, ryolite, and sandstones (Koinov et al., 1998; Shishkov and Kolev, 2014).

Soils developed under shrub vegetation in the subalpine areas of the Rila and Pirin mountains are defined as dark mountainous forest soils and are correlated to Humic Cambisols in the FAO classification from 1988 (Ninov, 2002). Their thick humic horizon is rich in raw humus, while the cambic B horizon is relatively thin. Soils developed in the highest mountain areas above the tree line are defined as mountainous meadow soils in the Bulgarian classification system and are correlated to Umbrisols. The organic content in these soils is high (organic C [C_{org}] between 8% and 17% or humus between 14% and 30%) and the soil reaction is acid (pH between 4.2 and 5.8) (Ninov, 2002).

6.2 DESCRIPTION OF THE PROFILES STUDIED

6.2.1 CAMBISOL (LIGHT BROWN FOREST SOIL), PROFILE DESIGNATION: T

Situated close to Turjan village, Smolyan district (Pirin mountain, South Bulgaria)

Location: N 41°33′29.0″; E 24°41′47.5″, h = 1100 m a.s.l.

Present-day climate conditions: MAT = 8.8°C; MAP = 891 mm, vegetation cover: coniferous forest; relief: moderate slope

Profile description: general view and details from the horizons are shown in Figs. 6.2.1–6.2.6.

0–9 cm: humic A_h horizon, light-brown, friable, humus-rich

9–21 cm: humic A horizon, dark brown-red color, dense, clayey, vegetation roots present, gley spots at the bottom

21–32 cm: illuvial B1 horizon, yellowish-brown, white talc spots, gley features, sharp upper boundary with A horizon

32–56 cm: B_{g2} illuvial horizon, yellowish-brown, abundant gray-blue gley spots, talc stripe at ~40–46 cm depth

FIGURE 6.2.1

View of the profile of Cambisol, T profile.

FIGURE 6.2.2

Cambisol profile "T." A_h horizon—detail.

FIGURE 6.2.3

Cambisol profile "T." A horizon—detail.

FIGURE 6.2.4

Cambisol profile "T." B1 horizon—detail.

FIGURE 6.2.5

Cambisol profile "T." BC_g horizon—detail.

FIGURE 6.2.6

Cambisols in the Smolyan area (Rila Mountain). View of the location of T profile.

56–80 cm: BC_g horizon, orange to dark-red, clayey, gley spots, rock fragments, and talc spots present

80–94 cm: C_g horizon, mottled, abundant rock pieces

Parent material: volcano sedimentary deposits

6.2.2 HUMIC CAMBISOL (DARK MOUNTAINOUS FOREST SOIL), PROFILE DESIGNATION: TR

Situated close to Treshtenik chalet (Rila Mountain, south Bulgaria)

Location: N 42°04′52.5″; E 23°37′39.9″, h = 1760 m a.s.l.

Present-day climate conditions: MAT = 0.1°C; MAP = 820 mm, vegetation cover: coniferous forest; relief: moderate slope

Profile description: general view and details from the horizons are shown in Figs. 6.2.7–6.2.10.

0–30 cm: A_h horizon, humus rich, dark brown, friable, vegetation roots

30–50 cm: A horizon, humus rich, brown, friable, sandy-loamy, gradual transition with A_h

50–80 cm: AC horizon, light brown, friable, sandy

FIGURE 6.2.7

View of the profile of Humic Cambisol, TR profile.

FIGURE 6.2.8

Humic Cambisol profile “TR.” A_h horizon—detail.

FIGURE 6.2.9

Humic Cambisol profile “TR.” AC horizon—detail.

FIGURE 6.2.10

Humic Cambisol profile "TR." C2 horizon—detail.

80–100 cm: C1 horizon, friable, brown-gray color, sandy
100–105 cm: C2 horizon, weathered granite, brown-gray
Parent material: granite

6.2.3 UMBRISOL (MOUNTAINOUS MEADOW SOIL), PROFILE DESIGNATION: GR

Situated close to Granchar chalet (Rila Mountain, south Bulgaria)

Location: N 42°07′21.1″; E 23°35′32.3″, h = 2190 m a.s.l.

Present-day climate conditions: MAT = 0.7°C; MAP = 900 mm, vegetation cover: alpine grassland; relief: flat

Profile description: general view, details from the horizons and surrounding landscape are shown in Figs. 6.2.11–6.2.15.

0–18 cm: A_h1 humic horizon, dark-brown to black color, organics and turf rich, friable.
18–56 cm: A_h2 humic horizon, organics rich, dark-brown, loamy-sandy, rock fragments present.
56 cm: C horizon, weathered granite.
Parent rock: granite

FIGURE 6.2.11

View of the profile of Umbrisol, GR profile.

FIGURE 6.2.12

Umbrisol profile GR. A_h1 horizon—detail.

FIGURE 6.2.13

Umbrisol profile GR. A_h2 horizon—detail.

FIGURE 6.2.14

View of the high alpine zone of the Rila mountain and the location of the Granchar chalet, where the GR profile is sampled.

FIGURE 6.2.15

High alpine zone in Rila Mountain.

6.3 TEXTURE, SOIL REACTION, AND MAJOR GEOCHEMISTRY OF CAMBISOLS AND UMBRISOLS

The mechanical composition of the "T" Cambisol profile is dominated by the silt fraction (Fig. 6.3.1a) comprising 50–70% in the different horizons. Sand and clay fractions make a relatively small contribution to the texture and display variations in the upper part of the profile. As is also seen from Figs. 6.2.1–6.2.5, this profile is strongly influenced by reduction processes, chalk deposition, and the presence of rock fragments. The soil reaction (pH) is slightly acid in the upper humic and illuvial horizon, while it drops to lower values (pH ~ 5.4) at the base of the soil profile (Fig. 6.3.1a). A similar pH pattern along the Cambisol profiles is also reported for representative Cambisols included in the Atlas of Soils in Bulgaria (Koinov et al., 1998) and the Reference Database for Soils in Bulgaria (Teoharov, 2009).

The silt fraction is the dominant mechanical fraction in the Humic Cambisol (TR) and the Umbrisol (GR) (Fig. 6.3.1b,c) and does not vary much along the soil profiles, again comprising about 60–70%. The clay fraction, although small in amount, is relatively enriched in the uppermost parts of these high mountain soils and rapidly decreases approaching the C horizon. Thus, the relatively weak textural

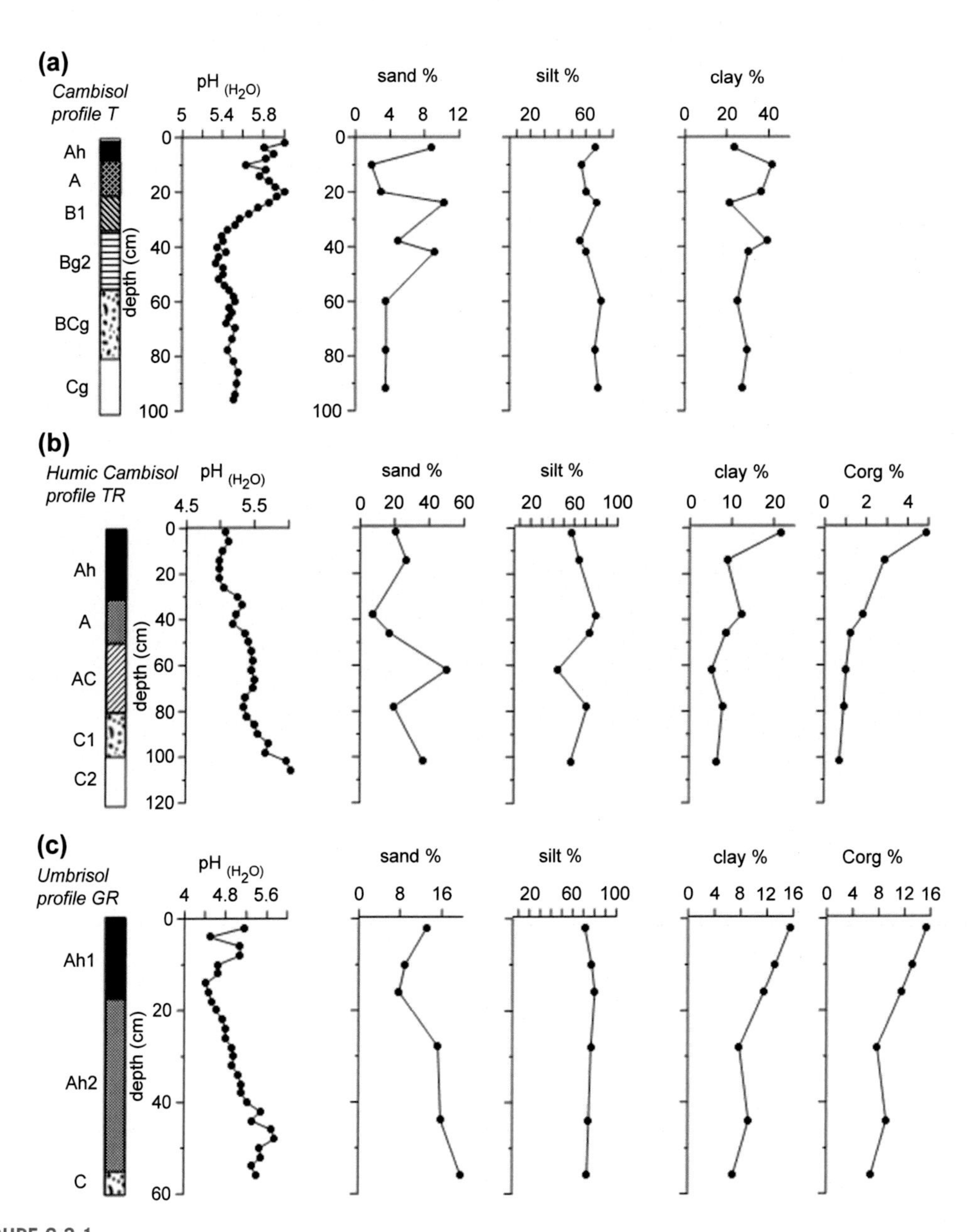

FIGURE 6.3.1

Soil reaction (pH) and texture (sand, silt, clay) for selected samples from the "T" Cambisol (a) Soil reaction (pH), texture (sand, silt, clay) and organic carbon content (C_{org}) for the Humic Cambisol (TR profile) (b) and Umbrisol (GR profile) (c).

differentiation in the Cambisol profiles is a reflection of their young age and early stages of weathering and pedogenic transformation (Schaetzl and Anderson, 2009). In contrast to the T profile, the soil reaction pH is strongly acid at the top of the TR and GR profiles and increases toward the C horizon (Fig. 6.3.1b,c). A comparison between the three profiles shows that the GR profile exhibits the most-acid pH values in the surface levels. This is most likely a reflection of the highest content of organic matter, as evidenced by the data for the TR and GR soils. In the Umbrisol profile (GR), C_{org} reaches 16% on the surface and retains relatively high values in the humic horizons, in agreement with its assignment to this soil order. The content of organic carbon in the Humic Cambisol (TR) is lower compared with the Umbrisol (GR), but is clearly enriched in the humic horizons.

The elemental composition studied in selected samples from the Humic Cambisol (TR) and the Umbrisol (GR) again evidences little variations and redistribution of the major soil constituents (Fig. 6.3.2). Silica inherited from the parent granitic rocks makes a major contribution to the elemental composition and comprises about 20%. Iron, aluminum, and titanium covary along the depth in both soil profiles, suggesting an intimate link also related to their lithogenic origin. The iron content

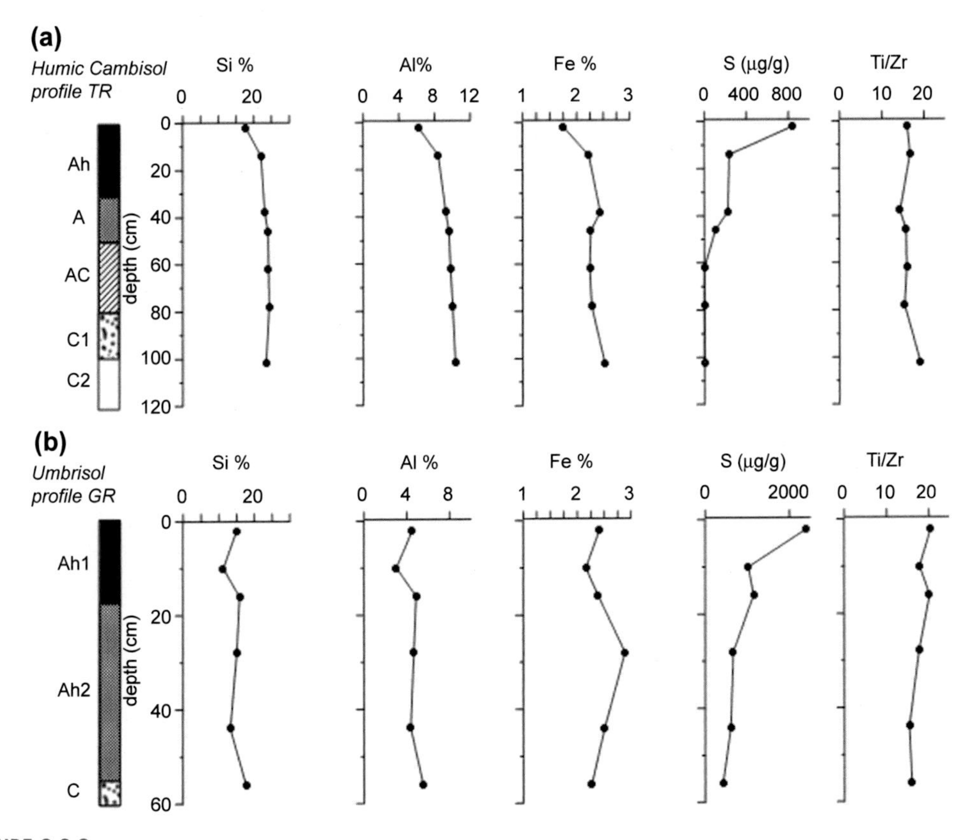

FIGURE 6.3.2

Content of some major and trace elements (Si, Al, Fe, S) and the ratio Ti/Zr for selected samples along the depth of profiles of (a) Humic Cambisol (TR) and (b) Umbrisol (GR).

accounts for ~2–2.5% of the bulk elemental composition and is similar in TR and GR profiles. Organic-rich humic horizons are enriched in sulfur (Fig. 6.3.2), likely of organic origin. It is exclusively enhanced in the Umbrisol humic horizon.

Uniformity in the parent material composition is evidenced in the constant Ti/Zr ratio values along the depth, as well as between the two profiles developed on similar parent material (Fig. 6.3.2). Calcium content (data not shown) is minor, varying between 0.4% and 1.8% in the two profiles. Thus, the elemental composition of the Humic Cambisol and the Umbrisol is dominated by the inherited rock minerals and weak weathering during their limited pedogenesis.

Selected samples from the TR and GR profiles are subjected to acid oxalate– and dithionite-selective chemical extractions to evaluate the relative contribution of amorphous and crystalline Fe and Al forms along the depth. Variations in the extracted Fe, Al, and the ratios Fe_o/Fe_d and Fe_d/Fe_t are shown in Fig. 6.3.3. In both profiles, a significant amount of oxalate extractable Al, Fe, and Mn (Al_o, Fe_o, Mn_o) is present ,which is a typical characteristic of genetically young soils (e.g., Schwertmann, 1988), suggesting the presence of a large amount of noncrystalline pedogenic (hydr) oxides. There are, however, substantial differences between the changes in the extracted elements from the two profiles. The oxalate-extracted Fe (Fe_o) in the Humic Cambisol (TR) is decreasing in depth, while, in the Umbrisol, the tendency is the opposite. This observation suggests an occurrence of

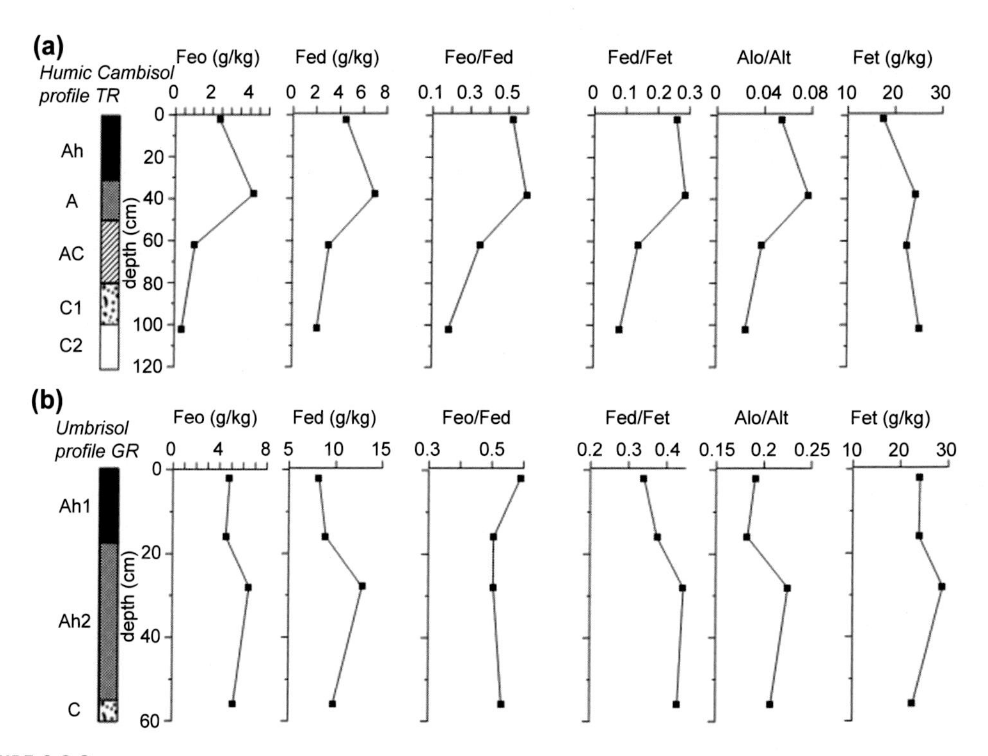

FIGURE 6.3.3

Variationsinoxalate-anddithionite-extractableiron(Fe_o and Fe_d) and total Fe (Fe_t) and ratios Fe_o/Fe_d, Fe_d/Fe_t, and Al_o/Al_t along the profiles of Humic Cambisol (a) and Umbrisol (b).

different processes in the TR and GR profiles, leading to the precipitation of amorphous pedogenic Fe and Al phases.

The dithionite-extractable iron (Fe_d) generally follows the variations obtained for Fe_o, and this causes a different depth behavior in the ratio Fe_o/Fe_d in the two profiles (Fig. 6.3.3). The changes in Fe_d/Fe_t also underline the differences obtained between the TR and GR profiles—it suggests a weak enhancement with pedogenic Fe and Al oxides in the humic horizon of the Humic Cambisol, while in the Umbrisol, the upper organic-rich horizon is relatively depleted. The differences obtained could be linked to a significantly larger amount of organics (and C_{org}) in the Umbrisol (Fig. 6.3.1c) and the related preferential formation of stable Fe–OM complexes (Cornell and Schwertmann, 2003).

6.4 MAGNETIC PROPERTIES OF CAMBISOLS, HUMIC CAMBISOLS, AND UMBRISOLS

Iron-containing minerals present in different depths of the studied profiles were identified through high-temperature magnetic susceptibility behavior. Representative examples from the Cambisol profile (T) are shown in Fig. 6.4.1. It is obvious that, in all samples, intense mineralogical transformations occur during heating, which lead to the creation of a new strongly magnetic phase on cooling.

The uppermost sample (T2) is characterized by an intense transformation at ~280°C and 500°C and a Curie temperature of T_c ~ 580°C (Fig. 6.4.1a). This behavior could be related to ferrihydrite transformation in the presence of organics (Campbell et al., 1997; Hanesch et al., 2006; Mitov et al., 2002), which maintains the reducing atmosphere during heating. The observed peaks in χ might be associated with ferrihydrite's transformation into maghemite and its subsequent inversion to hematite. Because of the reducing atmosphere, the hematite is further reduced to magnetite expressed by its Hopkinson peak and the T_c. The sample T38 from the B_g2 horizon is characterized by intense gleying (see the profile's description in part 6.2). In contrast to the surface sample (T2), the second transformation peak at ~500°C is missing in T38 and the magnetic susceptibility reaches zero at ~540°C

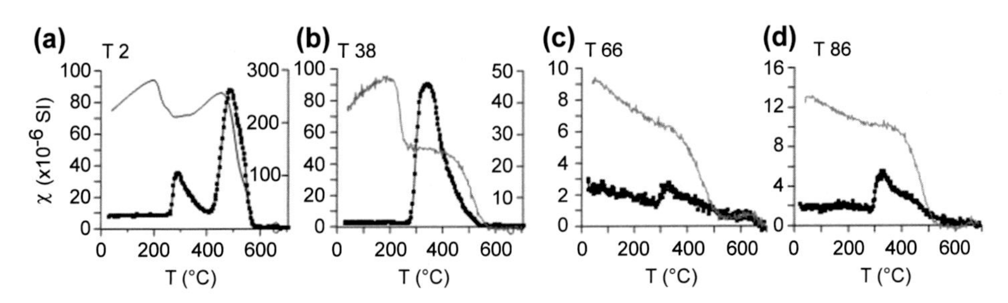

FIGURE 6.4.1

Thermomagnetic analysis of magnetic susceptibility for selected samples from different horizons of the "T" Cambisol. Thick black lines—heating run; thin gray line—cooling run. Heating in air, fast heating rate. The right vertical axis corresponds to the cooling curves for samples with exceptional mineralogical changes during heating. (a) Sample T2 from A_h horizon; (b) sample T38 from B1 horizon; (c) sample T66 from BC_g horizon; and (d) sample T86 from C_g horizon.

(Fig. 6.4.1b). The main transformation again occurs at ~280°C, suggesting that likely again maghemite is produced during heating, which further gradually inverts to hematite. It could be supposed that the initial Fe-containing mineral that exhibits these transformations is lepidocrocite (γ-FeOOH), taking into account Hanesch et al. (2006) observations. This iron hydroxide transforms into maghemite on heating even without organics to support reducing conditions. Lepidocrocite has also been detected in B horizons of Cambisols from Wales (Adams and Kassim, 1984). The sample from the BC_g horizon—T66—shows different behavior without the creation of a large amount of a strongly magnetic fraction on cooling (Fig. 6.4.1c). However, a weak alteration is obvious at ~300°C. Magnetic susceptibility further decreases, showing drops at magnetite's T_c (580°C) and at hematite's T_N ~ 700°C. Again, referring to model thermomagnetic curves for synthetic minerals reported in Hanesch et al. (2006), the behavior observed in the sample T66 could be related to ferrihydrite transformations but without organics. The deepest sample (T86) shows behavior similar to T66 (Fig. 6.4.1d) and thus analogous mineralogy could be supposed.

Thermomagnetic analyses for the Humic Cambisol (TR) are shown in Fig. 6.4.2a–c and demonstrate a distinct magnetite phase with a sharply expressed Curie temperature at 580°C in all samples. A weakly expressed kink at ~100°C is also observed, which could be ascribed to goethite's T_N (Dunlop and Özdemir, 1997) or to the effect of stress release during the heating of surface-maghemitized magnetite particles (van Velzen and Zijderveld, 1995). The thermomagnetic behavior

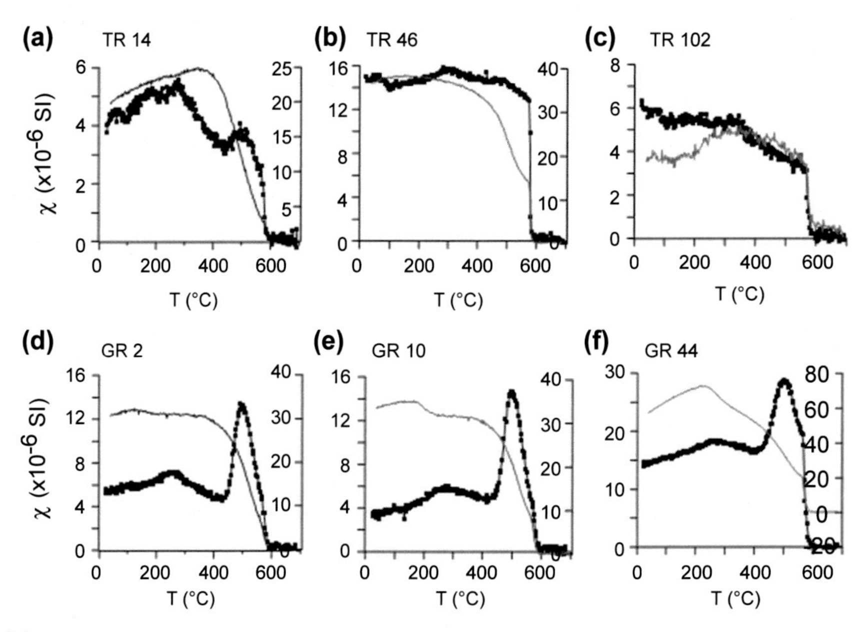

FIGURE 6.4.2

Thermomagnetic analysis of magnetic susceptibility for selected samples from different horizons of the "TR" Humic Cambisol (a–c) and "GR" Umbrisol (d–f). Thick black lines—heating run; thin gray line—cooling run. Cooling curves are shown on secondary Y-axis (except for sample TR102). Heating in air, fast heating rate.

in the interval 100−450°C is likely related to the presence of maghemite (already discussed in the previous chapters) with the biggest share in the near-surface sample TR14 (Fig. 6.4.2a). Significantly enhanced magnetic susceptibility on cooling for the samples TR14 and TR46 reflects the creation of a new magnetic phase due to abundant organics and its reducing effect during heating.

All samples studied from the Umbrisol profile display equivalent behavior during the thermomagnetic runs (Fig. 6.4.2d−f). As already discussed, such χ−T curves are characteristic for the presence of maghemite and organics in well-drained soils. An initial linear increase in χ up to 250°C and the subsequent drop to 400°C are explained by structural effects in maghemite during heating (e.g., Banfield et al., 2004), which is further inverted to hematite. In the strongly reduced atmosphere created by the abundant organics throughout the GR profile (Fig. 6.3.1b), this hematite is reduced to magnetite, detected through its clear T_c of 580°C.

The occurrence of magnetite and maghemite as major magnetic minerals in the TR and GR profiles is further supported by the results from the stepwise thermal demagnetization of composite IRM (Fig. 6.4.3). As is evident from the examples shown, the magnetically soft component dominates the remanence signal.

Two unblocking temperatures are characteristic for all the samples—at ~250°C and 600°C, related to maghemite and oxidized magnetite (Dunlop and Özdemir, 1997). The hard component is always very weak, and only for some samples (TR42, GR28, GR56) could goethite's presence be hypothesized.

The presence of high-coercivity magnetic minerals that, however, have a weak magnetization is further explored in the three profiles by the analysis of IRM acquisition curves in fields up to 5 T. Results obtained for selected samples from the T Cambisol are shown in Fig. 6.4.4. They demonstrate a very important content of high-coercivity fractions that begin to saturate in fields above 1 T and 2 T.

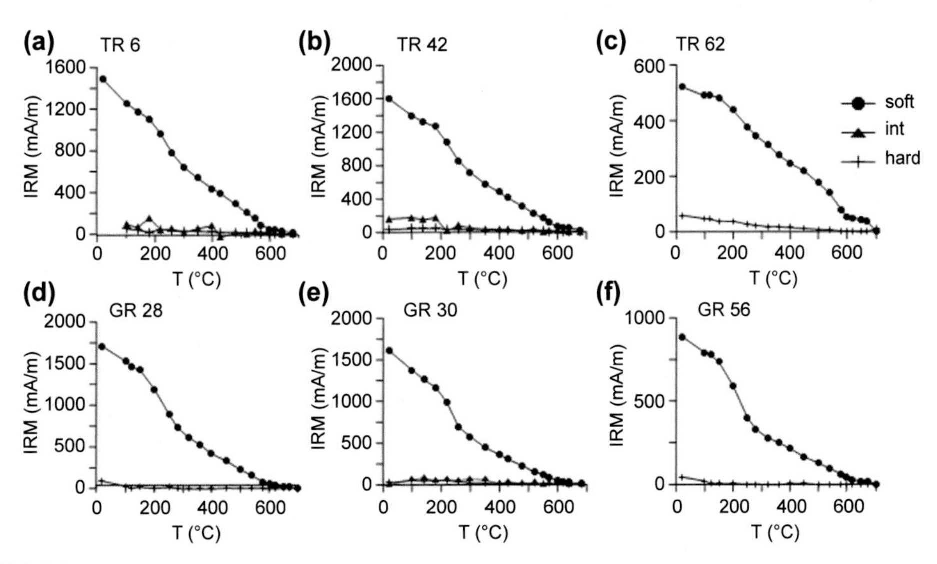

FIGURE 6.4.3

Stepwise thermal demagnetization of composite IRM for samples from TR profile (a−c) and GR profile (d−f). Magnetizing fields for composite IRM: soft fraction—acquired in a field of 0.23 T; intermediate fraction—acquired between 0.23 and 0.46 T; hard fraction—acquired between 0.46 and 2.0 T.

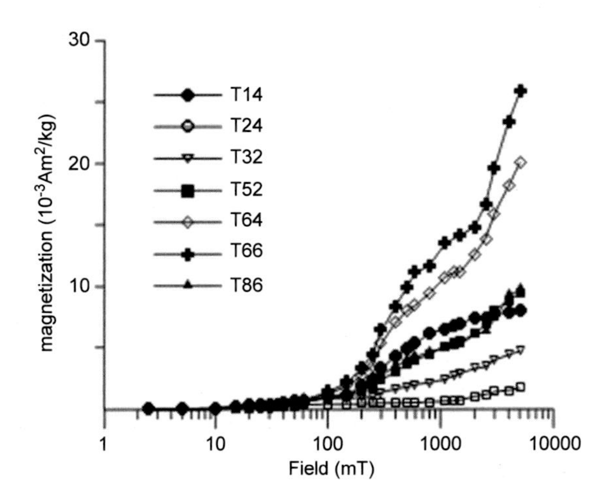

FIGURE 6.4.4

Stepwise acquisition of isothermal remanence (IRM) up to field of 5 T for selected samples from the "T" Cambisol profile.

As is evident from the acquisition curves, the share of the high-coercivity minerals increases with increasing depth and is best revealed in the samples from a 62-cm and 64-cm depth (Fig. 6.4.4). An analysis of coercivity components (Kruiver et al., 2001) reveals the presence of four coercivity fractions (Table 6.4.1), of which the IRM3 component is identified only in three samples from the B1 and B_g2 horizons (T24, T32, T52). The magnetically softest component (IRM1) is practically identical all along the profile, having $B_{1/2}$ ~ 28–29 mT but with a low contribution to the total IRM (Table 6.4.1). The second component (IRM2) is present only in the above-mentioned three samples and shows coercivity $B_{1/2}$ of 56 mT. The third component with a coercivity of about 300 mT has a relatively large contribution to the total IRM and accounts for about 20–40% of it. The most important phase contributing generally 50–70% to the IRM is the IRM4 with $B_{1/2}$ varying between 2000 and

Table 6.4.1 Summary parameters from coercivity analysis by fitting IRM acquisition curves with cumulative Gaussian functions, using IRM-CLG1.0 software (Kruiver et al., 2001) for selected samples from the "T" Cambisol

	IRM component 1			IRM component 2			IRM component 3			IRM component 4		
Depth (cm)	% from IRM_{total}	$B_{1/2}$ (mT)	DP	% from IRM_{total}	$B_{1/2}$ (mT)	DP	% from IRM_{total}	$B_{1/2}$ (mT)	DP	% from IRM_{total}	$B_{1/2}$ (mT)	DP
Profile T Cambisol												
14	11	28.2	0.40				71	354.8	0.28	18	2818.4	0.30
24	12	30.9	0.33	13	57.5	0.45				75	2290.9	0.29
32	8	28.2	0.30	10	56.2	0.33	20	316.2	0.10	62	1995.3	0.30
52	4	28.2	0.30	5	56.2	0.33	29	281.8	0.18	62	2089.3	0.32
64	4	28.2	0.40				37	288.4	0.22	59	2454.7	0.26
66	2	28.8	0.30				41	288.4	0.22	57	3020.0	0.25
86	5	28.8	0.36				38	288.4	0.22	57	3020.0	0.25

The relative intensity (% from IRM_{total}), median acquisition field ($B_{1/2}$), and dispersion parameter (DP) for each coercivity component are shown.

3000 mT (Table 6.4.1). Considering the obtained coercivities of the components identified, it could be supposed that the magnetically soft fractions IRM1 and IRM2 are related to coarse magnetite and to pedogenic maghemite, respectively (e.g., Egli, 2004). The third component, IRM3, could be linked to hematite (Özdemir and Dunlop, 2014) and the fourth, to the goethite. The intensity of the goethite's component is increasing downward, suggesting that it is of lithogenic origin.

In contrast to the complex mixture of mineral coercivity phases found in the Cambisol profile (T), the Humic Cambisol (TR) and the Umbrisol (GR) show the dominant role of the magnetically soft fraction (Fig. 6.4.5).

As is seen from the results of the coercivity analysis (Table 6.4.2), a soft coercivity phase with $B_{1/2}$ of 30 mT dominates the magnetization and a weak higher coercivity fraction with $B_{1/2}$ of 500 mT in the upper profile part and $B_{1/2}$ of 300 mT at the bottom accounts for about 20% of the total IRM. These two components could be linked to magnetite and hematite, respectively. In the Umbrisol profile (Table 6.4.3), the soft fraction is dominant. Only in the upper 14 cm is a minor high-coercivity phase

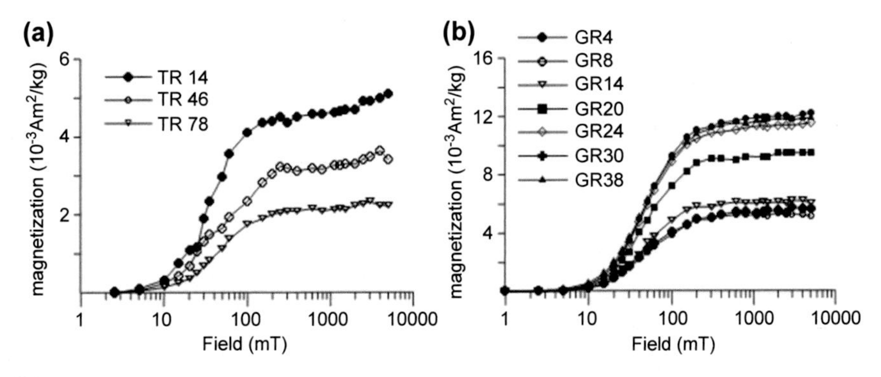

FIGURE 6.4.5

Stepwise acquisition of isothermal remanence (IRM) up to field of 5 T for selected samples from the "TR" Humic Cambisol profile (a) and "GR"Umbrisol (b).

Table 6.4.2 Summary parameters from coercivity analysis by fitting IRM acquisition curves with cumulative Gaussian functions, using IRM-CLG1.0 software (Kruiver et al., 2001) for selected samples from the "TR" Humic Cambisol

	IRM component 1				IRM component 2			
Sample depth (cm)	Intensity ($10^{-6}Am^2/kg$)	% from IRM_{total}	$B_{1/2}$ (mT)	DP	Intensity ($10^{-6}Am^2/kg$)	% from IRM_{total}	$B_{1/2}$ (mT)	DP
Profile TR Humic Cambisol								
14	5.882	85	29.20	0.32	0.750	15	501.20	0.50
46	2.700	77	33.10	0.35	0.800	23	501.20	0.55
78	1.750	76	33.90	0.31	0.550	24	316.20	0.55

The intensity, relative contribution (% from IRM_{total}), median acquisition field ($B_{1/2}$), and dispersion parameter (DP) for each coercivity component are shown.

Table 6.4.3 Summary parameters from coercivity analysis by fitting IRM acquisition curves with cumulative Gaussian functions, using IRM-CLG1.0 software (Kruiver et al., 2001) for selected samples from the "GR" Umbrisol

	IRM component 1				IRM component 2			
Sample depth (cm)	Intensity ($10^{-6}Am^2/kg$)	% from IRM_{total}	$B_{1/2}$ (mT)	DP	Intensity ($10^{-6}Am^2/kg$)	% from IRM_{total}	$B_{1/2}$ (mT)	DP
Profile GR Umbrisol								
4	5414.00	98	37.2	0.39	106.10	2	398.1	0.50
6	4748.10	96	33.9	0.38	193.78	4	398.1	0.50
8	5161.90	98	37.2	0.39	101.20	2	398.1	0.50
12	4609.22	96	29.9	0.33	200.40	4	398.1	0.50
14	5876.33	97	31.6	0.33	202.63	3	398.1	0.50
18	8703.87	100	31.6	0.35				
20	9443.86	100	39.8	0.39				
22	10,585.40	100	39.8	0.39				
24	11,244.98	100	37.2	0.39				
26	11,377.71	100	37.2	0.39				
30	12,158.06	100	37.2	0.41				
32	12,126.11	100	37.2	0.40				
34	11,811.02	100	37.2	0.40				
36	11,345.22	100	37.2	0.40				
38	11,578.95	100	37.2	0.38				

The intensity, relative contribution (% from IRM_{total}), median acquisition field ($B_{1/2}$), and dispersion parameter (DP) for each coercivity component are shown.

with hematite's characteristics identified, accounting for merely 3–4% of the IRM. The coercivity of the magnetically soft fraction in the two profiles is similar, being a little bit higher in the Umbrisol ($B_{1/2}$ of 30–37 mT), while in the Humic Cambisol, it is ~33 mT.

6.4.1 VARIATIONS OF MAGNETIC PARAMETERS ALONG DEPTH OF THE PROFILES

Cambisol (profile T)

The complex magnetic mineralogy in the Cambisol profile (T) is reflected in the depth variations of concentration- and grain-size–related magnetic characteristics along the profile. Magnetic susceptibility is generally low and varies between 10 and 60×10^{-8} m^3/kg (Fig. 6.4.6a) with two maxima—in the surface layer at 0–10 cm and at a depth interval of (55–70) cm. These two maxima are seen in the saturation magnetization M_s and the remanent magnetizations ARM and IRM_{1T} as well (Fig. 6.4.6c,d). Superparamagnetic magnetite/maghemite particles are present mostly in the upper A_h plus A horizons, where χ_{fd} shows maximum values. Single-domain (SD) magnetite particles are also abundantly found in this depth interval since the ARM is strongly enhanced there, while in the

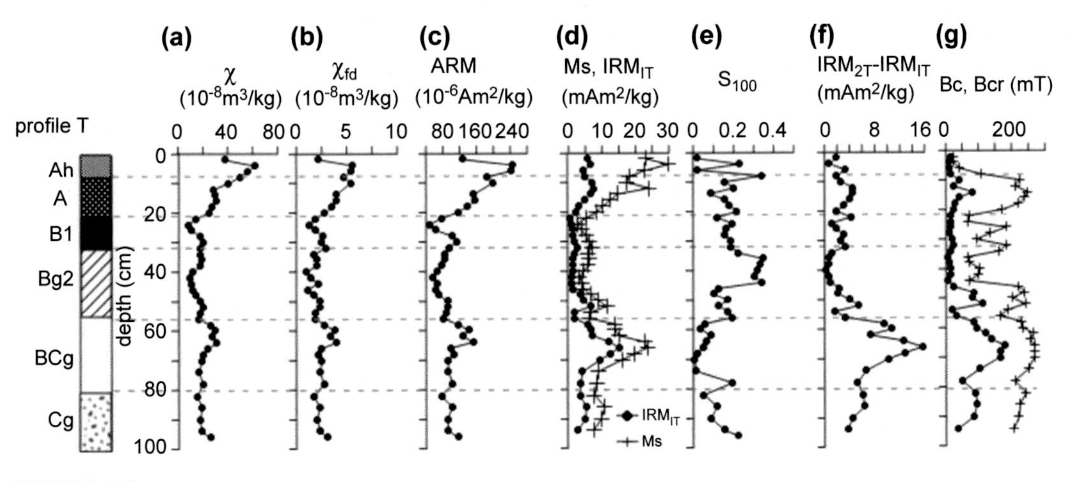

FIGURE 6.4.6

Depth variations of magnetic parameters and ratios along the "T" Cambisol profile: (a) magnetic susceptibility (χ); (b) frequency dependent magnetic susceptibility (χ_{fd}); (c) anhysteretic remanence (ARM); (d) saturation magnetization (M_s) (crosses) and isothermal remanence acquired in 1 T field (IRM_{1T}); (e) S ratio ($S = IRM_{0.1T}/IRM_{2T}$); (f) "hard" IRM obtained as a difference between IRM_{2T} and IRM_{1T}; (g) coercivity parameters B_c (full dots) and B_{cr} (crosses).

BC_g horizon, its increase is moderate at the expense of IRM, which shows the most pronounced maximum in the deeper part of the profile (Fig. 6.4.6c,d).

In addition, variations in the saturation magnetization (M_s) and IRM_{1T} (Fig. 6.4.6d) reveal a small difference between these two characteristics in comparison with soils dominated by strongly magnetic ferrimagnetic minerals. This is due to the significant amount of weakly magnetic antiferromagnetic minerals of high coercivity, as demonstrated by the exclusively low S ratios (Fig. 6.4.6e) reaching zero in the BC_g horizon. The isothermal remanence is dominated by the high-coercivity minerals, which is apparent in the variations of $IRM_{2T}-IRM_{1T}$ (Fig. 6.4.6f). The mixture of magnetically soft minerals and a significant amount of high-coercivity antiferromagnetic minerals is revealed in the hysteresis loops and parameters (Figs. 6.4.6g and 6.4.7).

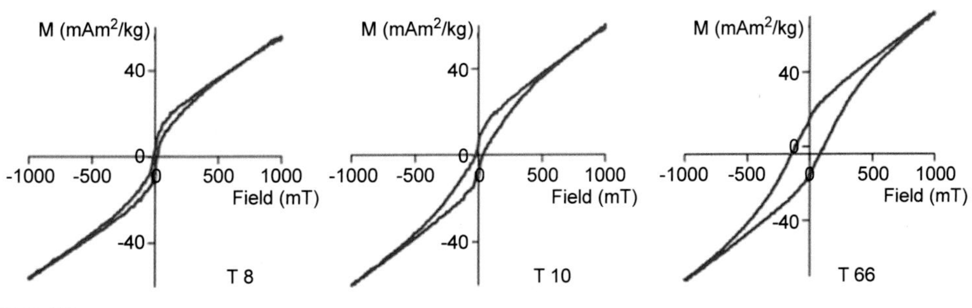

FIGURE 6.4.7

Hysteresis loops for soil samples from three depth intervals (8, 10, and 66 cm) of Cambisol profile ("T").

Samples from the upper humic horizons where magnetically soft SD and SP grains are evidenced (Fig. 6.4.6a–c) show well-expressed wasp-waisted hysteresis loops as a result of the mixture of two magnetic phases with contrasting coercivities (Tauxe et al., 1996). At the bottom part of the profile (sample T66 from the BC_g horizon), the antiferromagnetic fraction dominates the iron oxide mineralogy and the hysteresis loop is characteristic for high-coercivity minerals such as hematite and goethite. The mixture of different magnetic phases is also revealed in the B_c and B_{cr} variations along the profile (Fig. 6.4.6g). As seen there, in the upper part of the profile where wasp-waisted hysteresis loops are obtained (Fig. 6.4.7), the difference between the two coercivity parameters is large as a result of the specific constricted shape of the loop, while downward, where high-coercivity minerals dominate, both B_c and B_{cr} are high.

Variations in the grain-size–dependent magnetic ratios for the T Cambisol are shown in Fig. 6.4.8. Despite the indication obtained above that the SP magnetite/maghemite fraction is concentrated in the upper humic horizons of the soil, the SP-proxy ratio $\chi_{fd}\%$ exhibits an initial increase from 6% at the surface until ~12% at a ~10-cm depth and remains more or less constant up to the C horizon (Fig. 6.4.8a). Taking into account all the above results suggesting practically an absence of a strongly magnetic phase in the deeper parts of the profile, it is illogical to interpret the $\chi_{fd}\%$ behavior in terms of the SP-magnetite/maghemite fraction present. Thus, the $\chi_{fd}\%$ behavior obtained needs further consideration, and this will be addressed in Chapter 8.

In contrast to the $\chi_{fd}\%$ variations, the other SP-grain size proxy ratio— χ/M_s (Fig. 6.4.8b)— suggests relative enhancement with the SP fraction in the A-B1-B_g2 horizons. Since the two magnetic parameters involved, χ and M_s, depend in the same way on the concentration of the magnetic minerals (e.g., Peters and Dekkers, 2003), the χ/M_s ratio could be considered as a more reliable SP-grain size indicator for the T profile. Since the magnetic susceptibility and its frequency dependence (e.g., grain size) are obviously biased by the complex magnetic mineralogy in the profile, the ratios related to the

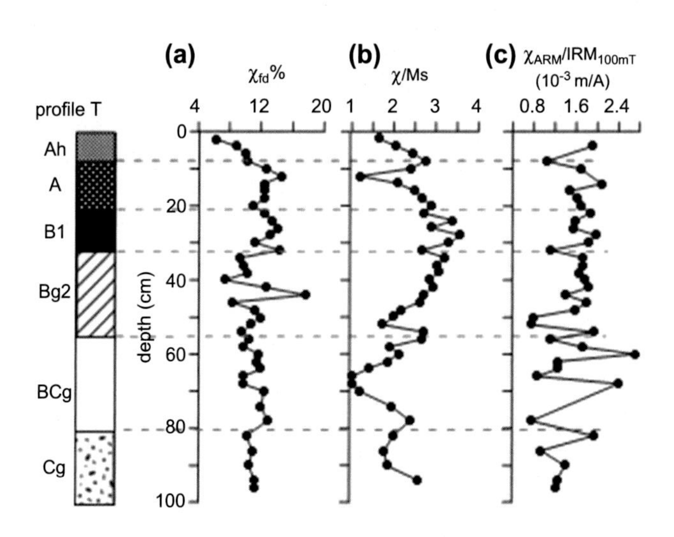

FIGURE 6.4.8

Magnetic proxy parameters for superparamagnetic [$\chi_{fd}\%$ (a), χ/M_s (b)] and single-domain [χ_{ARM}/IRM_{100mT} (c)] grain size along the depth of "T" Cambisol.

SD grain size fraction that incorporate χ (χ_{ARM}/χ, χ_{ARM}/χ_{fd}) could not be reliably used in this case. That is why we used the ratio χ_{ARM}/IRM_{100} as it mostly involves the ferrimagnetic low coercivity fraction (Fig. 6.4.8c). However, little well-resolved variations are seen in χ_{ARM}/IRM_{100} along the profile. Nevertheless, the high values of the ratio, reaching 1.6−1.8 in the upper part of the profile, suggest that stable SD maghemite particles could be present there (Geiss et al., 2008).

Humic Cambisol (profile TR)

Depth variations of different magnetic characteristics along the Humic Cambisol profile are shown in Fig. 6.4.9. The magnetic susceptibility of the Humic Cambisol is higher than that of the Cambisol T, but χ does not show well-constrained changes in relation to the genetic soil horizons (Fig. 6.4.9a). Moreover, a relative minimum in χ at the surface is observed, which could reflect the influence of significant organic matter content (Fig. 6.3.1b) in the A_h horizon and its diamagnetic signal (e.g., Oldfield, 1999). Frequency-dependent susceptibility suggests enrichment with ultrafine SP magnetite/maghemite grains mostly at the surface with a subsequent fast decrease deeper in the solum (Fig. 6.4.9b), which generally follows the pattern of C_{org} decrease (see Fig. 6.3.1b). Both χ_{fd} and $\chi_{fd}\%$ show similar variations, thus proving the exclusive control of the SP-magnetite concentration and its contribution to the bulk magnetic mineralogy. Effectively, $\chi_{fd}\%$ also distinguishes A_h plus A horizons as enriched in strongly magnetic minerals, while in the deeper parts this parameter has very low values indicating minor SP content (Dearing et al., 1996). In contrast, the remanent magnetizations—ARM and IRM_{2T}—very clearly separate the humic horizons (A_h plus A) from the transitional and the C horizons through a strong magnetic enhancement (Fig. 6.4.9c,d). Therefore, the concentration-dependent magnetic characteristics support the strong pedogenic enhancement of humic horizons

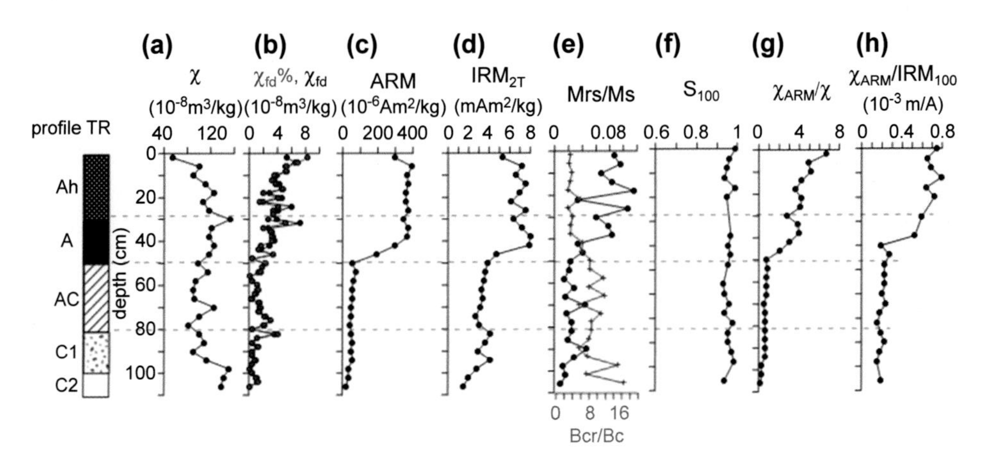

FIGURE 6.4.9

Depth variations of magnetic parameters and ratios along the "TR" Humic Cambisol profile: (a) magnetic susceptibility (χ); (b) frequency-dependent magnetic susceptibility (χ_{fd}) (full dots) and percent frequency-dependent magnetic susceptibility ($\chi_{fd}\%$) (crosses); (c) anhysteretic remanence (ARM); (d) saturation isothermal remanence acquired in 2 T (IRM_{2T}); (e) M_{rs}/M_s ratio; (f) S ratio ($IRM_{0.1T}/IRM_{2\,T}$); (g) ratio between anhysteretic susceptibility and magnetic susceptibility (χ_{ARM}/χ); (h) $\chi_{ARM}/IRM_{0.1T}$ ratio.

with dominantly magnetically stable (remanence-carrying) minerals throughout and an exponentially decreasing in depth concentration of the SP fraction.

As seen from Fig. 6.4.9c,d, IRM is more enhanced in the deeper soil horizons in comparison with the ARM, suggesting the presence of coarse (lithogenic) magnetite grains, while the upper pedogenically enhanced horizons contain predominantly finer SD grains. The same conclusion follows from the variation in the hysteresis ratios M_{rs}/M_s and B_{cr}/B_c (Fig. 6.4.9e). The dominant role of magnetically soft minerals in the bulk iron oxide mineralogy is confirmed by the S ratio (Fig. 6.4.9f) showing values close to 1.0 all along the depth. The grain-size proxy ratio χ_{ARM}/IRM_{100} suggests the finest SD grain size of the magnetite fraction in the upper A_h horizon (Fig. 6.4.9h), increasing toward coarse lithogenic MD grains at the bottom. The other ratio, χ_{ARM}/χ (Fig. 6.4.9g), is most likely influenced by the high organic content lowering the χ values in the uppermost parts of the profile.

Umbrisol (GR profile)

The Umbrisol profile is characterized by depleted magnetic susceptibility and saturation magnetization (M_s) in the upper organic-rich horizon (Fig. 6.4.10a), which could be ascribed to the presence of a significant amount of diamagnetic organic matter, as discussed before. Deeper in the A_h2 horizon, where C_{org} is lower (5–6%) (Fig. 6.3.1c), χ is increasing up to high values of $\sim 90-100 \times 10^{-8}$ m^3/kg. However, the magnetic remanences, ARM and IRM_{2T}, show the same character of depth variations (Fig. 6.4.10c,d), and these parameters are not influenced by the diamagnetic minerals; thus, the above supposition could not be accepted. Therefore, most probably the uppermost organics-enriched horizon is magnetically depleted while the underlying A_h2 horizon is enhanced in magnetically stable SD and coarser grains. Maxima in the concentrations of these stable fractions are linked to the upper-middle part of the A_h2 horizon (Fig. 6.4.10). The content of superparamagnetic grains approximated through the χ_{fd} and $\chi_{fd}\%$ parameters (Fig. 6.4.10b) suggests a constant low share of such particles in the bulk magnetic composition. The hysteresis ratios M_{rs}/M_s and B_{cr}/B_c (Fig. 6.4.10f,g) exhibit a consistent

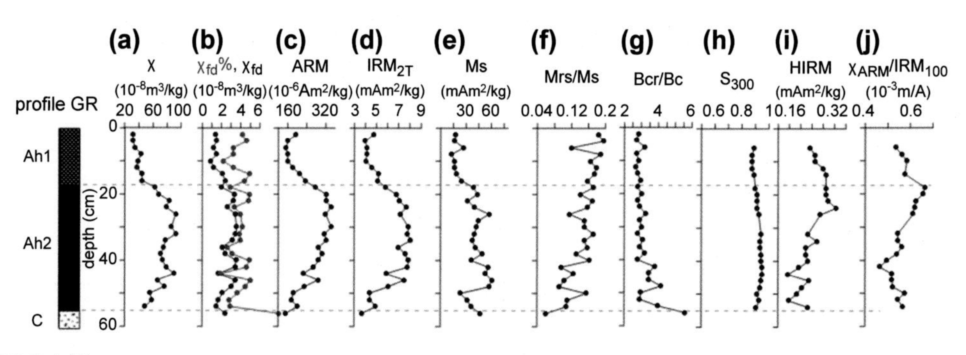

FIGURE 6.4.10

Depth variations of magnetic parameters and ratios along the "GR" Umbrisol profile: (a) magnetic susceptibility (χ); (b) frequency-dependent magnetic susceptibility (χ_{fd}) (full dots) and percent frequency-dependent magnetic susceptibility ($\chi_{fd}\%$) (crosses); (c) anhysteretic remanence (ARM); (d) saturation isothermal remanence acquired in 2 T (IRM_{2T}); (e) saturation magnetization (M_s); (f) M_{rs}/M_s ratio; (g) B_{cr}/B_c ratio; (h) S ratio ($IRM_{0.3T}/IRM_{2T}$); (i) hard isothermal remanent magnetization [HIRM = 0.5*(IRM_{2T}—$IRM_{0.3T}$); Liu et al., 2002]; (j) ratio between anhysteretic susceptibility and magnetic susceptibility (χ_{ARM}/χ).

trend suggesting a gradual coarsening of the magnetic grain size from the top toward the bottom of the Umbrisol profile. The grain-size–related variations are supported by the high S ratio all along the depth (Fig. 6.4.10h), although as is also shown by the IRM acquisition curves and their analysis (Table 6.4.3), a minor hematite component is present in the uppermost horizon. This result is corroborated by the HIRM variations (Fig. 6.4.10i) suggesting enhancement with high-coercivity hematite in the A_h1 and upper half of A_h2 horizons. The magnetic proxy ratio for the SD magnetite/maghemite χ_{ARM}/IRM_{100} shows a maximum at the upper boundary of the A_h2 horizon, implying the finest magnetite grain size there. Inside the A_h1, an increasing χ_{ARM}/IRM_{100} indicates refinement of the grain size, while starting from the A_h2, the magnetic grain size coarsens until a 40-cm depth followed again by a slight refinement.

6.5 PEDOGENESIS OF IRON OXIDES IN CAMBISOLS AND UMBRISOLS, AS REFLECTED IN SOIL MAGNETISM

Cambisols are widespread soils developed on different lithologies that play a substantial role in the pedogenic development of these relatively young soils. Pedogenic processes in high mountain forest soils are strongly influenced by the coniferous vegetation, relatively low mean annual temperatures, and high humidity, which favor the formation of secondary iron oxyhydroxides—ferrihydrite, goethite, and lepidocrocite. Most of them are amorphous low-crystalline phases extracted by the acid oxalate treatment (Cornel and Schwertmann, 2003). Data available for selective extractions on the Bulgarian Cambisols confirm the presence of a significant amount of oxalate-extractable iron (Jokova et al., 1991; Georgieva and Jokova, 1990; Jokova, 1999). The magnetic data of the Cambisol (T profile) shows complex iron oxide mineralogy (Figs. 6.4.1, 6.4.4 and 6.4.6–6.4.8) dominated by the antiferromagnetic minerals—goethite and hematite (Table 6.4.1). Hematite is not a typical pedogenic mineral in high mountain forest soils because of the high humidity and precipitation favoring goethite formation (Schwertmann, 1988). Also, taking into account the results for the coercivity components identified in the IRM acquisition curves (Table 6.4.1), it could be supposed that hematite and goethite are of lithogenic origin in the deeper parts of the T profile, formed as a result of the chemical weathering of the parent material. Despite the dominant role of the lithology on the magnetic signature of the "T" Cambisol, pedogenic iron oxyhydroxides are detected in the upper A_h-A-B1 horizons, represented by a small fraction of stable SD-like magnetite/maghemite (corresponding to the IRM2 component from Table 6.4.1) and a weak SP-component as deduced by the χ_{fd} variations (Fig. 6.4.6).

Our results are in agreement with other environmental magnetic studies on Cambisols from different regions. Anastacio et al. (2005) have studied the iron oxide mineralogy of the B horizons of a Cambisol from Brazil, and they found maghemite and SD goethite as being dominant phases. In high mountain Cambisols from Mexico (Rivas et al., 2012), the main contribution to the magnetic properties give lithogenic titanomagnetite and titanomaghemite as well as a small amount of pedogenic SP and SD magnetite. A Cambisol profile from Germany (Hanesch and Petersen, 1999) contained SP and SD pedogenic magnetite/maghemite in its B horizon. In a Cambisol profile from Corsica (France) (Stanjek, 1987), chemical and XRD analyses reveal the occurrence of maghemite, hematite, lepidocrocite, and goethite as well. However, Stanjek (1987) links maghemite and hematite to the influence of a forest fire that induced the thermal transformation of lepidocrocite to produce strongly magnetic maghemite and hematite in the upper soil levels. Brown earth profiles from the United Kingdom, studied by Maher (1986), show enhanced magnetic parameters (magnetic susceptibility, ARM, IRM) in their upper horizons with a maximum in the middle of the A horizons. Simultaneously, the percentage of frequency-dependent magnetic susceptibility was relatively low and uniform along the depth, as also observed for the T profile in our study. Cambisols from the Lublin Upland Region in Poland (Alekseev et al., 2002) show relatively low magnetic susceptibilities (χ between 5 and 30×10^{-8} m^3/kg) with no significant variations among the different grain-size fractions and inside the profile. A detailed magnetic, geochemical, and DRS study of Cambisols developed in tidal flat deposits in the coastal plain of Jiangsu Province, China (Dong et al., 2014), also revealed a complex magnetic mineralogy, composed of magnetite, maghemite, hematite, and goethite. A coercivity analysis of IRM acquisition curves was also applied for magnetic mineral identification. It is remarkable that the identified coercivities of the four components obtained by Dong et al. (2014) roughly coincide with those obtained for the Cambisol profile T. The authors found $B_{1/2}$ for the soft component 30 mT (or 50 mT in some profiles), related to coarse (fine) magnetite, $B_{1/2}$ of 100 mT, related to maghemite, $B_{1/2}$ of 300—400 mT linked to hematite and $B_{1/2}$ of 1600 mT and higher, associated with goethite. Similar to the T profile, goethite concentration was found to be higher than hematite's. The formation pathway of these iron oxide phases cannot be directly compared with our Cambisols because the tidal flat sediments are strongly influenced by periodic redox changes related to fluctuations in the water table inside the sediments. Nevertheless, all studies register a complicated mixture of lithogenic and pedogenic iron oxides in Cambisols. This could be related to the combination of lithogenically inherited (titano)magnetite/maghemite coarse particles, subjected to intense weathering during recent pedogenesis since Cambisols are considered to be young soils (Schaetzl and Anderson, 2009). The pedogenic formation of strongly magnetic phases is impeded by a commonly humid climate and low temperatures, which favor the formation of goethite and lepidocrocite (Cornell and Schwertmann, 2003). As Dahms et al. (2012) demonstrate, in soils of alpine chronosequence from the European Alps and the Wind River Range (USA), pedogenically formed stocks of Fe_o, Al_o, Si_o, and Mn_o increase considerably with time, but the formation rates distinctly decrease with the increasing age of the soil.

During the Pleistocene, most of the mountain ranges in southeastern Europe and the Balkans were covered by glaciers, which reached altitudes of ~1000 m below the contemporary snow line. In the highest parts of the Bulgarian mountains (Rila, Pirin, the Rhodopes, the Balkans), a number of geomorphological signs evidence the spread of glaciers during the last glacial epochs (Velchev, 1995). According to Velchev (1995), after the last glacial epoch (the Würm glacial) and the glaciers' retreat, intense weathering and the formation of soil cover began. The two high alpine soils from this study—the Humic Cambisol (TR) and the Umbrisol (GR)—were sampled in the eastern part of Rila Mountain. The TR profile was located in the coniferous belt, where the dominant species are *Pinus sylvestris* and *P. abies* (Bozilova and Tonkov, 2000). However, a number of pollen and radiocarbon studies show that in the Rila and Pirin mountains during the late glacial (~10,000—6700 BC) era, in the altitudinal belt above 1300 m, grassland and shrub vegetation dominated (Tonkov et al., 2008, 2011). Only during the Holocene climatic optimum (~6700 BC) did favorable conditions (higher humidity and precipitation) for the vertical migration of the coniferous species and the development of soils with a humic genetic horizon appear (Tonkov and Marinova, 2005). Therefore, the two soil profiles (TR and GR) are genetically young, most likely of Holocene age. This supposition is supported by the selective extraction data (Fig. 6.3.3), pointing to the presence of a significant amount of amorphous Fe, Al, and

Mn phases formed during recent intense weathering and pedogenesis. The pedogenesis of the Humic Cambisol developed under coniferous forest is strongly influenced by the specifics of the coniferous biomass and its decomposition. Coniferous litter decay is characterized by the formation of fulvic acids that promote faster weathering of the primary minerals and the formation of secondary (pedogenic) iron (hydr)oxides (Schaetzl and Anderson, 2009) which, under an alpine climate, often lead to podzol formation (Egli et al., 2008; Mourier et al., 2008; Dahms et al., 2012). In Bulgaria, however, podzol formation is not evidenced (Shishkov and Kolev, 2014).

Studies of alpine soils from Switzerland, Austria, and Italy (Egli et al., 2008; Dahms et al., 2012) show that the most intense processes of weathering and transformations of the primary rock-forming minerals occur in the altitudinal zone at the boundary between the coniferous belt and the high alpine grassland area. Soils developed in this area are characterized by a maximum concentration of the amorphous forms of Fe and Al, with an Al level that is always larger than that of Fe (Egli et al., 2008). The Umbrisol profile (GR) was sampled in the equivalent zone of Rila Mountain. This tendency in the acid oxalate–extractable Fe and Al (Fe_o, Al_o) is evidenced in the GR profile as well. A major role in the pedogenesis of the Umbrisols is played by the amount and the type of organic matter. In high alpine soils developed under rich peat grass vegetation, a large amount of poorly humified organic matter is accumulated. In such conditions, most of the iron becomes bound to the organic matter, which impedes its further transformation into other pedogenic Fe (hydr)oxides (Cornell and Schwertmann, 2003) and, very often, only pedogenic ferrihydrite is found (Schwertmann and Murad, 1988). As was discussed in the previous section, effective magnetic depletion was observed in the uppermost peaty horizon in the GR profile (Fig. 6.4.10), while magnetic enhancement with stable SD/PSD magnetite grains is linked to the underlying A_h2 horizon. Its formation may be facilitated by the increasing pH gradient observed in the sub-surface soil horizons of both TR and GR profiles (Fig. 6.3.1). However, along with this magnetic fraction, a significant amount of low-crystalline Fe minerals is also present.

6.6 FLUVISOLS: MAIN CHARACTERISTICS, FORMATION PROCESSES AND DISTRIBUTION

Fluvisols develop on alluvial deposits near river beds in lacustrine and marine deposits (FAO, 2001). The dynamics of the sediment deposition near large rivers is controlled by the complex interplay among the rocks, relief, and climate. The dynamics of alluvial fans strongly influence the temporal and spatial distribution of the sedimentation processes and, consequently, the soil distribution in the river plain (do Nascimento, 2013). Overbank flows and avulsion processes considerably affect soil properties by the formation of a textural gradient where fine sediments of floodplains are overlaid with sediments deposited during a rapid change in the course of the river bed (Farrell, 2001). Alternating horizons with different textures may arise when overbank sediments are formed by successive floods (do Nascimento, 2013). This is the reason why Fluvisols exhibit characteristics reflecting sedimentary transport/deposition as well as in situ pedogenesis. The diverse character of the sedimentary stratification in the different parts of the river bed is a major factor that determines the subsequent pedogenic changes. According to Gerrard (1993), homogenization in the Fluvisol profile is expressed in a disturbance of the sedimentary layering under the influence of the biological activity of vegetation roots, animals, and microorganisms. Deeper soil horizons are frequently prone to water saturation cycles under the influence of the high level of the groundwater table (Penkov, 1983; Gerrard, 1993).

Therefore, most Fluvisols display signs of gleying. Soils developed on low river terraces are subjected to seasonal water-logging from both rainfall and the fluctuating groundwater table (Gerrard, 1993). Commonly, the soil horizons in Fluvisols comprise (1) a humic (A) horizon with partially oxidative conditions and high organic matter content; (2) a mottled horizon formed under alternating oxidative and reducing conditions; and (3) a horizon with permanently reducing conditions, frequently gaining a blue-green coloring. In the deeper soil horizons with alternating redox conditions, Iron and Manganese precipitate as orange diffuse masses when a weak and slowly diffusing oxygen flow is present, while during fast oxygen diffusion, Fe-Mn concretions form (Gerrard, 1993). Alluvial and delluvial soils are considered as genetically young soils because their development is related to contemporary river terraces. Depending on the thickness of the humic horizon, the amount of organic matter, the presence/absence of carbonates, etc., Fluvisols show specific peculiarities. Often in the Fluvisol profile, separate layers corresponding to older buried humic horizons can be found (Penkov, 1983; Feakes and Retallack, 1988; Shishkov and Kolev, 2014).

Fluvisols occur on all continents and in all climates around rivers, lakes, in deltaic areas, and in areas of recent marine deposits (FAO, 2001) (Fig. 6.6.1). One special case of Fluvisols are the so-called Thionic Fluvisols—the acid sulfate soils - in which the parent material contains pyrite formed during sediment deposition.

In the Bulgarian classification system, alluvial-meadow soils are correlated to Fluvisols (Shishkov and Kolev, 2014). They are widespread in the river valleys in flat and semimountainous areas of the country. The most widespread are the Fluvisols near the Danube River and the other major rivers (Iskar, Lom, Ogosta, Kamchia, Yantra, etc.).

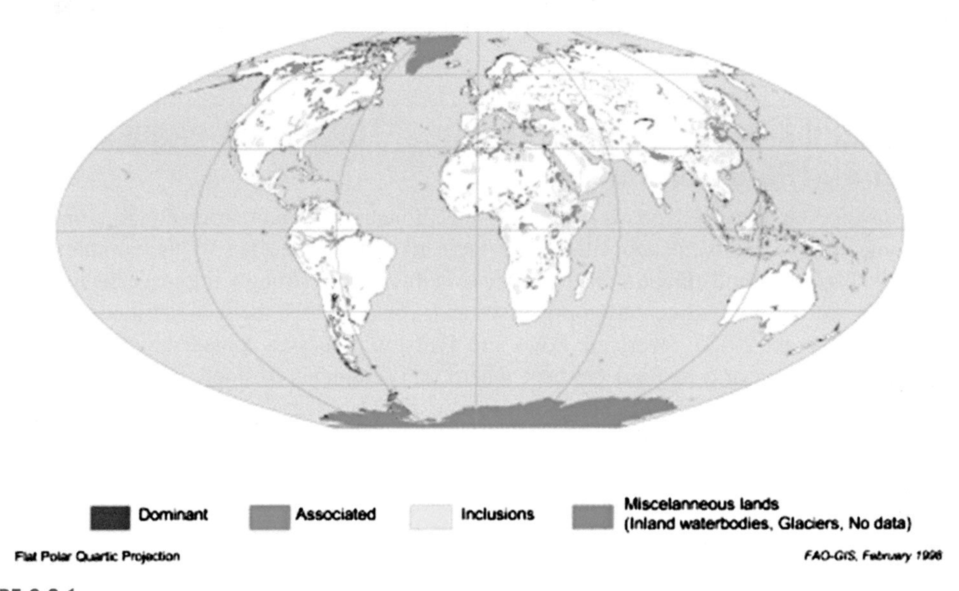

FIGURE 6.6.1

Spatial distribution of Fluvisols worldwide (FAO, 2001). Lecture notes on the major soils of the world. http://www.fao.org/docrep/003/y1899e/y1899e12.htm#P353_47671.

6.7 DESCRIPTION OF THE PROFILE STUDIED, TEXTURE, SOIL REACTION, AND MAJOR GEOCHEMISTRY

6.7.1 FLUVISOL (ALLUVIAL MEADOW SOIL), PROFILE DESIGNATION: AL

Situated close to Izgrev village, Varna district, in the river valley of Stara Reka River

Location: N 43°17′48.0″; E 27°42′45.4″, h = 307 m a.s.l.

Present-day climate conditions: MAT = 11.5°C; MAP = 522 mm, vegetation cover: grassland; relief: flat

Profile description: general view and details from the horizons are shown in Figs. 6.7.1–6.7.5.

0–45 cm: A_h horizon, dark brown, rich in organics, carbonates, friable, sandy-loamy

45–73 cm: transitional AC horizon, brown-gray, clayey-loamy, dense

73–120 cm: 1C horizon, brown-gray, clayey-loamy, gravel layer between 90 and 103 cm

120–185 cm: 2C horizon, sandy, gray-blue, mottled with gley spots

Parent material: alluvial deposits

The soil reaction and mechanical composition obtained for selected depth levels from the AL Fluvisol are shown in Fig. 6.7.6. The soil reaction is alkaline throughout the profile with a slight increase from the surface down to the 1C horizon and a lower pH with a larger scatter in the values measured at the bottom of the 2C horizon (Fig. 6.7.6a). A well expressed enhancement with silt and

FIGURE 6.7.1

View of the profile of Fluvisol (AL profile).

FIGURE 6.7.2

Fluvisol profile AL. A horizon—detail.

FIGURE 6.7.3

Fluvisol profile AL. AC horizon—detail.

FIGURE 6.7.4

Fluvisol profile AL. 1C horizon—detail.

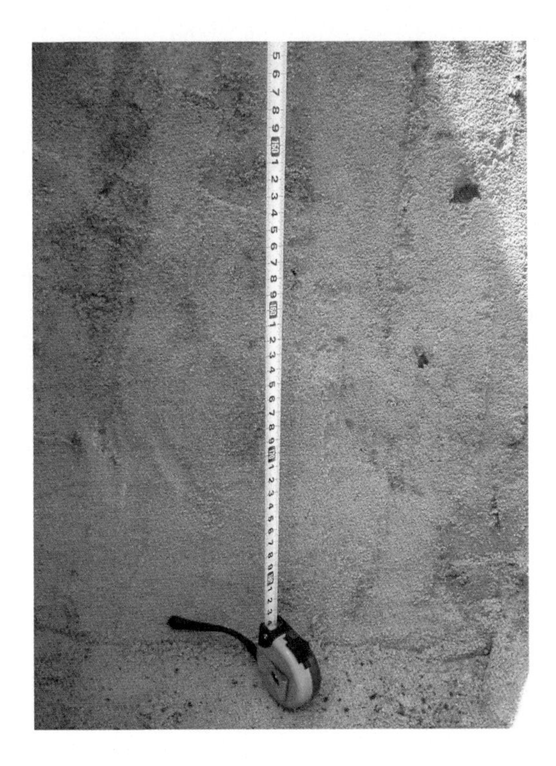

FIGURE 6.7.5

Fluvisol profile AL. 2C horizon—detail.

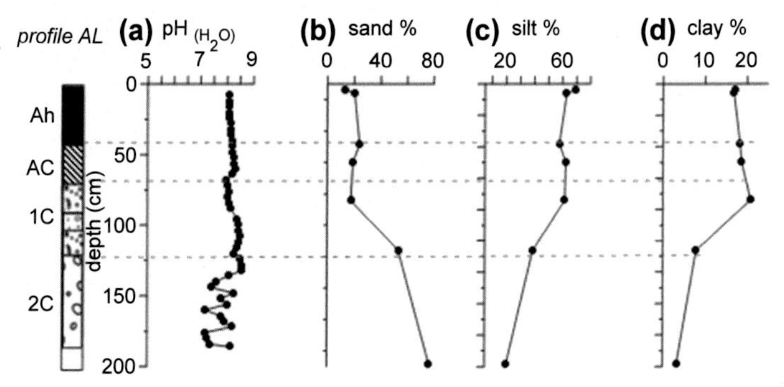

FIGURE 6.7.6

Soil reaction pH (a) and texture [sand (b), silt (c), clay (d)] for the AL Fluvisol profile.

clay is seen in the A_h and AC horizons of the profile (Fig. 6.7.6c–d), while the 2C horizon is dominated by the sand fraction. This textural differentiation clearly separates at least two major sedimentation events encompassed in the AL profile.

The major geochemistry as revealed by the variations in the elemental composition (Fig. 6.7.7), determined for the selected depth levels, confirms the observed differences in the mechanical composition. Most of the major elements are enhanced in the upper soil horizons—Ca, Al, Fe, Ti, Mn—while others (Si, K, Na) show a higher concentration in the *2C* horizon (Fig. 6.7.7). The uppermost A_h and AC horizons are also enriched in Cl and S (graphs not shown), which often arises from organic matter accumulation (Schaetzl and Anderson, 2009).

The iron content varies between 2% and 2.4% in the upper part of the profile (Fig. 6.7.7d), which is relatively low compared with other soil types. Its nondifferentiated variations inside the A_h-AC also support the initial stages of pedogenic transformations in Fluvisols.

Results from the acid oxalate extractions on selected samples are shown in Table 6.7.1. It is evident that the Fe_o extracted from the samples, belonging to the A_h-AC horizons is similar, as well the Fe_o/Fe_t

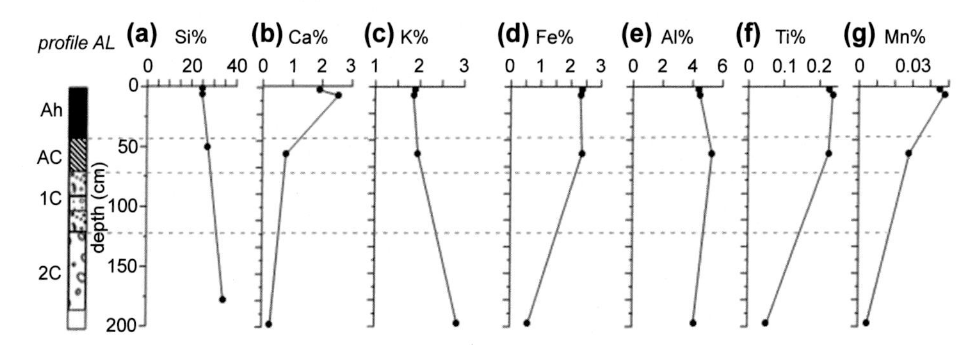

FIGURE 6.7.7

Content of some major and trace elements (Si, Ca, K, Fe, Al, Ti, Mn) (a–g) for selected samples along the depth of AL profile.

Table 6.7.1 Chemical extraction data for selected samples from AL Fluvisol profile

Sample depth (cm)	Al_o (g/kg)	Fe_o(g/kg)	Mn_o (g/kg)	Fe_t (g/kg)	Fe_o/ Fe_t	Al_t(g/kg)	Al_o/ Al_t
2	0.9526	1.057	0.267	23.93	0.044	44.34	0.021
6	0.944	1.078	0.273	23.06	0.047	44.81	0.021
50	1.1203	1.083	0.156	23.6	0.046	53.03	0.021
178	0.1653	0	0	5.861	0	40.88	0.004

Oxalate-extractable Al, Fe and Mn (Al_o, Fe_o, Mn_o), total Fe content (Fe_t), total Al content (Al_t), and the ratios Fe_o/ Fe_t and Al_o/ Al_t.

ratio, which varies between 4.4% and 4.7% (Table 6.7.1). Weakly crystalline and amorphous phases like ferrihydrite, lepidocrocite, and green rust are usually extracted through the oxalate (McKeague and Day, 1966; Cornell and Schwertmann, 2003). The amount of oxalate-extracted Al (Al_O) is also similar in the upper part of the profile, thus suggesting that Fe and Al are likely included in the same amorphous phase. This phenomenon is frequently observed in soils (Schwertmann, 1988).

6.8 MAGNETIC PROPERTIES OF FLUVISOLS AND THEIR RELATION TO PEDOGENIC PROCESSES

The high-temperature behavior of the magnetic susceptibility and remanences (ARM and IRM) (Fig. 6.8.1) is used for the identification of the main magnetic minerals in the AL Fluvisol. Thermomagnetic analysis of the sample from the A_h horizon (Fig. 6.8.1a) shows important thermal transformations occurring during heating, which leads to the formation of a new strongly magnetic phase on cooling. As already discussed in other chapters, such high-temperature behavior is frequently observed in samples containing Fe-hydroxides in the presence of organic matter (Hanesch et al., 2006). The thermomagnetic heating curve for the sample AL4 is characterized by the presence of two maxima—at 250°C and 500°C (Fig. 6.8.1a). The first peak could be related to the presence of ferrihydrite undergoing transformation to maghemite at ~250°C in the presence of organics. The second higher temperature peak at ~500°C likely reflects the reductive formation of magnetite from the hematite—the inverted product of the maghemite's transformation (Dunlop and Özdemir, 1997). An alternative possible explanation of the observed behavior is the transformation of lepidocrocite (γ-FeOOH) into maghemite (Hanesch et al., 2006; Mitov et al., 2002).

As seen from the χ–T heating curve (Fig. 6.8.1a), a small part of the magnetic susceptibility persists until 640°C, which could be indicative for the presence of maghemite that is stable on heating (Dunlop and Ödemir, 1997). The sample from the deeper part of the profile (Fig. 6.8.1b) shows completely different thermomagnetic curves, indicating the presence of magnetite through the T_c ~ 580°C and hematite through a T_N ~ 680°C. Due to intense mineral transformations, magnetic susceptibility on cooling strongly increases.

Stepwise thermal demagnetization of laboratory-induced remanences (ARM and IRM) helps in determining the stable remanence-carrying magnetic fraction. Representative results are shown in Fig. 6.8.1c–h. The anhysteretic remanence carried preferentially by SD grains (e.g., Dunlop and

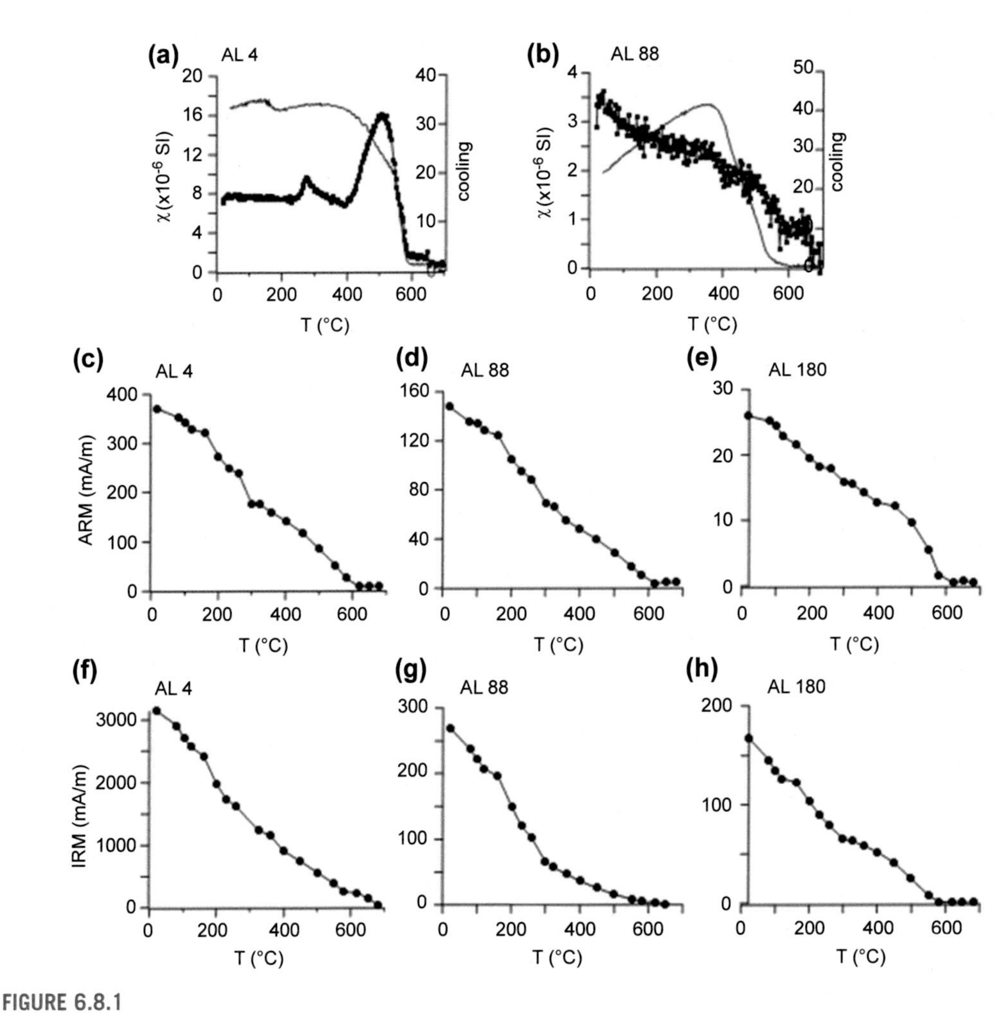

FIGURE 6.8.1

Magnetic mineral identification in AL profile: thermomagnetic analysis of magnetic susceptibility for two samples (a, b), heating curves—thick black line, cooling curves—thin gray line. Stepwise thermal demagnetization of ARM (c–e) and stepwise thermal demagnetization of IRM_{2T} (f–h).

Özdemir, 1997) shows a convex demagnetization curve with two more clearly expressed unblocking temperatures—at ~300°C and 580°C (Fig. 6.8.1c,d). These could be related to the presence of maghemite/maghemitized magnetite (e.g., van Velzen and Zijderveld, 1995) and magnetite, respectively. The most clearly expressed T_c of magnetite is seen for the deepest sample (AL180, Fig. 6.8.1e), thus revealing the narrowest grain-size distribution of the magnetite phase. The isothermal remanence displays a more concave shape for the demagnetization curve (Fig. 6.8.1f–h), underlining the prevailing contribution of coarser magnetic grains. Nevertheless, IRM is completely demagnetized at ~600°C, and only in the uppermost sample is a small tail extending to 680°C seen. Thus, the

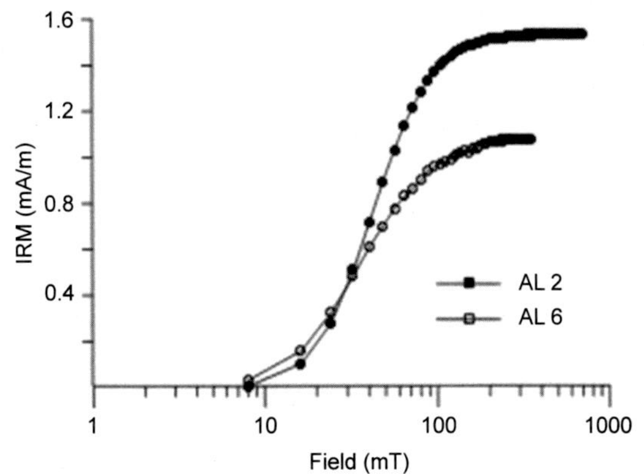

FIGURE 6.8.2

Stepwise acquisition of isothermal remanence (IRM) up to field of 0.8 T for selected samples from the AL Fluvisol.

dominant magnetite contribution to the remanent magnetic signal is confirmed. Sample AL88 from the 1C horizon (Fig. 6.8.1d, g) exhibits the widest grain size distribution of the magnetite fraction since both the ARM and IRM demagnetization curves are concave.

A stepwise acquisition of IRM was carried out for only two near-surface samples (Fig. 6.8.2). The curves obtained show the dominant role of the magnetically soft fraction. An analysis of the coercivity components (Kruiver et al., 2001) confirm this observation, revealing that the soft component accounts for 98−99% of the total IRM (Table 6.8.1).

The coercivity of the dominant component is slightly higher in the surface level (AL2) than in the underlying sample and $B_{1/2}$ of 36−43 mT could be related to magnetite of a relatively fine grain size (Egli, 2004). A very weak higher coercivity component, accounting for only 1−2% of the IRM (Table 6.8.1), may be attributed to fine-grained hematite (Özdemir and Dunlop, 2014).

Depth variations of different concentration- or grain-size−dependent magnetic parameters along the AL Fluvisol are shown in Fig. 6.8.3. The major feature that is obvious at a glance is the very

Table 6.8.1 Summary parameters from coercivity analysis by fitting IRM acquisition curves with cumulative Gaussian functions, using IRM-CLG1.0 software (Kruiver et al., 2001) for selected samples from the "AL" Fluvisol

	IRM component 1				IRM component 2			
Sample depth (cm)	Intensity (mA/m)	% from IRM_{total}	$B_{1/2}$ (mT)	DP	Intensity (mA/m)	% from IRM_{total}	$B_{1/2}$ (mT)	DP
Profile AL Fluvisol								
2	1.5	99	42.7	0.27	0.02	1	316.2	0.22
6	1.06	98	36.3	0.35	0.02	2	199.5	0.22

The intensity, relative contribution (% from IRM_{total}), median acquisition field ($B_{1/2}$), and dispersion parameter (DP) for each coercivity component are shown.

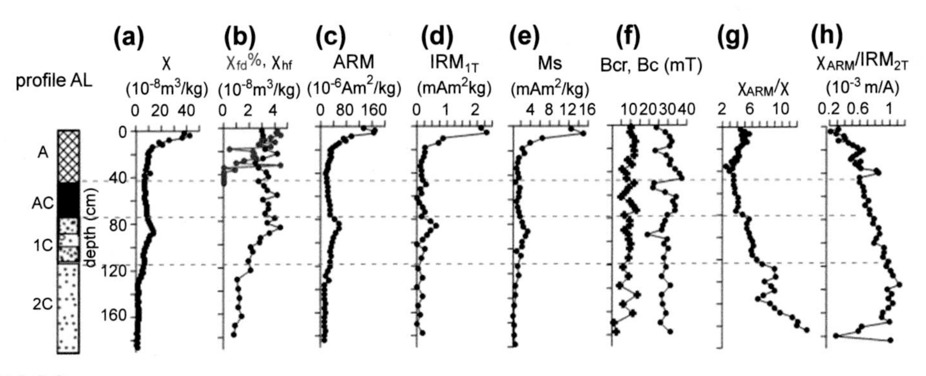

FIGURE 6.8.3

Depth variations of magnetic parameters and ratios along the "AL" Fluvisol profile: (a) magnetic susceptibility (χ); (b) percent frequency-dependent magnetic susceptibility ($\chi_{fd}\%$) (red crosses) and high field susceptibility (χ_{hf}) (full dots); (c) anhysteretic remanence (ARM); (d) saturation isothermal remanence acquired in 1 T (IRM_{1T}); (e) saturation magnetization (M_s); (f) B_c (crosses) and B_{cr} (dots); (g) ratio between anhysteretic susceptibility and magnetic susceptibility (χ_{ARM}/χ);(h) ratio between anhysteretic susceptibility and IRM_{2T} (χ_{ARM}/IRM_{2T}).

restricted near surface depth interval where an enhanced concentration of magnetically strong minerals is detected. Despite the visual (by a much darker color) well-expressed pedogenically transformed upper A_h and AC horizons (see Figs. 6.7.1–6.7.3), only the uppermost ~15 cm from the profile are magnetically enhanced, while downward all concentration-dependent magnetic characteristics (χ, M_s, IRM, ARM) show significantly lower values (Fig. 6.8.3). Weaker, but still well expressed, is a second maximum in the interval 80–100 cm. Magnetic susceptibility values are relatively low, varying generally between 5 and 20 ($\times 10^{-8}$ m^3/kg) except in the uppermost levels where χ reaches 40×10^{-8} m^3/kg (Fig. 6.8.3a).

Frequency-dependent magnetic susceptibility is reliably determined for only the upper ~40 cm (Fig. 6.8.3b) and shows low $\chi_{fd}\%$ at the surface (~4%) and a fast decrease down to the bottom of A_h. Thus, a weak magnetic enhancement with strongly magnetic SP grains is restricted to the uppermost well-aerated part of the Fluvisol profile. The magnetic remanences ARM and IRM also have maxima in the A_h horizon (Fig. 6.8.3c,d), suggesting that the pedogenic ferrimagnetic fraction has a wide grain size distribution, including SP as well as stable SD and coarser grains. On the other hand, high-field magnetic susceptibility (χ_{hf}) shows consistently higher values in the upper A_h-AC-*1C* part of the profile and, after a step-like decrease, low values in the deeper *2C* horizon (Fig. 6.8.3b). This behavior closely matches the obtained variations in the silt and clay mechanical fractions (Fig. 6.7.6), which are enhanced in this upper part of the profile, while downward, the sand fraction predominates. As is shown, commonly, χ_{hf} reflects the concentration of paramagnetic Fe-containing clay minerals (Oldfield, 1999) and the close correlation obtained evidences this relationship in the AL Fluvisol as well. The relatively sharp transition observed in the mechanical composition as well as in χ_{hf} most likely reflects different stages in the alluvial sedimentation. The coercivity parameters B_c and B_{cr} are practically constant along the depth (Fig. 6.8.3f), suggesting the dominant influence of the low coercivity magnetite fraction detected in the mineral magnetic identification data (Figs. 6.8.1 and

6.8.2). The grain-size–sensitive magnetic proxies χ_{ARM}/χ and χ_{ARM}/IRM_{2T} (Fig. 6.8.3g, h) reveal a steady tendency for refining the magnetic grain size toward the SD threshold at the middle of the *2C* horizon. Looking closer at the variations of these two parameters, it could be seen that in the uppermost A_h horizon, χ_{ARM}/χ and χ_{ARM}/IRM_{2T} have opposite trends—a decrease and an increase, respectively. This could be due to the effect of increased SP magnetite/maghemite in the uppermost soil depths, which enhances χ, while IRM is influenced only by coarser remanence carrying grains. In the deeper part of the solum, both grain-size proxy ratios show a decrease toward the SD grain size. This assumption is supported by the magnetic susceptibility measured after the dithionite extraction in the selected samples (Fig. 6.8.4).

It is evident from Fig. 6.8.4 that the biggest χ decrease after DCB treatment is obtained in the surface sample and in the subsurface sample at a 6-cm depth, while for the deeper samples, almost no change in χ occurs. Thus, the uppermost soil levels are the most enhanced with a fine-grained pedogenic strongly magnetic fraction that rapidly decreases in depth.

As discussed, the pedogenic processes in Fluvisols are relatively young and at an initial stage. Thus, their share should be separated from the contribution of the sediment-inherited ferrimagnetic grains. As demonstrated by the depth variations of the main magnetic parameters (Fig. 6.8.3), the AL profile is magnetically enhanced with SP, SD, and coarser pedogenic grains just in the uppermost ~20 cm. This near surface part of the profile is characterized by relatively good aeration and rich meadow organics that favor pedogenic precipitation of ferrimagnetic iron oxides (Schwertmann, 1988; Dearing et al., 1996). The magnetic signature in the deeper soil horizons, however, could not be explained simply as reflecting the lithological changes in the alluvial deposits because the observed magnetic grain-size refinement is in contradiction with the sand-dominated mechanical composition (Fig. 6.7.6). Therefore, the stable SD-like magnetite fraction in the *1C* and *2C* horizons (see Figs. 6.8.2 and 6.8.3) is most likely of pedogenic or diagenetic origin. As explained in part 6.7, the bottom parts of the Fluvisol profiles are strongly influenced by the fluctuating groundwater table and related periodic

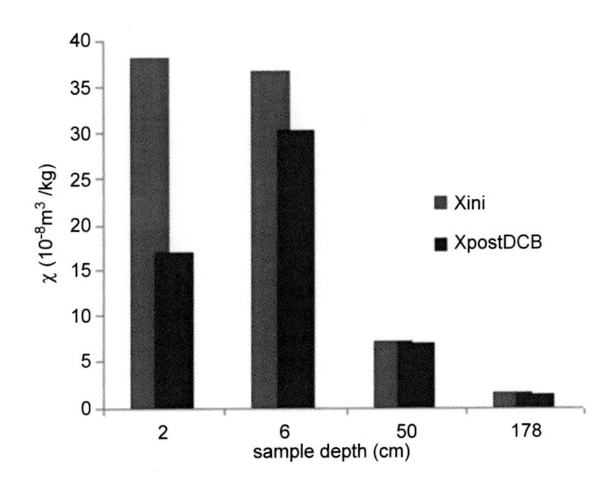

FIGURE 6.8.4

Magnetic susceptibility before (χ_{ini}—red columns) and after dithionite extraction ($\chi_{postDCB}$—blue columns) for selected samples from AL Fluvisol.

redox changes. Reductive dissolution of iron-containing minerals is thus possible, creating amorphous phases such as ferrihydrite, which in turn may undergo reductive dissolution, and secondary magnetite can precipitate under favorable conditions (Hansel et al., 2003). This hypothesis was discussed in more detail in other chapters related to soils undergoing periodic reduction–oxidation changes and will not be repeated here. Based on the observed regularities in the magnetic signature of these soils, it could be hypothesized that the deeper horizons of Fluvisols affected by fluctuations in the groundwater table (e.g., changing redox state) are enhanced with secondary stable magnetite grains formed as a result of ferrihydrite transformations.

Published magnetic studies on Fluvisols and other soils developed on alluvial deposits all deal with investigations of soil chronosequences and the relation between the obtained magnetic signature and soil age (White and Walden, 1997; Van Dam et al., 2006; Zielhofer et al., 2009; Quinton et al., 2011; Stinchcomb and Peppe, 2014). White and Walden (1997) have found a good correlation between the amount of Fe_d extracted from A_h horizons of Fluvisols developed in an alluvial fan in southern Tunisia and the magnetic susceptibility in this horizon despite the very low χ values reported ($\sim 2.3 \times 10^{-8}$ m^3/kg in the upper profile part). Thus, a direct link is assumed between the degree of the pedogenic development of alluvial soils and their magnetic enhancement with strongly magnetic minerals. The authors also underlined that the rates of magnetic enhancement vary with time and more rapid enrichment was observed in Holocene soils than in the Pleistocene soils. Zielhofer et al. (2009) report only field (volume) magnetic susceptibility measurements along their profiles, so a direct comparison with the other data is impossible. Quinton et al. (2011) carried out detailed magnetic measurements on five soils developed in coarse-grained fluvial deposits in the terraces in Red Canyon, Wyoming (USA). All soils exhibit relatively weak magnetic enhancement in the uppermost A_h horizon with a fast decrease downward, similar to the results obtained for the AL Fluvisol in our study (Fig. 6.8.3). Magnetic susceptibility in the A_h horizons varied in the range $40–50 \times 10^{-8}$ m^3/kg and is also compatible with the values obtained for the AL profile. Similarities also arise when comparing the results from the coercivity analysis of the IRM acquisition curves of Fluvisols from the Red Canyon and the AL profile. Two major coercivity components are found in the soils studied by Quinton et al. (2011)—a magnetically soft one with $\log B_{1/2}$ of 1.5 and a high-coercivity component with $\log B_{1/2}$ of 2.6–2.7, while for the AL profile the respective coercivities are $\log B_{1/2}$ of 1.5–1.6 for the soft component and $\log B_{1/2}$ of 2.3–2.5 for that of the high coercivity component. Considering the magnetic properties of soils in relation to the terrace's age, Quinton et al. (2011) have found a loss of the ferrimagnetic fraction and an increase in hematite content with increasing soil age. The authors assume that this supports the hypothesis of Barrón and Torrent (2002) that the pedogenic strongly magnetic maghemite is an intermediate product in the transformation of ferrihydrite to hematite. In the Fluvisol from Bulgaria, the hematite's high coercivity component is very weak, likely indicating the very young soil age according to the assumptions by Quinton et al. (2011). Fluvisol developed on the T_0 river terrace of the Delaware River (Stinchcomb and Peppe, 2014) displays a low concentration of the strongly magnetic fraction, expressed in low χ, IRM, and ARM values and scattered $\chi_{fd}\%$ reaching a 5% maximum. Detailed mineral magnetic investigations reveal the dominant role of the magnetically soft magnetite/maghemite component and a weak contribution of the higher coercivity fraction. Considering the magnetic signature of all soil profiles studied (comprising Fluvisol, Arenosol, Cambisol, as well as buried soils in the older terraces) in the alluvial plain of the river, Stinchcomb and Peppe (2014) show that the soil's magnetic properties are influenced by the weathering duration (e.g., the time factor), which agrees well with Boyle et al. (2010) chemical kinetic theory for the pedogenic magnetic enhancement of a soil. Thus the authors conclude that magnetic susceptibility measurements are a good indicator of weathering duration for soils that have developed for <16 ka in periglacial and alluvial deposits. Van Dam et al. (2006) studied young (16–413 ka) alluvial fan soil chronosequence along the Mojave section of the San Andreas fault. In contrast to the chronosequences discussed above, Van Dam et al. (2006) demonstrate that the magnetic susceptibility in these soils is predominantly correlated with the type of parent material and less with age or landscape position. This is likely due to the fact that the fan sediments contained significant amounts of mafic material and the inherited lithogenic magnetic signal overwhelms magnetic susceptibility and remanences.

The magnetic properties of Fluvisols are therefore conditioned first by the lithogenic contribution from the alluvial/delluvial deposits (the parent material) and, second, by the duration of the pedogenesis. The pedogenic magnetic fraction is most abundant in the surface layer (A_h) and rapidly diminishes in depth. It is dominated mainly by SP and stable SD-like grains of magnetite/maghemite. In older Fluvisols, hematite is also detected, most likely resulting from the oxidation of magnetite/maghemite.

REFERENCES

Adams, W., Kassim, J., 1984. Iron oxyhydroxides in soils developed from Lower Paleozoic sedimentary rocks in mid-Wales and implications for some pedogenic processes. J. Soil Sci. 35 (1), 117–126.

Alekseev, A., Alekseeva, T., Sokolowska, Z., Hajnos, M., 2002. Magnetic and mineralogical properties of different granulometric fractions in the soils of the Lublin Upland Region. Int. Agrophysics 16, 1–6.

Anastacio, A., Fabris, J., Stucki, J., Coelho, F., Pinto, I., Viana, J., 2005. Clay fraction mineralogy of a Cambisol in Brazil. Hyperfine Interact. 166, 619–624.

Banfield, J., Wasilewski, P., Veblen, D., 1994. TEM study of relationships between the microstructures and magnetic properties of strongly magnetized magnetite and maghemite. Am. Mineral. 79, 654–667.

Barrón, V., Torrent, J., 2002. Evidence for a simple pathway to maghemite in Earth and Mars soils. Geochim. Cosmochim. Acta 66 (15), 2801–2806.

Boyle, J., Dearing, J., Blundell, A., Hannam, J., 2010. Testing competing hypotheses for soil magnetic susceptibility using a new chemical kinetic model. Geology 38, 1059–1062.

Bozilova, E., Tonkov, S., 2000. Pollen from Lake Sedmo Rilsko reveals southeast European postglacial vegetation in the highest mountain area of the Balkans. New Phytol. 148, 315–325.

Campbell, A., Schwertmann, U., Campbell, P., 1997. Formation of cubic phases on heating ferrihydrite. Clay Miner. 32, 615–622.

Cornell, R., Schwertmann, U., 2003. The Iron Oxides. Structure, Properties, Reactions, Occurrence and Uses. Weinheim, New York.

Dahms, D., Favilli, F., Krebs, R., Egli, M., 2012. Soil weathering and accumulation rates of oxalate-extractable phases derived from alpine chronosequences of up to 1 Ma in age. Geomorphology 151–152, 99–113.

Dearing, J.A., Dann, R.J.L., Hay, K., Lees, J.A., Loveland, P.J., Maher, B.A., O'Grady, K., 1996. Frequency-dependent susceptibility measurements of environmental materials. Geophys. J. Int. 124, 228–240.

do Nascimento, A., Furquim, S., Couto, E., Beirigo, R., de Oliveira Júnior, J., de Camargo, P., Vidal-Torrado, P., 2013. Genesis of textural contrasts in subsurface soil horizons in the Northern Pantanal-Brazil. Rev. Bras. Ciênc. Solo. 37, 1113–1127.

Dong, Y., Zhang, W., Dong, C., Ge, C., Yu, L., 2014. Magnetic and diffuse reflectance spectroscopic characterization of iron oxides in the tidal flat sequence from the coastal plain of Jiangsu Province, China. Geophys. J. Int. 196, 175–188.

Dunlop, D., Özdemir, O., 1997. Rock magnetism. Fundamentals and frontiers. In: Edwards, D. (Ed.), Cambridge Studies in Magnetism. Cambridge University Press.

Egli, R., 2004. Characterization of individual rock magnetic components by analysis of remanence curves. 3. Bacterial magnetite and natural processes in lakes. Phys. Chem. Earth 29, 869–884.

Egli, M., Fitze, P., Mirabella, A., 2001. Weathering and evolution of soils formed on granitic, glacial deposits: results from chronosequences of Swiss alpine environments. Catena 45, 19–47.

Egli, M., Mirabella, A., Sartori, G., Zanelli, R., Bischof, S., 2006. Effect of north and south exposure on weathering rates and clay mineral formation in Alpine soils. Catena 67, 155–174.

Egli, M., Mirabella, A., Sartori, G., 2008. The role of climate and vegetation in weathering and clay mineral formation in late Quaternary soils of the Swiss and Italian Alps. Geomorphology 102, 307–324.

Egli, M., Sartori, G., Mirabella, A., Giaccai, D., 2010. The effects of exposure and climate on the weathering of late Pleistocene and Holocene Alpine soils. Geomorphology 114, 466–482.

Egli, M., Dahms, D., Norton, K., 2014. Soil formation rates on silicate parent material in alpine environments: different approaches–different results? Geoderma 213, 320–333.

FAO World soil resources reports, 2001. Lecture Notes on the Major Soils of the World. ISSN: 0532-0488. http://www.fao.org/DOCREP/003/Y1899E/Y1899E00.HTM.

Farrell, K., 2001. Geomorphology, facies architecture, and high-resolution, non-marine sequence stratigraphy in avulsion deposits, Cumberland Marshes, Saskatchewan. Sediment. Geol. 139, 93–150.

Feakes, C., Retallack, G., 1988. Recognition and chemical characterization of fossil soils developed on alluvium; A Late Ordovician Example. Geol. Soc. America, Spec. Publ. 216, 35–48.

Geiss, C.E., Egli, R., Zanner, C.W., 2008. Direct estimates of pedogenic magnetite as a tool to reconstruct past climates from buried soils. J. Geophys. Res. 113, B11102. http://dx.doi.org/10.1029/2008JB005669.

Georgieva, J., Jokova, M., 1990. Valence state and mineral forms of iron in brown forest soils from Western Stara Planina. For. Sci. 1, 68–73 (in Bulgarian).

Gerrard, A., 1993. Soil Geomorphology. An Interaction of Pedology and Geomorphology. Springer, Netherlands, ISBN 9780412441806, 272 pp.

Hanesch, M., Petersen, N., 1999. Magnetic properties of recent parabrown-earth from Southern Germany. Earth Planet. Sci. Lett. 169, 85–97.

Hanesch, M., Stanjek, H., Petersen, N., 2006. Thermomagnetic measurements of soil iron minerals: the role of organic carbon. Geophys. J. Intern. 165, 53–61.

Hansel, C., Benner, S., Netss, J., Dohnalkova, A., Kukkadapu, R., Fendorf, S., 2003. Secondary mineralization pathways induced by dissimilatory iron reduction of ferrihydrite under advective flow. Geochim. Cosmochim. Acta 67 (16), 2977–2992.

Jokova, M., Georgieva, J., Trichkov, L., 1991. Mobile compounds of Fe, Al and Mn in brown forest soils from Batak ridge of Western Rhodopes. For. Sci. 4, 59–66 (in Bulgarian).

Jokova, M., 1999. Extractable compounds of Fe, Al and Mn as participants and indicators of soil processes. Soil Sci. Agro-Chemistry Ecol. XXXIV (4–5), 95–102.

Koinov, V., Kabakchiev, I., Boneva, K., 1998. In: Koinov, V., Boneva, K. (Eds.), Atlas of Soils in Bulgaria. Zemizdat, Sofia.

Kruiver, P., Dekkers, M., Heslop, D., 2001. Quantification of magnetic coercivity components by the analysis of acquisition curves of isothermal remanent magnetization. Earth Planet. Sci. Lett. 189, 269–276.

Liu, Q., Banerjee, S., Jackson, M., Zhu, R., Pan, Y., 2002. A new method in mineral magnetism for the separation of weak antiferromagnetic signal from a strong ferrimagnetic background. Geophys. Res. Lett. 29 (12), 1565. http://dx.doi.org/10.1029/2002GL014699.

Maher, B., 1986. Characterization of soils by mineral magnetic measurements. Phys. Earth Planet. Inter. 42, 76–92.

McKeague, J.A., Day, J.H., 1966. Dithionite and oxalate extractable Fe and Al as aids in differentiating various classes of soils. Can. J. Soil Sci. 46, 13–22.

Mitov, I., Paneva, D., Kunev, B., 2002. Comparative study of the thermal decomposition of iron oxyhydroxides. Thermochim. Acta 386, 179–188.

Mourier, B., Poulenard, J., Chauvel, C., Faivre, P., Carcaillet, C., 2008. Distinguishing subalpine soil types using extractible Al and Fe fractions and REE geochemistry. Geoderma 145, 107–120.

Ninov, N., 2002. Soils. In: Kopralev, I., Yordanova, M., Mladenov, C. (Eds.), Geography of Bulgaria. Institute of Geography – Bulg. Acad. Sci., ForCom, Sofia, pp. 277–317.

Oldfield, F., 1999. The rock magnetic identification of magnetic mineral and magnetic grain size assemblages. In: Walden, J., Oldfield, F., Smith, J. (Eds.), Environmental Magnetism: A Practical Guide. ISSN: 0264-9241. Quaternary Research Association, Technical Guide No. 6, London, ISBN 0 907780 42 3, pp. 98–112.

Özdemir, Ö., Dunlop, D., 2014. Hysteresis and coercivity of hematite. J. Geophys. Res. Solid Earth 119, 2582–2594. http://dx.doi.org/10.1002/2013JB010739.

Penkov, M., 1983. Soils in Bulgaria. Conservation and Amelioration. Nauka I Izkustvo, Sofia, 221 pp. (in Bulgarian).

Peters, C., Dekkers, M.J., 2003. Selected room temperature magnetic parameters as a function of mineralogy, concentration and grain size. Phys. Chem. Earth 28, 659–667.

Quinton, E., Dahms, D., Geiss, C., 2011. Magnetic analyses of soils from the Wind River Range, Wyoming, constrain rates and pathways of magnetic enhancement for soils from semiarid climates. Geochem. Geophys. Geosyst. 12 http://dx.doi.org/10.1029/2011GC003728. Q07Z30.

Rivas, O., Guerrero, B., Rebolledo, E., Sedov, S., Perez, S., 2012. Mineralogia magnetica de suelos volcanicos en una toposecuencia del valle de Teotihuacan. Bol. Soc. Geol. Mex. 64 (1), 1–20.

Schaetzl, R., Anderson, A., 2009. Soils. Genesis and Geomorphology. Cambridge Univ. Press, UK, ISBN 978-0-521-81201-6.

Schwertmann, U., 1988. Occurrence and formation of iron oxides in various pedoenvironments. In: Stucki, J., Goodman, B., Schwertmann, U. (Eds.), Iron in Soils and Clay Minerals. NATO ASI Series: Series C: Mathematical and Physical Sciences, 217. Reidel Publ. Company, pp. 267–308.

Schwertmann, U., Murad, E., 1988. The nature of an iron oxide – organic iron association in a peaty environment. Clay Miner. 23, 291–299.

Shishkov, T., Kolev, N., 2014. The soils of Bulgaria. In: Hartemink, A.E. (Ed.), World Soils Book Series. Series. Springer Science+Business Media, Dordrecht, ISBN 978-94-007-7784-2, p. 205.

Stanjek, H., 1987. The formation of maghemite and hematite from lepidocrocite and goethite in a Cambisol from Corsica, France. Pflanzenernahr. Bodenk. 150, 314–318.

Stinchcomb, G., Peppe, D., 2014. The influence of time on the magnetic properties of late Quaternary periglacial and alluvial surface and buried soils along the Delaware River, USA. Front. Earth Sci. 2 http://dx.doi.org/10.3389/feart.2014.00017 article 17.

Tauxe, L., Mullender, T., Pick, T., 1996. Potbellies, wasp-waists, and superparamagnetism in magnetic hysteresis. J. Geophys. Res. 101 (B1), 571–583.

Teoharov, M., 2009. Systematic Database for the Soils in Bulgaria. Agricultural Academy. Inst. of Soil Sci., Agrotechnologies and Plant Protection "N. Poushkarov", Sofia, ISBN 978-954-9467-26-0 (in Bulgarian).

Tonkov, S., Marinova, E., 2005. Pollen and plant macrofossil analyses of radiocarbon dated mid-Holocene profiles from two subalpine lakes in the Rila Mountains, Bulgaria. The Holocene 15 (5), 663–671.

Tonkov, S., Bozilova, E., Possnert, G., Velcev, A., 2008. A contribution to the postglacial vegetation history of the Rila Mountains, Bulgaria: the pollen record of Lake Trilistnika. Quat. Int. 190, 58–70.

Tonkov, S., Possnert, G., Bozilova, E., 2011. The Lateglacial in the Rila mountains (Bulgaria) revisited: the pollen record of lake Ribno (2184 m). Rev. Palaeobotany Palynology 166, 1–11.

Van Dam, R., Harrison, J., Rittel, C., Hendrickx, J., Borchers, B., 2006. Magnetic soil properties at two arid to semi-arid sites in the Western United States. Proc. SPIE - Int. Soc. Opt. Eng. 6217 I http://dx.doi.org/10.1117/12.665625 [62170O].

van Velzen, A., Zijderveld, J., 1995. Effects of weathering on single domain magnetite in Early Pliocene marine marls. Geophys. J. Int. 121, 267–278.

Velcev, A., 1995. Pleistocene Glaciations in Bulgarian Mountains, vol. 87. Annuaire de l'Universite de Sofia "St. Kliment ohridski", Faculte de Geologie et Geographie, Book 2, pp. 53–64 (in Bulgarian).

White, K., Walden, J., 1997. The rate of iron oxide enrichment in arid zone alluvial fan soils, Tunisian southern atlas, measured by mineral magnetic techniques. Catena 30, 215–227.

Zanelli, R., Egli, M., Mirabella, A., Giaccai, D., Abdelmoula, M., 2007. Vegetation effects on pedogenetic forms of Fe, Al and Si and on clay minerals in soils in southern Switzerland and northern Italy. Geoderma 141, 119–129.

Zielhofer, C., Espejo, J., Granados, M., Faust, D., 2009. Durations of soil formation and soil development indices in a Holocene Mediterranean floodplain. Quat. Int. 209 (1–2), 44–65.

CHAPTER

MAGNETISM OF SOILS FROM THE ANTARCTIC PENINSULA

7

7.1 SOILS IN ANTARCTICA: DISTRIBUTION, FORMATION PROCESSES, AND SPECIFIC CHARACTERISTICS

Soil investigations in Antarctica started after the first expeditions to the McMurdo area during the British Antarctic Expedition at the beginning of the 20th century (1907–09) (Ugolini and Bockheim, 2008). The first publication on Antarctic soils appeared in 1916, authored by Jensen (Jensen, 1916), in which the term "soil" is shown in quotation marks. Detailed information about the history of soil research during this century and major advances in soil investigations in Antarctica can be found in Ugolini and Bockheim (2008), Bockheim (2015). Only 0.35% of Antarctica is ice-free, and this area is covered mainly by the Transantarctic Mountains and the mountains along the Antarctic Peninsula (Bockheim, 2015 and references therein). The main ice-free areas investigated by different expeditions are depicted in Fig. 7.1.1, according to Bockheim et al. (2015).

According to Bockheim and Hall (2002), Antarctica can be divided into three ecoclimatic regions: the Antarctic Peninsula and its offshore islands (c. 61–72°S), maritime East Antarctica (c. 66–71°S), and the Transantarctic Mountains (c. 72–87°S).

The geology of Antarctica is relatively well studied despite being limited by the ice cover and outcropping rocks. East Antarctic regions (e.g., Fig. 7.1.1) are dominantly covered by Precambrian gneisses and schists, and the Transantarctic mountains are covered by sediments cut by granite intrusions. Volcanic rocks and granites are frequently found in the Antarctic Peninsula and Marie Byrd Lands (Bockheim et al., 2015 and references therein). Since the ice-free areas are close to the glaciers, the soil parent materials are mainly of glacial origin—till, outwash, colluvium, etc. In the Antarctic Peninsula and along the coastal areas of continental Antarctica, debris flows and gelifluction deposits frequently occur. Because of the presence of active and past volcanoes, volcanic ash and tephra layers are common in many regions in the Antarctic Peninsula, East Antarctica, and Transantarctic Mountains (Narcisi et al., 2010; Patrick and Smellie, 2013). They are extensively used for establishing the tephrostratigraphy and for chronological constraint and the correlation of the tephra layers in Antarctic ice cores (e.g., Dunbar and Kurbatov, 2011).

Due to the extreme climate conditions with very low temperatures, humidity, and precipitation, accompanied by strong katabatic winds, soil formation in Antarctica is at a very early stage, dominated by physical weathering mainly through permafrost and related freeze–thaw cycles (Grosse et al., 2016). Commonly these soils have poor internal development and lack layering. Antarctic soil science, however, is becoming increasingly important because of the established high sensitivity of the Antarctic environment to changing climate conditions and human impact (Ugolini and Bockheim, 2008;

Soil Magnetism. http://dx.doi.org/10.1016/B978-0-12-809239-2.00007-3

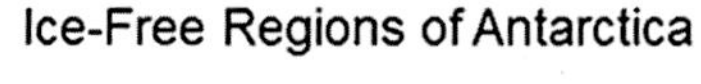

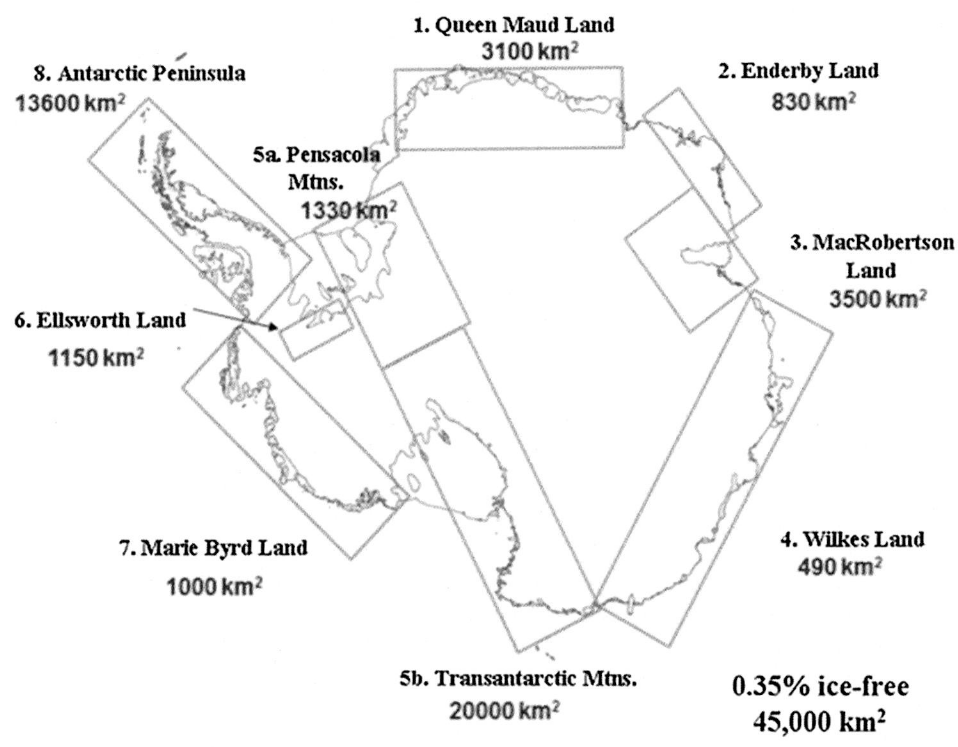

FIGURE 7.1.1

Ice-free regions of Antarctica.

Reprinted from Bockheim, J., Lupachev, A., Blume, H.-P., Bölter, M., Simas, F., McLeode, M., 2015. Distribution of soil taxa in Antarctica: A preliminary analysis. Geoderma 245–246, 104–111, with permission from Elsevier.

Bockheim, 2015). Therefore, Antarctic soils could be used as a baseline environment in the studies of global climate change (Smykla et al., 2015).

Antarctica experiences three main types of climate: (1) a humid-maritime climate along the coast of the West Antarctic Peninsula; (2) a dry maritime climate along the coast of East Antarctica; and (3) the most extreme hypercold and hyperarid climate of inland Antarctica (Bockheim et al., 2015). Because of these extreme conditions, the permafrost is continuous on the Antarctic continent and discontinuous at the South Shetland Islands where the mildest climate conditions occur. Plant communities are restricted to lichens, mosses, and algae. As demonstrated recently by Newsham et al. (2016), the air temperature is the main driving factor for the soil's fungal diversity and model predictions suggest an increase in the number of fungal species in maritime Antarctica of ~20% until the year 2100. On the other hand, the climate of continental Antarctica is different. A number of studies suggest a lack of air warming there, but ecosystem changes occur rapidly because of the increasing thickness of the active layer (Guglielmin et al., 2014), which produces a decrease in ground water content and related drying. The active layer has been defined as the surficial layer above the permafrost that thaws during summer (e.g., Burn, 1998). Thus, the surface albedo is an important factor for the active layer's thickness and its temporal dynamics. Pedogenesis of soils from the Antarctic Peninsula, its offshore islands, and maritime East Antarctica is related to the dynamics of the active layer. The thickness of the active layer is largest in the Antarctic islands where it can reach 150 cm. In inland Antarctica, the active layer and near-surface permafrost have a very low moisture content ($< 3\%$) so that the permafrost is "dry" (Bockheim and Hall, 2002).

Pedological studies of permanently frozen, desert ice-free areas in continental Antarctica show the presence of poorly developed soils with a significant accumulation of water-soluble salts (Bockheim, 1997). The warmer climate with higher water availability in maritime Antarctica results in a completely different soil formation, with the biggest soil and faunal diversity found there. The relatively mild temperatures compared with continental Antarctica intensify hydrological and biological cycles, so that weak chemical weathering is suggested to play a role in the pedogenesis (Simas et al., 2007; Haus et al., 2016). One important factor in the soil formation in maritime Antarctica is the higher nutrient inputs from marine animals, compared with continental Antarctica (Bockheim, 2015). Ornithogenic soils form in the weathered rock substrates due to the influence of flying birds (such as skuas, giant petrels, and seagulls) and nesting penguins. They commonly occur in coastal ice-free areas as discontinuous patches covered with abundant vegetation (mostly lichens, mosses, and, in some cases, flowering plants) (Simas et al., 2008). Well-developed ornithogenic soils in old penguin colonies have a thick organic horizon enriched in organic carbon and nitrogen. Intense cryoturbation and water percolation cause strong phosphorus leaching from the guano. Leached phosphorus further reacts with the mineral surfaces, leading to "phosphatization" (Simas et al., 2007 and references therein). Ornithogenic soils are acid because of the formation of sulfuric and nitric acids during the biological stabilization of guano (excrement) and have anomalously high total and bioavailable P levels (Simas et al., 2007; Emslie et al., 2014). The main soil types recognized in Antarctica are Gelisols, Cryosols, and ornithogenic soils (Bockheim et al., 2015), but controversies still exist between the correlation of Antarctica's soil classification according to *Soil Taxonomy* (Soil Survey Staff, 2014) and the WRB (IUSS Working Group WRB, 2014).

7.2 MAGNETIC STUDIES OF SOILS FROM THE ANTARCTIC PENINSULA

The Bulgarian national Antarctic Program started in 1987 when a refuge was established on Livingston Island (South Shetland Islands) on a spot located on the northeast side of the South Bay. In the period between 1993 and 2015, Bulgaria organized 22 successive Antarctic campaigns with scientific research focused on Earth and Life sciences (Pimpirev, 2015). In the several Antarctic expeditions (2005–08), a collection of topsoil samples from the area near the Bulgarian Antarctic Base was gathered for magnetic studies.

Livingston Island is the second largest in the South Shetlands archipelago, and its geology, tectonic evolution, and relationships among the main structural units are well studied by different geological expeditions because of its relatively large ice-free area (Kamenov, 2015 and references therein). Several sedimentary and igneous rock sequences are described in Livingston Island. The sediments from the Miers Bluff Formation are considered to be host rocks of Hesperides Point Pluton. A number of lava flows, subordinate pyroclastic deposits, isolated exposures of smaller plutonic bodies, and numerous dykes are also described (Kamenov, 2015; Bonev et al., 2015). The area close to the Bulgarian Antarctic Base is occupied by rocks of Hesperides Pluton, sill-like bodies related to it, and the sedimentary complex of the Miers Bluff Formation (Bonev et al., 2015).

It is supposed that soils in the South Shetland Islands have formed since the last deglaciation (ca. 9500–6000 BP) (Navas et al., 2008) as a result of a warmer climate and the related longer thaw period, and a larger amount of percolating water during the summer. Twenty-three topsoil samples were gathered along the coastline near the Bulgarian and Spanish Antarctic Bases. Samples LB1–L16 were gathered along the beach in front of the Bulgarian Antarctic Base (BAB), and samples L17 and L18 were taken at Caleta Argentina. Another four samples were taken from Sally Rocks and Glacier Rocosso. One sample from King George Island (taken close to the FERRAZ base) is also included in the study. Detailed sample locations and sampling depths are summarized in Table 7.2.1. Figs. 7.2.1–7.2.5 illustrate typical environments at several sampling locations. Soils from Livingston Island sampled close to the Bulgarian Antarctic Base were developed in till intermixed with frost-shattered rock and tephra (Haus et al., 2016). The main vegetation is represented by mosses, lichens, and grass (*Deschampsia antarctica*).

Table 7.2.1 Locations and sampling depths of the collection of topsoils from Livingston Island

Sample	Latitude (°S)	Longitude (°W)	Location	Sampled depth interval	Seabird/penguin influence
Sally rocks 2	62°42′07.7″	60°25′06.8″	Sally rocks	0–5 cm	
Sally rocks 3	62°42′45.6″	60°24′31.1″	Sally rocks	0–8 cm	
Sally rocks 4	62°42′45.6″	60°24′31.1″	Sally rocks	0–8 cm	
Rokosso	62°68′45.6″	60°42′16.2″	Glacier Rokosso	0–5 cm	
Buena nueva	62°42′15.7″	60°23′21.3″	Caleta Buena Nueva	0–5 cm	
LB-1	62°38′04.3″	60°20′55.2″	BAB	0–7 cm	
LB-2	62°38′04.3″	60°20′57.0″	BAB	0–8 cm	
LB-3	62°38′04.8″	60°20′59.2″	BAB	0–10 cm	
LB-4	62°38′06.2″	60°23′21.3″	BAB	0–4 cm	
LB-5	62°38′06.4″	60°21′15.0″	BAB	0–10 cm	
LB-6	62°38′07.9″	60°21′22.2″	BAB	0–8 cm	Skua nests
LB-7	62°38′07.1″	60°21′22.2″	BAB	0–10 cm	Skua nests
LB-8	62°38′07.1″	60°21′22.2″	BAB	0–8 cm	
LB-9	62°38′35.8″	60°22′14.8″	BAB	0–5 cm	
LB-10	62°38′14.3″	60°21′36.6″	BAB	0–7 cm	
LB-11	62°38′24.6″	60°21′55.6″	BAB	0–10 cm	
L-12	62°38′35.8″	60°22′14.8″	BAB	0–11 cm	
L-13	62°38′35.7″	60°22′24.5″	BAB	0–7 cm	
L-14	62°38′36.1″	60°22′16.7″	BAB	0–8 cm	
L-15	62°38′40.1″	60°22′12.9″	BAB	0–7 cm	
L-16	62°38′49.0″	60°22′17.9″	BAB	0–5 cm	
L-17	62°40′05.5″	60°23′39.3″	Caleta Argentina	0–3 cm	Penguins
L-18	62°40′04.0″	60°24′06.8″	Caleta Argentina	0–6 cm	Penguins
Ferraz			King George island—Admiralty Bay base FERRAZ		

FIGURE 7.2.1

Locality at Caleta Argentina, where sample L18 was gathered.

FIGURE 7.2.2

Caleta Argentina—general view.

FIGURE 7.2.3

Sampling locality for sample Rokosso.

Most of the sampling sites are characterized by the presence of a thick layer of mosses/lichens (Figs. 7.2.1–7.2.5), so that the sampling interval encompasses the organic-rich part of the soil. Sampling locations L1–L14 close to the BAB have altitudes between 3 and 10 m a.s.l., and sites L15 and L16 are situated at only ~1 m a.s.l.

The soil reaction (pH) is nearly neutral, varying generally between 5.0 and 6.0 (Fig. 7.2.6).

These values are in agreement with the data reported by Haus et al. (2016) for the uppermost layers of the profiles sampled near the BAB in their study. Such relatively low values are reported as characteristic for ornithogenic soils because of the acidifying action of the acids H_2SO_4 and HNO_3 produced during guano decomposition (Simas et al., 2007; Moura et al., 2012).

The topsoils studied contain between 6% and 7% of iron (Fig. 7.2.7), while the Al content varies in a wider range. Comparing the contents of these two elements in soils and rocks taken from the same sampling locations, it appears that for some soils the Fe content is much higher than that in the

FIGURE 7.2.4

Landscape around the Bulgarian Antarctic Base—general view.

FIGURE 7.2.5

Soil sampling near the Bulgarian Antarctic Base (2008).

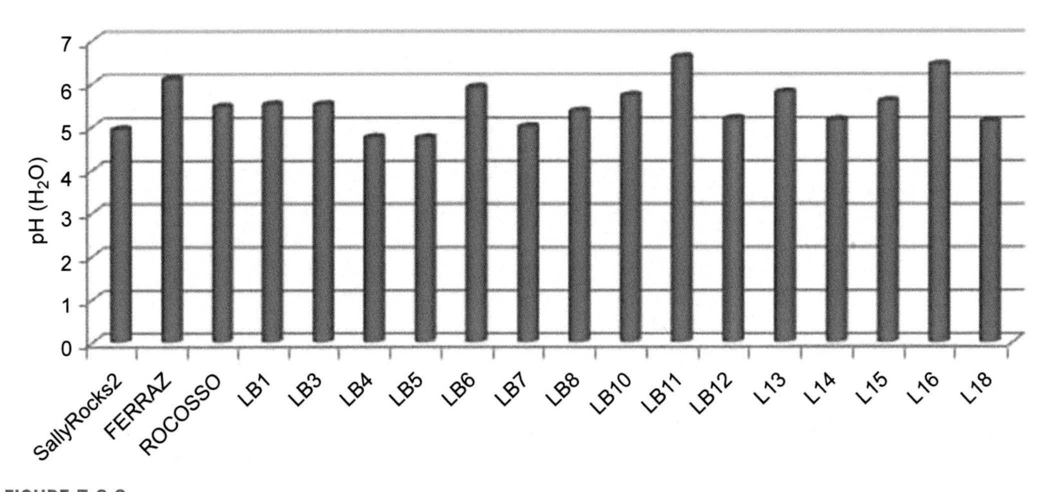

FIGURE 7.2.6

Soil reaction (pH) measured in water (1:5 soil:water ratio) for a set of samples.

corresponding bedrock. This, however, is most probably due to the well-established fact that rock debris and tephra are abundant in these soils (Bockheim et al., 2015; Haus et al., 2016).

An extremely high P content (in comparison with other soil varieties) is typically obtained in ornithogenic soils due to the high P supply from the guano decomposition (Moura et al., 2012; Emslie et al., 2014). Soils in the vicinity of the Bulgarian Antarctic Base also exhibit high P concentrations (Fig. 7.2.8), between 0.2% and 0.5% with the exception of one sample where P reaches 0.8%. It is considered that soils showing P values $> 0.5\%$ are strongly ornithogenic (Haus et al., 2016). For the topsoil collection studied, such high values are obtained for samples from the Rokosso Glacier and the L18 site. In the rest of the locations the influence of seabirds and penguins on the soil formation is less, but still important.

The magnetic susceptibility of the soil samples studied is depicted in Fig. 7.2.9. As can be seen, χ is relatively high and varies in the interval $(100-200) \times 10^{-8}$ m^3/kg; only the sample from King George Island (Ferraz) is higher. Frequency-dependent magnetic susceptibility (χ_{fd}) and the percent χ_{fd} ($\chi_{fd}\%$) suggest the presence of a relatively low amount of fine-grained superparamagnetic minerals (Fig. 7.2.9a–c).

A few samples show $\chi_{fd}\% > 5\%$ (Fig. 7.2.9c), thus indicating a more important relative content of nanosized magnetic minerals. Their magnetic properties are further investigated in more detail. The Ferraz sample from King George Island again underlines its different origin showing a very low $\chi_{fd}\%$, while at the same time its χ is the highest among the other samples (Fig. 7.2.9a). Thus, even the simplest magnetic characteristic like magnetic susceptibility evidences its high resolution power for the classification and characterization of the specific soils from Antarctica.

The magnetic mineralogy of the topsoils is studied through the temperature dependence of magnetic susceptibility, using both high-temperature and low-temperature cycling (down to liquid nitrogen temperature ($\sim$77 K)) to obtain information about possible structural magnetic transitions (e.g., Dunlop and Özdemir, 1997). Several representative examples of the investigations carried out are shown in Fig. 7.2.10.

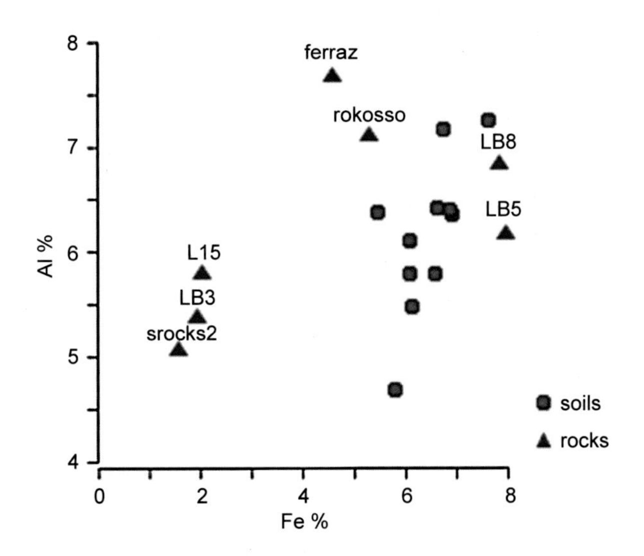

FIGURE 7.2.7

Biplot of Fe versus Al contents (in wt%) for the studied topsoils (*circles*) and rocks (*triangles*) from Livingston Island.

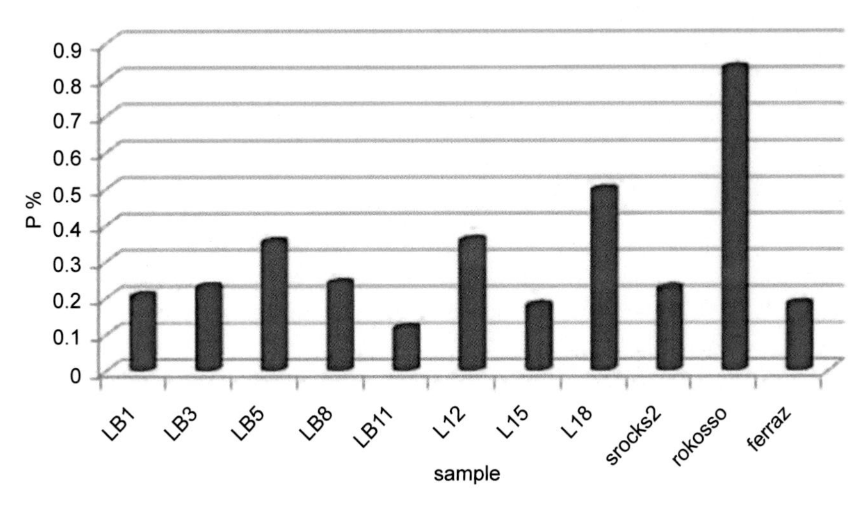

FIGURE 7.2.8

Phosphorous content (wt%) in the studied topsoils. Relatively high values are due to the P input from guano.

The high-temperature behavior of magnetic susceptibility reveals the presence of magnetite through the observed T_c of ~580°C for all samples (Fig. 7.2.10). Its proportion in the total magnetic assemblage varies, as can be deduced by the different amplitude of the sharp χ decrease just below the T_c. The Ferraz sample again shows a distinctly different χ–T curve, characterized by a near

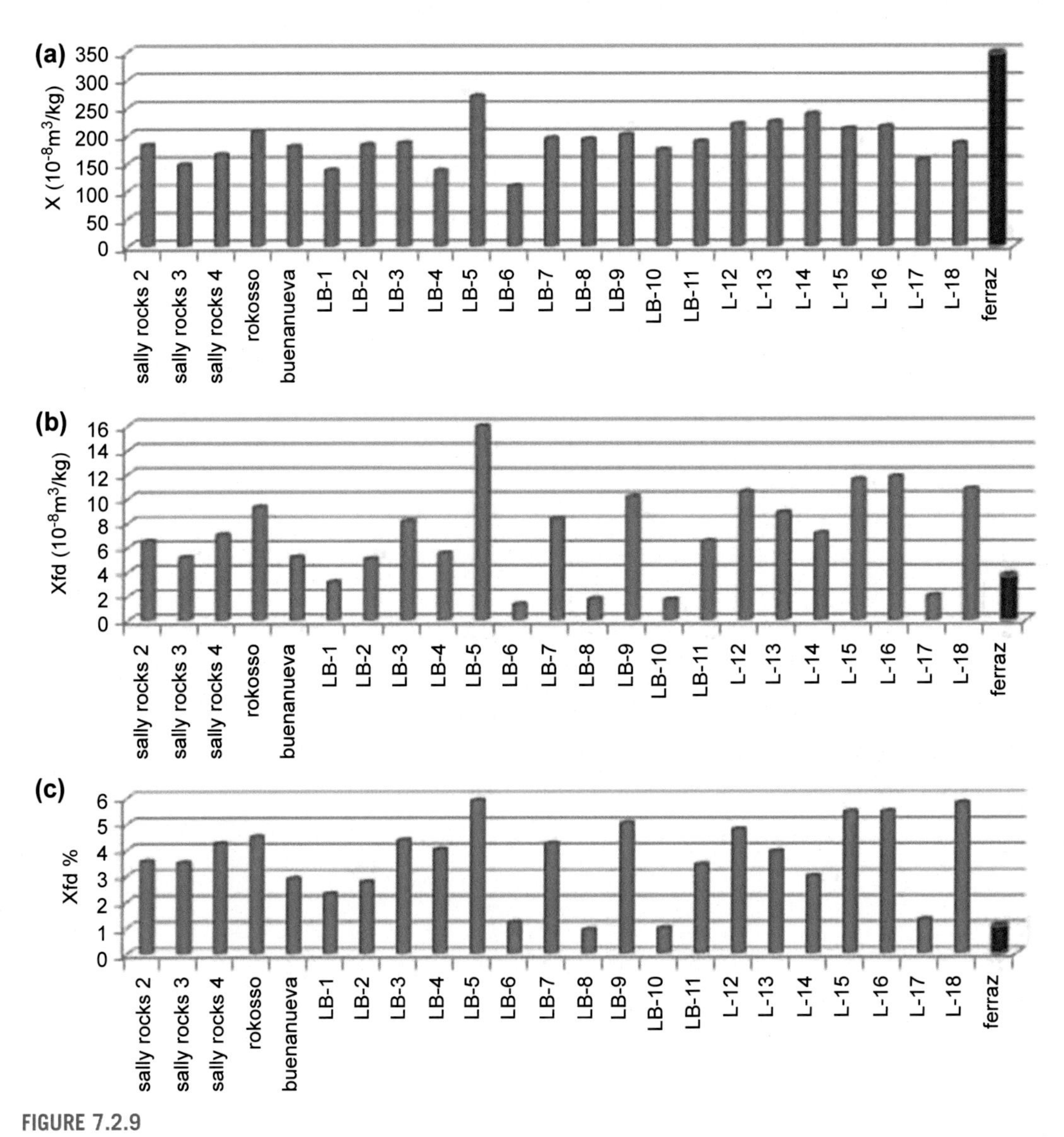

FIGURE 7.2.9

Mass-specific magnetic susceptibility (a), frequency-dependent magnetic susceptibility (χ_{fd}) (b), and precent frequency-dependent susceptibility (χ_{fd}%) (c) of the soil samples from Livingston Island. Sample from the King George Island is shown in red.

rectangular shape (Fig. 7.2.10d), suggesting the prevailing influence of coarse-grained magnetite. In contrast, sample L12 shows the most concave heating χ–T curve (Fig. 7.2.10c), implying the presence of a wide grain-size distribution of the ferrimagnetic fraction. A specific feature in the temperature interval 100–200°C is observed for the SallyRocks2 and L12 samples, expressed in a well-underlined maximum (Fig. 7.2.10a–c), which is preserved on the cooling curve as well. This behavior could not

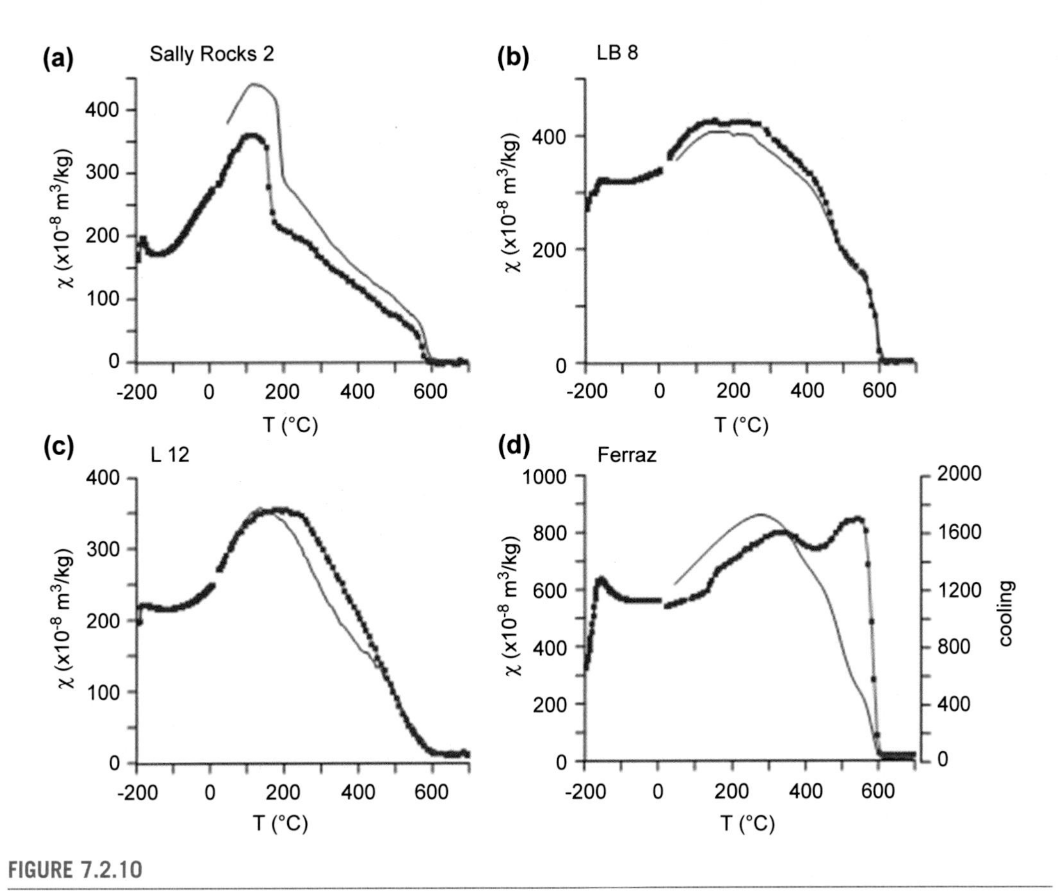

FIGURE 7.2.10

Examples of low- and high-temperature thermomagnetic analysis of magnetic susceptibility for soil samples from Livingston Island. *Heating curve* is indicated in red; cooling—in blue. Heating from room temperature up to 700°C is carried out in air at heating rate of 11°C/min.

be ascribed to a thermal transformation but rather to the presence of a different mineral magnetic phase or reversible structural transition. As evident from the geological information available (Kamenov, 2015 and references therein), the parent material for the soil formation often contains weathering products from the host rock (sediments from the Miers Bluff formation) but also detritus from widely spread plutonic and volcanic outcrops. Thus, it could be supposed that the observed magnetic phase with T_c ~150°C is linked to the presence of hemoilmenites (Dunlop and Özdemir, 1997). The low-temperature behavior of χ clearly underlines the presence of magnetite through the observed peak on the χ–T curve near ~ −180°C, related to the Morin transition in magnetite. Therefore, the dominant presence of coarse magnetite grains in the topsoils inherited mostly from the plutonic rocks of the Hesperides Pluton (Jordanova et al., 2015) and the additional tephra fraction from the Deception island suggest that the magnetic properties of Antarctic soils are strongly influenced by the iron oxides in the parent material.

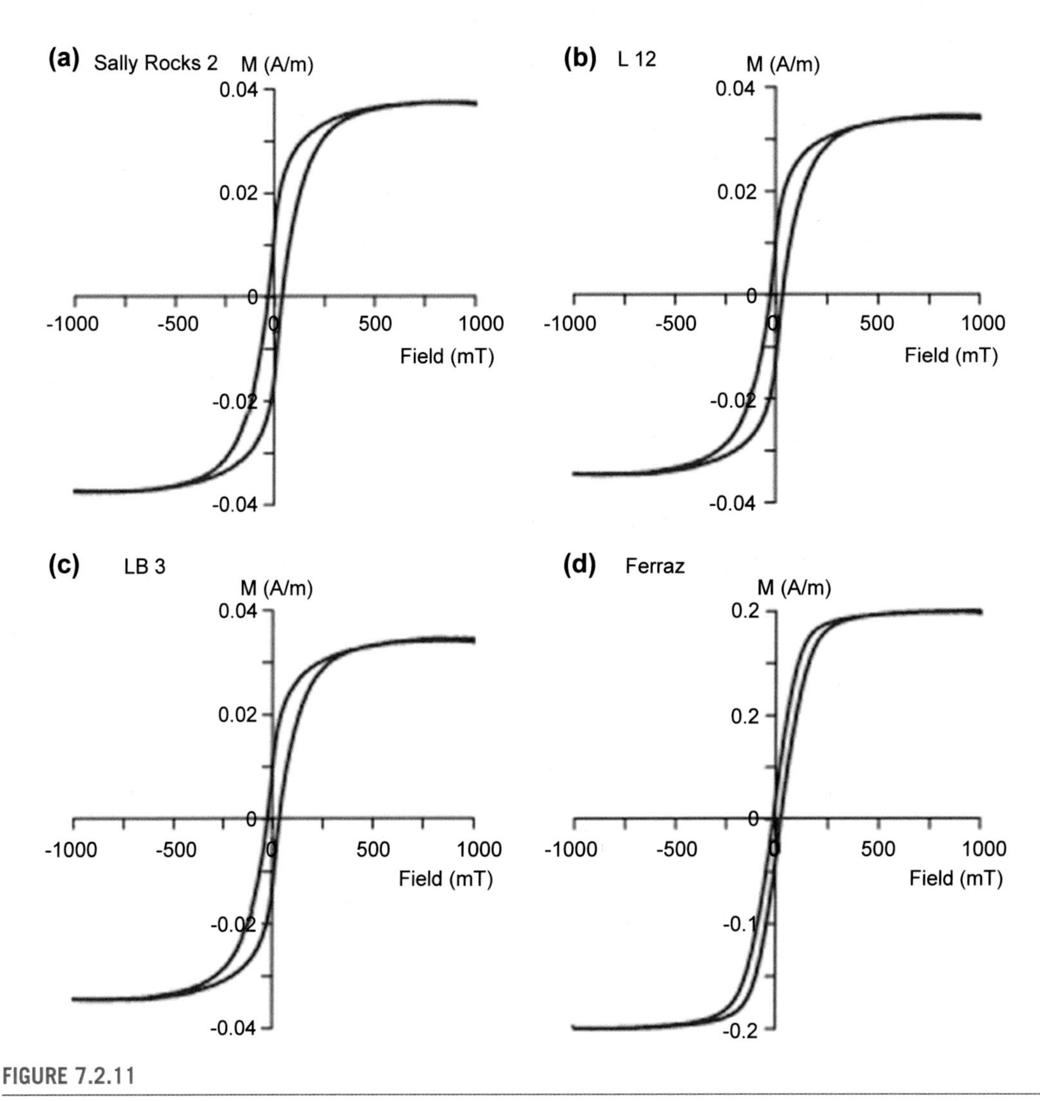

FIGURE 7.2.11

Hysteresis loops for selected samples from the studied topsoils from Livingston Island (a–c) and King George Island (d).

Hysteresis curves were measured for selected samples up to 1 T field. Several examples of the obtained loops are shown in Fig. 7.2.11. All samples from Livingston Island display hysteresis curves evidencing the dominant content of a stable remanence carrying ferrimagnetic fraction with appreciable coercivity (Fig. 7.2.11a–c). A very weak constriction of the loops can be also observed for these samples. Since no high-coercivity magnetic phase was detected (Fig. 7.2.11), the observed feature could be assigned to the effect of the mixture of different grain-size fractions of magnetite (e.g., Tauxe et al., 1996). The soil sample from King George Island (Ferraz) has a different hysteresis loop

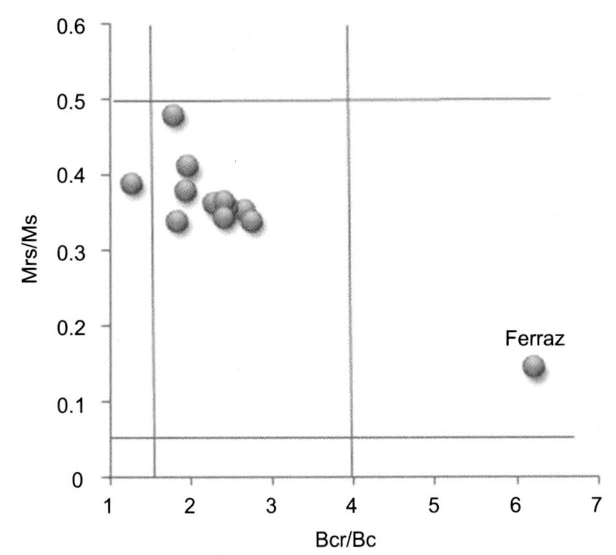

FIGURE 7.2.12

Day plot (Day et al., 1977) for the studied soil samples.

(Fig. 7.2.11d) characterized by a more rectangular shaped curve and lower coercivity. Thus, its generally different origin is again manifested.

The hysteresis parameters of the soil samples plotted on a Day plot (Day et al., 1977) cluster in the pseudo-single-domain (PSD) area (Fig. 7.2.12), except for the Ferraz sample. Relatively high coercivity and remanent magnetizations result in grouping the samples into the upper edge of the PSD area.

Laboratory remanences [anhysteretic remanence (ARM) and isothermal remanence (IRM)] of the soil samples were used to infer the relative content of stable coarse (through IRM) and finer SD/PSD grains (carrying the most effectively ARM). High values of both remanences confirm the abundance of strongly magnetic remanence-carrying iron oxides in the soil collection. A linear dependence between the ARM and IRM (Fig. 7.2.13) suggests that coarser and finer magnetically stable grains originate from the same process/source. The deviating sample point represents the data for the Ferraz sample, demonstrating its different composition.

As suggested by Oldfield (1994) the logarithmic biplot of the ratios χ_{ARM}/χ and χ_{ARM}/χ_{fd} could be used to assess the extent to which the magnetic assemblage is dominated by viscous/SP grains (e.g., lower χ_{ARM}/χ_{fd}) or stable SD magnetite, usually characteristic of bacterial magnetite. As shown in Fig. 7.2.14, the magnetic data on the soils from Livingston Island show a uniform χ_{ARM}/χ ratio (except the Ferraz sample) between 7 and 9, while the χ_{ARM}/χ_{fd} varies in a wider interval spanning one order of magnitude. Therefore, the SP fraction, although in a smaller relative amount, is a distinguishing factor in the magnetic signature of soils from Livingston Island. The obtained values of both ratios are characteristic of soils and fine sediments, while for bacterially derived magnetite, both ratios are much higher, grouping the data in the upper right corner of the biplot (Oldfield, 1994; Dearing, 1999; Evans and Heller, 2003).

An analysis of the data presented leads to the conclusion that topsoil samples from Livingston Island most probably contain a mixture of a magnetite-like ferrimagnetic fraction with a different grain size—from single-domain to large multidomain grains—giving the high magnetic susceptibility of the

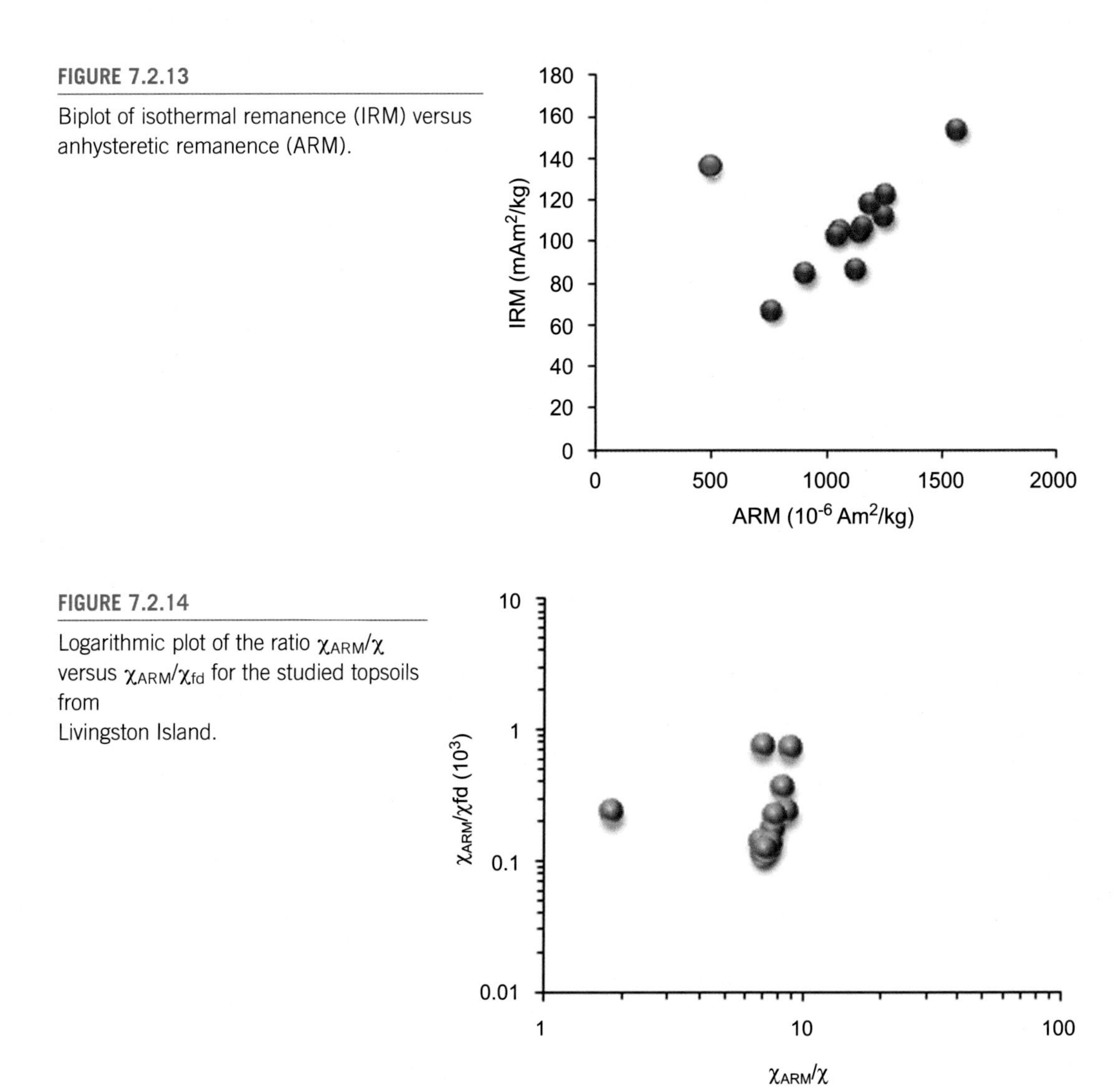

FIGURE 7.2.13

Biplot of isothermal remanence (IRM) versus anhysteretic remanence (ARM).

FIGURE 7.2.14

Logarithmic plot of the ratio χ_{ARM}/χ versus χ_{ARM}/χ_{fd} for the studied topsoils from
Livingston Island.

samples (Fig. 7.2.9), as well as a higher coercivity fraction of remanence carrying magnetite. In addition, as evidenced by the frequency-dependent magnetic susceptibility (Fig. 7.2.9), a small fraction of SP grains is also present.

It is generally considered that soil formation in Antarctica proceeds via physical weathering (rock disintegration and mechanical grinding), while the chemical weathering is very limited (Ugolini and Bockheim, 2008) because of the extremely low temperatures and limited precipitation. As a result, simple shallow soils form with an A-C horizonation (Beyer et al., 1999). Soils in the Antarctic Peninsula experience stronger chemical weathering (Haus et al., 2016) compared with continental Antarctica due to the milder climate and, thus, careful consideration of pedogenic transformations is needed. Navas et al. (2008) studied soil characteristics in the samples of South Shetland Islands and concluded that the soils from the Hurd Peninsula (where our collection is also gathered) are more immature than the soils in the Byers Peninsula since the mineralogical composition is dominated by silicate minerals. The elemental composition of topsoil samples H1 and H3 from Navas et al. (2008) correlate very well with the data obtained for samples L17, L18, and SallyRocks2, respectively, which have identical locations in both studies. Based on the results from the statistical evaluation of the elemental and chemical composition of soils, Navas et al. (2008) conclude that the bedrock type has the strongest influence and, thus, physical weathering is the main driving force of pedogenesis in Antarctica. However, Haus et al. (2016) prove that chemical weathering is also important in soils from Livingston Island. The authors suggest that the main driving factors of the chemical weathering are enhanced tephra depositions from Deception Island and the guano deposition in coastal areas. Rapid tephra weathering and enhanced ion complexation due to guano decomposition and related acid production govern the processes of chemical weathering (Haus et al., 2016). As shown, magnetic data on soils from Livingston Island (Fig. 7.2.9) suggest the presence of a nonnegligible superparamagnetic fraction in some of the soils. Pedogenic synthesis of nanosized superparamagnetic grains in soils is evidenced in their magnetic signature, commonly by an enhanced frequency-dependent magnetic susceptibility (Mullins and Tite, 1973; Dearing et al., 1996; Worm, 1998). Favorable conditions for SP magnetite/maghemite growth involve good aeration, mild precipitation, near neutral pH, and a warm climate (Cornell and Schwertmann, 2003; Maher, 1998; Barrón and Torrent, 2002). Obviously, these conditions are not met in the Antarctic environment and thus the evidence obtained for the superparamagnetic fraction in soils from Livingston Island could not be reliably related to the pedogenic growth of the nanosized ferrimagnetic phase. In such settings, only a restricted amount of short-ranged amorphous iron hydroxides are likely to exist (Cornell and Schwertmann, 2003; Simas et al., 2008; Yesavage et al., 2015; Haus et al., 2016). However, the identification and allocation of different (magnetic) grain-size fractions in the soils of Livingston Island are hampered by the abundance of allogenic materials, eolian additions of tephra brought from the neighboring Deception Island (Narcisi et al., 2010; Haus et al., 2016), erosional transport from the outcropping plutonic and volcanic bodies (Kamenov, 2015), etc. Refinement of particles in soils dominated by physical weathering is attributed to cryoclastic particle size reduction (Simas et al., 2008), although the wetting–drying cycles are also shown to play a substantial role in weathering in Arctic/Antarctic environments (Yesavage et al., 2015). In a detailed magnetic study of rocks and sediments from Southern Victoria Land (continental Antarctica), Ohneiser et al. (2015) found evidence for the dominant influence of a mixture of SP plus SD magnetite fractions on sediment magnetic properties, independent of the depositional settings. The authors suggest that the superparamagnetic magnetite fraction originates from glacial processes and/or aeolian transport, and magnetite is further prevented from oxidation by the hypercold and hyperarid conditions in continental Antarctica (Ohneiser et al., 2015). The magnetic characteristics of topsoils from Livingston Island also suggest the presence of a significant portion of a stable SD/PSD magnetite fraction in a mixture with SP particles (Fig. 7.2.9; Fig. 7.2.13; Fig. 7.2.14) and, therefore, one possible origin of these magnetic fractions is the mechanical disintegration and sorting (physical weathering). However, indications regarding chemical weathering provided in the study by Haus et al. (2016) should also be considered as a possible mechanism, provoking the pedogenic synthesis of fine-grained magnetic fractions. Since, as pointed out by these authors, these processes are not well studied from both the geochemical and environmental magnetic points of view, further more-detailed and focused investigations on Antarctic soils are needed to clarify the pedogenesis in such extreme environmental settings.

REFERENCES

Barrón, V., Torrent, J., 2002. Evidence for a simple pathway to maghemite in Earth and Mars soils. Geochimica et Cosmochimica Acta 66, 2801–2806.

Beyer, L., Bockheim, J., Campbell, I., Claridge, G., 1999. Genesis, properties and sensitivity of Antarctic Gelisols. Review. Antarctic Sci. 11 (4), 387–398.

Bockheim, J., Hall, K., 2002. Permafrost, active-layer dynamics and periglacial environments of continental Antarctica. South African J. Sci. 98, 82–90.

Bockheim, J., 1997. Properties and classification of cold desert soils from Antarctica. Soil Sci. Soc. Am. J. 61, 224–231.

Bockheim, J., 2015. Soils of Antarctica: history and challenges. World Soils Book Series. In: Bockheim, G. (Ed.), The Soils of Antarctica. Springer International Publishing, Print, ISBN 978-3-319-05496-4, pp. 1–3.

Bockheim, J., Lupachev, A., Blume, H.-P., Bölter, M., Simas, F., McLeode, M., 2015. Distribution of soil taxa in Antarctica: a preliminary analysis. Geoderma 245–246, 104–111.

Bonev, K., Dimov, D., Georgiev, N., 2015. Explanatory notes to the geological map of Hurd Peninsula, Livingston island. In: Pimpirev, C.H., Chipev, N. (Eds.), Bulgarian Antarctic Research: A Synthesis. "St. Kliment Ohridski" University Press, Sofia, ISBN 978-954-07-3939-7, pp. 29–32.

Burn, C., 1998. The active layer: two contrasting definitions. Permafr. Periglac. Process. 9 (4), 411–441.

Cornell, R., Schwertmann, U., 2003. The Iron Oxides. Structure, Properties, Reactions, Occurrence and Uses. Weinheim, New York.

Day, R., Fuller, M., Schmidt, V.A., 1977. Hystereis properties of titanomagnetites—grain size and compositional dependence. Phys. Earth Planet. Inter. 13, 260–267.

Dearing, J., 1999. Magnetic susceptibility. In: Walden, J., Oldfield, F., Smith, J.P. (Eds.), Environmental Magnetism: A Practical Guide. Technical Guide No 6. Quaternary Research Association, London, pp. 35–63.

Dearing, J.A., Dann, R.J.L., Hay, K., Lees, J.A., Loveland, P.J., Maher, B.A., O'Grady, K., 1996. Frequency-dependent susceptibility measurements of environmental materials. Geophys. J. Int 124, 228–240.

Dunbar, N., Kurbatov, A., 2011. Tephrochronology of the Siple Dome ice core, West Antarctica: correlations and sources. Quat. Sci. Rev. 30, 1602–1614.

Dunlop, D., Özdemir, Ö., 1997. Rock magnetism. Fundamentals and frontiers. In: Edwards, D. (Ed.), Cambridge Studies in Magnetism. Cambridge University Press.

Emslie, S., Polito, M., Brasso, R., Patterson, W., Sun, L., 2014. Ornithogenic soils and the paleoecology of pygoscelid penguins in Antarctica. Quat. Int. 352, 4–15.

Evans, M., Heller, F., 2003. Environmental Magnetism: Principles and Applications of Enviromagnetics. Academic Press, San Diego, CA.

Grosse, G., Goetz, S., McGuire, A., Romanovsky, V., Schuur, E., 2016. Changing permafrost in a warming world and feedbacks to the Earth system. Environ. Res. Lett. 11, 040201.

Guglielmin, M., Fratte, M., Cannone, N., 2014. Permafrost warming and vegetation changes in continental Antarctica. Environ. Res. Lett. 9, 045001.

Haus, N., Wilhelma, K., Bockheim, J., Fournelle, J.b, Miller, M., 2016. A case for chemical weathering in soils of Hurd Peninsula, Livingston Island, South Shetland Islands, Antarctica. Geoderma 263, 185–194.

IUSS Working Group WRB, 2014. World Reference Base for Soil Resources 2014. International Soil Classification System for Naming Soils and Creating Legends for Soil Maps. World Soil Resources Reports No. 106. FAO, Rome.

Jensen, H., 1916. Report on Antarctic soils. Repts. Sci. Invest. Brit. Antarct. Exped.1907–1909. Part IV. Geology 2, 89–92.

Jordanova, D., Jordanova, N., Dimov, D., 2015. Paleomagnetic and mineral magnetic studies on rock formations from Livingston Island, Antarctica. In: Pimpirev, C.H., Chipev, N. (Eds.), Bulgarian Antarctic Research: A Synthesis. "St. Kliment Ohridski" University Press, Sofia, ISBN 978-954-07-3939-7, pp. 208–220.

Kamenov, B., 2015. Magmatism in Hurd Peninsula, Livingston island. In: Pimpirev, C.H., Chipev, N. (Eds.), Bulgarian Antarctic Research: A Synthesis. "St. Kliment Ohridski" University Press, Sofia, ISBN 978-954-07-3939-7, pp. 70–176.

Maher, B., 1998. Magnetic properties of modern soils and Quaternary loessic paleosols: paleoclimatic implications. Palaeogeogr. Palaeoclimat. Palaeoecol. 137, 25–54.

Moura, P., Francelino, M., Schaefer, C., Simas, F., de Mendonça, B., 2012. Distribution and characterization of soils and landform relationships in Byers Peninsula, Livingston Island, maritime Antarctica. Geomorphology 155–156, 45–54.

Mullins, C., Tite, M., 1973. Magnetic viscosity, quadrature susceptibility and frequency dependence of susceptibility in single-domain assemblies of magnetite and maghemite. J. Geophys. Res. 78 (5), 804–809.

Narcisi, B., Petit, J., Delmonte, B., 2010. Extended East Antarctic ice-core tephrostratigraphy. Quat. Sci. Rev. 29, 21–27.

Navas, A., López-Martínez, J., Casas, J., Machín, J.a, Durán, J., Serrano, E., Cuchi, J.-A., Mink, S., 2008. Soil characteristics on varying lithological substrates in the South Shetland Islands, maritime Antarctica. Geoderma 144, 123–139.

Newsham, K., Hopkins, D., Carvalhais, L., Fretwell, P., Rushton, S., O'Donnell, A., Dennis, P., 2016. Relationship between soil fungal diversity and temperature in the maritime Antarctic. Nat. Climate Change 6 (2), 182–186.

Ohneiser, C., Wilson, G., Cox, S., 2015. Characterisation of magnetic minerals from southern Victoria Land, Antarctica. N.Z. J. Geol. Geophys. http://dx.doi.org/10.1080/00288306.2014.990044.

Oldfield, F., 1994. Toward the discrimination of fine grained ferrimagnets by magnetic measurements in lake and near-shore marine sediments. J. Geophys. Res. 99, 9045–9050.

Patrick, M., Smellie, J., 2013. Synthesis. A spaceborne inventory of volcanic activity in Antarctica and southern oceans, 2000–10. Antarctic Sci. 25 (4), 475–500.

Pimpirev, C., 2015. Bulgarian antarctic research: a synthesis. Editorial notes. In: Pimpirev, C.H., Chipev, N. (Eds.), Bulgarian Antarctic Research: A Synthesis. "St. Kliment Ohridski" University Press, Sofia, ISBN 978-954-07-3939-7, pp. 7–12.

Simas, F., Schaefer, C., Albuquerque Filho, M., Francelino, M.R., Filho, E., da Costa, L., 2008. Genesis, properties and classification of Cryosols from Admiralty Bay, maritime Antarctica. Geoderma 144, 116–122.

Simas, F., Schaefer, C., Melo, V., Albuquerque-Filho, M., Michel, R., Pereira, V., Gomes, M., Da Costa, L., 2007. Ornithogenic Cryosols from Maritime Antarctica: phosphatization as a soil forming process. Geoderma 138, 191–203.

Smykla, J., Drewnik, M., Szarek-Gwiazda, E., Hii, Y.S., Knap, W., Emslie, S., 2015. Variation in the characteristics and development of soils at Edmonson Point due to abiotic and biotic factors, northern Victoria Land, Antarctica. Catena 132, 56–67.

Soil Survey Staff. In: Keys to Soil Taxonomy, twelfth, 2014. United States Department of Agriculture, Natural Resources Conservation Servic.

Tauxe, L., Mullender, T., Pick, T., 1996. Potbellies, wasp-waists, and superparamagnetism in magnetic hysteresis. J. Geophys. Res. 101 (B1), 571–583.

Ugolini, F., Bockheim, J., 2008. Antarctic soils and soil formation in a changing environment: a review. Geoderma 144, 1–8.

Worm, H.-U., 1998. On the superparamagnetic - stable single domain transition for magnetite, and frequency dependence of susceptibility. Geophys. J. Int. 133, 201–206.

Yesavage, T., Thompson, A., Hausrath, E., Brantley, S., 2015. Basalt weathering in an Arctic Mars-analog site. Icarus 254, 219–232.

CHAPTER

THE DISCRIMINATING POWER OF SOIL MAGNETISM FOR THE CHARACTERIZATION OF DIFFERENT SOIL TYPES

8

8.1 MAJOR FEEDBACK AND FEATURES OF THE MAGNETIC SIGNATURE OF SOIL PROFILE DEVELOPMENT

As shown in Chapters 2–7, the magnetic properties of soil profiles may provide valuable information about the degree of pedogenic development and different destruction processes (such as illuviation, podzolization, etc.). It is therefore important not only to compile the data obtained but also to summarize and capitalize on the observed relations and features according to different soil characteristics. This could further serve as a basis of magnetic discrimination among fine-scale changes in soil characteristics toward degradation and related environmental threats.

The group of Chernozems. As demonstrated in Chapter 2, Chernozem soils exhibit similar features worldwide and are generally characterized by significant magnetic enhancement in the upper soil horizons. When detailing the magnetic proxy record through the investigation of Chernozems of different degrees of textural differentiation (Chapter 2), the following major relations, as schematically drafted in Fig. 8.1.1a, appear:

1. Increased leaching and illuviation accompanying the transition from Haplic Chernozem to Luvic Phaeozem lead to a well-expressed split between the maximum concentration of pedogenic strongly magnetic superparamagnetic (SP) and single-domain (SD) fractions. Synchronous variations are linked to a Haplic variety, while increasing profile differentiation "pushes" the SD fraction to the near surface horizon, while the SP fraction is enhanced in the deeper AC or B horizons.
2. The major pedogenic strongly magnetic mineral is maghemite.
3. Increasing profile degradation and differentiation (e.g., development of Luvic Phaeozems) result in a prevailing role of the SD pedogenic fraction, while during the soil magnetic enhancement (e.g., Haplic Chernozems), the SP fraction dominates.
4. The observed depth differentiation in the magnetic grain size fractions for texturally differentiated profiles suggests that grain growth is the dominant process in the humic horizon, while in the deeper parts where the SP fraction has maximum concentration, grain nucleation is occurring. Therefore, the studied Chernozems exhibit top-down pedogenesis.

The group of Luvisols. Since Luvisols have a well-developed illuvial horizon as a major pedogenic feature, the "starting" point, compared with Chernozems, is different. The magnetic characteristics

Soil Magnetism. http://dx.doi.org/10.1016/B978-0-12-809239-2.00008-5

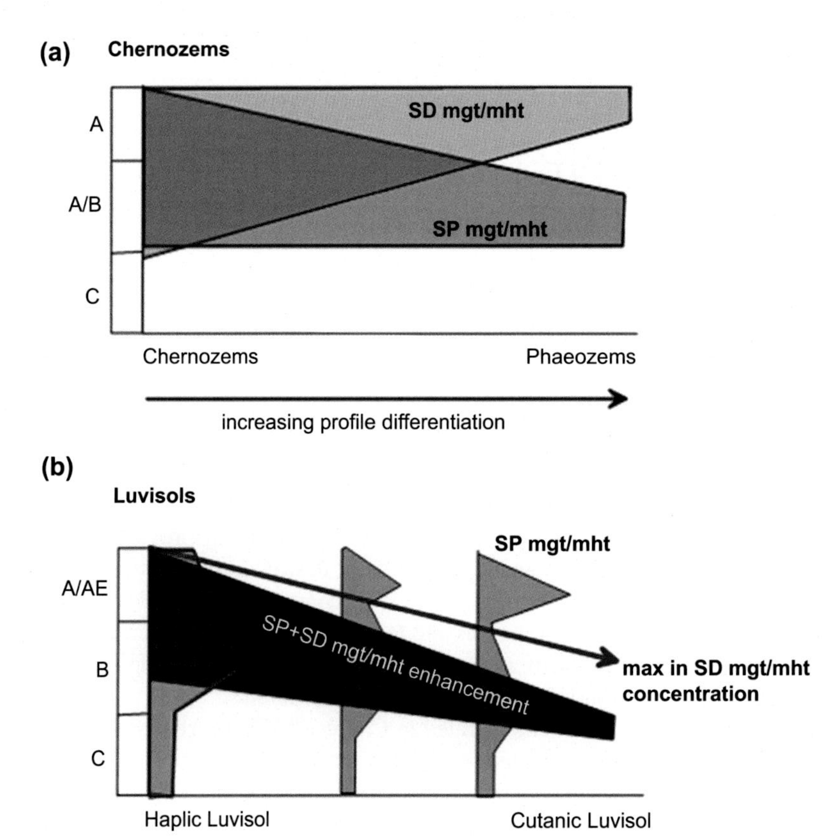

FIGURE 8.1.1

Schematic summary of the depth distribution of pedogenic superparamagnetic magnetite/maghemite (SP mgt/mht) and single-domain magnetite/maghemite (SD mgt/mht) fractions in well-aerated soils: Chernozem-Phaeozem soils with related increasing profile differentiation (a) and Luvisols with increasing eluviation and podzolization (b). Summary based on the magnetic properties of soil profiles described in Chapters 2 and 3.

obtained for a sequence of Luvisols exhibiting varying degrees of profile degradation (eluviation, weak podzolization) were discussed in detail in Chapter 3. Considering the obtained concentration and grain size related magnetic proxies, the following chain of relations depicted in Fig. 8.1.1b could be proposed:

1. Haplic Luvisols not affected by eluviation exhibit synchronous enhancement with an SP and SD strongly magnetic fraction throughout the A and B horizons, but with maxima linked to the upper/middle part of the B horizon.
2. Increasing profile degradation leads to a decrease in the absolute amount and a split in the SP fraction distribution into the eluvial and illuvial horizons (Fig. 8.1.1b). The development of a strongly eluviated (weakly podzolized) horizon (the third stage in Fig. 8.1.1b) leads to an increased SP fraction there, while the SP share in the B horizon diminishes and deepens.

3. Increased Luvisol degradation pushes the maxima in the SD strongly magnetic fraction deeper into the illuvial horizon.
4. The obtained increase in the coercivity of the magnetically soft IRM fraction from the Haplic Luvisol toward the Cutanic Luvisol (see Chapter 3) suggests the presence of smaller magnetite/maghemite grains or surface maghemitized particles in the most strongly degraded Luvisol variety. This would suggest impeded grain growth due to increased clay illuviation, gleying, and groundwater saturation.

The group of Planosols. The Planosols' magnetic signature was discussed in detail in Chapter 4. These soils are strongly affected by iron oxide dissolution because of oscillating reduction–oxidation conditions inside the profile. As outlined in Chapter 4, the magnetic properties along the depth of the Planosols indicate an important enhancement of the deeper soil parts with stable SD magnetite, while the SP fraction is small and also related to the B/C horizons. The major findings from the detailed magnetic characterization of three Planosols can be summarized schematically in Fig. 8.1.2, as follows:

1. Eluvial and illuvial horizons in Planosols are enriched with stable near SD magnetite, probably formed as a result of ferrihydrite transformations under repeating oxidative-reducing conditions (e.g., Hansel et al., 2003). This suggests relatively good crystallinity, hypothesized as being due to the effect of multiple redox cycles (Thompson et al., 2006).
2. Illuvial horizons of Planosols are enhanced with hematite, while magnetite is detected in the eluvial and C horizons (Fig. 8.1.2).
3. Frequency-dependent magnetic susceptibility in Planosols correlates well with the sand fraction along the depth (Chapter 4), thus suggesting the existence of iron oxides as a precipitation film on the sand particles, confirming its pedogenic origin.

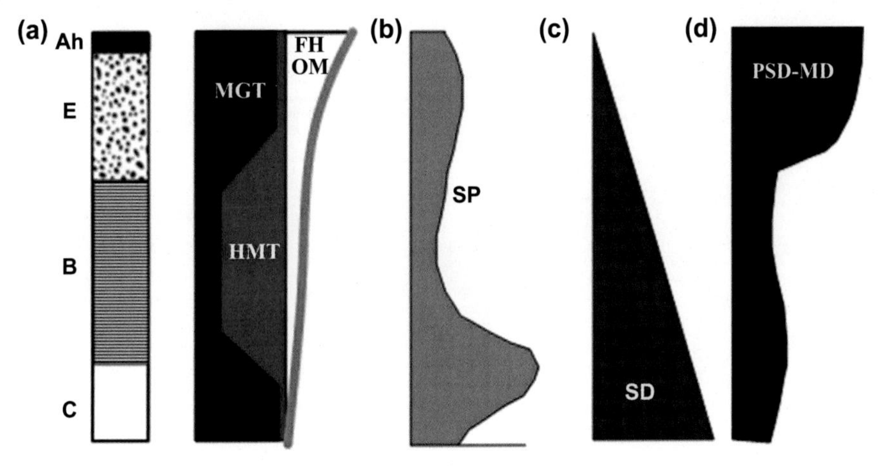

FIGURE 8.1.2

Schematic illustration of the major trends and relations among the magnetic mineralogy and magnetic grain size variations along the Planosol profiles, as deduced from the magnetic measurements. (a) Magnetite (MGT)-hematite (HMT) and ferrihydrite/organic matter (FH, OM) distribution along depth; (b) variations in concentration of the superparamagnetic (SP) fraction along depth; (c) variations in concentration of the single-domain (SD) fraction along depth; (d) variations in concentration of the pseudo-single-domain/multi-domain (PSD-MD) fractions along depth.

The group of Podzols. Podzols are strongly affected by the development of a thick organic layer of coniferous needles, which, in combination with cool humid climates, promotes iron complexation with the organic matter and its further translocation into the illuvial horizon, leaving a whitish strongly bleached eluvial horizon above. A detailed discussion is provided in Chapter 4. Because of the above mentioned specific combination of environmental factors, iron oxides in Podzols are mostly presented by weakly crystalline iron (oxy)hydroxides, mainly ferrihydrite (Schaetzl and Anderson, 2009). The magnetic signature in Podzols shows that:

1. The magnetic susceptibility in Podzols is very low and a relative maximum is concentrated in a small peak in the organic horizon just above the eluvial one.
2. The magnetic characteristics indicate the presence of a small amount of stable SD magnetite in the illuvial horizon as well as in the eluvial one.
3. Hematite is detected in the deeper part of Podzols and a presence of nanosized goethite (not detectable through room temperature magnetic measurements) is supposed.

The group of Vertisols. Vertisols are a specific soil type characterized by the dominant presence of smectite clays, displaying a strong shrinking—swelling capacity under repeated dry—wet seasonal cycles (Schaetzl and Anderson, 2009). As discussed in detail in Chapter 5, these soils are strongly influenced by the reductive dissolution of Fe-containing minerals and subsequent related environmental reactions (e.g., Van Breemen, 1988; Hansel et al., 2003; Usman et al., 2012). The magnetic signature of the three Vertisol profiles from Bulgaria (Chapter 5) can be summarized by the following general observations (Fig. 8.1.3):

1. The magnetic susceptibility in Vertisols is weak, supporting the dominant role of reductive dissolution of strongly magnetic Fe-containing (lithogenic) coarse grains.

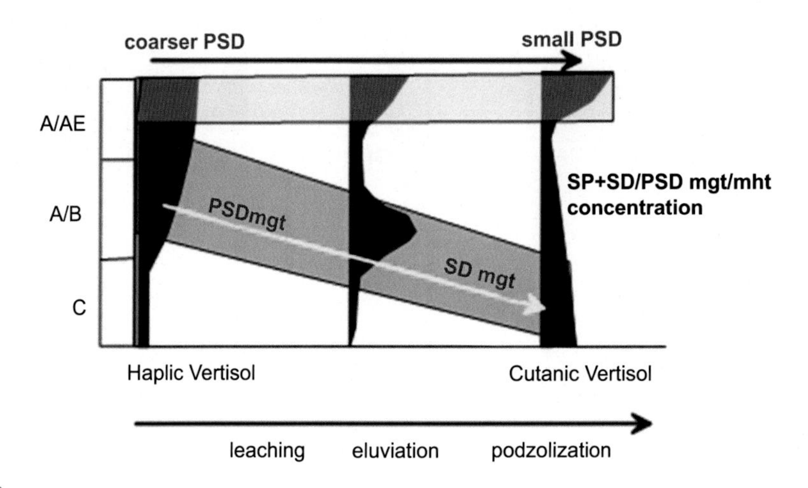

FIGURE 8.1.3

Schematic illustration of the evolution in distribution of the pedogenic magnetic grain size fractions in Vertisols with increasing profile degradation (eluviation, podzolization). Summary based on the magnetic properties of soil profiles described in Chapter 5.

2. All profiles of Vertisols having different degree of profile degradation (leaching, eluviation) are characterized by synchronous variations in the concentration of both the stable SD/pseudo-SD (PSD) and the SP fraction. This implies that the pedogenic component is formed by simultaneous nucleation and grain growth.
3. Increasing profile differentiation and degradation causes the appearance of two maxima with abundant SD/PSD and SP fractions (Fig. 8.1.3). In Vertisols influenced by leaching, eluviations, or podzolization, the SD/PSD plus SP fractions are concentrated in the uppermost A/AE horizon and in the illuvial or deeper horizon.
4. Increasing profile differentiation results in the decreasing grain size of the magnetically stable fraction, such as from PSD toward SD magnetite characteristics (Fig. 8.1.3).
5. As discussed in detail in Chapter 5, the origin of the stable magnetite fraction is related to the processes of reductive dissolution of ferrihydrite and its pedogenic transformations under repeated redox cycles.

The group of Leptosols. Leptosols are shallow soils that are strongly influenced by the nature and properties of the parent material. As examined in Chapter 5, profiles of Rendzic Leptosols (Rendzinas) and a Leptosol developed on highly magnetic volcanoclastic sediments showed the following major relations:

1. The whole soil depth (up to the parent rock) is influenced by the pedogenic changes.
2. Leptosols developed on weakly magnetic sedimentary materials (e.g., limestones in the case of Rendzina) are strongly enhanced with a pedogenically formed SP and SD magnetite fraction. The SP and the stable (SD/PSD) magnetic fractions covary along the depth, suggesting the combined nucleation and growth of the pedogenic iron oxides.
3. Leptosols developed on rocks containing a large amount of primary strongly magnetic iron oxides, show an effective magnetic depletion. Nevertheless, the frequency-dependent magnetic susceptibility (χ_{fd}) tangibly distinguishes a weak enhancement in the uppermost soil part with very fine pedogenic grains.

The group of Cambisols and Umbrisol. Cambisols are soils with little or no profile differentiation (IUSS Working Group WRB, 2014), and their magnetic signature is also significantly affected by the parent rock's mineralogy, the altitude of the site, and the organic matter content and type. The magnetic properties of Cambisols developed in different altitudinal zones were presented in Chapter 6. Their major magnetic characteristics are schematically depicted in Fig. 8.1.4.

1. Pedogenic magnetic enhancement with SP and stable SD/PSD grains exhibits clear altitudinal separation. SP magnetite particles are in a small amount and detected only in forest soils. The Umbrisol developed immediately above the timber line contains only coarser SD/PSD/MD particles.
2. With increasing altitude, the humic horizons in the Cambisol become progressively more enhanced with stable remanence carrying grains (Fig. 8.1.4).
3. The uppermost peat-enhanced humic horizon in the Umbrisol is much depleted in iron and practically no strongly magnetic minerals are detected there.
4. Intense gleying in the bottom parts of Cambisols situated at lower altitudes leads to the formation of a complex mixture of primary (lithogenic) and secondary (pedogenic) Fe-containing minerals. In the case of the "T" profile (Chapter 6) a large amount of goethite is detected in the subsoil horizons (Fig. 8.1.4).

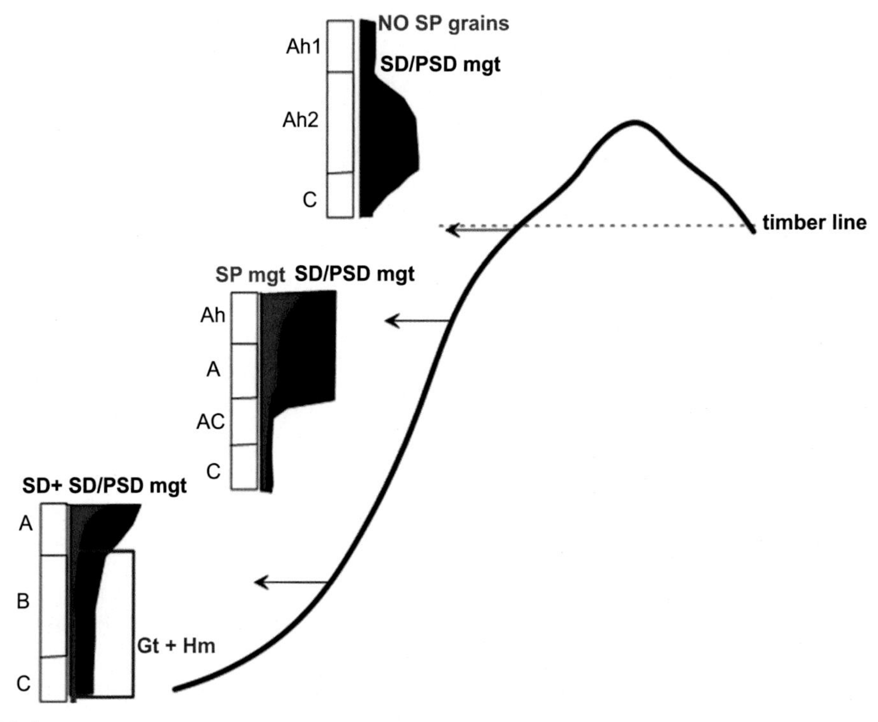

FIGURE 8.1.4

Schematic illustration of the evolution in distribution of the pedogenic grain size fractions magnetite [superparamagnetic (SP) (denoted in red) and stable single-domain/pseudo-single-domain SD/PSD mgt (denoted in black)] with increasing altitude of the soil profile. Summary based on the magnetic properties of soil profiles described in Chapter 6.

5. It could be supposed that grain growth is the dominant process of pedogenic enhancement in high mountain soils.

8.2 STATISTICAL ANALYSIS OF SOIL PROFILE DATA ON MAGNETIC PARAMETERS, MECHANICAL COMPOSITION, AND IRON CONTENT

The observed relationships and statistical classification of the magnetic characteristics described in the following Chapter 9, reveal the spatial association and discrimination of the topsoil samples, while vertical differentiation and variations in the properties according to the genetic horizons need separate analysis. The latter can be accomplished by considering the relationships between the measured soil properties along the depth of the studied soil profiles, described in Chapters 2–7. For that purpose, all samples from the studied soil profiles that have been analyzed for grain size (sand, silt, clay) and iron content (oxalate-extractable (Fe_o), dithionite-extractable (Fe_d), total iron (Fe_{tot}), together with the magnetic parameters, such as magnetic susceptibility, frequency-dependent magnetic susceptibility, percentage of frequency-dependent magnetic susceptibility (χ, χ_{fd}, $\chi_{fd}\%$), anhysteretic remanence and isothermal remanence acquired in 2-T field (ARM, $SIRM_{2T}$), saturation magnetization (M_s), ratios between anhysteretic susceptibility (χ_{ARM}) and magnetic susceptibility (χ_{ARM}/χ), χ_{ARM} and IRM acquired in a weak magnetic field (IRM_{soft}) (χ_{ARM}/IRM_{soft}), χ_{ARM} and χ_{fd} (χ_{ARM}/χ_{fd}), "hard" IRM (HIRM) (e.g., Liu et al., 2012), were included in a database used for statistical analysis. As far as a normal distribution is required for the variables used in the subsequent factor and cluster analyses, only the parameters holding such normality have been used further—silt, clay, silt/clay, Fe_d, Fe_{tot}, χ_{fd}%, and $SIRM_{2T}$ and the ratios χ_{ARM}/χ, χ_{ARM}/IRM_{soft}, and χ_{ARM}/χ_{fd}. After the application of the standardization procedure, factor analysis was carried out, excluding Fe_d and Fe_o, because the lack of such data for some profiles greatly reduces the number of valid cases. The results for the calculated Eigenvalues and the contribution of the extracted factors for the explanation of the data variability are shown in Table 8.2.1.

Four factors with eigenvalues higher than 1 explain 87% of data variability, the first and the second factor having a 35% and 24% contribution, respectively (Table 8.2.1). The other two factors are of lower importance, explaining 14–15% of the data variability. The variables having the highest contribution to each of the four factors are shown in bold in Table 8.2.2.

A major portion of data variability is ascribed to the first factor, in which the loadings from the clay content and the silt/clay ratio dominate. These parameters are essential indicators of pedogenic development (Schaetzl and Anderson, 2009). On the other hand, the second factor of main importance is dominated by the loadings of magnetic parameters which are proxies for the presence of fine-grained pedogenic iron oxides – χ_{ARM}/χ, χ_{ARM}/IRM_{soft}, and χ_{fd}%. Consequently, textural differentiation accompanied by the pedogenic formation of secondary iron oxides within the soil profiles are the main processes determining the pedogenic horizons' variability in the soil database. The change in the concentration of strongly magnetic iron oxides (through χ and $SIRM_{2T}$) is of secondary importance, as defined by the contribution of the third factor. The importance of the χ_{ARM}/χ_{fd} ratio (representing the only significant loading in factor 4) should also be emphasized as an explanation of soil variability. It is predetermined by the sensitivity of the ratio to the relative prevalence of SD or SP fractions in the magnetic signal of the samples, which may be an indication of soil burning or the presence of extracellular magnetite (Oldfield, 2013). This rating of the different magnetic parameters in the explanation of the variability along the depth of soil profiles is in contrast to that obtained for the database on the magnetism of the topsoil samples (Chapter 9). In the latter, concentration-dependent magnetic variables dominate.

Using the results from the factor analysis, the following independent variables were chosen for running cluster analysis—clay content (%), χ_{fd}%, $SIRM_{2T}$, and χ_{ARM}/χ_{fd} (see Table 8.2.2). A k-means clustering algorithm with a 100-fold cross-validation procedure was carried out using STATISTICA 8.0 software. Two categorical variables have been introduced—*soil order* and *conditions* (oxidative and reducing). The introduction of a third categorical variable—*the soil horizon*—caused a significantly higher error in the model, so it was not incorporated in the analysis. A cross-validation

Table 8.2.1 Eigenvalues and the corresponding variance of the four factors, explaining the variability of the magnetic data for different soil profiles

Factor	Eigenvalue	% Total Variance	Cumulative Eigenvalue	Cumulative%
1	3.2	35.3	3.2	35.3
2	2.1	23.6	5.3	59.0
3	1.3	14.6	6.6	73.5
4	1.2	13.5	7.8	87.1

Table 8.2.2 Contribution of the continuous variables, included in the factor analysis to each of the four factors

Variable	Factor 1	Factor 2	Factor 3	Factor 4
Silt	0.524	0.019	0.483	0.299
Clay	**−0.945**	0.114	−0.006	0.095
Silt/clay	**0.938**	0.077	0.22	0.02
χ	0.056	−0.078	**0.927**	−0.174
$\chi_{fd}\%$	0.095	**−0.827**	0.301	−0.346
$SIRM_{2T}$	0.191	−0.146	**0.933**	0.075
χ_{ARM}/χ	0.226	**−0.829**	0.193	0.335
χ_{ARM}/IRM_{soft}	−0.199	**−0.877**	−0.116	−0.116
χ_{ARM}/χ_{fd}	−0.022	0.075	−0.046	**0.958**

procedure allowed the estimation of the optimum number of clusters – four, as evidenced by the graph of cluster costs (Fig. 8.2.1).

The flattening of the curve after the fourth cluster indicates that introduction of more clusters (five or more) does not lead to any significant improvement of the model. Cluster means for each numerical variable differ significantly, as evidenced in Fig. 8.2.2.

The distribution of the categories of the "*soil type*" categorical variable among the clusters is presented in Table 8.2.3. The first cluster includes all samples from the profiles of Planosols (OK, PR, ZL) (Chapter 4) and the salt-affected soil (S) (Chapter 5), as well as samples from the C horizon of the Alisol profile (YPS) (Chapter 3) and the AC plus C horizons of the Humic Cambisol TR (Chapter 6). These are actually the weakly magnetic profiles (Planosols and the salt-affected soil) and the bottom parts of the YPS and TR profiles (also exhibiting weak magnetism and restricted content of strongly magnetic pedogenic iron oxides). All samples from the Vertisol profiles

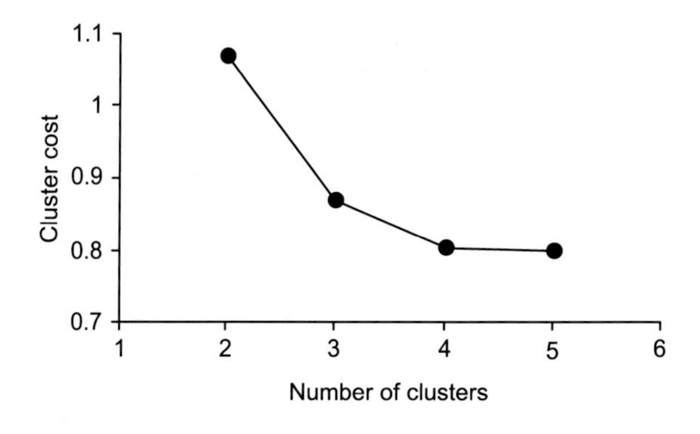

FIGURE 8.2.1

Graph of cluster cost as a function of the number of clusters estimated through 100-fold cross-validation algorithm of the k-means clustering. Training cases: 132; training error: 0.791.

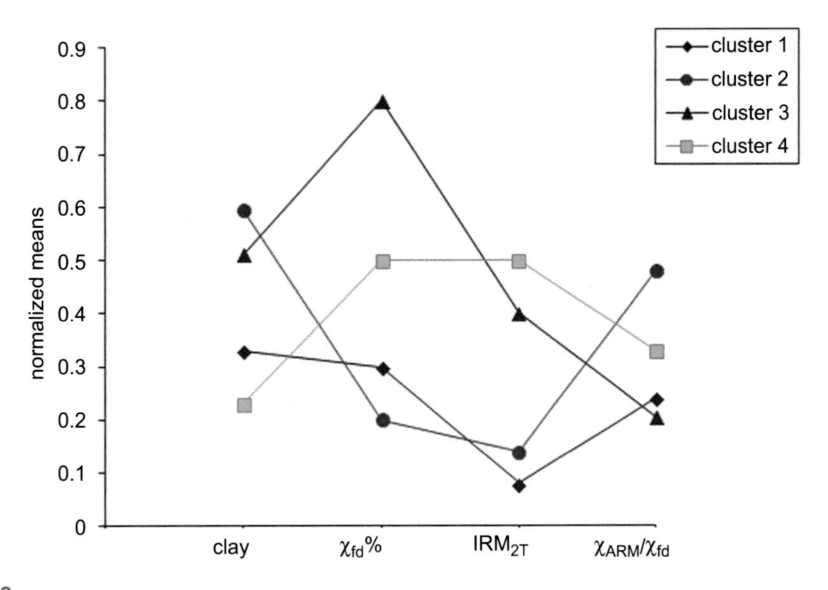

FIGURE 8.2.2

Cluster means of the continuous variables included in the k-means clustering of the soil profiles' data.

(VR, JAS, SM) (Chapter 5) and the samples from the E1 and B_t horizons of the Alisol form the second cluster. Therefore, it includes profiles consisting of a couplet of the eluvial horizon and the heavy clayey illuvial horizon with slickensides inside. Samples from the Luvisols (CIN and RED) (Chapter 3) belong to the third cluster, with a few samples from the A and C horizons of a Luvic Phaeozem GF (Chapter 2) and Leptosol R_Z (Chapter 5) included. This cluster combines the most strongly magnetic profiles possessing extreme pedogenic enhancement with strongly magnetic iron oxides, as well as significant hematite content. The fourth cluster includes samples from variable soil

Table 8.2.3 Distribution of cases of the categorical cariable "soil type" among different clusters

	Cluster 1	Cluster 2	Cluster 3	Cluster 4	Total
Luvisol	0	0	**14**	0	14
Alisol	2	6	0	1	9
Chernozem	0	0	0	**17**	17
Phaeozem	0	0	**2**	**5**	7
Planosol	**28**	1	0	0	29
Vertisol	0	**28**	0	0	28
Salt-affected soil	**10**	0	0	0	10
Leptosol	0	0	**2**	**3**	5
Cambisol	3	0	0	**4**	7
Umbrisol	0	0	0	**6**	6

Table 8.2.4 Distribution of cases of the categorical variable *"soil conditions"* (oxidative, oxidative/reducing and reducing) among the four clusters

	Cluster 1	Cluster 2	Cluster 3	Cluster 4	Total
Oxidative	**20**	**12**	**18**	**32**	82
Oxidative–reducing	**10**	**15**	0	0	25
Reducing	**13**	**8**	0	**4**	25

orders, including all samples from the Chernozems (profiles TB and OV) (Chapter 2), B_t horizons of the Luvic Phaeozem (GF), and A horizons of the profiles of Humic Cambisol (TR), Umbrisol (GR) and Leptosol (R_Z) (Chapter 5). Considering the most characteristic properties of these soils/horizons, it appears that cluster 4 joins the soil profiles/horizons with a significant accumulation of organic matter, also exhibiting good aeration and thus creating favorable conditions for pedogenic magnetic enhancement.

The distribution of categories of the *"conditions"* categorical variable among the four clusters is not so straightforward (Table 8.2.4). Clusters 3 and 4 contain nearly only samples with *oxidative* conditions, except the samples from the B_t horizon of the Luvic Phaeozem (GF), which is described in the field as possessing reductive features. Clusters 1 and 2 incorporate samples with all types of conditions, as far as samples from all genetic horizons of the Vertisols (cluster 2) and Planosols (cluster 1) are characterized by varying conditions depending on the horizon. Consequently, for clusters 1 and 2, other factors than reducing–oxidative conditions play a major role in the establishment of their magnetic signal. In contrast, it could be supposed that the clusters of soils with oxidative conditions (clusters 3 and 4) are governed by pedogenic processes related to good aeration and drainage, which ensure the dominance of oxidative conditions.

Thus, a statistical analysis of the magnetic and textural signature in soil profiles evidences the power of soil magnetism in revealing the general characteristics of the major soil types and its high potential applicability in pedology.

8.3 PERCENT FREQUENCY-DEPENDENT MAGNETIC SUSCEPTIBILITY (χ_{FD}%) AS A PROXY FOR THE CONTENT OF PEDOGENIC MAGNETITE/MAGHEMITE GRAINS IN RED-COLORED SOILS

Frequency-dependent magnetic susceptibility is widely used for the detection of fine nanometer-sized ferrimagnetic particles in environmental and rock magnetism (Mullins and Tite, 1973; Thompson and Oldfield, 1986; Maher, 1986; Dearing et al., 1996; Evans and Heller, 2003; Liu et al., 2012). This magnetic parameter is extensively used, especially in soil and sediment magnetic studies (e.g., Maher, 1998; Liu et al., 2012; Roberts, 2015), since pedogenesis in soils and diagenesis in sediments produce fine-grained submicrometer and nanometer-sized particles, usually magnetite and/or maghemite. In contrast, lithogenic, rock inherited particles are commonly coarse grained and thus predominantly display multidomain (MD) properties. Therefore, magnetic grain-size–sensitive proxy parameters successfully discriminate pedogenic/diagenetic from lithogenic magnetic fractions. Theoretical studies on SP-SD transition in magnetite and the frequency-dependence of magnetic susceptibility (Mullins and Tite, 1973; Eyre, 1997; Worm, 1998) provide the physical basis of the experimentally obtained data on different natural materials. The empirically derived relations between the content of the fine-grained magnetite/maghemite and the $\chi_{fd}\%$ value (Dearing et al., 1996, 1997) in different soil types are used as a standard fast and nondestructive test for the evaluation of the relative contribution of the SP strongly magnetic fraction into the measured bulk (low field) magnetic susceptibility. The "threshold" $\chi_{fd}\%$ values proposed by Dearing (1999) suggest that, if $\chi_{fd}\% > 10\%$ (between 10% and 14%), the SP assemblage dominates the magnetic susceptibility (SP content >75%); medium range: $2\% < \chi_{fd}\% < 10\%$ shows an admixture of SP and coarser grains, while $\chi_{fd}\% < 2\%$ is obtained when SP grains are practically missing.

These empirical considerations, theories, and experimental data on the reliability of $\chi_{fd}\%$ as a proxy for the content of the SP magnetite/maghemite fraction in soil are developed and based on the studies of soils from the temperate climatic belt. In these conditions, magnetite/maghemite is the pedogenic magnetic mineral that dominates magnetic susceptibility, while the antiferromagnetic fractions (hematite, goethite), although in a much larger amount, provide a minor contribution to the measured signal (e.g., Dearing, 1999). However, as discussed in Chapter 4.13, tropical strongly weathered soils (Ferralsols) as well as Mediterranean red-colored soils (*terra rossa*) (Chapter 3) display contradictory results for the $\chi_{fd}\%$ depth variations. This contradiction arises when comparing the χ, χ_{fd}, and $\chi_{fd}\%$ depth variations. Examples of this problem are illustrated by the soil profiles of the Rhodic Luvisol (profile RED, described in Chapter 3) and the andic soil (profile BP described in Chapter 4) (Fig. 8.3.1) from Bulgaria.

As seen from Fig. 8.3.1, the mass-specific frequency-dependent magnetic susceptibility χ_{fd} is following the behavior of the bulk χ, thereby suggesting enhanced SP ferrimagnetic content in the upper soil horizons. Deeper soil parts are magnetically depleted, but at the same time the $\chi_{fd}\%$ is steadily increasing toward the bottom of the profile, which, according to the criteria described here earlier, would stand for an increasing share of the SP magnetite/maghemite grains. As already pointed out in Chapter 3, similar $\chi_{fd}\%$ behavior is also reported for red Mediterranean soils from Spain (Liu et al., 2010a). Based on the data provided in Liu et al. (2010a), Fig. 8.3.2 shows the obtained χ, χ_{fd}, and $\chi_{fd}\%$ for the three Luvisol profiles from Spain.

In the same way as for the profile RED (Fig. 8.3.1), the χ and χ_{fd} covary and generally decrease toward the C horizon, but $\chi_{fd}\%$ remains very high or even increases in depth. Analogous behavior is observed in lateritic soil profiles from India (Ananthapadmanabha et al., 2014) (see Fig. 4.13.1 from Chapter 4), and the authors interpret it in the usual way by applying Dearing (1999) threshold $\chi_{fd}\%$ values (i.e., an increasing share of SP magnetite/maghemite). Still another example that probably best illustrates the contradiction discussed comes from the study by Long et al. (2015), who present detailed magnetic investigations on three red Ferralsols from Hainan Island in South China. All three soils were developed on the same parent material but under significantly different precipitation (varying between 1440 and 1650 mm/year). A very high magnetic susceptibility ($800-3200 \times 10^{-8}$ m^3/kg) is obtained in the upper soil horizons. In the first profile (under the highest MAP), both χ and $\chi_{fd}\%$ show a decrease with increasing depth, a feature usually observed in temperate soils. However, in the second profile developed under lower MAP, χ decreases in depth, while $\chi_{fd}\%$ exhibits a constant high value of $\sim 15\%$ all along the profile. In the third profile developed in the lowest MAP, χ is the highest in the near surface horizon but also decreases toward the bottom of the profile, while $\chi_{fd}\%$ increases, reaching 20% (Long et al., 2015). The increasing contribution of hematite in the order of the first to the third profile is evidenced through a strong

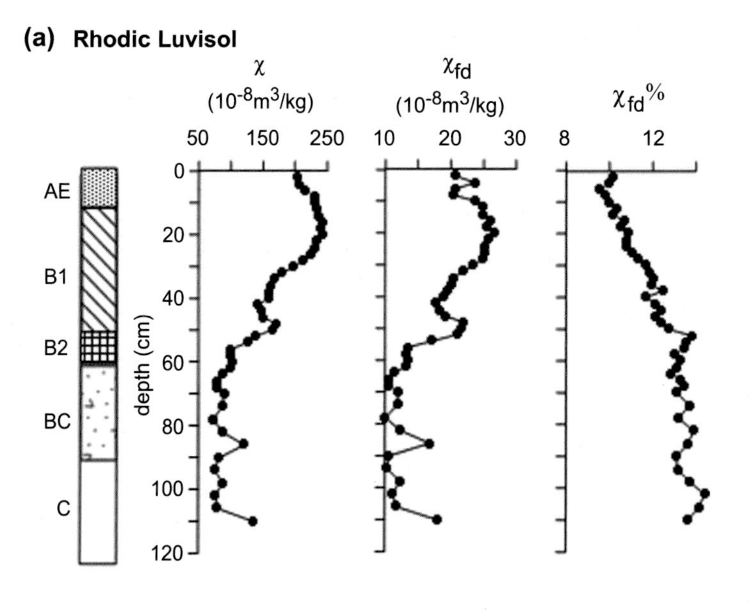

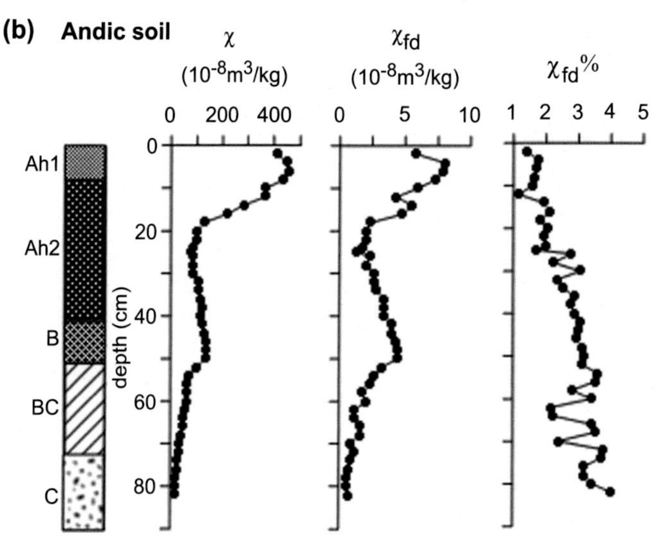

FIGURE 8.3.1

Magnetic susceptibility (χ), frequency-dependent magnetic susceptibility (χ_{fd}), and percent frequency-dependent magnetic susceptibility ($\chi_{fd}\%$) for profiles of (a) Rhodic Luvisol (RED) (Chapter 3) and (b) andic soil (BP profile) (Chapter 5). χ_{fd} and $\chi_{fd}\%$ considered together imply controversial interpretation concerning SP fraction contribution along depth of the profile. See the text for more details.

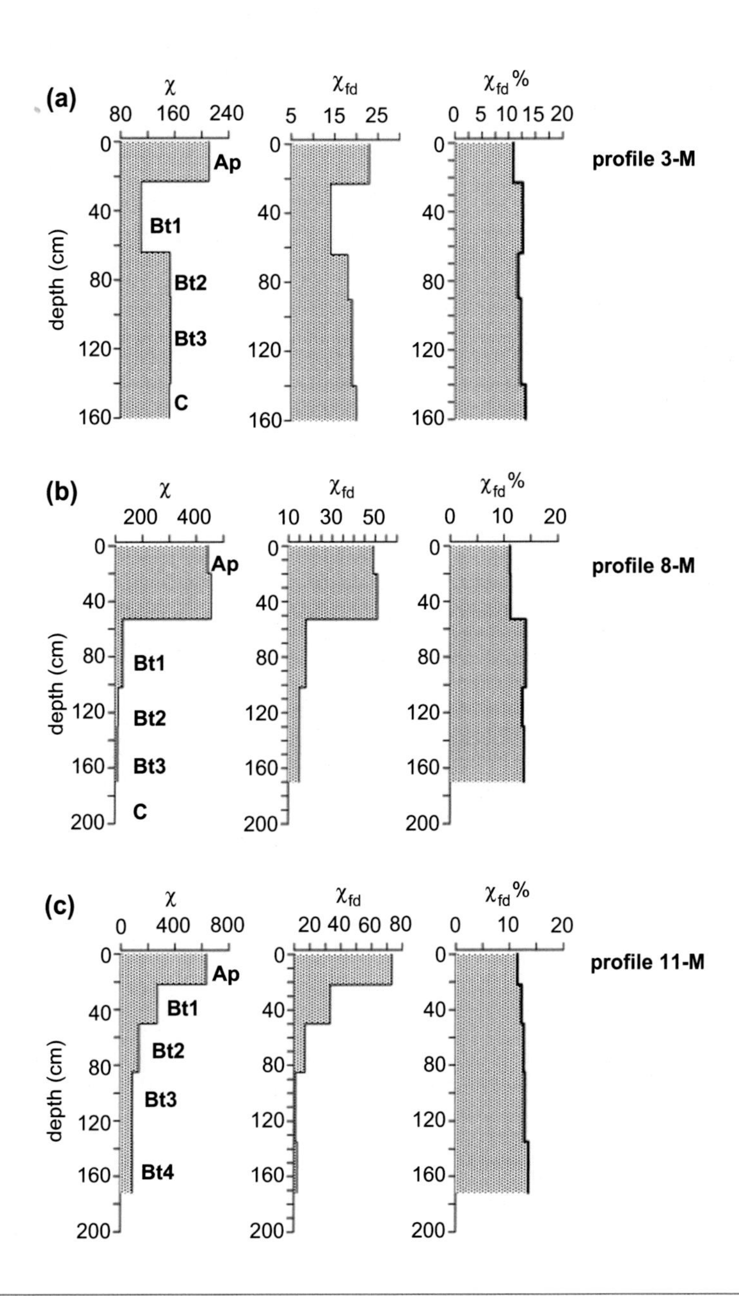

FIGURE 8.3.2

Magnetic susceptibility (χ), frequency-dependent magnetic susceptibility (χ_{fd}), and percent frequency-dependent magnetic susceptibility (χ_{fd}%) for profiles of three red Mediterranean soils from Spain. Magnetic susceptibility is shown in 10^{-8} m^3/kg units.

Data compiled from Liu, Q., Hu, P., Torrent, J., Barrón, V., Zhao, X., Jiang, Z., Su, Y., 2010a. Environmental magnetic study of a Xeralf chronosequence in northwestern Spain: indications for pedogenesis. Palaeogeogr. Palaeoclimatol. Palaeoecol. 293, 144–156.

increase in the Hm/(Hm + Gt) ratio. However, the authors do not comment on the reasons for the obtained pattern and the extremely high $\chi_{fd}\%$ obtained (which, according to Dearing et al. (1996) analyses, as well as Worm (1998), theoretical considerations should indicate an erroneous measurement). A very similar situation is reported in Wang and Lu (2014) in a study of red paleosols in southwest China, where only in the first paleosol are covarying χ and $\chi_{fd}\%$ obtained, while in all older paleosols, $\chi_{fd}\%$ retains very high values of ~12–14% despite the diminished χ. Analogous contradictory data are reported by Li et al. (2011), again for tropical Ferralsols from Hainan Island. Therefore, considering the compiled data altogether, the following facts appear common:

1. Opposing χ and $\chi_{fd}\%$ depth variations are observed in highly weathered red soils from the tropics and the Mediterranean areas.
2. A significant amount of pedogenic hematite is identified in all these soils.
3. The pedogenic iron oxides are a complex mixture of maghemite, hematite, and goethite.
4. The parent material is enriched in easily weatherable Fe-bearing silicates (basalts, shists, loess, etc.).
5. Hematite is supposed to represent the final product in pedogenic iron oxide formation (e.g., Barrón and Torrent, 2002).

Thus, it seems reasonable to hypothesize that the increased $\chi_{fd}\%$ observed in the deeper soil horizons, despite the diminished concentration of the ferrimagnetic strongly magnetic fraction, is due to the specific properties of the abundant pedogenic hematite.

There are numerous studies on the magnetic properties of hematite elucidating its coercivity, Morin transition temperature, Neel temperature, grain size and foreign ion substitution effects, etc. (Dunlop and Özdemir, 1997 references therein; Kletetschka and Wasilewski, 2002; Liu et al., 2010b; Özdemir and Dunlop, 2014). There is, however, only one study (Wells et al., 1999) (to our knowledge) reporting measurements of room temperature frequency-dependent magnetic susceptibility ($\chi_{fd}\%$) for well-defined (by means of composition and grain size) synthetic hematites. According to the data from Wells et al. (1999), Al-substituted hematites, which are of nanometer size and thus in an SP domain state (Dunlop and Özdemir, 1997), show very high $\chi_{fd}\%$ values (Table 8.3.1).

Table 8.3.1 Magnetic susceptibility (χ) and percent frequency-dependent magnetic susceptibility ($\chi_{fd}\%$) for synthetic hematites with different degree of Al-substitutions

Al substitution (%)	MCD a (nm)	MCDc (nm)	χ (10^{-8} m^3/kg)	$\chi_{fd}\%$
4.6	31.3	38.9	183.8	18.9
8.3	27.6	21.2	99.3	28.1
13.4	22.1	7.9	47.4	100.0
15.0	18.7	5.2	41.9	100.0

Mean crystallite dimensions along the hematite's a- (MCDa) and c- (MCDc) directions are also shown.
Data according toWells, M., Fitzpatrick, R., Gilkes, R., Dobson, J., 1999. Magnetic properties of metal-substituted haematite. Geophys. J. Int. 138, 571–580.

As the data demonstrate, nanosized highly substituted hematites show anomalously high $\chi_{fd}\%$ values, reaching 100%. Larger, tens-of-nanometer–sized Al-substituted particles show very high $\chi_{fd}\%$ as well, which is unusual even for SP ferrimagnetic magnetite/maghemite. The lowering of the absolute value of the bulk magnetic susceptibility (Table 8.3.1) with increasing Al substitutions is logically explained by the diluting effect of the Al ions, but the extremely high $\chi_{fd}\%$ could not be explained solely as a grain size effect. Wells et al. (1999) propose that the Al substitutions break down the antiferromagnetic coupling between the sublattices and effectively create smaller SP particles that exhibit magnetostatic interactions. It is now well proved that nanoparticles could display magnetic properties completely different from the microsized and coarser analogs of the respective minerals (Papaefthymiou, 2009; Shcherbakov et al., 2012; Guo and Barnard, 2013). Mørup et al. (2007) discuss in detail the magnetic properties of antiferromagnetic nanoparticles and the effects of magnetic interactions. It is stressed that the SP relaxation of nanoparticles is very sensitive to interparticle interactions. This is well demonstrated in the study by Reufer et al. (2011), where a comparative investigation of interacting and noninteracting nanohematites nominally in SD-range grain size is reported. It has been evidenced that interacting, Si-coated nanohematite particles exhibit frequency-dependent magnetic susceptibility (both in in-phase and out-of-phase components) at room temperature, while a noninteracting sample showed negligible frequency-dependence. This SP relaxation at room temperature (much higher than the blocking temperature of nanosized hematite) is explained by the role of exchange coupling between the grains. Reufer et al. (2011) conclude that the nanohematite showing exchange interactions behaves as clusters of interacting SP particles. Furthermore, Hansen et al. (2000) measured the SP blocking temperature of noninteracting hematite particles and demonstrated that it is always $<$ 250 K. Thus, the authors attribute the observed relaxation at room temperature for the interacting hematite particles to the "superferromagnetic" phenomenon (Mørup et al., 1983). It implies that interparticle interactions could lead to "superferromagnetic" ordering of the sublattice magnetization directions of the particles at temperatures when the noninteracting particles would exhibit fast SP relaxation.

The presence of highly Al-substituted hematites is one of the typical features of the red tropical soils (Cornell and Schwertmann, 2003), indicating their advanced stage of mineral weathering. Therefore, considering the experimental data from Wells et al. (1999), and the magnetic studies of red-colored soils summarized here earlier, it can be supposed that the unusual $\chi_{fd}\%$ increase accompanied by a decreasing concentration of the strongly magnetic ferrimagnetic fraction reflects the presence of highly Al-substituted hematite nanoparticles. Their occurrence becomes "visible" at the expense of diminished ferrimagnetic concentration and is expressed in the $\chi_{fd}\%$ parameter. The measured bulk χ values in these bottom parts of the profiles comply well with the values reported for fine-grained hematite (Dearing, 1999; Liu et al., 2010b). Another issue worth commenting on is as to why the $\chi_{fd}\%$ increase in the weakly magnetic soil parts is not always observed in red soils [as discussed for the examples in Long et al. (2015), Wang and Lu (2014), etc.]. The answer may reside in the assumption that the observed anomalous $\chi_{fd}\%$ is characteristic only for highly substituted hematites, behaving effectively as clusters of SP grains, while nonsubstituted hematites probably would exhibit a "normal" relation (e.g., negligible frequency-dependence). Thus, a future challenging task in the mineral magnetic characterization of soils would be a theoretical and experimental study on the frequency-dependent properties of fine-grained hematites.

REFERENCES

Ananthapadmanabha, A., Shankar, R., Sandeep, K., 2014. Rock magnetic properties of lateritic soil profiles from southern India: evidence for pedogenic processes. J. Appl. Geophys. 111, 203–210.

Barrón, V., Torrent, J., 2002. Evidence for a simple pathway to maghemite in Earth and Mars soils. Geochim. Cosmochim. Acta 66 (15), 2801–2806.

Cornell, R., Schwertmann, U., 2003. The Iron Oxides. Structure, Properties, Reactions, Occurrence and Uses (Weinheim, New York).

Dearing, J.A., Dann, R.J.L., Hay, K., Lees, J.A., Loveland, P.J., Maher, B.A., O'Grady, K., 1996. Frequency-dependent susceptibility measurements of environmental materials. Geophys. J. Int. 127, 228–240.

Dearing, J., 1999. Environmental Magnetic Susceptibility. Using the Bartington MS2 System. Bartington Instruments Ltd, ISBN 0 9523409 0 9. OM0409 ISSUE 8.

Dearing, J., Bird, P., Dann, R., Benjamin, S., 1997. Secondary ferrimagnetic minerals in Welsh soils: a comparison of mineral magnetic detection methods and implications for mineral formation. Geophys. J. Int. 130, 727–736.

Dunlop, D., Özdemir, Ö., 1997. Rock magnetism. Fundamentals and frontiers. In: Edwards, D. (Ed.), Cambridge Studies in Magnetism. Cambridge University Press.

Evans, M., Heller, F., 2003. Environmental Magnetism: Principles and Applications of Enviromagnetics. Academic Press, San Diego, CA.

Eyre, J.K., 1997. Frequency dependence of magnetic susceptibility for populations of single-domain grains. Geophys. J. Int. 129, 209–211.

Guo, H., Barnard, A., 2013. Naturally occurring iron oxide nanoparticles: morphology, surface chemistry and environmental stability. J. Mater. Chem. A 2013 (1), 27–42.

Hansel, C., Benner, S., Netss, J., Dohnalkova, A., Kukkadapu, R., Fendorf, S., 2003. Secondary mineralization pathways induced by dissimilatory iron reduction of ferrihydrite under advective flow. Geochim. Cosmochim. Acta 67 (16), 2977–2992.

Hansen, M., Bender Koch, C., Mørup, S., 2000. Magnetic dynamics of weakly and strongly interacting hematite nanoparticles. Phys. Rev. B (Condens. Matter Mater. Phys.) 62 (2), 1124–1135.

IUSS Working Group WRB, 2014. World Reference Base for Soil Resources 2014. International Soil Classification System for Naming Soils and Creating Legends for Soil Maps. World Soil Resources Reports No. 106. FAO, Rome.

Kletetschka, G., Wasilewski, P., 2002. Grain size limit for SD hematite. Phys. Earth Planet. Inter. 129, 173–179.

Li, D., Yang, Y., Guo, J., Velde, B., Zhang, G., Hu, F., Zhao, M., 2011. Evolution and significance of soil magnetism of basalt-derived chronosequence soils in tropical southern China. Agric. Sci. 2, 536–543. http://dx.doi.org/10.4236/as.2011.24070.

Liu, Q., Barrón, V., Torrent, J., Qin, H., Yu, Y., 2010b. The magnetism of micro-sized hematite explained. Phys. Earth Planet. Inter. 183, 387–397.

Liu, Q., Hu, P., Torrent, J., Barrón, V., Zhao, X., Jiang, Z., Su, Y., 2010a. Environmental magnetic study of a Xeralf chronosequence in northwestern Spain: indications for pedogenesis. Palaeogeogr. Palaeoclimatol. Palaeoecol. 293, 144–156.

Liu, Q., Roberts, A., Larrasoaña, J., Banerjee, S., Guyodo, Y., Tauxe, L., Oldfield, F., 2012. Environmental magnetism: principles and applications. Rev. Geophys. 50, RG4002.

Long, X., Ji, J., Balsam, W., Barrón, V., Torrent, J., 2015. Grain growth and transformation of pedogenic magnetic particles in red. Ferralsols. Geophys. Res. Lett. 42, 5762–5770. http://dx.doi.org/10.1002/2015GL064678.

Maher, B., 1986. Characterization of soils by mineral magnetic measurements. Phys. Earth Planet. Inter. 42, 76–92.

Maher, B.A., 1998. Magnetic properties of modern soils and Quaternary loessic paleosols: paleoclimatic implications. Palaeogeogr. Palaeoclimatol. Palaeoecol. 137, 25–54.

Mørup, S., Madsen, D., Franck, J., Villadsen, J., Bender Koch, C., 1983. A new interpretation of Mössbauer spectra of microcrystalline goethite: "Super-ferromagnetism" or "super-spin-glass" behaviour? J. Magn. Magn. Mater. 40 (1–2), 163–174.

Mørup, S., Madsen, D., Frandsen, C., Bahl, C., Hansen, M., 2007. Experimental and theoretical studies of nanoparticles of antiferromagnetic materials. (topical review). J. Phys. Condens. Matter 19, 213202, 31pp.

Mullins, C., Tite, M., 1973. Magnetic viscosity, quadrature susceptibility and frequency dependence of susceptibility in single-domain assemblies of magnetite and maghemite. J. Geophys. Res. 78 (5), 804–809.

Oldfield, F., 2013. Mud and magnetism: records of late Pleistocene and Holocene environmental change recorded by magnetic measurements. J. Paleolimnol. 49, 465–480.

Özdemir, Ö., Dunlop, D., 2014. Hysteresis and coercivity of hematite. J. Geophys. Res. Solid Earth 119, 2582–2594. http://dx.doi.org/10.1002/2013JB010739.

Papaefthymiou, G.C., 2009. Nanoparticle magnetism. (review). Nano Today 4, 438–447.

Reufer, M., Dietsch, H., Gasser, U., Grobety, B., Hirt, A., Malik, V., Schurtenberger, P., 2011. Magnetic properties of silica coated spindle-type hematite particles. J. Phys. Condens. Matter 23 (6), 065102.

Roberts, A.P., 2015. Magnetic mineral diagenesis. Earth-Sci. Rev. 151, 1–47.

Schaetzl, R., Anderson, A., 2009. Soils. Genesis and Geomorphology. Cambridge University Press, UK, ISBN 978-0-521-81201-6.

Shcherbakov, V., Fabian, K., Sycheva, N., McEnroe, S., 2012. Size and shape dependence of the magnetic ordering temperature in nanoscale magnetic particles. Geophys. J. Int. 191, 954–964.

Thompson, A., Chadwick, O., Rancourt, D., Chorover, J., 2006. Iron oxide crystallinity increases during soil redox oscillations. Geochim. Cosmochim. Acta 70, 1710–1727.

Thompson, R., Oldfield, F., 1986. Environmental Magnetism. Allen&Unwin, London.

Usman, M., Abdelmoula, M., Hanna, K., Grégoire, B., Faure, P., Ruby, C., 2012. FeII induced mineralogical transformations of ferric oxyhydroxides into magnetite of variable stoichiometry and morphology. J. Solid State Chem. 194, 328–335.

Van Breemen, N., 1988. Long-term chemical, mineralogical and morphological effects of iron-redox processes in periodically flooded soils. In: Stucki, J., Goodman, B., Schwertmann, U. (Eds.), Iron in Soils and Clay Minerals. NATO ASI Series. D. Reidel Publishing Company.

Wang, S., Lu, S., 2014. A rock magnetic study on red palaeosols in Yun-Gui Plateau (Southwestern China) and evidence for uplift of Plateau. Geophys. J. Int. 196, 736–747.

Wells, M., Fitzpatrick, R., Gilkes, R., Dobson, J., 1999. Magnetic properties of metal-substituted haematite. Geophys. J. Int. 138, 571–580.

Worm, H.-U., 1998. On the superparamagnetic – stable single domain transition for magnetite, and frequency dependence of susceptibility. Geophys. J. Int. 133, 201–206.

CHAPTER

9

THE MAPPING OF TOPSOIL MAGNETIC PROPERTIES: A MAGNETIC DATABASE FOR BULGARIA—STATISTICAL DATA ANALYSIS AND THE SIGNIFICANCE FOR SOIL STUDIES

9.1 INTRODUCTION

The soil is a heterogeneous media that is characterized by temporal, lateral, and vertical non-homogeneity. This is a natural response to the changing environmental (external) conditions and the existing vertical geochemical gradients in the soil's body (Lin et al., 2005; Lawrence et al., 2013). The mapping of soil spatial variability is of great importance, not only for the construction of soil cover maps but also to provide critical information needed for modeling the landscape-scale processes and evolution. Soil magnetic studies can be used as a rapid tool for detailed spatial reconnaissance of soil cover, as far as topsoil magnetism reflects4 the specific characteristics of the various soil types, related to their genesis and soil-forming factors (see Chapters 2–7).

Large-scale mapping of magnetic susceptibility and frequency-dependent magnetic susceptibility by using national soil data sets from England and Wales, Austria, Bosnia-Herzegovina, France, and other countries with partial coverage have been published (Dearing et al., 1996a; Hanesch et al., 2007; Hannam and Dearing, 2008; Blundell et al., 2009; Thiesson et al., 2012; Bian et al., 2014). Most of the cited works use magnetic susceptibility and its frequency dependence as variables for revealing the relationship among soil magnetism, other soil physical properties, and the factors of soil formation as defined by Jenny (1941). The commonly used data sets cover the national territories through applied regular sampling grids, including natural, agricultural, and anthropogenically affected sites. This results in highly variable and complicated data, where many external factors interact and the resulting soil response is highly dependent on the location's attributes. A specific topic of magnetic mapping for revealing the degree of soil pollution deals with smaller sampling areas, representing particular case studies (Wang, 2013; Sarris et al., 2009; Kapicka et al., 2008; Chaparro et al., 2007; Magiera et al., 2007; Hanesch and Scholger, 2002).

Similar to other earth and environmental studies, different processes and factors govern the lateral changes of a physical property, depending on the spatial scale, sampling density, and design (Benayas et al., 2004). As emphasized by Hannam and Dearing (2008), a soil's magnetic properties and characteristics are influenced by different factors, acting at different scales (Fig. 9.1.1).

Soil Magnetism. http://dx.doi.org/10.1016/B978-0-12-809239-2.00009-7

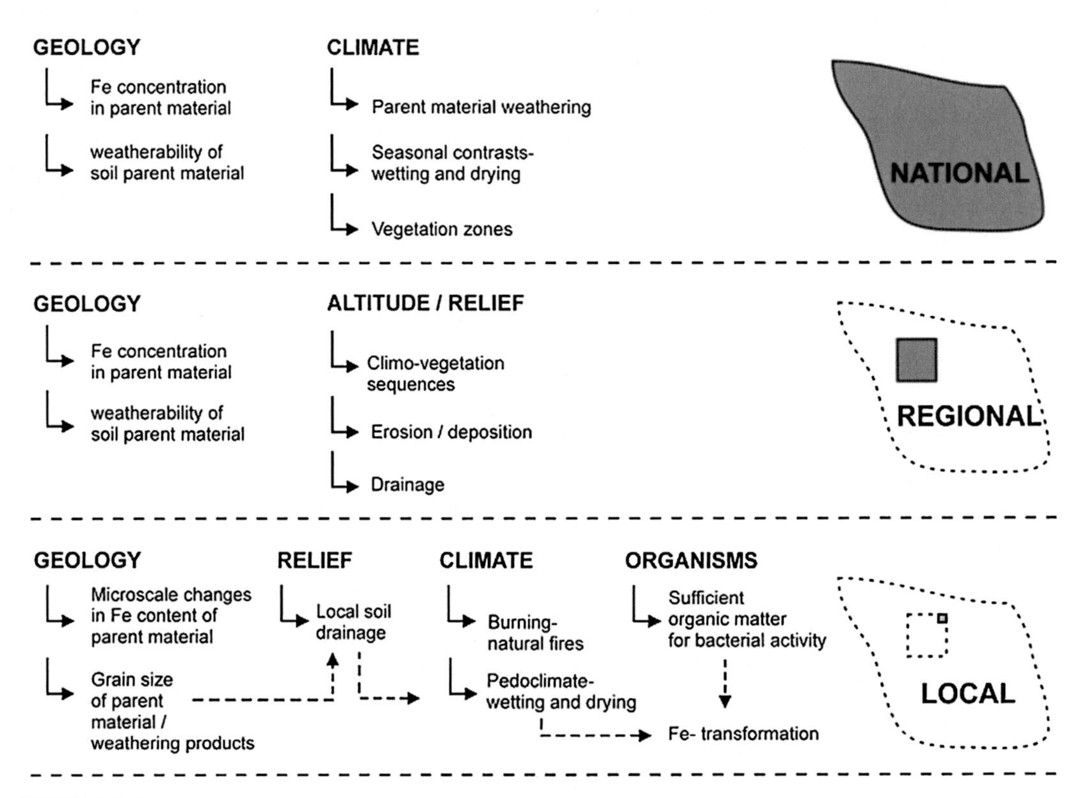

FIGURE 9.1.1

Scale-dependent range of the factors, determining the magnetic signal of topsoils.

From Hannam, J.A., Dearing, J.A., 2008. Mapping soil magnetic properties in Bosnia and Herzegovina for landmine clearance operations. Earth Planet. Sci. Lett. 274, 285–294 with permission from Elsevier.

The Bulgarian data set for the topsoils from the whole territory of the country ($\sim 110{,}000\ km^2$) has been acquired in a different manner. The aim of the study was to identify the role of natural soil forming processes on the magnetic signature of unpolluted soils, developed on different parent materials. According to this general objective, soil sampling was carried out on an irregular grid, including only nonpolluted and nonagricultural sites. The latter requirement was difficult to follow for the soils from the northern part of the country, covered by Chernozems, which are actively used for corn and food production. In such cases, locations outside the main agricultural fields have been chosen, away from the direct influence of anthropogenic activities (roads, industries, urban areas, etc.). The uppermost 20 cm of the soils were sampled at five points (square ends and the center) of the $2 \times 2\ m^2$ surface according to the protocol for standard geochemical soil sampling (Darnley et al., 1995; Tóth et al., 2013). The material was mixed and assigned to a single point with georeferenced coordinates, taken with a portable GPS (Garmin Vista). As a result, a bulk composite sample of ~ 1.5 kg was gathered from each location. For each sampling site, a soil type was verified and assigned, using the 1:200,000 *Soil Map of Bulgaria* (Tanov et al., 1956) and the *Soil Atlas of Bulgaria* (Koinov et al., 1998). Information on parent rock lithology was obtained from the 1:100,000 geological map of Bulgaria for each sampling location. The sampling grid had a mean size of 12 km, resulting in 511 single locations being sampled. Soils were classified in terms of main soil groups (orders). A more detailed classification was not attempted because of the grid density (1 point per 144 km^2), which is not sufficient for applying sub-order classification. The soil classification is presented according to the WRB system, and the correspondence with the Bulgarian soil classification system is given according to Ninov (2000).

Sample preparation and laboratory procedures follow the same protocol as described in Chapter 1 related to the sampling and analyses of the soil profiles. In addition, field magnetic susceptibility per locality has been measured on the sampled square grid, cleaned from vegetation, taking 10 independent measurements and calculating the average (K_{field}). A field kappamater KT5 (Satis Geo, Czech Republic) was used, having a sensitivity of 1×10^{-5} SI.

The statistical analyses of the database (descriptive statistics, factor analysis, k-means clustering) were performed by using STATISTICA 8 (StatSoft Inc.) software. The variogram analysis, data gridding, and plotting of the maps were performed by using the SURFER 11 software package (Golden Software Ltd.). The meteorological information was provided by the National Institute of Meteorology and Hydrology (Bulgarian Academy of Sciences) covering the mean annual precipitation (MAP) and the mean annual temperature (MAT) data from a 20-year observation period for 102 permanent meteorological stations in Bulgaria.

9.2 FIELD AND MASS-SPECIFIC MAGNETIC SUSCEPTIBILITY OF THE TOPSOILS FROM BULGARIA

There is a good agreement between the measured field and laboratory values of the topsoil's magnetic susceptibility (Fig. 9.2.1). It should be mentioned that some deviations may occur due to the differences in the soil's volume included in the sensor's space. During the field measurements, only the uppermost 2–3 cm provide the main contribution to the susceptibility values measured, while laboratory measurements are conducted on the homogenized material gathered from the upper 20 cm of the topsoil layer. Thus, the latter may include Fe minerals that are not typically found at the surface in some particular soil types with a strong vertical differentiation of the profile.

On the other hand, the relationship between the measured mass-specific magnetic susceptibility of the topsoils and the signal of outcropping parent rock samples is much less consistent (Fig. 9.2.2). There is no clear dependence for the weakly magnetic sedimentary rocks, while some positive relationship appears for the strongly magnetic volcanic and intrusive (granitic) rocks. The delineation of different linear trends is obviously dictated by the different Fe oxide content of the rocks from different geological formations on the territory of Bulgaria.

9.3 MAGNETIC CHARACTERISTICS OF THE TOPSOILS FROM MAJOR SOIL ORDERS

9.3.1 MAGNETIC MINERALOGY

The magnetic mineralogy of the topsoils studied has been deduced from the thermomagnetic analysis of magnetic susceptibility. Generally, the strongly magnetic ferrimagnetic phases dominate the signal and the minor presence of weakly magnetic hematite is difficult to detect (Frank and Nowaczyk, 2008; Petrovský and Kapicka, 2006). Fig. 9.3.1 shows typical examples of the obtained data for various soil types.

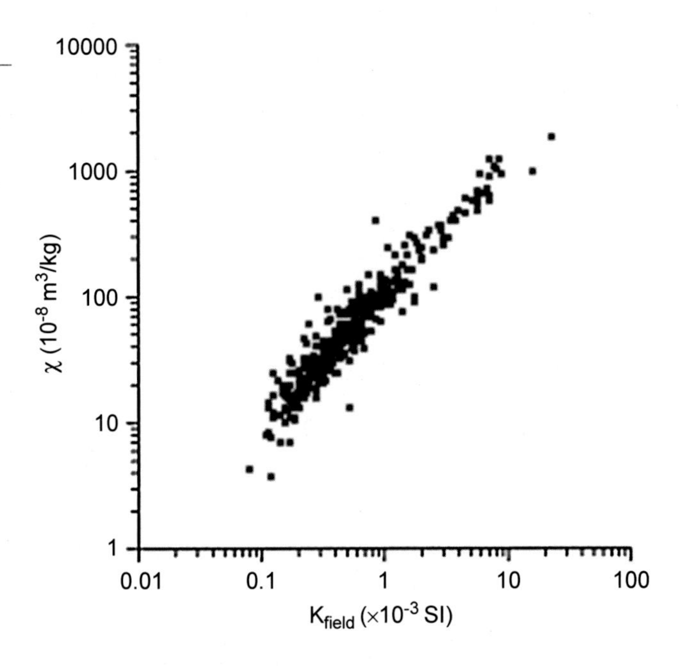

FIGURE 9.2.1

Dependence between the field-measured volume magnetic susceptibility (K_{field}) and the laboratory mass-specific magnetic susceptibility (χ) of topsoils from Bulgaria.

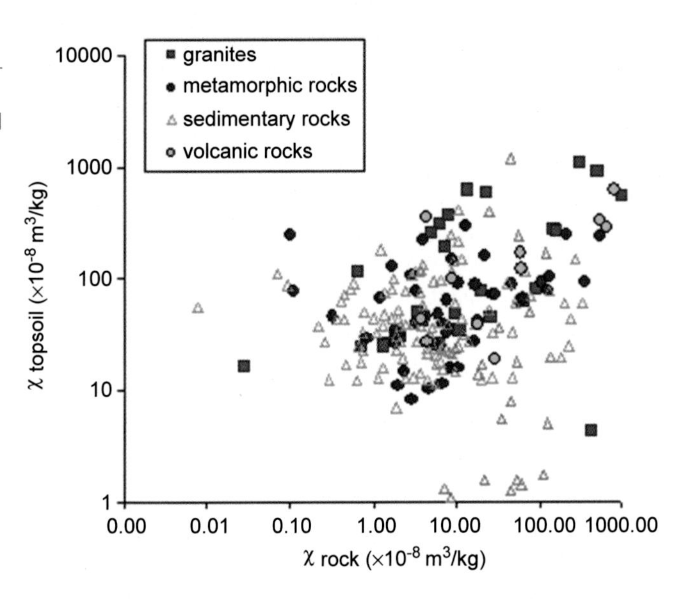

FIGURE 9.2.2

Relationship between mass-specific magnetic susceptibility of the topsoils and the magnetic susceptibility of the parent rocks (where available for sampling).

Almost all soil samples show strong irreversible changes of magnetic susceptibility during heating–cooling cycles. This is common phenomena, resulting from the presence of instable clay minerals and organic matter in the soils. The different temperatures at which the most significant mineralogical changes occur may give clues for the initial mineralogy of the soils. The typical shape of

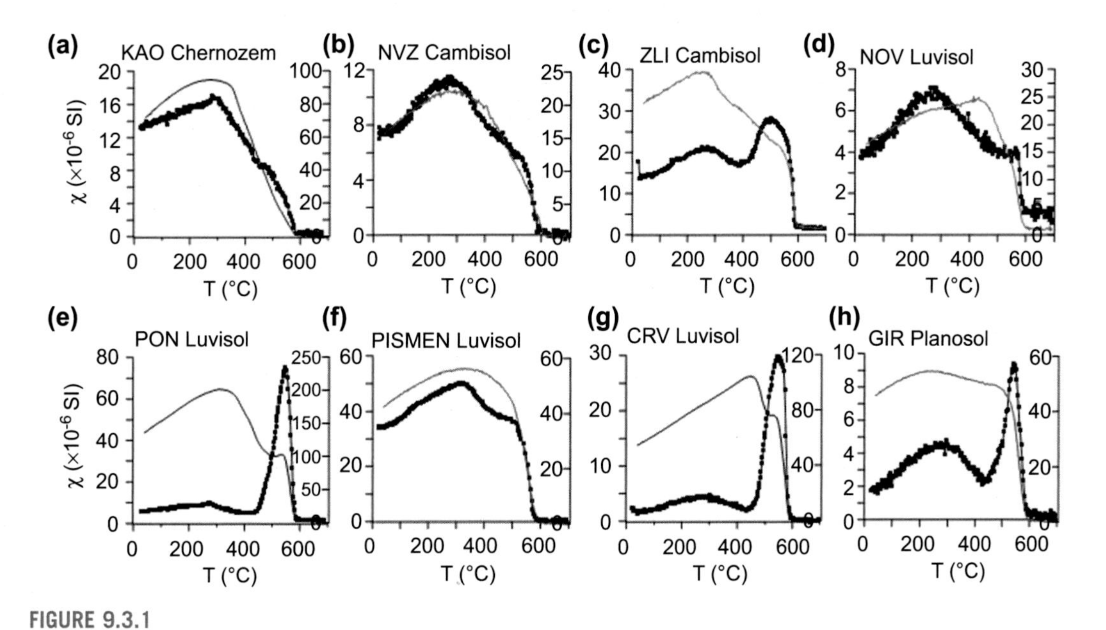

FIGURE 9.3.1

Examples of thermomagnetic analysis of magnetic susceptibility for topsoils from different soil orders, included in the magnetic database for Bulgaria. Heating curve is indicated as *thick line*; cooling—a thin line. Analyses done in air with heating rate of 11°C/min. For samples, exhibiting strong mineralogical transformations, the cooling curve is referenced by the right *y*-axis.

the heating curve obtained for the Chernozem soils (Fig. 9.3.1a) suggests the presence of maghemite, which is transformed to hematite after 300°C, causing the significant decrease of the signal after that temperature. Upon heating, a new strongly magnetic phase appears, probably resulting from hematite reduction after 600°C or the formation of new magnetite from Fe, released during the thermal destruction of clay minerals (Murad and Wagner, 1998; Murad et al., 2002; Jelenska et al., 2010). Similar mineralogical transformations possibly occur in the Cambisol (NVZ) developed on alluvium (Fig. 9.3.1b). The initial increase of susceptibility from room temperature up to 300°C is probably related to the unblocking of fine grains at the superparamagnetic (SP)–single-domain (SD) boundary (Liu et al., 2005; Wang and Løvlie, 2008). Another Cambisol developed on dioritic parent material (sample ZLI) shows different thermal behavior, primarily determined by the transformations of the lithogenic magnetic minerals of the diorite. As a result, much less-significant thermal transformations are observed as far as the mineralogy of the diorites crystallized from magma cooling down from high temperatures. The examples from Luvisols include topsoils developed on limestones, tuffs, breccia, and sandstones (Fig. 9.3.1d–g). The most thermally instable samples are the Luvisols developed on limestone and red sandstone, where secondary magnetite with a T_c of 580°C is formed during heating after 450°C. A similar mineralogy is revealed for the Planosol sample GYR (Fig. 9.3.1h). The initial presence of a magnetite-like fraction on the topsoils from Bulgaria is further evidenced by the unblocking of laboratory remanences (ARM, SIRM) during thermal demagnetization experiments (Fig. 9.3.2).

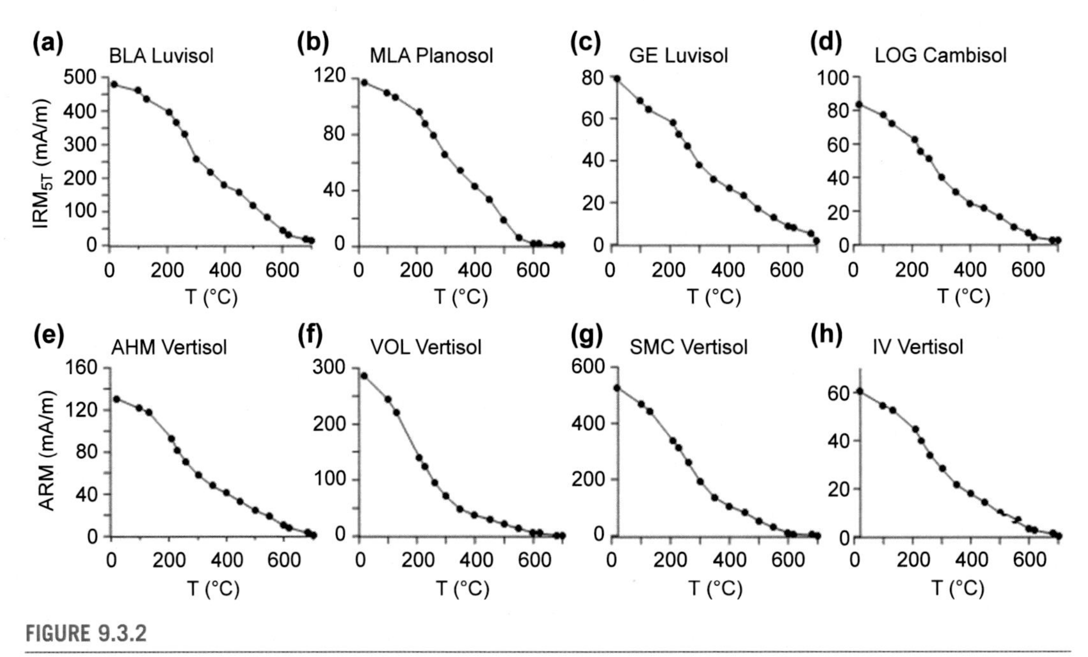

FIGURE 9.3.2

Examples of stepwise thermal demagnetization of isothermal remanence (IRM_{5T}) (a–d) and anhysteretic remanence (ARM) (e–h) for different soil types.

The presence of a viscous (fine) pedogenic fraction causes the systematic smooth unblocking of the laboratory remanences in the temperature range 200–400°C. Final T_{ub}s of (oxidized) magnetite (600°C) and, in some samples, hematite (700°C) are clearly expressed.

9.3.2 MASS-SPECIFIC MAGNETIC SUSCEPTIBILITY AND FREQUENCY-DEPENDENT MAGNETIC SUSCEPTIBILITY

The topsoil's mass-specific magnetic susceptibility of the Bulgarian data set varies in the range from -0.67×10^{-8} m^3/kg to a maximum of 2731.6×10^{-8} m^3/kg. The two extremes correspond to pure diamagnetic material with negative susceptibility, and a strongly magnetic material containing ~4–5 wt% magnetite (Hunt et al., 1995). We consider the magnetic susceptibilities of the major soil orders, assuming that the mass-specific susceptibility of the topsoils of a particular type is determined by the simultaneous action of all soil-forming factors. Thus, median susceptibilities per soil type with the corresponding statistical estimates (median, range, quartiles, outliers) are presented in Fig. 9.3.3a. The highest variability of the mass-specific magnetic susceptibility is typical for the soil orders that do not depend on climate–Leptosols, Fluvisols, and Vertisols. Their magnetic signal is greatly influenced by the lithogenic magnetic minerals inherited from the parent material. Although Vertisols are characterized by a thick solum (see Chapter 5), their topsoil susceptibility is highly dependent on the parent rock mineralogy because of the weak pedogenic enhancement typical for these soils. The contrast is due to the two major types of lithologies on which Bulgarian Vertisols are developed—weakly magnetic Pliocene clays in western Bulgaria and strongly magnetic volcano-sedimentary rock formations in

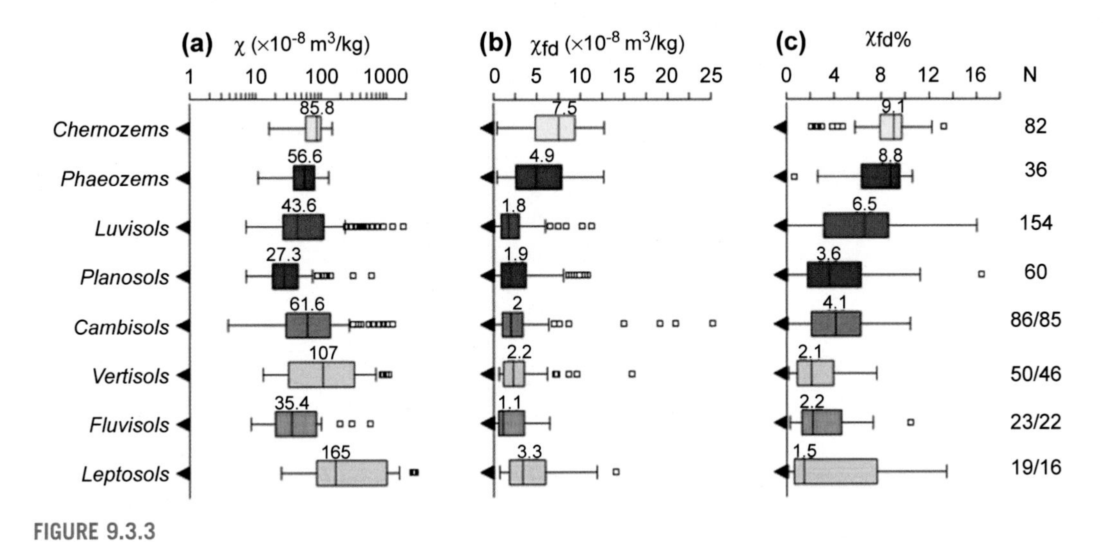

FIGURE 9.3.3

Box-and-whisker plots of (a) magnetic susceptibility; (b) frequency-dependent magnetic susceptibility; and (c) percent frequency-dependent magnetic susceptibility per soil type. Number of samples (N) is given in the right column. The second values for Cambisols, Vertisols, Fluvisols, and Leptosols correspond to the number of samples included in χ_{fd} and χ_{fd}% statistics, where very low susceptibilities are not included because of the uncertain measurements. Outliers are represented by circles, and median values are indicated in the figure.

southeastern Bulgaria. The median χ value for Vertisols is calculated for two groups—the weakly magnetic soils (with $\chi < 100 \times 10^{-8}$ m^3/kg) and the strongly magnetic soils ($\chi > 100 \times 10^{-8}$ m^3/kg). The resulting parameters indicate a normal distribution of values for the weakly magnetic Vertisols with $\chi_{median} = 31.8 \times 10^{-8}$ m^3/kg and a scattered distribution of higher susceptibilities with $\chi_{median} = 331 \times 10^{-8}$ m^3/kg. Consequently, a reliable statistical estimate of mean magnetic susceptibility for the soil order of Vertisols could be established only for those developed on weakly magnetic parent material, while for the others, the topsoil's susceptibility is strongly site-specific and depends on the local concentration of strongly magnetic minerals inherited from the parent rock.

A high variability is also observed within the Cambisols group. Similar to the strongly magnetic Vertisols, Cambisols developed on strongly magnetic intrusive rocks from northwest Bulgaria show extremely high values of topsoil magnetic susceptibility. The distributions close to normal are characteristic for the Chernozem and Phaeozem orders. The distribution of topsoil susceptibilities of the Chernozems is characterized by the presence of a group of consistently low values, falling in the lower quartile ($16 < \chi < 58 \times 10^{-8}$ m^3/kg) (Fig. 9.3.4). Spatially, these locations with low magnetic susceptibilities are situated in northwest Bulgaria and northeast from the city of Varna (northeast Bulgaria). On the soil maps (1:100,000), these Chernozems are further classified as strongly eroded. Recently, Christov and Teoharov (2008) suggested that such Chernozems belong to the order of weakly developed Leptosols. Magnetic susceptibility data clearly discriminate this subdivision.

The frequency-dependent magnetic susceptibility (χ_{fd}) is proportional to the amount of ultrafine nanometer-sized pedogenic strongly magnetic Fe minerals within the size range of 10–16 nm (Dunlop and Özdemir, 1997; Worm, 1998). The values vary between 0 and 25×10^{-8} m^3/kg. Significant

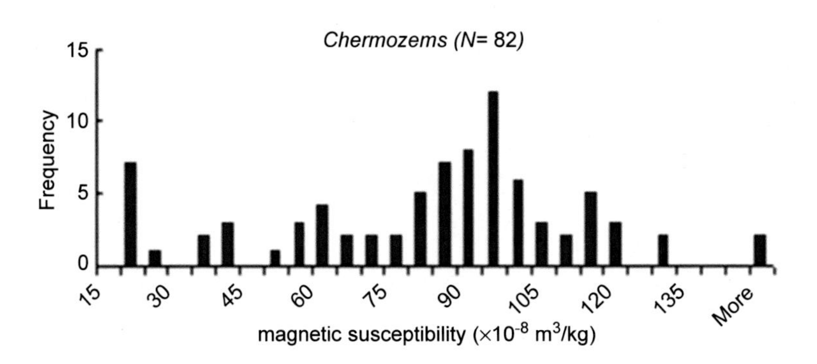

FIGURE 9.3.4

Histogram showing the distribution of values of magnetic susceptibility of the Chernozems. *N*, number of samples.

differences exist among the different soil groups (Fig. 9.3.3b). The highest χ_{fd} values are obtained for the Chernozems and Phaeozems, while systematically lower median values are obtained for the other soil types. Outliers with extremely high values are found in the group of the Cambisols (Fig. 9.3.1b). Minimum values are observed for Fluvisols and Planosols, reflecting the effects of the initial stages of pedogenesis in Fluvisols and the leaching of Fe^{2+} in Planosols. The percent frequency-dependent magnetic susceptibility ($\chi_{fd}\%$) (Fig. 9.3.3c) represents the relative contribution of the pedogenic SP fraction in the total magnetic susceptibility signal (Dearing et al., 1996b). The highest values are typical for the Chernozems and Phaeozems, which are developed on weakly magnetic loess and/or sediments, followed by Luvisols. The latter, however, show a wide distribution of values, which is a reflection of the variability in the mineralogy of the parent materials.

9.3.3 HYSTERESIS PARAMETERS

Descriptive statistics for the hysteresis parameters coercive force (B_c), coercivity of remanence (B_{cr}), saturation magnetization (M_s), isothermal remenence (M_{rs}), and high-field magnetic susceptibility (χ_{hf}) are presented in Table 9.3.1. The median values of the coercivity parameters B_c and B_{cr} per soil group are very similar, while the coefficient of variation (CV%) depends on the variability in the properties of the parent material. This is deduced by the very small CV% for the Chernozem and Phaeozem groups, which are developed on spatially homogeneous parent material (mainly loess sediments). The latter has been shown to possess the lowest spatial variability compared with the other materials such as alluvium, pyroclastic rocks, and other various disturbed materials (Drees and Wilding, 1973). Median B_{cr} values among soil types vary in the range of 23–28 mT and coercive force B_c in the range of 6.8–8.5 mT, which are very narrow intervals. It suggests that pedogenic Fe oxides in different soil types dominate the coercivity distributions. The latter is an indication of the uniformity of the grain size and phase compositions, irrespective of the peculiarities of the pedogenesis in each soil order.

The obtained median B_{cr} values per soil type are shown in Fig. 9.3.5. Lower B_{cr} values for Chernozems and Phaeozems are due to the significant contribution of viscous SD grains of pedogenic origin. On the other hand, similarly low B_{cr} values for Leptosols can be explained by the dominance of lithogenic coarse MD grains, having in mind the observed low χ_{fd} and $\chi_{fd}\%$ values and that these soils are young (e.g., did not reach maturity).

Table 9.3.1 Descriptive statistics of the hysteresis parameters [saturation magnetization M_s, isothermal remanence (M_{rs}), coercive force (B_c), coercivity of remanence (B_{cr}), and high-field susceptibility (χ_{hf})] per soil type

Soil Type	*N*	Mean	Median	SD	Variance	Minimum	Maximum	Skewness	Kurtosis	CV (%)
M_s (mAm^2/kg)										
Chernozems	82	45.5	49.4	21.0	443.5	5.86	133.36	0.59	2.90	46.2
Phaeozems	35	33.4	34.3	18.5	341.8	5.09	81.78	0.56	−0.23	55.3
Luvisols	153	143.8	29.5	347.6	120,795.2	2.0	3080.0	5.33	36.7	241.7
Planosols	59	56.3	17.0	187.4	35,103.7	4.2	1422.0	6.96	50.9	332.9
Cambisols	86	184.9	41.9	374.4	140,201.2	1.9	2051.0	3.49	12.6	202.5
Vertisols	46	237.0	46.9	519.5	269,861.4	7.82	2909.0	4.09	19.41	219.2
Leptosols	20	705.7	118.4	1104.8	1,220,566.0	12.4	3365.0	1.62	1.25	156.6
Fluvisols	16	74.7	19.4	172.5	29,749.1	1.4	677.7	3.5	12.68	230.9
M_{rs} (mAm^2/kg)										
Chernozems	82	6.3	6.8	2.8	8.1	0.76	19.7	0.84	5.12	44.8
Phaeozems	35	4.8	5.2	2.6	6.7	0.84	9.96	0.14	−1.1	53.9
Luvisols	153	13.9	3.9	34.5	1190.6	0.38	333.5	6.35	51.1	248.2
Planosols	59	4.9	2.6	11.2	125.6	0.71	86.0	6.8	49.4	230.0
Cambisols	86	11.7	5.5	19.3	372.1	0.16	113.8	3.37	12.8	165.0
Vertisols	46	13.8	5.2	26.0	673.7	1.08	146.6	4.04	19.0	188.4
Leptosols	20	33.5	12.2	45.0	2026.7	1.7	149.0	1.77	1.94	134.3
Fluvisols	16	4.6	1.9	8.3	69.6	0.19	33.5	3.35	12.0	180.9
B_c (mT)										
Chernozems	82	8.0	7.7	1.1	1.2	6.24	12.2	2.06	5.25	13.6
Phaeozems	35	7.6	7.4	0.9	0.9	6.1	10.0	0.60	0.375	12.1
Luvisols	153	9.1	8.2	4.7	22.0	2.9	48.6	4.67	34.1	51.6
Planosols	58	9.4	8.5	4.4	19.1	3.76	28.1	2.36	8.15	46.5
Cambisols	86	9.0	7.8	6.2	38.1	2.19	41.4	2.6	9.5	68.4
Vertisols	38	7.4	7.5	3.2	10.3	2.25	14.7	0.25	−1.05	43.8
Leptosols	20	6.8	6.8	2.3	5.5	3.21	12.2	0.38	−0.33	34.3

Continued

Table 9.3.1 Descriptive statistics of the hysteresis parameters [saturation magnetization M_s, isothermal remanence (M_{rs}), coercive force (B_c), coercivity of remanence (B_{cr}), and high-field susceptibility (χ_{hf})] per soil type—cont'd

Soil Type	*N*	Mean	Median	SD	Variance	Minimum	Maximum	Skewness	Kurtosis	CV (%)
Fluvisols	15	7.8	8.3	3.3	10.7	3.02	16.8	1.31	3.25	41.9
B_{cr} (mT)										
Chernozems	82	25.7	25.1	3.4	11.8	18.7	40.38	1.92	4.95	13.4
Phaeozems	35	24.6	24.2	2.7	7.1	18.1	31.3	0.40	0.88	10.8
Luvisols	153	34.5	28.2	27.2	742.1	15.6	264.8	6.02	43.1	78.8
Planosols	59	33.5	26.3	37.5	1407.8	13.4	234.3	5.03	25.1	111.9
Cambisols	86	37.8	27.7	37.1	1455.1	16.1	292.0	5.0	28.2	98.1
Vertisols	38	26.2	27.4	7.4	54.7	10.4	45.0	0.22	−0.06	28.2
Leptosols	20	24.2	23.0	6.71	45.0	15.0	42.8	1.54	2.58	27.7
Fluvisols	15	30.9	26.4	12.8	164.5	12.7	67.7	1.71	4.29	41.6
χ_{hf} ($\times 10^{-8}$ m^3/kg)										
Chernozems	82	8.0	8.3	1.3	1.6	3.4	11.1	−1.14	2.73	16.2
Phaeozems	35	7.2	7.5	1.7	2.9	2.9	10.0	−0.76	0.21	23.6
Luvisols	153	8.5	7.7	6.1	37.1	1.0	63.3	5.32	43.69	71.8
Planosols	59	5.5	6.0	2.3	5.3	1.0	10.5	−0.26	−0.48	41.8
Cambisols	85	6.6	5.5	3.2	10.5	1.6	18.7	1.19	1.91	48.5
Vertisols	38	10.1	9.3	4.0	16.1	2.6	22.2	1.19	2.04	39.6
Leptosols	20	11.2	8.8	7.9	63.0	3.7	32.2	1.5	1.66	70.5
Fluvisols	15	6.3	6.2	3.1	10.0	2.7	14.2	1.3	1.77	49.2

CV, *coefficient of variation;* N, *number of samples.*

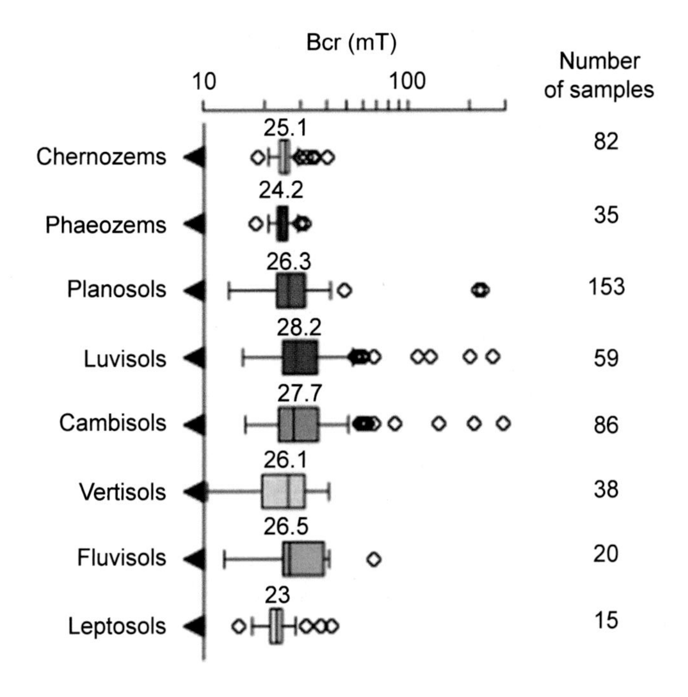

FIGURE 9.3.5

Box-and-whisker plot of coercivity of remanence (B_{cr}) per soil type. Outliers are represented by open circles; median values per soil type are indicated.

9.3.4 ANHYSTERETIC AND ISOTHERMAL REMANENCES AND RATIOS

The anhysteretic remanence (ARM) is widely used in environmental magnetism studies as a proxy for the concentration of the stable SD ferrimagnetic fraction (Maher, 1988). In soil magnetic studies of well-aerated magnetically enhanced soils, this fraction of SD grains is usually considered to be of pedogenic origin (Maher, 1986; Geiss and Zanner, 2006; Evans and Heller, 2003). This is particularly well established for the Chernozem-like soils developed on loess parent material (Maher, 1998; Liu et al., 2004; Geiss et al., 2008; Jordanova et al., 2010). The genesis of this SD fraction is more difficult to reveal when dealing with soils developed on parent materials that already contain an SD-like ferrimagnetic fraction. Examples of such materials are limestones and marls (Jackson and Swanson-Hysell, 2012), metamorphic rocks, and volcanic (volcano-sedimentary) facies (Dunlop and Özdemir, 1997). The topsoil samples from Bulgaria show a wide variability in ARM intensity. The calculated anhysteretic susceptibility (χ_{ARM}) per soil type shows that Chernozems and Phaeozems exhibit consistently high values with small variations and median values of 50.1 and 42.2 × 10^{-8} m^3/kg, respectively. A higher median value is obtained for Leptosols (χ_{ARM} = 72.1 × 10^{-8} m^3/kg) due to the contribution of Leptosols developed on volcanic/volcano-sedimentary rocks from southeast Bulgaria.

Fig. 9.3.6 shows the calculated median values of the ratio χ_{ARM}/SIRM, which was proposed by Maher (1986) as a measure of the grain size of pedogenic magnetite grains.

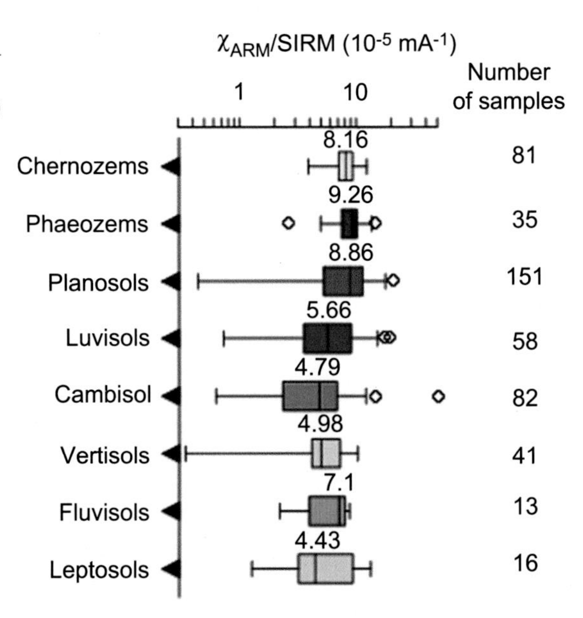

FIGURE 9.3.6

Box-and-whisker plot of the ratio of anhysteretic susceptibility to saturation remanence (χ_{ARM}/SIRM) per soil type. Outliers are represented by *open circles*; median values per soil type are indicated.

Similar to the relationships shown in Fig. 9.3.3, Chernozems, Phaeozems, and Planosols are again characterized by systematically higher values, pointing out the importance of the pedogenic SD fraction. Analogous behavior shows the ratio χ_{ARM}/χ.

9.4 RELATIONSHIPS AMONG MAGNETIC PARAMETERS FOR THE WHOLE TOPSOIL DATABASE

9.4.1 INTERPLAY BETWEEN THE PEDOGENIC MAGNETIC ENHANCEMENT AND LITHOGENIC MAGNETIC SIGNAL

Soil development under aerobic well-drained conditions results in the formation of a pedogenic strongly magnetic fraction of SP-SD magnetite (maghemite) (see Chapters 2 and 3). This magnetic enhancement is usually expressed by the increased values of the percent frequency-dependent magnetic susceptibility $\chi_{fd}\%$ (Dearing et al., 1996b, 1997). Fig. 9.4.1 shows the observed relationship between mass-specific magnetic susceptibility χ and the percent frequency-dependent magnetic susceptibility $\chi_{fd}\%$ for the Bulgarian topsoils. Two clearly defined trends are evident—a linear increase of $\chi_{fd}\%$ with increasing χ (for $\chi \leq 100 \times 10^{-8}$ m^3/kg) and an opposite trend for $\chi > 100 \times 10^{-8}$ m^3/kg.

The two tendencies reflect the competing effects of the pedogenic enhancement (e.g., the formation of an SP fraction of magnetite/maghemite) and the lithogenic magnetic minerals in the parent material. Soils, developed on strongly magnetic parent materials, have magnetic susceptibilities primarily determined by the concentration of the lithogenic (titano)magnetites. Even though the pedogenic magnetic fraction appears during soil formation, its relative contribution to the total susceptibility is

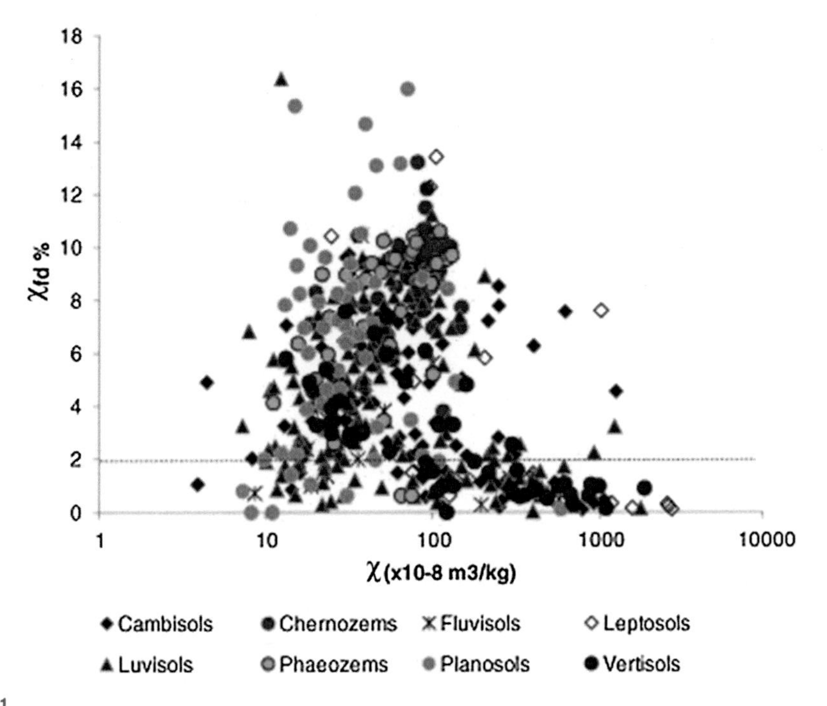

FIGURE 9.4.1

Relationship between mass-specific magnetic susceptibility of the topsoils and the percent frequency-dependent magnetic susceptibility.

From Jordanova, N., Jordanova, D., Petrov, P., 2016. Soil magnetic properties in Bulgaria at a national scale—Challenges and benefits. Glob. Planet. Change 137, 107–122 with permission from Elsevier.

low. The decreasing $\chi_{fd}\%$ with an increase of χ is related to the change in the relative importance of the two magnetic components (pedogenic and lithogenic). The most strongly magnetic topsoils from the Bulgarian database are soils from southeast Bulgaria, developed on the intrusive and volcanic rocks from the area. Plotting the data according to the soil type does not show systematic dependence, which suggests that the major role is played by the parent rock's magnetic signal.

According to the phenomenological model by Dearing et al. (1996a), in soils with $\chi_{fd}\% > 10\%$, almost all pedogenic strongly magnetic minerals are in an SP state. From this supposition it follows that for such an SP assemblage at $\chi_{fd} = 0$, $M_s = 0$. Selecting the soils with $\chi_{fd}\% > 10\%$ from the Bulgarian database, we obtain a good linear relationship between χ_{fd} and saturation magnetization M_s (Fig. 9.4.2).

From the regression equation, we obtain $M_s/\chi_{fd} = 0.4739 * 106$ A/m. If we suppose that the strongly magnetic SP fraction is represented by maghemite with $M_s = 60$ Am^2/kg (Dunlop and Özdemir, 1997), then $\chi_{fd} = 126.6 \times 10^{-6}$ m^3/kg. This value is comparable with the measured frequency-dependent susceptibility of synthetic maghemites with a mean grain size of d = 0.016 μm (Dearing et al., 1996b). Calculating χ_{fd} for magnetite ($M_s = 90$ Am^2/kg), the values are much higher than the measured χ_{fd} for synthetic well-defined SP magnetites. Consequently, the pedogenic strongly magnetic fraction in the Bulgarian topsoils is maghemite.

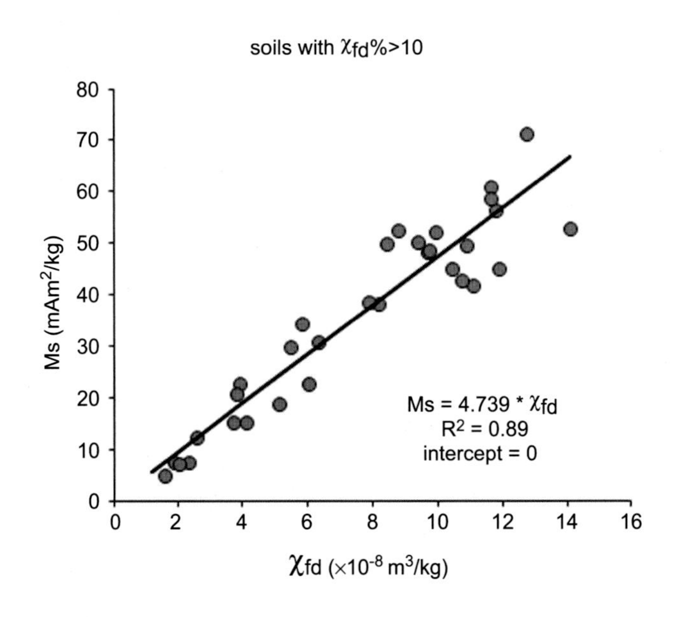

FIGURE 9.4.2

Relationship between saturation magnetization (M_s) and the frequency-dependent magnetic susceptibility (χ_{fd}) for topsoil samples with dominant SP fraction (e.g., $\chi_{fd}\% > 10\%$). Regression equation is given as well.

From Jordanova, N., Jordanova, D., Petrov, P., 2016. Soil magnetic properties in Bulgaria at a national scale—Challenges and benefits. Glob. Planet. Change 137, 107–122 with permission from Elsevier.

In order to verify if the SD fraction responsible for the increased values of the ratio χ_{ARM}/SIRM originates from the pedogenic in situ formation, it is supposed that the grain sizes of the pedogenic fraction span the SP-SD range, thus leading to a proportional increase in SP and SD content in the soils. This supposition would result in a linear relationship between χ_{fd} and χ_{ARM}/SIRM. The biplot of the two parameters (Fig. 9.4.3) reveals the presence of several trends that are related to the nature of the parent material.

The most prominent SP enhancement is evident for the group of soils developed on loess and limestones. These materials ensure a high pH and, thus, facilitate the synthesis of fine-grained magnetite (Cornell and Schwertmann, 2003). Very low values of χ_{ARM}/SIRM are obtained for the soils developed on the intrusive rocks in southwest Bulgaria. The latter lithologies are coarse grained and contain large magnetite crystals (Georgiev et al., 2014). Soils developed on other parent materials are characterized by a lower content of the SP fraction but a higher proportion of SD grains (e.g., higher χ_{ARM}/SIRM values). This shift could be either explained by an inherited SD fraction from the parent rock or by a skewed grain size distribution of the pedogenic fraction. The latter implies the existence of some peculiar factors that would lead to such truncated distribution. There also exists a third group of topsoils, developed on marls and marly limestones, which show an almost zero frequency-dependent susceptibility and very high χ_{ARM}/SIRM values. These topsoils could be considered as being dominated by an SD magnetite fraction from the parent material.

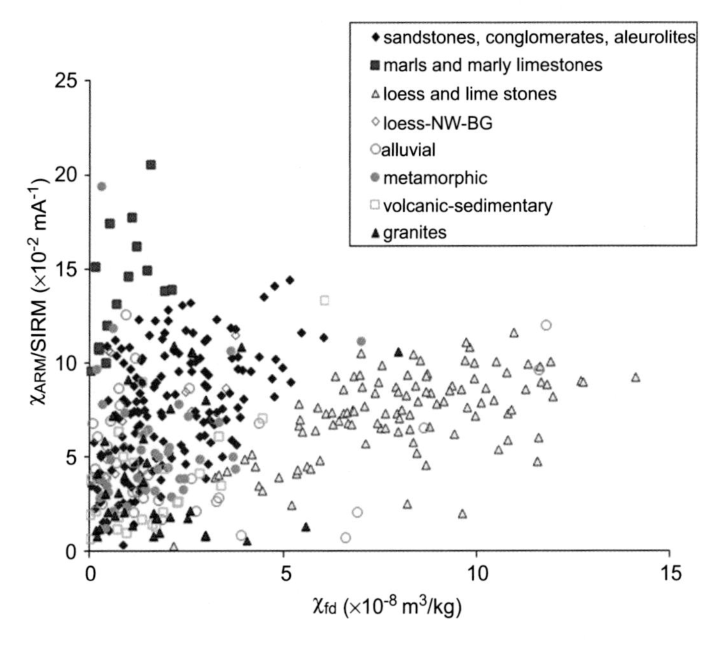

FIGURE 9.4.3

Biplot of frequency-dependent magnetic susceptibility (χ_{fd}) versus ratio of anhysteretic susceptibility to saturation remanence (χ_{ARM}/SIRM) of topsoils, developed on different parent rock lithology.

9.5 STATISTICAL DATA ANALYSIS

Data distributions of the concentration-dependent magnetic characteristics—χ, χ_{fd}, M_s, M_{rs}, χ_{ARM}—are skewed toward the low values because of the prevailing number of topsoil samples with a low magnetic signal, such as Luvisols and Cambisols, which represent 50% of the database. Concerning the χ distribution, a second smaller maximum appears in the interval 80–90 × 10^{-8} m^3/kg, corresponding to the contribution of Chernozems and Phaeozems (in total being 24% from the database). Extreme values in each variable were evaluated and considered as outliers if they had values higher than 1.5 times the interquartile range (IQR). These have been removed from the data set before further statistical analysis. The total number of outliers varies for each particular parameter, being usually between 20 and 25 single samples. The skewed distributions of the concentration-dependent magnetic parameters were log-normally transformed in order to apply statistical analyses and kriging interpolation to normally distributed variables. Table 9.5.1 presents the skewness and kurtosis of the original and log-normally transformed variables. As a measure of normality, values less than ± 1 are considered as sufficient to accept the hypothesis that the distribution is close to normal.

The obtained log-normal distributions are typical for natural systems where multiplicative variations around the mean lead to an evolution toward more probable states. This evolution is accompanied by a decrease in free energy and an increase in enthropy (Grönholm and Annila, 2007). Considering soil formation from a thermodynamics point of view, the establishment of soil horizons is governed by the

Table 9.5.1 Skewness and kurtosis of the original and log-normally transformed parameters for the whole topsoil database

			Initial data		Ln-transformed data	
Parameter	N	CV%	Skewness	Kurtosis	Skewness	Kurtosis
χ	492	209	5.415	36.39	0.317	−0.156
χ_{fd}	502	95.6	1.646	4.08	0.168	−0.887
$\chi_{fd}\%$	494	67	0.324	−0.78	0.303	−0.78
χ_{ARM}	476	118.2	4.354	30.69	−0.366	0.382
χ_{ARM}/χ	474	49.1	0.092	−0.503		
M_{rs}	461	226.5	6.833	66.43	0.348	−0.144
M_s	465	271.2	5.342	32.78	0.21	0.032
ARM / SIRM	475	54.8	0.619	0.84	−0.46	−0.56
MDF_{ARM}	475	18.8	0.519	2.968	−0.855	*2.91*
MDF_{IRM}	469	30.5	2.864	17.56	−0.564	*2.236*
B_c	465	49.4	3.938	26.82	−1.337	*2.78*
B_{cr}	462	83.2	6.812	53.03	*0.95*	*5.35*
χ_{ARM} / χ_{fd}	429	56.8	1.459	1.728		

CV, coefficient of variation; N, number of samples without outliers; Values indicated in italics mark the persistent deviation from log-normal distribution of the transformed variables.
From Jordanova, N., Jordanova, D., Petrov, P., 2016. Soil magnetic properties in Bulgaria at a national scale—Challenges and benefits. Glob. Planet. Change, 137, 107–122, with permission from Elsevier.

processes in an open system that leads to a decrease in the enthropy of the soil and an increase in the enthropy of the surroundings. The formation of secondary oxides during weathering (including Fe oxides) is related to a decrease in enthropy, being more and more significant in soil orders with a high degree of weathering (e.g., Oxisols) (Smeck et al., 1983). Thus, the obtained log-normal distributions of the concentration-dependent magnetic characteristics result from the universal laws of thermodynamics.

Distributions of values of the grain-size—sensitive parameters and ratios ($\chi_{fd}\%$, χ_{ARM}/SIRM, χ_{ARM}/χ) show a skewness and kurtosis close to zero (Table 9.5.1), thus allowing us to accept them as normally distributed. However, closer inspection reveals the presence of two maxima that are related to the magnetic signals from two different origins: very strongly magnetic topsoils, developed on intrusive and volcanogenic rocks, which, however, are dominated by a coarse-grained lithogenic magnetic fraction; and soils with a strong pedogenic magnetic enhancement, developed on weakly magnetic parent material. The magnetic parameter, related to the coercivity and internal microstructure of the ferromagnetic carriers [e.g., B_c, B_{cr}, median destructive field of ARM (MDF_{ARM}), median destructive field of IRM (MDF_{IRM})], do not transform into a normal distribution because of the higher kurtosis of the transformed distributions (Table 9.5.1). This is probably related to the presence of a strongly magnetic pedogenic component with a uniform and narrow grain-size distribution.

9.5.1 FACTOR ANALYSIS

Because of the observed high correlation between the concentration-dependent variables ($R^2 = 0.7–0.9$) (Jordanova et al., 2016), factor analysis is used for reducing the number of variables,

Table 9.5.2 Results from the factor analysis with Eigenvalues and explained variance for each of the four factors

	Factor 1	Factor 2	Factor 3	Factor 4
Eigenvalue	4.605	3.124	1.721	0.987
% Total variance	38.37	26.03	14.36	8.24
Cumulative Eigenvalue	4.605	7.729	9.45	10.437
Cumulative %	38.373	64.405	78.751	86.976
Variable	Factor loadings			
$\text{Ln}(\chi)$	*0.934*	−0.150	0.258	−0.029
$\text{Ln}(\chi_{fd})$	*0.736*	0.530	0.085	−0.197
$\text{Ln}(M_s)$	*0.889*	−0.316	0.236	0.022
$\text{Ln}(M_{rs})$	*0.970*	−0.123	0.069	0.018
$\text{Ln}(\chi_{ARM})$	*0.857*	0.431	0.180	0.058
$\text{Ln}(B_c)$	−0.145	0.369	*−0.780*	0.091
$\text{Ln}(B_{cr})$	−0.129	−0.340	*−0.861*	0.036
$\text{Ln}(\text{MDF}_{ARM})$	−0.001	−0.045	−0.114	*0.967*
$\text{Ln}(\text{MDF}_{IRM})$	−0.267	−0.029	*−0.802*	0.052
Ln(ARM / SIRM)	−0.102	*0.879*	0.247	0.008
χ_{ARM}/χ	−0.077	*0.889*	−0.074	0.116
$\chi_{fd}\%$	0.068	*0.839*	−0.124	−0.207
Expl.Var	4.008	3.126	2.254	1.048

Varimax-normalized factor loadings of the magnetic variables are presented. Loadings >0.7 are indicated in italics.
From Jordanova, N., Jordanova, D., Petrov, P., 2016. Soil magnetic properties in Bulgaria at a national scale—Challenges and benefits. Glob. Planet. Change 137, 107–122, with permission from Elsevier.

explaining the variability in the database. Log-transformed and standardized by Z-score scaling variables with outliers removed were used as input data. The results indicate that four factors may explain 87% of the total variance (Table 9.5.2).

The first factor is responsible for 38% of the total variance, and major loadings in this factor are the concentration-dependent variables χ, M_s, M_{rs}, χ_{ARM}, and χ_{fd}. The second factor explains 26% of the total variance, and it is related to the grain-size proxies for SP and SD pedogenic fractions ($\chi_{fd}\%$, χ_{ARM}/χ, ARM/SIRM). The third factor explains 14% of the total variance originating from the variability of coercivities (B_c, B_{cr}, and MDF_{IRM}). Thus, as is expected as a general consideration, the combination of concentration, grain size, and coercivity of the mineral magnetic assemblages in the topsoils defines the variability of their magnetic signature.

9.5.2 CLUSTER ANALYSIS

Further clarification of the effects of geology and soil type on the spatial pattern of soil magnetic properties is obtained through k-means clustering using three independent variables, representing the ones with the highest loadings in the extracted factors (M_{rs}, χ_{ARM}/χ, and B_{cr}) together with the two categorical variables introduced—the "*soil type*" and the "*lithology of the parent material*." The same major soil groups are considered, as in Section 9.3—Chernozems, Phaeozems, Luvisols, Planosols, Cambisols, Vertisols, Fluvisols, and Leptosols. The parent material lithology is represented by the following classes—alluvial-delluvial materials, loess, sediments (limestones, marls, sandstones, conglomerates, aleurolites), granites, metamorphic rocks, and volcanic/volcano-sedimentary rocks. The results after using a 100-fold cross-validation algorithm indicate that the topsoils could be subdivided into five clusters. The samples in the first cluster are mainly Luvisols, developed on volcanic/volcano-sedimentary rocks. The rest of the Luvisols, developed on sedimentary, delluvial, and partly granitic parent materials, are separated in cluster 2. In contrast, the grouping of the samples in the third cluster occurs mainly according to the "*soil type*" category, including all Cambisols in the database and part of the Vertisols. The fourth cluster is very well separated and incorporates Chernozems and Phaeozems, developed on loess. The fifth cluster is variable and includes Luvisols, Planosols, and some Phaeozems, developed on sedimentary rocks. Consequently, two of the clusters are governed by the soil type (Cambisols and Chernozems), and the other three clusters are affected by the properties of the parent material (volcanic/volcanogenic sediments; weakly magnetic sediments, etc.).

9.6 SPATIAL DISTRIBUTION OF TOPSOIL MAGNETIC PROPERTIES

Spatial maps and analysis of the observed pattern of soil magnetism variability have been explored by the application of a kriging procedure for the construction of the interpolated maps of the magnetic parameters (original or Ln-transformed) resembling a normal distribution. This limited the number of magnetic characteristics to the following set: $Ln(\chi)$, $Ln(M_{rs})$, $Ln(M_s)$, $Ln(\chi_{fd})$, $Ln(\chi_{ARM})$, $\chi_{fd}\%$, χ_{ARM}/χ, χ_{ARM}/χ_{fd}, and ARM/SIRM.

The spatial variability of the different parameters has been estimated by constructing the experimental variograms, and their interpolation with standard functions has been carried out using nested models (McBratney and Webster, 1986). Further details on the applied variogram analysis can be found in Jordanova et al. (2016). Using the modeled variograms, a kriging method of interpolation was used to construct the interpolated maps. The obtained spatial distributions of the concentration-dependent magnetic parameters $Ln(\chi)$, $Ln(\chi_{fd})$, $Ln(\chi_{ARM})$, and $Ln(M_{rs})$ are depicted on Fig. 9.6.1. The mapped variabilities of mass-specific magnetic susceptibility and isothermal remanence correspond closely to the delineation of the main lithologies and units of the surface geology on the Bulgarian territory (Fig. 9.6.2). It reflects the role of lithologic magnetic fraction, inherited from the parent material on the bulk magnetic signature of the topsoils. On the other hand, frequency-dependent susceptibility and anhysteretic susceptibility do not reveal such systematic dependence on the parent rock type and composition (e.g., geology). The observed consistent tendency for a systematic increase of the χ_{fd} values toward a northeast direction reflects the dominance of the pedogenic SP fraction in the magnetic signal of the loess sediments from the low Danube area. The obtained minimum in the most southwest part of Bulgaria is related to the dominance of Luvisols developed on weakly magnetic metamorphic rocks from the Ograzden and western Rhodopes units and weakly magnetic alluvium along the Struma River Valley. The other magnetic parameter—anhysteretic susceptibility—shows a different lateral pattern, related to the content of stable SD

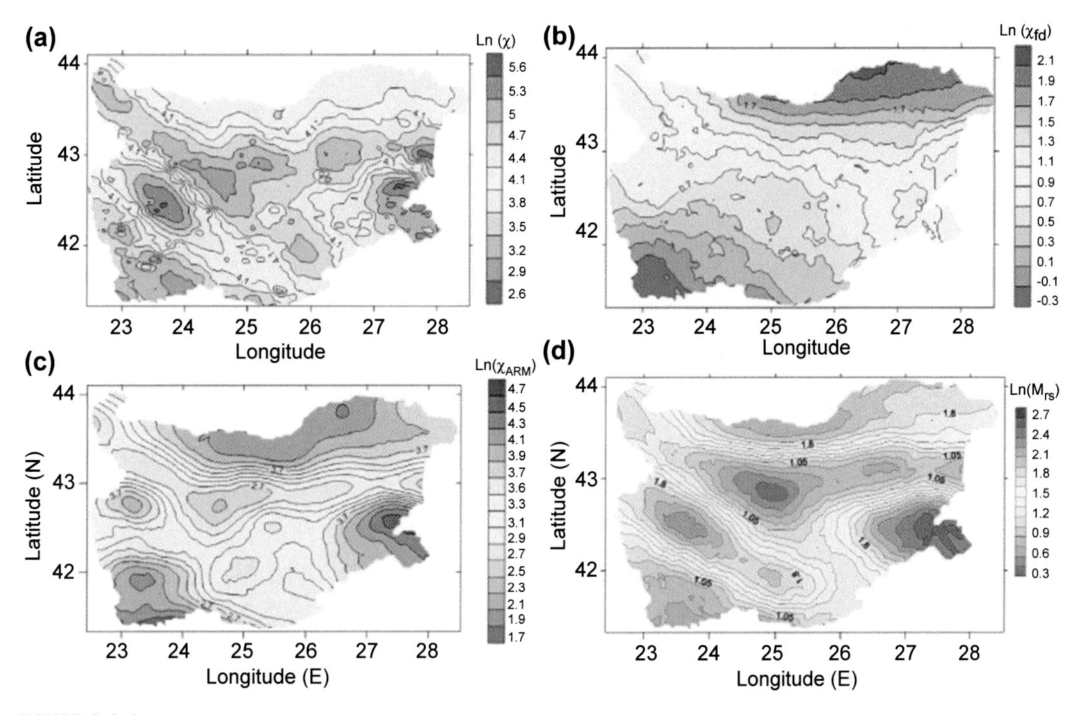

FIGURE 9.6.1

Interpolated maps prepared by using ordinary kriging of the Ln-transformed concentration-dependent magnetic parameters for the topsoils samples from Bulgaria: (a) Ln(χ), (b) Ln(χ_{fd}), (c) Ln(χ_{ARM}), and (d) Ln(M_{rs}).

ferrimagnetic fraction in the topsoils. The map of χ_{ARM} exhibits strong maxima in southeast and west Bulgaria, linked to the outcropping volcanic rocks, as well as a broad maxima in the loess area of north Bulgaria. Thus, having also in mind the relationships envisaged in Fig. 9.4.3, we infer that the stable SD fraction is related to the pedogenic enhancement in north Bulgaria and to the inherited lithogenic SD fraction, in the southeastern and western regions.

Kriged maps of the spatial pattern of the grain-size–sensitive parameters $\chi_{fd}\%$ and χ_{ARM}/χ are presented in Fig. 9.6.3a,b, and the point maps of the coercivities B_{cr} and MDF_{ARM} in Fig. 9.6.3c,d. Comparing the maps of $\chi_{fd}\%$ and χ_{ARM}/χ in north Bulgaria, where most of the sampling locations are in the loess area, we could infer an offset of the maxima for SD and SP contents, which suggests that the grain size of the pedogenic fraction is sensitive to the combination of the climate factors (temperature and precipitation). Looking at the distribution of mean coercivities of the ARM (Fig. 9.6.3d), it is evident that the Chernozems from north–northeast Bulgaria show systematically lower coercivities compared with the rest of the sampled locations, except for the coarse-grained granites from northwest Bulgaria. This is caused by the fraction of viscous SD-SP grains, which contribute significantly to the magnetic enhancement of the Chernozems (Jordanova et al., 2001). Most of the topsoils developed on sediments from central Bulgaria are characterized by MDF_{ARM} in the range of 15–20 mT, indicating the combination of pedogenic and lithogenic SD fractions. The coercivity of

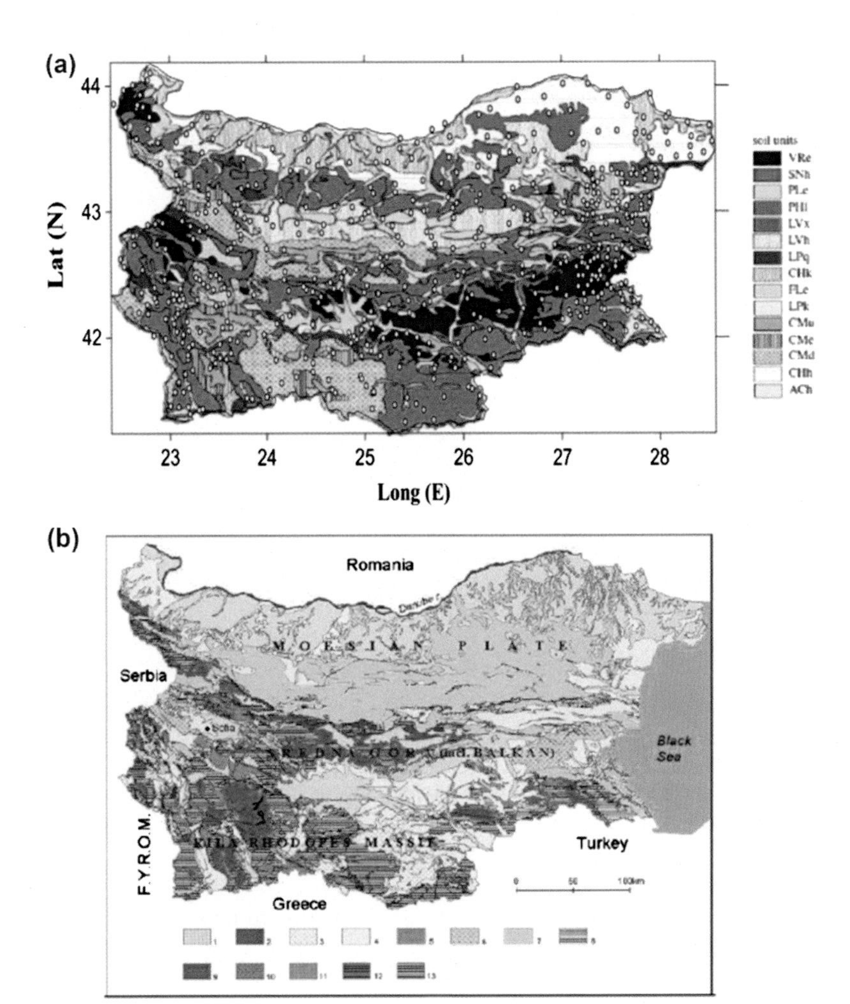

FIGURE 9.6.2

(a) Soil map of Bulgaria after Shishkov and Kolev (2014) with sampling sites of the present study indicated as *dots*. Abbreviations of the soil units are according to the FAO–UNESCO 1990 classification: VRe, Eutric Vertisols; SNh, Haplic Solonetzs; PLe, Eutric Planosols; PHl, Luvic Phaeozems; LVx, Chromic Luvisols; LVh, Haplic Luvisols; LPq, Lithic Leptosols; CHk, Calcic Chernozems; Fle, Eutric Fluvisols; LPk, Rendzic Leptosols; CMu, Humic Cambisols; CMe, Eutric Cambisols; CMd, Dystric Cambisols; CHh, Haplic Chernozems; ACh, Haplic Acrisols. (b) Simplified geological map of Bulgaria (compilation 1989). Legend: 1, sedimentary deposits, including loess sediments in North Bulgaria (Quaternary); 2, intrusive rocks (Neozoic); 3, volcanic and volcano-sedimentary rocks (Neozoic); 4, sedimentary rocks (Neozoic); 5, intrusive rocks (Mesozoic); 6, volcanic and volcano-sedimentary rocks (Mesozoic); 7, sedimentary rocks (Mesozoic); 8,metamorphic rocks (Mesozoic); 9, intrusive rocks (Paleozoic); 10, volcanic and volcano-sedimentary rocks (Mesozoic); 11, sedimentary rocks (Paleozoic); 12, metamorphic rocks (Paleozoic); 13, metamorphic rocks (pre-Paleozoic).

From Jordanova, N., Jordanova, D., Petrov, P., 2016. Soil magnetic properties in Bulgaria at a national scale—Challenges and benefits. Glob. Planet. Change 137, 107–122 with permission from the publisher.

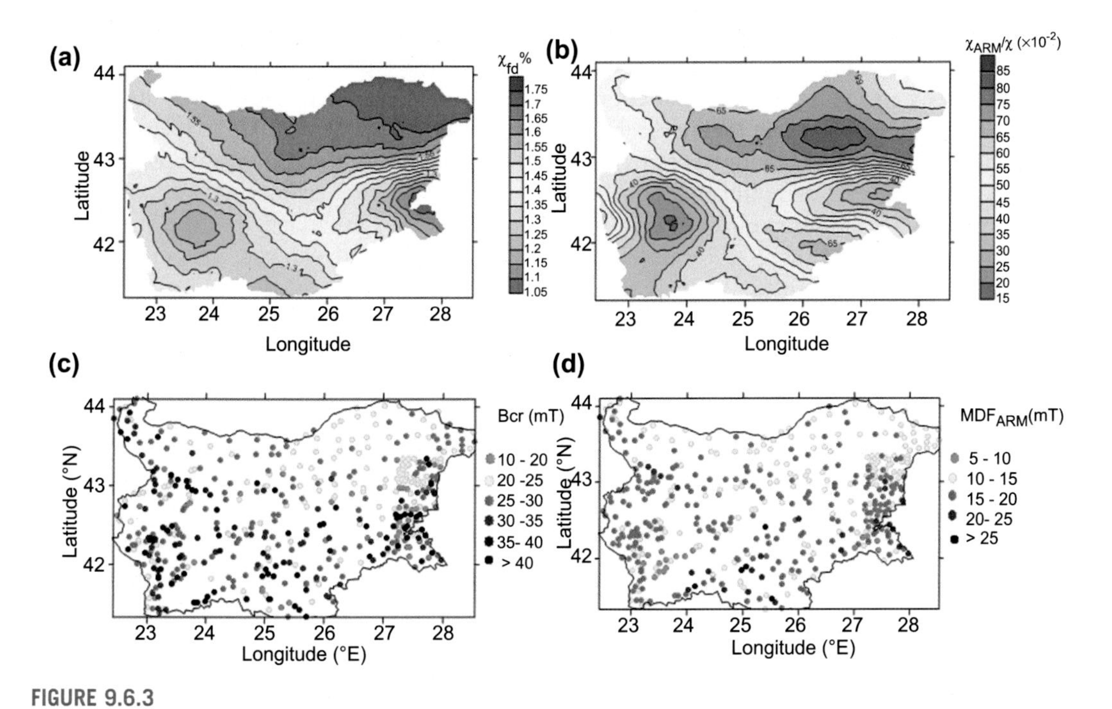

FIGURE 9.6.3

Interpolated maps prepared using ordinary kriging of spatial distribution of grain-size–sensitive magnetic parameters of the topsoils: (a) percent frequency-dependent magnetic susceptibility and (b) ratio of anhysteretic to low field magnetic susceptibility. Raw data for the coercivity of remanence (B_{cr}) (c) and the median destructive field of ARM (MDF_{ARM}) (d).

isothermal remanence B_{cr} shows slightly different spatial variation (Fig. 9.6.3c). The sensitivity of IRM coercivity to both grain size and internal domain structures and defects (Dunlop and Özdemir, 1997) is manifested through higher contrasts in the coercivity of topsoils from adjacent areas, such as very low and very high values for the locations in southeast Bulgaria, depending on the exact lithology—volcanic/volcano-sedimentary or intrusive outcropping rocks.

9.7 TOPSOIL MAGNETIC SUSCEPTIBILITY DATABASES: OVERVIEW AND DISCUSSION

Analyses of the magnetic signature of topsoil samples at national scales are rare, as far as a large effort of sampling and the following analyses are required. Examples of detailed mapping of the magnetic properties of topsoils from the available archives of National Soil surveys are the publications concerning topsoils from England and Wales (Dearing et al., 1996a; Blundell et al., 2009), Austria (Hanesch et al., 2007), Bosnia-Herzegovina (Hannam and Dearing, 2008), and France (Thiesson et al., 2012). Generally, magnetic susceptibility and frequency-dependent magnetic susceptibility are measured and interpreted in terms of major soil-forming factors, playing a role for the establishment of the soil magnetic signature. There is a common agreement on the leading role of lithology of the parent material for the observed lateral pattern of soil magnetic susceptibility. Recently, reports on the magnetic susceptibility of agricultural soils at a European scale from the GEMAS survey also revealed the major effect of geology on the magnetic signature of agricultural topsoils on a continental scale (Fabian and Reimann, 2014; Kuzina et al., 2015). Other sources of information on topsoil magnetic susceptibility are publications dealing with establishment of a "background magnetic susceptibility" value for soil pollution research using magnetic proxies (see Chapter 10). In a large study of forest soils from natural reserves in Poland, Lukasic et al. (2016) reported data on topsoil susceptibility in the forest soils from Poland, showing systematically low values due to the dominant weakly magnetic parent materials. On the other hand, rock magnetic studies on soils, developed on strongly magnetic igneous (basaltic) rocks, show high susceptibilities (Shenggao, 2000; Van Dam et al., 2008; Su et al., 2015). Numerous topsoil magnetic susceptibility data on Chernozem-like soils from the Chinese Loess Plateau have been published, but unfortunately without numerical mean (or median) values per region, as far as correlations with climatic variables (MAT, MAP) were sought (Song et al., 2014). Generally, the main part of the χ values fall in the range from 80 to 110×10^{-8} m^3/kg, which is consistent with the data for the Chernozem soils from Bulgaria with $\chi_{median} = 85.8 \times 10^{-8}$ m^3/kg. The effect of the parent rock lithology and soil type on topsoil magnetic susceptibility according to the published research is synthesized in Table 9.7.1.

Table 9.7.1 Compilation of the published data on magnetic susceptibility of topsoils, developed on various parent materials

Soil Type	Country	*N*	Parent material	χ_{median} ($\times 10^{-8}$ m^3/kg)	References
Chernozems	Bulgaria	55	Loess	92.1	This study
		25	Calcareous sediments	70.9	
		2	Alluvium	74.8	
Chernozems	Austria	169	Loess	77.0	Hanesch and Scholger (2005)
		37	Quaternary sediments	75.0	
		12	Calcareous alluvium	57.0	
		57	Old alluvium calcareous	64.0	
Cambisols	Bulgaria	35	Sediments	37.4	This study
		27	Metamorphic rocks (schists, gneiss, etc.)	69.3	
		10	Granites (Rila, Plana)	364.9	
		6	Alluvium	58.2	
Cambisols	Austria	100	Granites (Bohemia)	20.0	Hanesch and Scholger (2005)

Table 9.7.1 Compilation of the published data on magnetic susceptibility of topsoils, developed on various parent materials—cont'd

Soil Type	Country	*N*	Parent material	χ_{median} ($\times 10^{-8}$ m^3/kg)	References
		81	Schists, gneiss	22.0	
		25	Calcareous Alps	23.0	
		56	Central Alps	33.0	
		46	loess	39.0	
Cambisols	Poland	19	Various	18.9	Lukasik et al. (2016)
Cambisols/ surface water gleys/FAO correlation according to Keay et al. (2009)	England and Wales	1204	Various	27.0	Blundell et al. (2009)
Cambisols	Bosnia-Herzegovina	116	Various	29.6	Hannam and Dearing (2008)
Luvisols	Bulgaria	10	Alluvium	39.6	This study
		15	Granites Sakar, Sredna gora	41.7	
		21	Metamorphic rocks	37.4	
		83	Sediments	37.7	
		17	Volcanic and volcano-sedimentary rocks	331.9	
Luvisols	Bosnia-Herzegovina	24		210.8	Hannam and Dearing (2008)
Luvisols	Poland	5	Various	18.8	Lukasic et al. (2016)
Luvisols/brown soils/FAO correlation according to Keay et al. (2009)	England and Wales	1846	Various	51.0	Blundell et al. (2009)
Vertisols	Bulgaria	24	Pliocene sediments	31.6	This study
Vertisols	Bulgaria	25	Sedimentary and volcanic/ volcano-sedimentary from southeast Bulgaria	331.0	This study

N, *number of samples. Where published data do not contain information about the parent material, "various" is indicated in the table.*

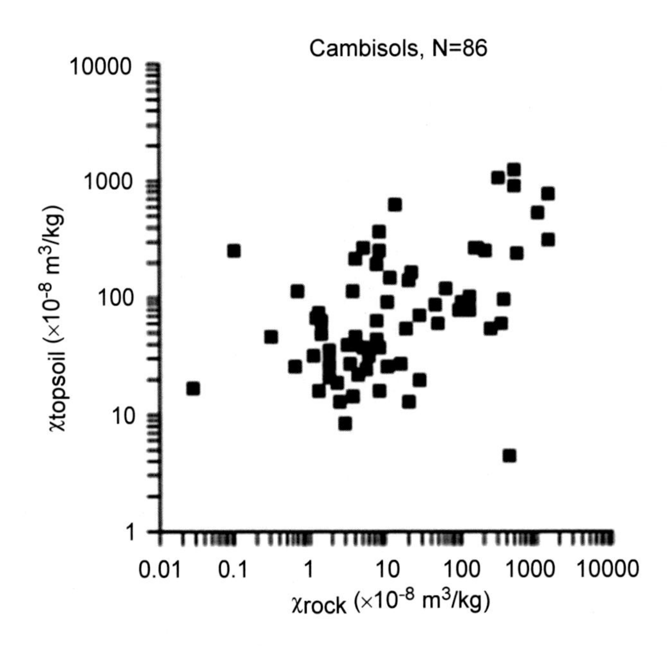

FIGURE 9.7.1

Linear relationship observed between topsoils' and parent rocks' magnetic susceptibilities for Cambisols from Bulgarian database.

It reveals the existence of high magnetic susceptibility of Chernozems developed on weakly magnetic loess and sediments, which again proves the high capacity of the loess material for weathering and neoformation of secondary strongly magnetic iron oxides under typical well-drained oxic environment. On the other hand, similar and sometimes higher magnetic susceptibilities are found in Cambisols. The latter, however, show a much larger variability, which originates from the various magnetic signals of the parent rock lithology. This is further visualized by the obtained linear dependence between topsoil magnetic susceptibility of Cambisols and the magnetic susceptibility of the parent rocks for the soil samples from the Bulgarian database (Fig. 9.7.1). A similar wide variability is observed for Cambisols from Bosnia-Herzegovina (Hannam and Dearing, 2008). Leptosols, the typical shallow soils, also exhibit large variability due to the same underlying factor (see Fig. 9.3.3a and Hannam and Dearing, 2008).

Thus, topsoil susceptibility mapping projects are very successful in revealing the primary role of the parent rock lithology in the development of soil magnetic susceptibility. This observation is grounded in the existing direct link between weathering of the parent material and secondary Fe oxide synthesis under various combinations of the other factors of soil formation (climate, vegetation, time, relief).

The effects of climate (through MAP, MAT, and effective evapotranspiration) are shown to be controversial depending on the scale (local, regional, or global), drainage class and the exact values of MAP (see discussion in Chapter 10).

REFERENCES

Benayas, J.M.R., Sanchez-Colomer, M., Escudero, A., 2004. Landscape- and field-scale control of spatial variation of soil properties in Mediterranean montane meadows. Biogeochemistry 69, 207–225.

Bian, Y., Ouyang, T., Zhu, Z., Huang, N., Wan, H., Li, M., 2014. Magnetic properties of agricultural soil in the Pearl River Delta, South China—Spatial distribution and influencing factor analysis. J. Appl. Geophys. 107, 36–44.

Blundell, À., Dearing, J.A., Boyle, J.F., Hannam, J.A., 2009. Controlling factors for the spatial variability of soil magnetic susceptibility across England and Wales. Earth Sci. Rev. 95, 158–188.

Chaparro, M., Nunez, H., Lirio, H., Gogorza, C., Sinito, A., 2007. Magnetic screening and heavy metal pollution studies in soils from Marambio Station, Antarctica. Antarct. Sci. 19 (3), 379–393.

Christov, B., Teoharov, M., 2008. Genetic and diagnostic peculiarities of soils, defined as eroded Chernozems. Soil Sci. Agrochem. Ecol. 2, 37–44 (inBulgarian).

Cornell, R., Schwertmann, U., 2003. The Iron Oxides. Structure, Properties, Reactions, Occurrence and Uses (Weinheim, New York).

Darnley, A., Bjorklund, A., Bolviken, B., Gustavsson, N., Koval, P., Plant, J., Steenfelt, A., Tauchid, M., Xuejing, X., Garrett, R., Hall, G., 1995. A Global Geochemical Database for Environmental and Resource Management. Recommendations for International Geochemical Mapping. Final Report of IGCP Project 259. UNESCO Publishing, ISBN 92-3-103085-X. Earth Sciences 19.

Dearing, J., Hay, K., Baban, S., Huddleston, A., Wellington, E., Loveland, P., 1996a. Magnetic susceptibility of soil: an evaluation of conflicting theories using a national data set. Geophys. J. Int. 127, 728–734.

Dearing, J.A., Dann, R.J.L., Hay, K., Lees, J.A., Loveland, P.J., Maher, B.A., O'Grady, K., 1996b. Frequency-dependent susceptibility measurements of environmental materials. Geophys. J. Int. 127, 228–240.

Dearing, J., Bird, P., Dann, R., Benjamin, S., 1997. Secondary ferrimagnetic minerals in Welsh soils: a comparison of mineral magnetic detection methods and implications for mineral formation. Geophys. J. Int. 130, 727–736.

Drees, L.R., Wilding, L.P., 1973. Elemental variability within a sampling unit. Soil Sci. Soc. Am. Proc. 37, 82–87.

Dunlop, D., Özdemir, Ö., 1997. Rock magnetism. Fundamentals and frontiers. In: Edwards, D. (Ed.), Cambridge Studies in Magnetism. Cambridge University Press.

Evans, M., Heller, F., 2003. Environmental Magnetism: Principles and Applications of Enviromagnetics. Academic Press, San Diego, CA.

Fabian, K., Reimann, C., 2014. GEMAS: a unique data set to define magnetic susceptibility variability of European agricultural soil. EGU General Assembly 2014. Geophys. Res. Abstr. 16, EGU2014–7892.

Frank, U., Nowaczyk, N.R., 2008. Mineral magnetic properties of artificial samples systematically mixed from haematite and magnetite. Geophys. J. Int. 175, 449–461.

Geiss, Ch, Zanner, W., 2006. How abundant is pedogenic magnetite? Abundance and grain size estimates for loessic soils based on rock magnetic analyses. J. Geophys. Res. 111 (B12) http://dx.doi.org/10.1029/2006JB004564.

Geiss, C.E., Egli, R., Zanner, C.W., 2008. Direct estimates of pedogenic magnetite as a tool to reconstruct past climates from buried soils. J. Geophys. Res. 113, B11102. http://dx.doi.org/10.1029/2008JB005669.

Georgiev, N., Henry, B., Jordanova, N., Jordanova, D., Naydenov, K., 2014. Emplacement and fabric-forming conditions of plutons from structural and magnetic fabric analysis: a case study of the Plana pluton (Central Bulgaria). Tectonophysics 629, 138–154.

Grönholm, T., Annila, A., 2007. Natural distribution. Math. Biosci. 210, 659–667.

Hanesch, M., Scholger, R., 2002. Mapping of heavy metal loadings in soils by means of magnetic susceptibility measurements. Environ. Geol. 42, 857–870.

Hanesch, M., Scholger, R., 2005. The influence of soil type on the magnetic susceptibility measured throughout soil profiles. Geophys. J. Int. 161, 50–55.

Hanesch, M., Rantitsch, G., Hemetsberger, S., Scholger, R., 2007. Lithological and pedological influences on the magnetic susceptibility of soil: their consideration in magnetic pollution mapping. Sci. Tot. Environ. 382, 351–363.

Hannam, J.A., Dearing, J.A., 2008. Mapping soil magnetic properties in Bosnia and Herzegovina for landmine clearance operations. Earth Planet. Sci. Lett. 274, 285–294.

Hunt, C.P., Moskowitz, B.M., Banerjee, S.K., 1995. Magnetic properties of rocks and minerals. Rock physics and phase relations. A handbook of physical constants. AGU Ref. Shelf 3, 189–204.

Jackson, M., Swanson-Hysell, N.L., 2012. Rock magnetism of remagnetized carbonate rocks: another look. In: Elmore, R.D., Muxworthy, A.R., Aldana, M.M., Mena, M. (Eds.), Remagnetization and Chemical Alteration of Sedimentary Rocks, 371. The Geological Society of London. Geological Society, London, Special Publications.

Jeleńska, M., Hasso-Agopsowicz, A., Kopcewicz, B., 2010. Thermally induced transformation of magnetic minerals in soil based on rock magnetic study and Mössbauer analysis. Phys. Earth Planet. Inter. 179 (3–4), 164–177.

Jenny, H., 1941. Factors of soil formation. In: Amundson, R. (Ed.), A System of Quantitative Pedology. Dover Publication Inc., New York (edition 1994).

Jordanova, D., Yancheva, G., Gigov, V., 2001. Viscous magnetization of loess/palaeosol samples from Bulgaria. Earth Planets and Space 53, 169–180.

Jordanova, D., Jordanova, N., Petrov, P., Tsacheva, T., 2010. Soil development of three Chernozem-like profiles from North Bulgaria revealed by magnetic studies. CATENA 83 (2–3), 158–169.

Jordanova, N., Jordanova, D., Petrov, P., 2016. Soil magnetic properties in Bulgaria at a national scale—Challenges and benefits. Glob. Planet. Change 137, 107–122.

Kapicka, A., Petrovsky, E., Fialova, H., Podrazsky, V., Dvorak, I., 2008. High resolution mapping of anthropogenic pollution in the Giant mountains National park using soil magnetometry. Stud. Geophys. Geod 52, 271–284.

Keay, C.A., Hallett, S.H., Farewell, T.S., Rayner, A.P., Jones, R.J.A., 2009. Moving the national soil database for England and Wales (LandIS) towards INSPIRE compliance. Int. J. Spat. Data Infrastruct. Res. 4, 134–155.

Koinov, V., Kabakchiev, I., Boneva, K., 1998. Soil Atlas of Bulgaria. Zemizdat, Sofia (in Bulgarian).

Kuzina, D., Kosareva, L., Fattakhova, L., Fabian, K., Nourgaliev, D., Reimann, C., 2015. GEMAS: mineral magnetic properties of European agricultural soils. EGU General Assembly 2015. Geophys. Res. Abstr. 17, EGU2015–10756.

Lawrence, G.B., Fernandez, I.J., Richter, D.D., Ross, D.S., Hazlett, P.W., Bailey, S.W., Ouimet, R., Warby, R.A.F., Johnson, A.H., Lin, H., Kaste, J.M., Lapenis, A.G., Sullivan, T.J., 2013. Measuring environmental change in forest ecosystems by repeated soil sampling: a North American perspective. J. Environ. Qual. 42, 623–639.

Lin, H., Wheeler, D., Bell, J., Wilding, L., 2005. Assessment of soil spatial variability at multiple scales. Ecol. Model. 182, 271–290.

Liu, Q., Banerjee, S.K., Jackson, M.J., Maher, B.A., Pan, Y., Zhu, R., Deng, C., Chen, F., 2004. Grain sizes of susceptibility and anhysteretic remanent magnetization carriers in Chinese loess/paleosol sequences. J. Geophys. Res. 109, B03101. http://dx.doi.org/10.1029/2003 JB002747.

Liu, Q.S., Deng, Ch, Yu, Y., Torrent, J., Jackson, M., Banerjee, S., Zhu, R., 2005. Temperature dependence of magnetic susceptibility in an argon environment: implications for pedogenesis of Chinese loess/palaeosols. Geophys. J. Int. 161, 102–112.

Łukasik, A., Magiera, T., Lasota, J., Błońska, E., 2016. Background value of magnetic susceptibility in forest topsoil: assessment on the basis of studies conducted in forest preserves of Poland. Geoderma 264, 140–149.

Magiera, T., Strzyszcz, Z., Rachwal, M., 2007. Mapping particulate pollution loads using soil magnetometry in urban forests in the Upper Silesia Industrial Region, Poland. For. Ecol. Manage. 248, 36–42.

Maher, B., 1986. Characterization of soils by mineral magnetic measurements. Phys. Earth Planet. Inter. 42, 76–92.

Maher, B., 1988. Magnetic properties of some synthetic sub-micron magnetites. Geophys. J. R. Astr. Soc. 94, 83–96.

Maher, B., 1998. Magnetic properties of modern soils and Quaternary loessic paleosols: paleoclimatic implications. Palaeogeogr. Palaeoclimat. Palaeoecol. 137, 25–54.

McBratney, A.B., Webster, R., 1986. Choosing functions for semi-variograms of soil properties and fitting them to sampling estimates. J. Soil Sci. 37, 617–639.

Murad, E., Wagner, U., 1998. Clays and clay minerals: the firing process. Hyperfine Interact. 117, 337–356.

Murad, E., Wagner, U., Wagner, F.E., Hausler, W., 2002. The thermal reactions of montmorillonite: a Mossbauer study. Clay Miner. 37, 583–590.

Ninov, N., 2000. Taxonomic list of soils in Bulgaria according to FAO WRB. Probl. Geogr. 1–4, 38–43 (in Bulgarian).

Petrovský, E., Kapicka, A., 2006. On determination of the Curie point from thermomagnetic curves. J. Geophys. Res. 111, B12S27.

Sarris, A., Kokinou, E., Aidona, E., Kallithrakas-Kontos, N., Koulouridakis, P., Kakoulaki, G., Droulia, K., Damianovits, O., 2009. Environmental study for pollution in the area of Megalopolis power plant (Peloponnesos, Greece). Environ. Geol. 58, 1769–1783.

Smeck, N.E., Runge, E.C.A., Mackintosh, E.E., 1983. Dynamics and genetic modelling of soil systems. In: Wilding, L.P., Smeck, N.E., Hall, G.F. (Eds.), Pedogenesis and Soil Taxonomy. I. Concepts and Interactions. Elsevier, Netherlands, pp. 51–81.

Song, Y., Hao, Q., Ge, J., Zhao, D., Zhang, Y., Li, Q., Zuo, X., Lü, Y., Wang, P., 2014. Quantitative relationships between magnetic enhancement of modern soils and climatic variables over the Chinese Loess Plateau. Quat. Int. 334–335, 119–131.

Shenggao, L., 2000. Lithological factors affecting magnetic susceptibility of subtropical soils, Zhejiang Province, China. CATENA 40, 359–373.

Shishkov, T., Kolev, N., 2014. The soils of Bulgaria. In: World Soils Book Series. Springer. http://dx.doi.org/10.1007/978-94-007-7784-2.

Su, N., Yang, S.-Y., Wang, X.-D., Bi, L., Yang, C.-F., 2015. Magnetic parameters indicate the intensity of chemical weathering developed on igneous rocks in China. CATENA 133, 328–341.

Tanov, et al., 1956. Soil Map of Peoples' Republic of Bulgaria. 1:200 000. Central Bureau of Geodesy and Cartography. Sofia. (in Bulgarian).

Thiesson, J., Boulonne, L., Buvat, S., Jolivet, C., Ortolland, B., Saby, N., 2012. Magnetic properties of the French soil monitoring network: first results. In: Near Surface Geoscience 2012 – 18th European Meeting of Environmental and Engineering Geophysics Paris, France, 3–5 September 2012; Abstract Book.

Tóth, G., Jones, A., Montanarella, L., 2013. Lucas topsoil survey – methodology, data and results. JRC Tech. Rep. http://dx.doi.org/10.2788/97922.

Van Dam, R.L., Harrison, J.B.J., Hirschfeld, D.A., Meglich, T.M., Li, Y., North, R.E., 2008. Mineralogy and magnetic properties of basaltic Substrate soils: Kaho'olawe and Big Island. Hawaii Soil Sci. Soc. Am. J. 72 (1), 244–257.

Wang, R., Løvlie, R., 2008. SP-grain production during thermal demagnetization of some Chinese loess/palaeosol. Geophys. J. Int. 172, 504–512.

Wang, X.S., 2013. Magnetic properties and heavy metal pollution of soils in the vicinity of a cement plant, Xuzhou (China). J. Appl. Geophys. 98, 73–78.

Worm, H.-U., 1998. On the superparamagnetic – stable single domain transition for magnetite, and frequency dependence of susceptibility. Geophys. J. Int. 133, 201–206.

CHAPTER

APPLICATIONS OF SOIL MAGNETISM

10

10.1 LINK BETWEEN SOIL MAGNETIC PROPERTIES AND CLIMATE PARAMETERS

Since the pioneering work by Heller and Liu (1986), recognizing the potential of magnetic susceptibility variations along loess–paleosol sequences at the Chinese loess plateau for recording climate variations during the Quaternary, there have been numerous publications on this subject. Several reviews give detailed overviews of the major stages and developments of the ideas and results concerning the link between the soil's magnetic properties and climate variables (Maher and Thompson, 1995; Liu et al., 2007; Balsam et al., 2011; Orgeira et al., 2011; Maher, 2011; Maxbauer et al., 2016). Here, we briefly summarize the major points, while more detailed information can be found in the cited works.

The plausibility of using magnetic signal of paleosols for paleoclimate reconstructions is based on the observed relationships between the magnetic characteristics and climate variables (temperature, precipitation, evapotranspiration) for modern soils that are developed under known conditions. Thus, establishing robust relationships valid for a wide spatial coverage and soil varieties is an essential point. Special attention is given to the climate–magnetism studies of recent soils developed on loess from the Chinese loess plateau as a reflection of Asian monsoon dynamics and history (Maher, 1998; Bloemendal and Liu, 2005; Song et al., 2014; Nie et al., 2014). The "pedogenic magnetic enhancement" χ_{pedo} is defined as the magnetic susceptibility of the B horizon of the soil minus the magnetic susceptibility of the parent C horizon (Maher et al., 1994). The observed log-linear dependence between mean annual precipitation (MAP) and χ_{pedo} differ for modern soils from different regions (Fig. 10.1.1).

The various trends observed are tentatively explained by the effect of the soil moisture regime, which is determined by the effective evapotranspiration. Orgeira et al. (2011) applied a correction for this factor and showed that all previous scatter of different regional data sets is greatly reduced.

However, the obtained log linear MAP–χ_{pedo} relationship holds up to a certain threshold value of the MAP; afterward, a reverse relationship (a decrease in χ_{pedo} with a further increase in MAP) is established. Such phenomena are also observed in other soil properties, known as "pedogenic thresholds" (Chadwick and Chorover, 2001). Balsam et al. (2011) reviewed the existing data sets in relation to the exact MAP value, serving as a threshold for the dependence between MAP and $\chi_{pedo.}$

Long et al. (2011) suggested a new model for the observed climate–pedogenic magnetic signal, which considers the combined effects of temperature and precipitation on the amounts of pedogenic ferrimagnetic iron oxides. The underlying supposition is that hematite (Hm) and goethite (Gt) in the soils are mainly of pedogenic origin. The authors consider changes in the amount of dithionite-

Soil Magnetism. http://dx.doi.org/10.1016/B978-0-12-809239-2.00010-3

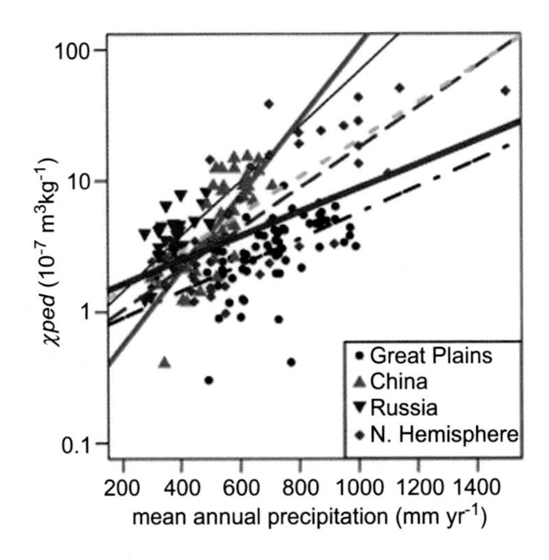

FIGURE 10.1.1

Relationship between pedogenic susceptibility (χ_{ped}) and mean annual precipitation. All lines represent simple linear regression models. All models were statistically significant ($p < .05$). *Thin blue line* = Russia ($R^2 = 0.35$), *double dashed black line* = Great Plains ($R^2 = 0.25$), *thick green line* = China ($R^2 = 0.61$), *coarse-dashed purple line* = Northern Hemisphere ($R^2 = 0.60$), *fine-dashed orange line* = Russia, China, and Northern Hemisphere data combined ($R^2 = 0.52$), *thick solid red line* = all data ($R^2 = 0.24$).

From Maxbauer, D.P., Feinberg, J.M., Fox, D.L., 2016. Magnetic mineral assemblages in soils and paleosols as the basis for paleoprecipitation proxies: a review of magnetic methods and challenges. Earth-Science Rev. 155, 28–48 with permission from Elsevier. Great Plains data from Geiss, C.E., Egli, R., Zanner, C.W., 2008. Direct estimates of pedogenic magnetite as a tool to reconstruct past climates from buried soils. J. Geophys. Res. 113, B11102 and Geiss, C.E., Zanner, C.W., 2007. Sediment magnetic signature of climate in modern loessic soils from the Great Plains. Quat. Int. 162–163, 97–110; China data from Porter, S.C., Hallet, B., Wu, X., An, Z., 2001. Dependence of near-surface magnetic susceptibility on dust accumulation rate and precipitation on the Chinese Loess Plateau. Quatern. Res., 55, 271–283 and Maher B.A., Thompson R., Zhou L.P., 1994. Spatial and temporal reconstructions of changes in the Asian palaeomonsoon: a new mineral magnetic approach. Earth Planet. Sci. Lett. 125, 461–471; Russia data from Maher, B.A., Alekseev, A., Alekseeva, T., 2002. Variation of soil magnetism across the Russian steppe: its significance for use of soil magnetism as a palaeorainfall proxy. Quat. Sci. Rev., 21, 1571–1576 and Alekseev, A.O., Alekseeva, T.V., Maher, B.A., 2003. Magnetic properties and mineralogy of iron compounds in steppe soils. Eurasian Soil Sci., 36, 59–70, and Northern Hemisphere data compiled in Maher, B.A., Thompson, R., 1995. Paleorainfall reconstructions from pedogenic magnetic susceptibility variations in the Chinese loess and paleosols. Quat. Int. 44, 383–391.

extractable iron (Fe_d) as reflecting the weathering intensity under the combined influence of temperature and precipitation. Thus, Fe_d increases with increasing mean annual temperature (MAT) and mean annual precipitation (MAP). On the other hand, the ratio Hm/(Hm + Gt) changes in an opposite manner compared with Fe_d (the ratio decreases with a decrease in temperature and an increase in precipitation). The relative changes in the two parameters depending on the particular combination of MAP and MAT defines the position of the precipitation threshold, above which magnetic enhancement starts to decrease with a further increase in precipitation. Via this model, Long et al. (2011) explain the various MAP thresholds obtained for data sets from different climate regimes (temperate, subtropical, tropical). The effects of the seasonal distribution of the MAP per location are considered essential for the absolute amount of pedogenic magnetic enhancement in temperate and tropical areas (Balsam et al., 2011). This is related to the establishment of most suitable microenvironments in the soil pores for the synthesis of magnetite (maghemite), which require frequent changes in oxidative–reductive conditions (Taylor et al., 1987) (see Chapter 1). A further statistical evaluation of the uncertainty of paleoprecipitation estimates based on magnetic susceptibility enhancement is presented in Maher and Posolo (2013) and Heslop and Roberts (2013).

The second group of magnetic proxies for paleoprecipitation considers ratios of different magnetic remanences, acquired in the laboratory. Geiss et al. (2008) propose as the most suitable parameter the ratio χ_{arm}/isothermal remanent magnetization $(IRM)_{100mT}$ (χ_{ARM}/IRM_{100mT}). The authors show that the pedogenic magnetic component in the loessic soils from the Midwestern United States possesses uniform values of this ratio, which are indicative of the presence of maghemites or magnetites of a single-domain (SD) grain size. The ratio of the magnetic susceptibility of the B horizon to the magnetic susceptibility of the C horizon is also considered suitable as paleo-rainfall proxy in cases where variations in the mineralogy of the parent material significantly influence the magnetic signature of the soils (Geiss and Zanner, 2007). Using these approaches, differences still exist between the precipitation–magnetic proxies relationships for different data sets (Geiss et al., 2008), which are tentatively ascribed to the differences in effective soil moisture, which controls the reaction rates in the soil's microenvironment. The model of magnetic enhancement proxy (MEP), which considers the effect of soil moisture on the production rate of the ferrimagnetic pedogenic fraction, is extensively described in Orgeira et al. (2011).

The third group of paleoprecipitation proxies using magnetic properties of soils proposes to use the estimates of hematite and goethite content as sensitive indicators of paleoenvironmental conditions. It is well known that hematite and goethite contents (as a weight %) are much higher than those of magnetite (maghemite) and are sensitive to the combination of temperature and precipitation, defining different climate regimes (Cornell and Schwertmann, 2003). As recently proposed by Hayland et al. (2015), the paleoprecipitation proxy uses the Gt/Hm ratio as a proxy for changes in precipitation values from 200 to 3000 mm/year for modern soils from different locations around the world. As far as goethite and hematite are the most stable iron oxide minerals over geological timescales, and the established proxy covers a wider MAP interval, this new approach is promising. However, significant drawbacks still exist in its application, since the use of the magnetic unmixing of IRM acquisition curves for estimating Hmt and Gt content supposes that all the pedogenic Hm and Gt are carrying remanence; that is, they all have grain sizes larger than the supraparamagnetic (SP) threshold. This is in conflict with the widely observed enhancement of soils from Mediterranean and tropical areas with SP hematite (Liu et al., 2010; Long et al., 2015; Li et al., 2011; Jordanova et al., 2013a). Further analysis can be found in Maxbauer et al. (2016). Another promising proxy parameter is the ratio χ_{fd}/HIRM, proposed by Liu et al. (2013) in a study of 10 modern soils in China spanning a MAP range of 300–1000 mm/year. The strong positive correlation between χ_{fd}/HIRM and MAP is explained by the change in the relative amount of the ferrimagnetic fraction (giving a contribution to χ_{fd}) and the antiferromagnetic (hematite + goethite) fraction contributing to HIRM.

10.2 EXPERIMENTAL DETERMINATION OF SOIL REDISTRIBUTION AND EROSION PATTERN IN AGRICULTURAL LANDS

Soil erosion causes a severe loss of biodiversity, soil fertility, and plant biomass productivity, along with indirect damage to the environment such as global warming and negative human health effects from the soil dust particulates (Pimentel, 2006). Agricultural activities increase the vulnerability of the soil to lateral displacements through tillage operations (Govers et al., 1999). The major natural phenomena triggering erosion are water, rain, and wind. Soil attributes such as mechanical composition, organic matter content, and water permeability, as well as external factors including climate conditions and the length and slope of the field, define the vulnerability of a field to soil erosion. Classic approaches for estimation and modeling of soil erosion and redistribution (^{137}Cs, empirical models such as the distributed RUSLE model) pose drawbacks because of the lack of site- and scale-specific input parameters and restriction in the temporal integration of the signal in case of radionuclide methods (Mabit et al., 2008; Vieira and Dabney, 2009, 2011). An experimental approach to soil erosion estimates is the use of various tracers (steel, magnetite, fly ash, aluminium, colored carbonate, or plastic grains) (Guzman et al., 2010; Logsdon, 2013; Olson et al., 2013). One shortcoming regarding this approach is the use of markers with a fixed grain size, usually several millimeters, while erosion rates depend on many factors, among which is the grain-size distribution of the original soil (de Lima et al., 2011). An alternative approach is proposed by Armstrong et al. (2012), who tested the efficiency and sensitivity of the "natural" tracer represented by heated soil (bulk and different size fractions). This heated soil material has significantly enhanced magnetic parameters, and when mixed with the original soil material, it provides the possibility of tracing the redistribution of the tracer by measuring the magnetic properties after simulated rain events.

The widely observed phenomenon of the magnetic susceptibility enhancement of soils has long been recognized as a suitable tool for mapping and delineation of soil loss and redistribution due to the inherent variability of magnetic susceptibility with depth (Dearing et al., 1986; de Jong et al., 1998). Royall (2001) proposed a tillage homogenization model for predicting the magnetic signal of the soil plow horizon and its subsequent erosion. In the next publications (Royall, 2004, 2007), the author introduced the effects of grain size and natural soil variability into the model. An additional magnetic parameter (the ratio of saturation isothermal remance to magnetic susceptibility (SIRM/χ) was used to eliminate the effects of twin values and outliers. Further improvement was achieved by using several undisturbed soil profiles for model calibration, and the results have been compared with the output of an independent RUSLE model (Royall, 2007). Using the simple tillage homogenization model proposed by Royall (2001), Jordanova et al. (2014) estimated soil loss and redistribution as a result of agricultural practices in an experimental area covered by Chernozems. Spatial variations of the topsoil magnetic parameters on the sloping land under agricultural use show a strong relation to the particular topographic features (summit, shoulder, backslope). Fig. 10.2.1 presents the obtained spatial pattern of mass-specific susceptibility, frequency-dependent susceptibility, and isothermal remanence of the topsoil samples along with the topography of the study area.

The minimum magnetic susceptibility values are found at the backslope, while maximum values appear at the summit and the margin between the shoulder and the backslope. The lower susceptibilities at Y-sampling lines 20–21 correspond to more abrupt changes in the slope at that position. The spatial distribution of topsoil frequency-dependent susceptibility (Fig. 10.2.1c) possesses higher variability and a number of small-scale structures. Such a funnel-like structure is delineated by a systematically lower χ_{fd} centered along the Y-line 9 and extended downward. Such a topographic feature is not recognized in the field, suggesting that the observed χ_{fd} pattern reflects the mechanical sorting of the ploughed soil material caused by water flow. This flow obviously affected the finest (clay) fraction, which also contains the superparamagnetic strongly magnetic fraction contributing to χ_{fd} (Mullins and Tite, 1973). The surface map of SIRM shown on Fig. 10.2.1d reveals a similar

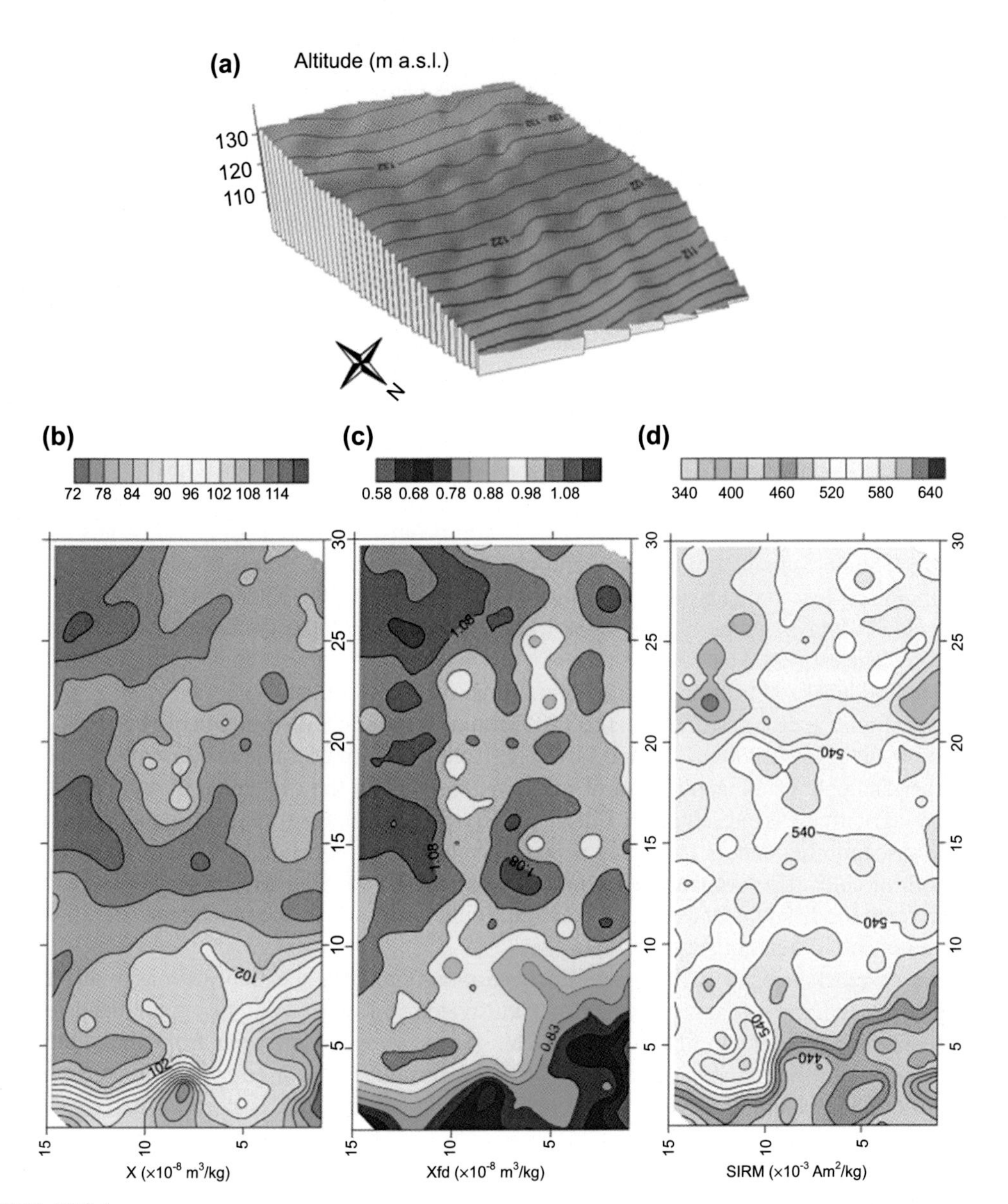

FIGURE 10.2.1

Spatial pattern of magnetic parameters of topsoils in a study site under agricultural use: (a) three-dimensional topography of the study area; (b) kriged map of mass-specific magnetic susceptibility of the topsoil samples (χ); (c) kriged map of frequency-dependent magnetic susceptibility (χ_{fd}); (d) kriged map of isothermal remanence (SIRM). The numbers on X- and Y-axes indicate the number of the sampling lines.

From Jordanova, D., Jordanova, N., Petrov, P., 2014. Pattern of cumulative soil erosion and redistribution pinpointed through magnetic signature of Chernozem soils. CATENA 120, 46–56 with permission from Elsevier.

pattern; however, the maxima are situated at lower elevations compared with those of susceptibility. This lateral shift again emphasizes the sorting of the magnetic and mechanical fractions during soil redistribution as far as grain-size dependence of SIRM for magnetites shows a maxima at 0.2- to 0.5-μm grain sizes (Dunlop and Özdemir, 1997). Consequently, χ_{fd} will mainly represent the redistribution of the fine clay fraction, while SIRM preferentially represents that of the fine silt fraction. This is in accordance with the findings that erosion is a grain-size–sensitive process (de Lima et al., 2011; Shi et al., 2012).

Using the approach by Royall (2001) for the construction of predictive curves for soil loss due to erosion, two estimates based on χ and χ_{fd} depth variations along nondisturbed reference depth profiles were carried out. The estimated soil redistribution within the experimental field is obtained. Fig. 10.2.2 shows the estimated depth of cumulative soil erosion using χ_{fd}.

Matching the two maps of the estimated depth of cumulative soil erosion using a χ_{fd} proxy and the generated map of the terrain slope (using Surfer 11 software) shows that the minimum soil loss is typical for local slopes of about 3–5°, occurring generally at higher elevations. The highest soil loss is observed at the backslope positions with high slope values (up to 16°), obviously reflecting the strong removal of soil material at these locations. This lateral pattern of soil redistribution is most probably related to the combined effect of tillage operations and water erosion. Tillage erosion is characterized by maximum soil loss at upper convex positions (De Alba et al., 2004; Van Oost et al., 2000), while water erosion is highest at the highest slope locations (Li et al., 2008; Quine et al., 1994; Van Oost et al., 2006). The estimated cumulative soil loss, calculated as a volume of the body, delineated by the surface and the level $z = 0$ from Fig. 10.2.2 is estimated at 394.1 t/ha.

Using a similar approach, Kapička et al. (2015) studied the soil redistribution of two areas covered by different soil types (Chernozems and Luvisols) from Southern Moravia and Central Bohemia.

Another example of the application of magnetic methods in the evaluation of soil redistribution caused by agriculture is presented in Liu et al. (2015). The authors compare depth variations of magnetic susceptibility and its frequency dependence along soil cores of 1-m length taken from two transects from cultivated farmland and a reforested area. The variations of the magnetic parameters along the depth for different landscape positions at a cultivated transect are much more significant than those at a reforested transect (Fig. 10.2.3).

The observed changes of $\chi_{fd}\%$ in a vertical cross-section along the slopes strongly differentiate between cultivated and reforested areas. In the cultivated soil, $\chi_{fd}\%$ is higher at the backslope and footslope, indicating that the finest SP magnetic grains were redistributed from the summit and shoulder toward the backslope and footslope. Similar results are reported by Mokhtari Karchegani et al. (2011) and Ayoubi et al. (2012) for cultivated and forested soils developed on calcareous material.

A detailed rock magnetic investigation of Calcisols developed on Tertiary clays and sandstones in Spain for revealing soil redistribution caused by cultivation practices and water erosion is presented in Quijano et al. (2014). The authors observed a strong correlation between the magnetic susceptibility (in situ field and laboratory measured) and the number of variables such as elevation, organic matter, content of clay and silt, slope, and negative correlation with the content of $CaCO_3$. They show that the spatial variability of the magnetic signal of the fine mechanical fraction is a good indicator for revealing the pattern of soil redistribution in the agroecosystems of the Mediterranean mountain environment.

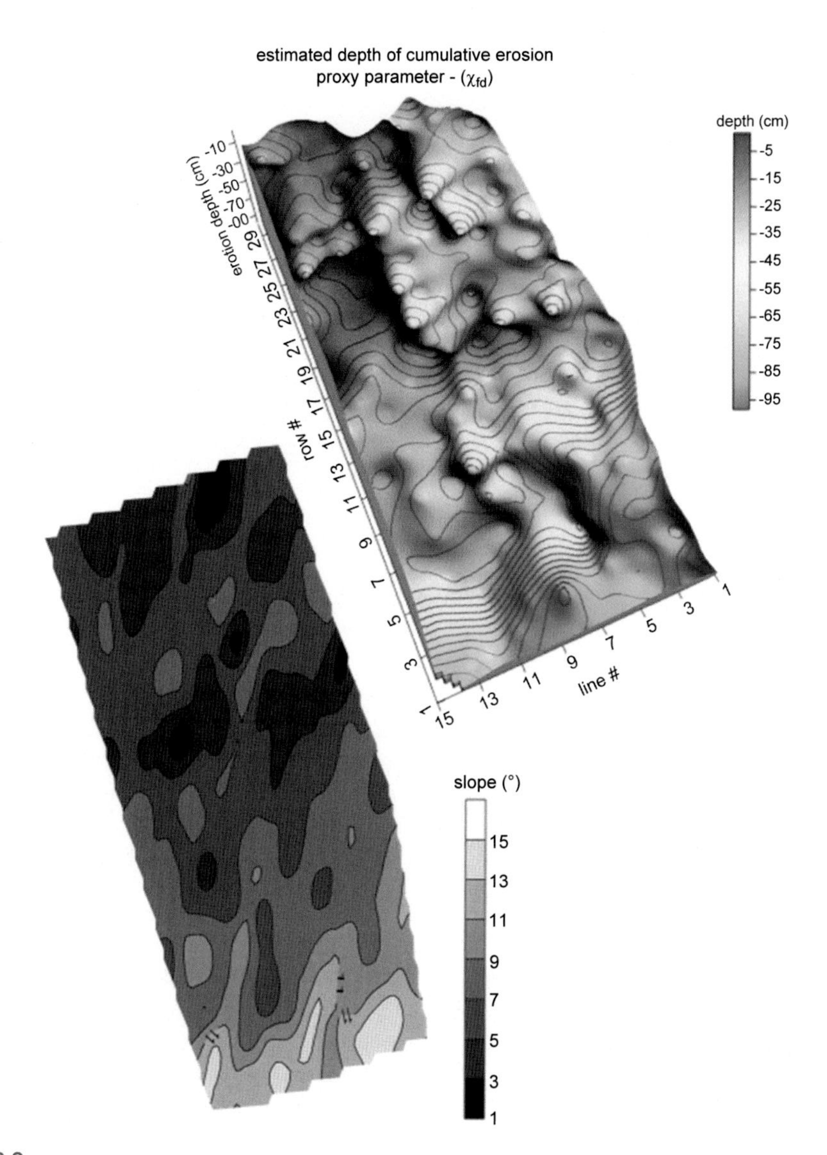

FIGURE 10.2.2

Distributed data of the estimated depth of cumulative soil loss obtained through comparison of surface frequency-dependent magnetic susceptibility of the topsoils with a modeled X_{fd} curve. Contour map of the terrain slope is shown on the left.

From Jordanova, D., Jordanova, N., Petrov, P., 2014. Pattern of cumulative soil erosion and redistribution pinpointed through magnetic signature of Chernozem soils. CATENA 120, 46–56 with permission from Elsevier.

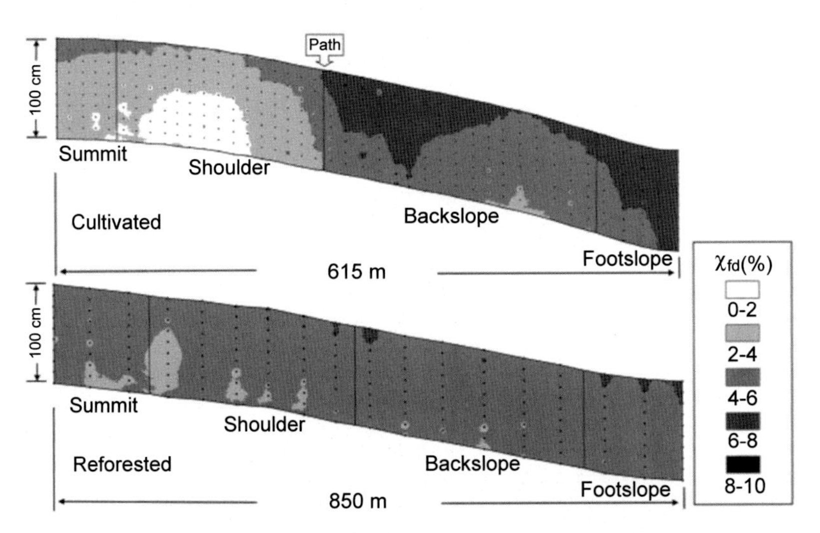

FIGURE 10.2.3

Downslope transect of the surface soil patterns of X_{fd}% for 0- to 100-cm depth in cultivated and reforested soils.

From Liu, L., Zhang, K., Zhang, Z., Qiu, Q., 2015. Identifying soil redistribution patterns by magnetic susceptibility on the black soil farmland in Northeast China. CATENA 129, 103–111 with permission from Elsevier.

10.3 MAGNETIC PROXIES FOR RELATIVE CHANGES IN THE FATE OF TOTAL CARBON AND NITROGEN IN DIFFERENT SOIL TYPES

Global cycling of carbon (C), phosphorous (P), and nitrogen (N) and their interactions play an essential role in the Earth's climate system and global elements cycling (Thornton et al., 2009; Schlesinger and Bernhardt, 2013; Xu et al., 2013; Achat et al., 2016). Phosphorous (P) is an essential nutrition element because of the relatively large amounts of P required for optimal plant growth. Phosphorous, which is present in different parent materials, is released by weathering and enters the biogeochemical cycle (Molina, 2009). Part of the P becomes bioavailable when it is assimilated by plants and microorganisms, while part of P adsorbs or precipitates as inorganic phosphates (Achat et al., 2016). Nitrogen in soils is derived from the decomposition of the plant litter and the organic matter (Thornton et al., 2009; Achat et al., 2016) and thus is intrinsically linked with the organic carbon (C_{org}) cycle. Besides, Achat et al. (2016) prove the major role of Al–Fe oxides and C_{org} in controlling the availability of soil inorganic P and thus the need to incorporate the P cycle into the global C–N cycle.

Iron (oxy)hydroxides are widespread in soils and control the fate and transport of many nutrients and contaminants via adsorption–desorption and oxidation–reduction (Borch and Fendorf, 2008 and references therein). Being highly redox sensitive, Fe oxides in soils are strongly influenced by changes in the oxygen availability in the different soil horizons. As discussed in the other chapters, Fe(III) (hydr)oxides can undergo reductive dissolution in anaerobic environments, resulting in a release of Fe(II), which may remain in solution, precipitate as Fe(II)-bearing or mixed valency iron phases, or be oxidized into ferrihydrite (Cornell and Schwertmann, 2003; Borch and Fendorf, 2008). Dissimilatory iron reduction by bacteria (e.g., *Shewanella* and *Geobacter*) strongly influences the reductive dissolution and mineralization of Fe(III) (hydr)oxides (Lovley et al., 1987; Zachara et al., 2002; Weber et al., 2006). The nature of the secondary iron minerals formed depends on various factors, but the most important are the presence and type of ligands and Fe(II) concentration (Hansel et al., 2003, 2005;

Borch and Fendorf, 2008; Kukkadapu et al., 2004). C_{org} is another major nutrient element in soil having prominent importance for soil functioning and fertility. On the other hand, magnetic enhancement in well-drained aerobic soils in the temperate climate is closely related to the abundance of organic matter (Maher and Thompson, 1995). Therefore, it is reasonable to seek a possible link between the magnetic minerals in soils and the contents of the major nutrients.

Phosphorous has a profound effect on the pathways of ferrihydrite transformations in soils, and many studies deal with laboratory investigations of its transformation products under different conditions. In series of laboratory experiments, Barrón and Torrent (2002) showed that two-line ferrihydrite with adsorbed phosphate ($P/Fe > 2.75\%$) aged at 150°C transforms into an intermediate "hydromaghemite" phase that further rapidly converts to hematite. These results are further accompanied by detailed magnetic investigations (Torrent et al., 2006; Liu et al., 2008) showing that the grain size of hydromaghemite particles increases on aging from the SP to the SD region before being converted into hematite. At smaller amount of adsorbed P, the ferrihydrite transforms directly into hematite (Barrón et al., 2003). Therefore, in well-aerated soils in the temperate climate, higher phosphate amount adsorbed on ferrihydrite promotes its transformation into (hydro)maghemite. In anaerobic soils and especially those experiencing oscillating reduction–oxidation changes, phosphate has a profound impact on the reduction and biotransformation of silica-ferrihydrite (Borch and Fendorf, 2008; Kukkadapu et al., 2004). Its presence leads to a decrease in magnetite precipitation and formation of carbonate green rust and vivianite. In contrast, Zachara et al. (1998) demonstrated that P inhibited or had no effect on the reducibility of synthetic and natural Fe(III) (hydr)oxides and only magnetite precipitates as a secondary transformation product. In addition, Adhikari et al. (2016), in reduction experiments of hematite with bound organic matter, demonstrated that the release of C was asynchronous as it was not linearly related to the reduction and release of Fe from the iron oxide. To check a possible link of soil magnetic properties and the P amount, a simple relation between the total P obtained via XRF analyses and the magnetic susceptibility of different depth intervals from various soil types (presented in detail in Chapters 2–6) is shown in Fig. 10.3.1. Only samples from the upper soil horizons (excluding the C horizons) are included in the plots, to avoid as much as possible the lithogenic P influence.

As can be seen from Fig. 10.3.1, no distinct relation is obtained for soils of Chernozem type (Fig. 10.3.1a) as well as for soils characterized by the presence of perched water table and seasonal surface waterlogging (Fig. 10.3.1b) (Planosols [profiles OK, PR, ZL], Stagnic Alisol [profile YPS], Alluvial soil [profile AL]). Two possible reasons could explain this observation—(i) most of the P is inherited from the parent rock and is not included in the biogeochemical cycle and precipitation of ferrimagnetic iron oxides and (ii) for these soil types, P does not directly influence the concentration of the ferrimagnetic fraction. However, for soil profiles showing important reduction–oxidation changes in their deeper horizons (Fig. 10.3.1c—Vertisols [JAS, SM], Gleysol [GL], salt-affected soil [S]), a well-expressed direct relationship between magnetic susceptibility and phosphorous content is observed. As discussed in detail in Chapters 4 and 5, magnetic properties of these soils suggest enhancement with low amount of SD-like magnetite fraction of their horizons, which are most affected by the redox changes. As was discussed, this fraction represents the secondary transformation product of ferrihydrite reductive dissolution. Therefore, the obtained relatively good relationship between P and χ is an indirect confirmation of this hypothesis. In contrast to all other soil profiles, the two high-mountain soil profiles (Humic Cambisol [TR] and Umbrisol [GR]) display an inverse relation between P and χ (Fig. 10.3.1d). As shown in Chapter 6, these profiles are characterized by the presence of a

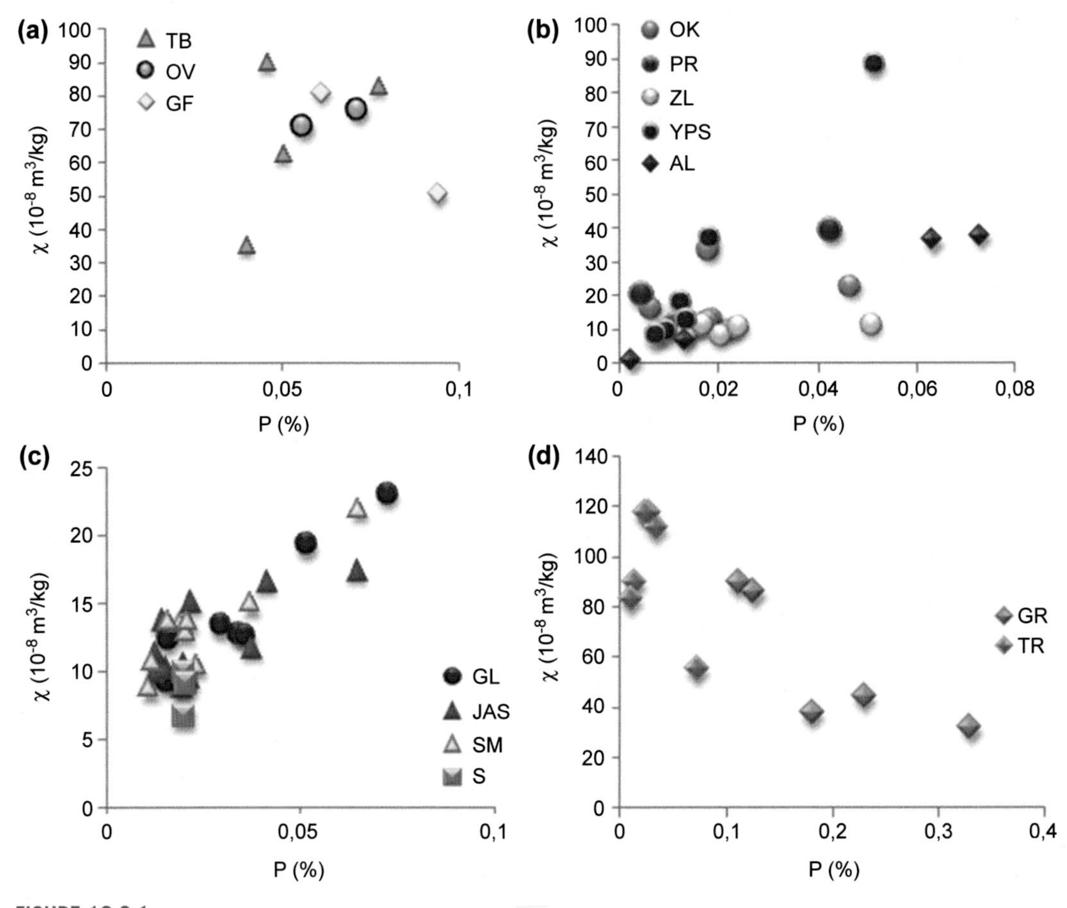

FIGURE 10.3.1

Relationships between bulk mass-specific magnetic susceptibility and the phosphorous content (in wt%) for samples from different depths and various soil types: (a) Chernozems and Phaeozems; (b) Planosols, Alisol, and Fluvisol; (c) Vertisols, Gleysol, and salt-affected soil; and (d) high mountain soils (Humic Cambisol and Umbrisol).

significant amount of organic matter in their uppermost horizon. Organic matter complexation with ferrihydrite is shown to inhibit its further transformation into more crystalline iron oxide minerals (Cornell and Schwertmann, 2003). Therefore, the obtained relation reflects this major characteristic of mountain soils developed in cold humid climate.

One of the most fertile soil types—the Chernozem—is frequently studied in environmental magnetism as part of the loess–paleosol sequences keeping a robust magnetic record of paleoclimate (Maher, 1998; Liu et al., 2007). Jakšik et al. (2016) studied topsoils of Haplic Chernozems in an area of ~100 ha where the C_{org} content varied between 0.3% and 2%. Three subsets of soil samples were investigated, and the correlation analysis carried out demonstrated a significant positive correlation between the low-field magnetic susceptibility and C_{org} with R^2 between 0.91 and 0.98. Models constructed for predicting the C_{org} content on the basis of magnetic susceptibility (χ_{lf}) resulted in the following pedotransfer function:

$$C_{org} = 0.3764 + 0.2030\,\chi_{lf} \text{ with } R^2 = 0.971 \text{ (Jakšik et al., 2016)}$$

Therefore, the magnetic susceptibility of the Haplic Chernozems is suggested as a good proxy parameter for the fast evaluation of C_{org} and thus provides further possibilities for practical applications of soil magnetism in agricultural science. Further support for the reliability of soil magnetic properties as proxies for C_{org} content is demonstrated in the study of Calcisols developed on Tertiary clays and sandstones from Spain (Quijano et al., 2014). The magnetic susceptibility at (0–30 cm) depth varied between 30 and 47×10^{-8} m^3/kg, while the $\chi_{fd}\%$ retained high values between 10% and 11%. The authors have found a direct link and high correlation between magnetic susceptibility and organic matter (OM) with a correlation coefficient $R^2 = 0.81$. Therefore, well-aerated soils showing moderate magnetic enhancement with ferrimagnetic magnetite/maghemite and enriched in fine SP particles are suitable targets for proxy estimation of C_{org} through magnetic methods.

A similar conclusion is provided by the data obtained from two soil collections, studied by the author. The first data set consists of soil samples gathered from a test area near Rosslau (situated in Saxony Anhalt, eastern Germany, 80 km north of Leipzig), where the major soils display considerable gleying. A more-detailed magnetic study is presented in Jordanova et al. (2013a). The magnetic susceptibility of the studied soils is relatively low, compared with the earlier-mentioned well-aerated soils but spans a much wider interval—from $\chi \sim 5 \times 10^{-8}$ m^3/kg up to $\sim 60 \times 10^{-8}$ m^3/kg. For the same soil samples, total carbon (C_{total} %) and total nitrogen (N_{total} %) have been determined according to standard methods DIN ISO 10,694 and DIN ISO 13878, 1998. Fig. 10.3.2 shows the obtained relation between χ and C_{total}, as well as between χ and N_{total}.

It should be taken into account that the relations in Fig. 10.3.2 link soil magnetic susceptibility and the total C, while the previous examples of Chernozems and Calcisols show relationships between χ

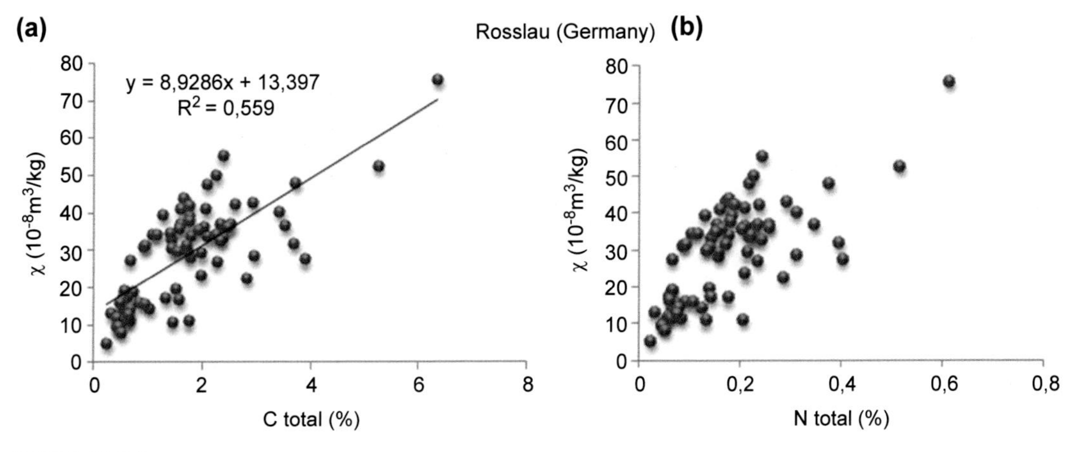

FIGURE 10.3.2

Relationship between mass-specific magnetic susceptibility and total carbon (a) and total nitrogen (b) content (in wt%) for soil samples from 0 to 10 cm and 10- to 30-cm depth intervals for Gleysols from Rosslau (Germany).

and C_{org} (Jakšik et al., 2016) or OM (Quijano et al., 2014). The organic pool in the total C is, however, calculated from the C_{total} making the correction for the calcium carbonate (DIN ISO 10694, 1995), and in the case of a very low amount of or missing $CaCO_3$, the C_{total} is equivalent to C_{org}. Nevertheless, even in the case of gleyed soil (Fig. 10.3.2), a relatively well-defined correlation is obtained between the magnetic susceptibility and the C_{total} content. As evidenced in Jordanova et al. (2013a), the multivariate statistical analysis of the data yields pedotransfer functions linking the C_{total} and N_{total} to the magnetic susceptibility but in combination with the soil reaction (pH). In contrast, for Chernozems from the Czech Republic (Jakšik et al., 2016), a straightforward $C_{org}-\chi$ link is obtained.

The second collection of soils comes from the CULS (Czech University of Life Sciences, Prague) farm test site Lany in central Bohemia, where the dominant soil type is Haplic Cambisol. Relationships equivalent to those shown for the Rosslau test site are presented in Fig. 10.3.3.

It is obvious that the linear regressions between the magnetic susceptibility and the C_{total} and N_{total} are better defined for the Cambisol topsoil samples. Considering deeper depth interval (10–30) cm, the

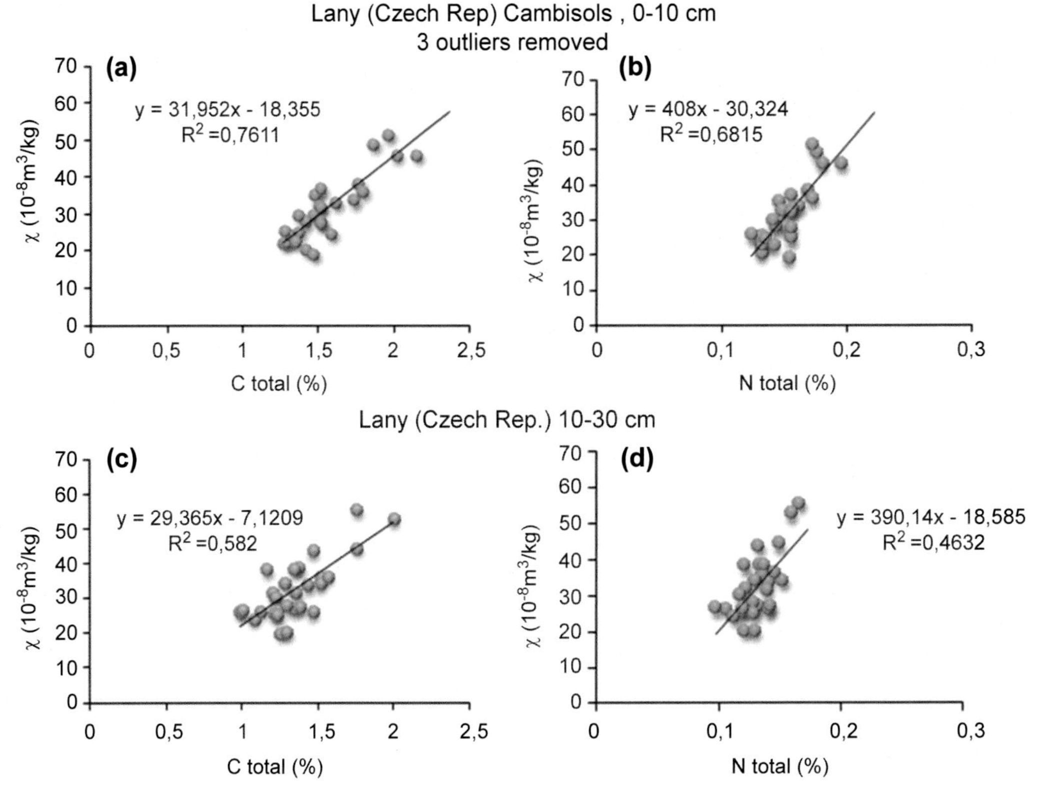

FIGURE 10.3.3

Relationship between mass-specific magnetic susceptibility and total carbon (a) and total nitrogen (b) content (in wt%) for soil samples from 0 to 10 cm and 10- to 30-cm depth intervals (c) and (d) for Cambisols from Lany (Czech Republic).

defined correlation is not so well expressed and the R^2 correlation coefficient is lower. Similar positive correlation between the magnetic susceptibility of Cambisols and the total carbon content has been reported by Chlupáčová et al. (2010) for soil collection from the Czech Republic. Thus, it could be supposed that the relation of magnetic susceptibility to the C_{total}, C_{org} (or OM) is more reliable for well-aerated soils with considerable topsoil magnetic enhancement, contrasting the deeper soil horizons. The link obtained is the best and simplest expressed in Chernozem-type soils (Jakšik et al., 2016), while a more complicated relation is found in soils with reducing conditions (such as the Gleysols from Rosslau, e.g., Jordanova et al., 2013a). The background of the obtained relation between the C_{total} or C_{org} and the magnetic susceptibility could be searched for in the well-known process of binding of the OM to mineral surfaces and especially to iron oxides (Lützow et al., 2006; Adhikari and Yang, 2015).

10.4 PRINCIPLES OF THE APPLICATION OF MAGNETIC METHODS FOR THE DETECTION OF SOIL CONTAMINATION AND PADDY MANAGEMENT

10.4.1 MAGNETOMETRY FOR EVALUATION OF SOIL POLLUTION

Knowledge regarding the magnetic signature of undisturbed nonpolluted soils is an important prerequisite for the application of magnetometric methods for the detection of anthropogenic soil pollution. The latter may have different origins—the atmospheric fallout of industrial and traffic emissions (dry deposition), soil contamination through polluted groundwater/surface water, or liquid effluents and slags from industrial processes (Alloway, 1995; Singh et al., 2011; Hu et al., 2014). The presence of a concomitant strongly magnetic fraction of iron oxides in the anthropogenic emissions causes a subsequent extra enhancement of the magnetic soil signature at the surface of the polluted media. This signature is registered by using fast magnetic susceptibility measurements (in situ in the field or using soil samples in the laboratory). Depending on the iron availability in the waste products, different degrees of magnetic enhancement are observed in the contaminated soils. Data on the mineralogical composition and magnetic signal of the major anthropogenic emissions (fly ashes, lagooned ashes, slags, cement dusts, etc.) can be found in a number of publications (Goluchowska, 2001; Veneva et al., 2004; Jordanova et al., 2006; Magiera et al., 2011; Howard and Orlicki, 2016). Recently, soil contamination by organic compounds (PAH, oil spills) has also been successfully studied with magnetic methods (Rijal et al., 2012; Wawer et al., 2015). Soil pollution caused by the accumulation of anthropogenic emissions has been considered a suitable target for the application of the magnetic methodology for the identification of environmental pollution since its establishment (Heller et al., 1998). The purpose of this section of the chapter is not to give a review of the most recent achievements in this field of research but rather to present a useful set of criteria and methodological approaches for outlining soil pollution at different spatial and temporal scales. There are several major issues that must be considered when planning, executing, and interpreting the magnetometric study of soil pollution:

Preliminary knowledge regarding the mineralogical characteristics and the relative importance of the magnetic fraction in the bulk (total) anthropogenic emissions from major pollution sources

The basic requirement for the application of magnetometry in soil pollution studies is that the bulk emission of the pollution source contains some fraction of strongly magnetic iron oxides (or pure iron). This would guarantee that the polluted media will be characterized by an anthropogenically mediated enhancement of its magnetic signal. According to the absolute amount and the phase composition of the Fe-containing fraction, this enhancement will vary accordingly. The highest percentage of a strongly magnetic component is typical for slags and fly ashes from metallurgical Fe-processing plants, the mining of Fe-bearing mineral deposits and the fly ashes from coal-burning power plants (Pacyna, 1987; Veneva et al., 2004; Yang et al., 2014; Jordanova et al., 2013b; Piatak et al., 2015). Less-magnetic minerals are present in the emissions from other industrial productions such as cement, wood processing, textile, etc. (Goluchowska, 2001; Szuszkiewicz et al., 2015; Lu et al., 2016). However, the relative importance of the Fe-bearing fraction does not correspond linearly to the absolute concentration of toxic heavy metals because of the source-specific amount of the magnetic fraction, as discussed. As a result, a complex pollution pattern originating from a combination of several sources is difficult to resolve by using magnetic proxy methods only. Except for classic soil contamination by heavy metals, emerging pollutants such as different organic compounds (PAHs, gasoline, oil spills, waste water, compost) have been shown to produce detectable magnetic enhancement (Zhang et al., 2013; Ameen et al., 2014; Porsch et al., 2014; Paradelo and Barral, 2014; Wawer et al., 2015).

Estimation of the extent of the vertical migration of the polluting fraction and affected soil horizons: factors affecting the depth distribution of anthropogenic iron oxides and the major heavy metals (pollutants)

A second important aspect of magnetometry applied to soil pollution is related to the detection and estimation of the naturally occurring processes of the vertical migration of pollutants within the soil profiles driven by geochemical gradients and reactions. A key challenge here is to assess the mobility, such as vertical migration of both the contaminants (toxic heavy metals) and the magnetic fraction. In cases where heavy metals are incorporated in the crystal structure of the Fe compounds or are adsorbed on the surfaces of the secondary (neoformed as a precipitation products) Fe (oxy)hydroxides, this problem is not relevant. However, in other cases, Fe-containing mineral grains in the anthropogenic emissions are physically separated from the other particulates of toxic compounds. Laboratory magnetic study on fly ash migration in a column simulation of rain events (Kapička et al., 2000) suggests that fly ashes deposited on top of sandy soils show remarkable stability and the maximum magnetic enhancement is detected at a few centimeters below the surface. On the other hand, different soil types exhibit different physical characteristics (clay content, acidity, organic material, cation exchange capacity) and depth variations of pedogenic magnetic enhancement, which result in various combinations of the superposition of the pedogenic and anthropogenic magnetic signature. In a number of publications, shallow forest soils from Poland, the Czech Republic, and Germany have been considered as representative for the anthropogenic magnetic signal of polluted soils (Klose et al., 2003; Magiera et al., 2007, 2008, 2015; Kapička et al., 2008). In these cases, the magnetic enhancement is only observed in the uppermost organic layers and the mineral soil up to depths of 5–10 cm and only exceptionally at higher depths. However, in thick Chernozem profiles, natural pedogenic magnetic enhancement is observed at much higher depths (60–80 cm) (Jordanova et al., 1997, 2010; Jelenska et al., 2004). In a comparison of depth variations of the magnetic signature of industrially polluted and nonpolluted Chernozems from Ukraine, Jelenska et al. (2004) found differences in mineralogy and the magnetic enhancement down to a 60-cm depth, reaching the transitional BC_k (B_k) horizons. Industrial pollution of soils from Argentina (Chaparro et al., 2006) is also identified in the surface as well as subsurface layers down to a 20-cm depth as a significant enhancement overlaid on the strongly magnetic signal of pedogenic origin.

Evaluation of the effect of the natural pedogenic magnetic enhancement of soils in the study area, which must be considered as a "background" in pollution studies

The problem of discrimination between natural pedogenic enhancement and the anthropogenic signal of polluted soils has been addressed in a number of publications (Magiera et al., 2006; Hanesch et al., 2007; Jordanova et al., 2008; Szuszkiewicz et al., 2016). The occurrence of strong magnetic enhancement in the uppermost organic layers of forest topsoils, low magnetic susceptibility down to a 20-cm depth, and increased heavy metal content, according to Magiera et al. (2006), is indicative of an anthropogenic origin for the magnetic signal. A more objective approach is presented in Hanesch et al. (2007), using a statistical analysis of a large data set of soils from eastern Austria. The thresholds of the background values of magnetic susceptibility for different soil types and parent rock lithologies have been calculated by using different methods (the median absolute deviation method, the boxplot method, and the population modeling method). The median absolute deviation (MAD) method is recommended in high-resolution mapping for the early detection of changes in the environment. The boxplot method is considered useful in most cases because it gives a useful overview of possible anthropogenically affected areas. The population modeling method sets the highest threshold and is appropriate for the rapid detection of "hot spots" of pollution where immediate intervention is necessary (Hanesch et al., 2007). A thorough discussion and analysis of the soil database for England and Wales (Blundell et al., 2009) allowed the authors to define the threshold values of χ and $\chi_{fd}\%$ ($\chi \leq 0.38 \times 10^{-6}$ m^3/kg and $\chi_{fd}\% < 3\%$) suitable for the discrimination of anthropogenically affected topsoils from the whole data set. They explore the effect of changing threshold values for the reliability of the polluted locations identified according to the distance from the source and the known historical pollution record. However, this approach is based on the assumption that strongly magnetic background resulting from a bedrock lithology having a high content of large multi-domain (MD) ferromagnetic grains is not present. This is not the case for other regions. For example, a large area in southeastern Bulgaria is covered by soils in which a lithogenic strongly magnetic component is represented by coarse-grained magnetites (see Chapter 9). It results in very high soil magnetic susceptibilities and low $\chi_{fd}\%$. Another approach for defining threshold values for the discrimination between nonpolluted and contaminated soils of different types in a special case of weakly magnetic parent rocks is presented in Jordanova et al. (2008). In this case, the effect of lithology is not significant; that is why only a correction for soil type has been applied. The field magnetic susceptibility (K), laboratory mass-specific susceptibility of the soil material (χ), and the percent frequency-dependent susceptibility ($\chi_{fd}\%$) for samples from the soil types presented in the study area were used in a fuzzy k-means cluster analysis with two clusters defined (representing natural and polluted samples). The corresponding mean values of the three variables (χ, $\chi_{fd}\%$, K) for the cluster containing samples with a lower magnetic enhancement are taken as "background" values (K_{bg} and χ_{bg}) (Table 10.4.1).

These background values of K and χ are used to normalize the corresponding measured field- and laboratory values for the soils of this type. Normalized values higher than 1 represent polluted sites, while values equal to or lower than 1 correspond to clean sites. The spatial maps of the corrected field- and mass-specific laboratory magnetic susceptibility values obtained are shown in Fig. 10.4.1.

The normalized magnetic susceptibility maps (from field measurements and the soil samples) depict areas influenced by industrial pollution from the Devnja complex and especially by the cement

Table 10.4.1 Calculated mean values of field (K_{bg} in 10^{-5} SI units) and mass-specific (χ_{bg} in 10^{-8} m^3/kg) background susceptibilities according to soil types in the study area, used for correction (e.g., normalization)

	Chernozems (N = 21/51)		Alluvial soils (N = 10/20)		Phaeozems (N = 11/25)		Luvisols (N = 23/48)		Rendzinas (N = 12/26)		Planosols (N = 3/4)	
	K_{bg}	χ_{bg}	K_{bg}	χ_{bg}	K_{bg}	χ_{bg}	K_{bg}	χ_{bg}	K_{bg}	χ_{bg}	K_{bg}	χ_{bg}
Mean	27.02	43.07	16.45	19.32	23.36	40.41	15.24	17.62	26.11	42.16	12.83	13.54
SD	6.79	13.39	2.78	6.40	6.33	12.84	3.42	5.25	18.02	22.83	1.46	7.19
Variance	46.13	179.2	7.72	40.95	40.12	164.93	11.7	27.61	324.79	521.28	2.13	51.6

The corresponding standard deviations (SD) and variances (Variance) are also given. Number of samples in the cluster and the total number of samples for each group (soil type) are indicated in the first row.

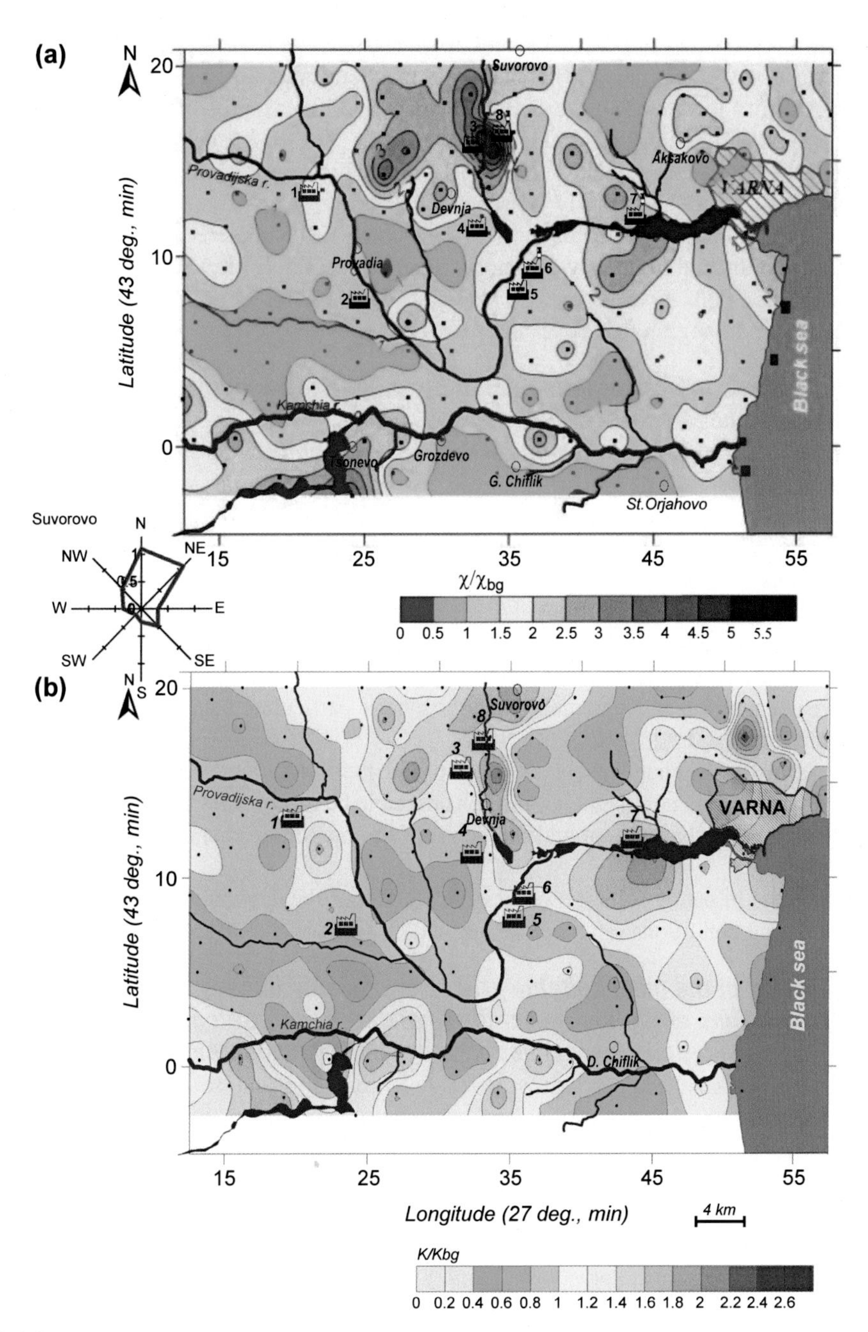

FIGURE 10.4.1

Spatial maps of normalized sample mass-specific magnetic susceptibility (χ/χ_{bg}) and field volume magnetic susceptibility (K/K_{bg}) of the mapped area around the Devnja industrial complex. Main sources of pollution are indicated: 1—Soda II; 2—polymer I; 3—Soda I; 4—fertilizer plant; 5—polymer II; 6—Beloslav power plant; 7—Varna power plant; 8—Devnja power plant and cement plant. Wind rose for the location close to Devnja is shown as well.

production factory and power plants (sources 6, 7, and 8 in Fig. 10.4.1). Obviously, wastes from chemical production (sources 1 and 3) contain less magnetic fraction, and this leads to the observed relative susceptibility lows around the Soda I factory. Field measurements at a dump site of waste material from soda production show values of only 0.23×10^{-5} SI in contrast to the susceptibility of the slag material from the power plants (225×10^{-5} SI), which has been dumped in the Beloslav lake.

According to the data, previously presented in Chapter 9, different soil types from Bulgaria developed on different parent materials show various magnetic characteristics and especially different magnetic susceptibility and frequency-dependent magnetic susceptibility. Thus, the obtained magnetic data on nonpolluted soils may serve as a valuable database for defining the threshold values in pollution studies in the region that cover different soil types.

Consideration of the effects of land use on the magnetic susceptibility enhancement observed in anthropogenically polluted areas

Magnetic susceptibility mapping projects for the identification of anthropogenic pollution covering areas characterized by various land use (forest, meadow, agricultural, urban, etc.) have faced the problem of a well-defined distinction between the measured topsoil susceptibilities for adjacent locations covered by the same soil type, which must exert a similar anthropogenic load (Magiera et al., 2006; Magiera and Zawadski, 2007; Yan et al., 2011). The observed land-use dependence of the soil magnetic susceptibility has been attributed to the effect of differences in the pollutants' accumulation by the dominant vegetation cover. An example of such land-use dependence of the observed anthropogenic enhancement is presented in Fig. 10.4.2. Deciduous forest litter is considered as the most effective in capturing and accumulating atmospheric particulate matter in the uppermost organic horizons of the forest soils.

Differences in the mechanisms of deposition and the vertical migration of the anthropogenic particulates (magnetic and nonmagnetic) are considered to be responsible for the observed contrasts in the absolute magnetic enhancement and the depth of the maximum magnetic signal (Magiera et al., 2006).

The special case of paddy soils shows the severe effect of the depletion of Fe oxides in paddy soils due to waterlogging conditions (Han and Zhang, 2013; Chen et al., 2015).

Advance planning of the most suitable grid size, mapping density, and sampling design, according to the aims of the study and the financial frame. Selection of field and laboratory instrumentations for magnetic susceptibility measurements

The planning of the size and grid density for field mapping and soil sampling are of primary importance for the successful application of magnetometry in pollution studies. A thorough analysis of the principles and the general rules of thumb for the selection of grid resolution in mapping and cartography (GIS) are presented in Hengl (2006). Here, we summarize the main recommendations given in this work, while a detailed analysis can be found in the cited reference.

For the sake of consistency, the mapped area should have equal sampling density. The cartographic rule used in soil mapping postulates to have minimum one observation at 1 cm^2 of the map at the chosen scale. The following three major issues should be considered in each mapping project:

- The selection of grid resolution should respect the principles of general statistics and information theory. According to the Nyquist frequency concept, in order to reconstruct a continuous signal

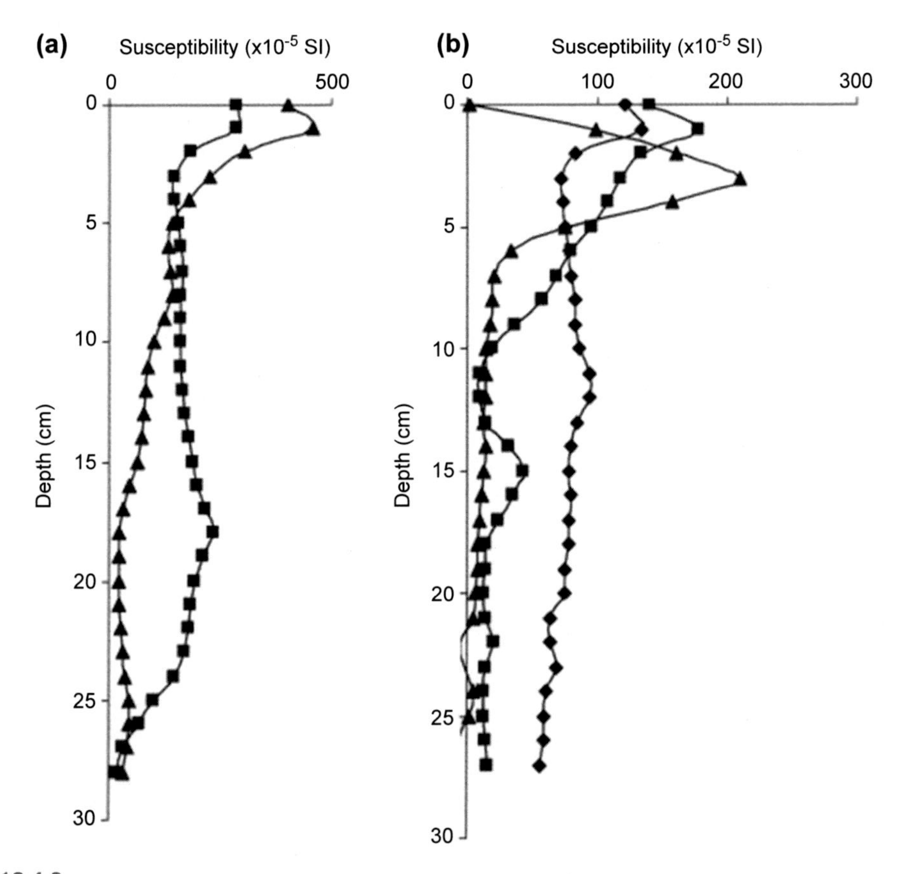

FIGURE 10.4.2

The influence of different land use on the vertical distribution of magnetic susceptibility in soil profiles from polluted areas. (a) Cambisol developed on sandstone from Ostrava region—Czech Republic (triangles—profile under forest, squares—profile under grassland). (b) Podzol developed on Quaternary sand from Upper Silesia—Poland (triangles—profile under forest, squares—profile under grassland, rhombus—arable field).

From Magiera, T., Strzyszcz, Z., Kapička, A., Petrovský, E., 2006. Discrimination of lithogenic and anthropogenic influences on topsoil magnetic susceptibility in Central Europe. Geoderma 130, 299–311 with permission from Elsevier.

from a number of discrete observations, the grid resolution should be at most half of the average spacing between the closest point pairs (e.g., the mean shortest distance).

- The following standard grid resolutions must be derived for each data set: (i) the coarsest legible grid resolution—the highest resolution that should be used, taking into account the specific scale, the accuracy of the positioning method and the size of the objects; (ii) the finest legible grid resolution e.g. the smallest resolution that confidently represents (at a 95% level) the spatial objects; and (iii) the recommended grid resolution e.g. the compromising resolution between the previous two.

- The choice of grid resolution should always take into account the inherent peculiarities and properties of the study area.

An example of a small-scale (e.g., large study area) mapping of soil pollution in central Europe is the magnetic susceptibility map of the forest soils covering an area of about 200,000 km^2 obtained from the MAGPROX project (Fig. 10.4.3).

The sampling grid of 10 km reveals the general pattern of topsoil field magnetic susceptibility with well-presented anomalous patterns related to the atmospheric deposition of strongly magnetic particulate matter from the major industrial centers in Europe (coal-burning power plants, steel works, cokeries, cement plants, etc., especially in Upper Silesia, northern Moravia and Bohemia, and eastern Saxony), as well as from a wide lithogenic strongly magnetic anomaly of central Bohemia (Fialova et al., 2006).

Another important aspect of planning the magnetic mapping of soils is the correct choice of field sensor. Knowledge of the penetration depth and geometry of the susceptibility sensors is necessary to interpret the data precisely. There are several field susceptibility sensors for measurements of magnetic susceptibility. The most widely used are the Bartington MS2D (Bartington Ltd, UK), KT-5 (Geofizika, Brno, Czech Republic), KT- 6 (Satis Geo, Czech Republic), and SM30 (AH Instruments, Czech Republic). A detailed examination of the effects of sensor geometry on the integrating volume of the material contributing the most to the susceptibility signal for each instrument is presented in Lecoanet et al. (1999) and Jordanova et al. (2003). The higher penetration depth of the MS2D sensor compared

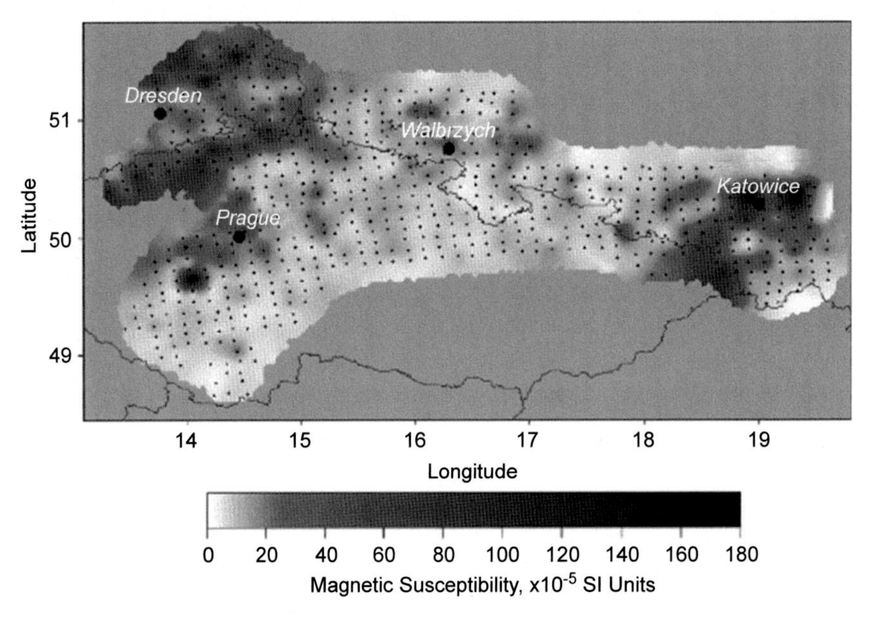

FIGURE 10.4.3

Map of magnetic susceptibility of forest topsoils in central Europe.

From Magiera, T., Strzyszcz, Z., Kapička, A., Petrovský, E., 2006. Discrimination of lithogenic and anthropogenic influences on topsoil magnetic susceptibility in Central Europe. Geoderma 130, 299–311 with permission from Elsevier.

with the other sensors should be carefully weighed against the higher portability and sensitivity of SM30 and facile KT-5 use. The methodology for field magnetic susceptibility mapping is described and evaluated in Schibler et al. (2001).

Planning and executing additional laboratory magnetic and heavy metal analyses

Mapping the field magnetic susceptibility of soils is the fastest initial step in the application of magnetometry for the evaluation of soil pollution. However, more robust and detailed information could be obtained by carrying out concomitant soil sampling (both topsoil and soil cores), which are subsequently used for laboratory magnetic and nonmagnetic analyses. This approach ensures a better evaluation of the origin of the magnetic enhancement (anthropogenic versus lithogenic) as well as further data on the effective magnetic grain size and mineralogical composition of the anthropogenic magnetic phases. The initial field mapping information is very useful in compiling the spatial pattern of topsoil magnetic susceptibility already on field maps. According to its peculiarities, nested grids of a higher resolution may be planned and additional mapping performed to get a better description of potential hot spots of anthropogenic pollution. Based on this initial overview, samples for further analysis of the content of heavy metals could be chosen by spotting the best representatives of magnetic susceptibility grids.

10.4.2 PADDY SOILS: A SPECIAL CASE OF AN ANTHROPOGENICALLY INFLUENCED SOIL

Paddy soils are strongly modified by the anthropogenic management during rice cultivation and represent the most widespread agricultural land in Asia, especially in China, India, Indonesia, and Bangladesh (Kögel-Knabner et al., 2010). This is the reason that they have been considered as an independent unit in the soil classification (IUSS Working Group WRB, 2006). Paddy management involves several usual practices, including the artificial submergence and drainage of the original soil, tillage of flooded soil, rice transplantation, and the following growth of the crop in a submerged environment. Several days before the harvest, the paddy field is drained to support easier rice manipulation (Kögel-Knabner et al., 2010 and references therein). Organic manuring, liming, and fertilization are also commonly applied. The interchanging submergence and drainage cause alternating fluctuations in the redox conditions in the plow pan. A high input of organic materials and their low decomposition rate under anaerobic conditions determine the accumulation of organic carbon in the uppermost horizon in the paddy soils (Huang et al., 2015). Waterlogging impedes oxygen supply, and continuing microbial activity causes anoxic conditions in the soil. As already discussed in the other chapters (Chapters 4 and 5), in such conditions the redox-sensitive Fe oxides are the major alternative electron acceptors for the metabolic decomposition of OM by bacteria and thus play a crucial role in paddy soil evolution (Van Breemen, 1988; Schwertmann, 1988; Winkler et al., 2016). In such environmental conditions, reductive iron oxide dissolution is the most prominent process in the iron-containing minerals (Van Breemen, 1988; Grimley and Arruda, 2007). Repeated plowing and puddling practices over long time periods lead to the formation of an impermeable plow pan, which reduces percolation and improves water and nutrient efficiency in the paddy fields (Huang et al., 2015). According to their hydrological regimes, paddy soils are grouped into three major categories—reducing paddy soil, oxidizing–reducing paddy soil, and oxidizing paddy soil. Reducing paddy soils are characteristic for the bog areas where the groundwater table is high and the whole solum is under reducing conditions. The oxidizing–reducing paddy soils are mostly spread in alluvial plains where groundwater level experiences seasonal fluctuations inside the soil profile. Oxidizing paddy soils are found in uplands where the groundwater table is deep, so that the soil is well drained and oxidized except in its uppermost submerged plow layer. A further detailed summary of paddy management practices can be found in Huang et al. (2015). A wealth of research is devoted to the biogeochemistry of paddy soils (e.g., Gong, 1986; Kögel-Knabner et al., 2010; Mueller-Niggemann et al., 2012; Kölbl et al., 2014) owing to their major importance for food production and ecosystem functioning. Strong management-induced changes in the morphological and bio-geochemical properties of paddy soils are extensively studied in model laboratory or field experiments that record soil changes over different timescales (Zhang and Gong, 2003; Hao et al., 2008; Huang et al., 2015 and references therein). Further on, biogeochemical processes, reactions, and implications for greenhouse gas emissions in paddy soils are well studied, and a number of books, reviews, and articles are available for more-detailed information (e.g., Conrad and Frenzel, 2002; Kirk, 2004; Kögel-Knabner et al., 2010). Macroscopic redox gradients between the anoxic plow layer and the oxic deeper horizons frequently cause a significant redistribution in the Fe (hydr)oxides along the paddy profile (Winkler et al., 2016 and references therein). The reductive dissolution of iron oxides in the anoxic upper part of the profile commonly leads to iron depletion in this part of the solum, while deeper soil parts become enriched in short-ranged Fe oxides (Kögel-Knabner et al., 2010). Since the anthropogenic management of paddy soils most strongly influences the redox state of the uppermost plow layer of the soil and thus, the iron oxide forms, the magnetic signature of paddy soils could be successfully used in their precise description and characterization.

As demonstrated in many environmental magnetic studies (Dearing et al., 1995; Maher, 1998; De Jong et al., 2005; Grimley and Arruda, 2007), reductive conditions and waterlogging significantly reduce the concentration of strongly magnetic iron oxides in the soils due to intense reductive dissolution. Paddy soils represent the ultimate example of waterlogged soils with alternation in the periods of aeration and anoxia. The strongly depleted magnetic susceptibility and frequency-dependent magnetic susceptibility in paddy soils compared with dryland soils developed on the same parent material are reported in studies of magnetic properties of subtropical soils (Yu et al., 1981; Lu, 2000). Paddy soils are used for rice cultivation during various time periods from very recently to hundreds to thousands of years B.P. (Chen et al., 2011). Paddy soil chronosequences are widely studied to understand the rates and directions of pedogenic processes and associated thresholds, nutrient dynamics, the impact of management practices on the temporal changes in paddy soils, etc. (Huang et al., 2015 and references therein). Magnetic susceptibility is used as a sensitive, fast, and cost-effective method for the evaluation of changes in crystalline and amorphous iron oxides in a paddy soil chronosequence located on a coastal plain in Cixi, Zhejiang Province, in China (Chen et al., 2011). The authors found a fast logarithmic decrease in magnetic susceptibility and the isothermal remanence (IRM) with increasing paddy cultivation age and a slower decrease in the concentration of the hard-coercivity IRM fraction. Thus, Chen et al. (2011) concluded that the magnetic minerals of different coercivities in paddy soils have a different response time to the cultivation period and maghemite (e.g., the soft IRM component) is more sensitive to waterlogging than the hard-coercivity minerals (e.g., goethite). Their results further demonstrate that magnetic susceptibility and IRM have a very fast response time to waterlogging since significant changes have been observed after only about 50 years of paddy cultivation. Similar results are reported in Chen et al. (2015), also in a study of paddy soil chronosequence. The authors conclude that the magnetic properties of paddy and nonpaddy soil derived from calcareous sediments are controlled by the changes in soil moisture regimes caused by land use, rather than the period of cultivation. A different response time regarding various paddy soil characteristics to paddy management is further investigated by Kolbl et al. (2014), who compared two chronosequences (paddy and non-paddy) from the coastal region of the Zhejiang Province, spanning about 2000 years. The authors suggest three distinct phases in paddy soil formation. The initial phase takes a few decades and is dominated by desalinization and the formation of a compacted plow-pan. During the second phase, accelerated carbonate losses and constantly increasing C_{org} concentrations are observed in paddy topsoils. In the third stage of paddy soil development ($\geq$700 years), a transformation and redistribution of iron oxides occurs. Based on the comparison between the paddy and nonpaddy chronosequences, Kolbl et al. (2014) demonstrate that pedogenesis under paddy management was accelerated relative to that under dryland cropping. The distinguishing power of the magnetic susceptibility of paddy soils related to cultivation age is further demonstrated in the study by Mueller-Niggemann et al. (2012) who also investigated paddy soil chronosequence in the coastal Cixi area (spanning ~200 years of cultivation). Using an advanced descriptive, explorative, and nonparametric statistical data analysis, Mueller-Niggemann et al. (2012) show that magnetic susceptibility as a conservative soil parameter is the second major factor in the principal component analysis, reflecting the minerogenic composition of soils. The authors conclude that the intrinsic heterogeneity of paddy soil organic and minerogenic components (including magnetic susceptibility) per field is smaller than between study sites. The highest intrinsic heterogeneity is found in the youngest paddy soils.

A further detailed magnetic study of paddy soils is reported by Lu et al. (2012) who investigated natural paddy/nonpaddy pairs of soils as well as performed laboratory incubation experiments to simulate paddy cultivation management. Soils subjected to laboratory waterlogging showed a fast decrease in magnetic susceptibility as well as in frequency-dependent magnetic susceptibility over time (Fig. 10.4.4).

Differences in the decrease rate of the magnetic enhancement depending on the parent material are obvious, as observed in Fig. 10.4.4. Lu et al. (2012) demonstrate that the stronger the magnetic susceptibility of dryland soil, the greater is the loss of magnetic susceptibility occurring after paddy cultivation. Paddy soils developed on well-aerated soils show magnetic susceptibility enhancement with SP/SD magnetite/maghemite in their subsoil horizons, related to the secondary precipitation of magnetic minerals due to iron dissolution and leaching from the uppermost reduced plow layer. The formation of amorphous iron oxyhydroxides in the anoxic topsoils is further demonstrated by thermomagnetic analyses of magnetic susceptibility, which display transformation behavior typical of ferrihydrite and/or pyrite. Ferrihydrite presence is strongly supported by the inhibiting role of P and the OM in its transformation to more crystalline phases (Cornell and Schwertmann, 2003).

Another important issue in the studies of paddy soils and their pedogenic evolution is whether and how the initial nonpaddy soil variety that converts to paddy soil influences the properties and characteristics of the paddy soil. This problem is addressed in the study by Winkler et al. (2016), who investigated the response of Vertisols, Andosols, and Alisols to paddy management. The authors show that the main characteristics of the initial soil type are preserved and not completely overridden by the paddy management. Similar to the other studies, the uppermost waterlogged layer in all initial soil types becomes enriched in organics and in amorphous iron oxides. Winkler et al. (2016) found that paddy management induced losses of Fe oxides in the topsoils that were the largest in the Alisols. According to the authors, the amount of crystalline and amorphous Fe oxides produced as a result of

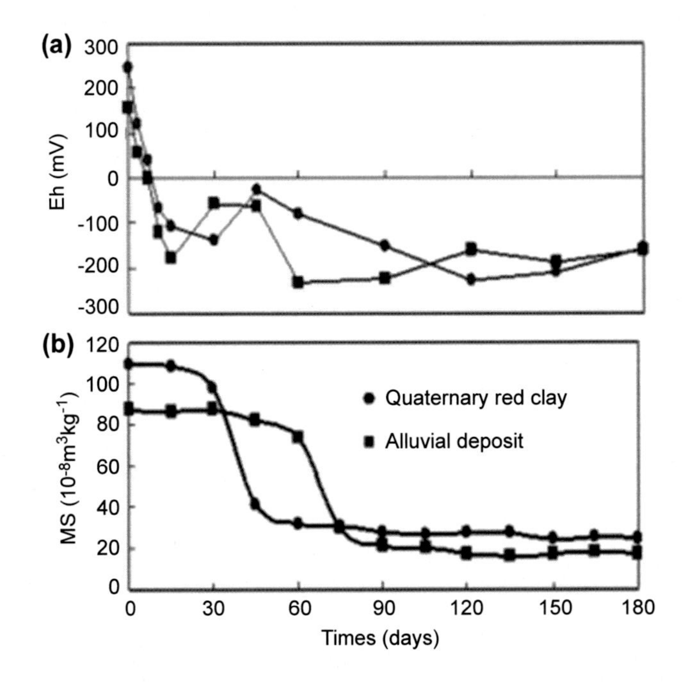

FIGURE 10.4.4

Changes in soil Eh (a) and magnetic susceptibility (MS) (b) during waterlogging incubation experiments.

From Lu, S.G., Zhu, L., Yu, J.Y., 2012. Mineral magnetic properties of Chinese paddy soils and its pedogenic implications. CATENA 93, 9–17 with permission from Elsevier.

paddy management depends mostly on the hydrological regimes of the corresponding initial soil (e.g., its texture).

Specific pedogenic transformations in the topsoil layer in paddy soils causing extensive magnetic depletion (e.g. Chen et al., 2011, 2015; Lu et al., 2012) make these anthropogenically influenced soils a suitable medium for the magnetic tracing of anthropogenic pollution. This is demonstrated in the study by Yan et al. (2011), presenting field magnetic susceptibility screening and laboratory magnetic investigations on paddy soils around the Meishan steel mill near Nanjing City (China). Yan et al. (2011) show that the background (nonpolluted) paddy soils display the described characteristic magnetic depletion in the plow layer, while closer to the major pollution source, the magnetic properties in this layer are enhanced with coarse, strongly magnetic particles of an anthropogenic origin. Anthropogenic dust emitted from the mill enters the topsoil layer and is homogenized there by the plowing practices in the upper approximately 20 cm. Still, an open question that needs further investigation is whether and how the ferromagnetic coarse dust particles from the anthropogenic emissions are influenced by the reductive dissolution in paddy soils, related time dependence, etc. Taking into account the coarse grain size of most of the anthropogenically derived strongly magnetic fractions, it could be supposed that this process is relatively slow because of the lower surface area available for reductive dissolution (compared with nano-sized particles).

10.5 EFFECTS OF WILDFIRES ON THE MAGNETISM OF SOILS: POSSIBLE APPLICATION IN SOIL SCIENCE AND PALEOENVIRONMENTAL STUDIES

Fire has played a significant role in the Earth's processes for millions of years (Bond, 2015). Its occurrence affected the composition of gases in the atmosphere, climate, weathering rates, erosion, and deposition processes, thus also influencing ecosystem evolution (Bodi et al., 2014). Severe forest fires induce changes in all soil properties including physical, chemical, mineralogical, and biological characteristics (Certini, 2005). Among other constituents, iron oxides also suffer significant changes, as far as they are very sensitive to any changes in the environment (Cornell and Schwertmann, 2003). Controlled (prescribed) fires, used in forest management, are usually with low-to-moderate severity and promote renovation of the dominant vegetation through removal of undesired species and temporary increase of pH and available nutrients (Certini, 2005; Santin and Doerr, 2016). Severe fires, like most wildfires, can cause irreversible damage to the soil through removal of organic matter, loss of nutrients, changes in texture, promoting erosion, and alteration in biological community (Certini, 2005). The heterogeneity of the soil cover, vegetation, and topography at a landscape scale lead to differences in the spatial position of the highest fire temperatures, on the one hand, and the highest temperatures, experienced by the soil, on the other (Stoof et al., 2013). The authors show in a catchment-scale fire experiment that soils are cooler where fuel load is high and fire is hot. This suggests that the most severe soil damages occur at locations where fuel load and fire intensity are low rather than high. Heterogeneity of the vegetation cover (e.g., fuel load) and weather conditions (wind speed, wind direction) lead to inhomogeneous heating of the soil beneath the fire (Gimeno Garcia et al., 2004).

Depending on burning conditions and, more specially, the abundance of O_2 supply, OM may burn completely and only ash is the residue. If the burning process is interrupted or O_2 supply is limited, charcoal is formed as a residue (Braadbaart and Poole, 2008). The chemical composition of the charcoal reflects the prevailing burning conditions, defined into three classes—grass and forest ground fires (reaching T_{max} of $285 \pm 143°C$), shrub fires (T_{max} of $503 \pm 211°C$), and domestic fires (T_{max} of $797 \pm 165°C$) (Wolf et al., 2013). Three main factors influence the charcoal formation: charring duration, temperature, and fuel. Usually, fire affects the topmost layers of soil up to the depth of 6–8 cm, thus also changing iron oxide mineralogy. Typically, pedogenic iron (oxy)hydroxides, which are synthesized at normal conditions near the Earth's surface, experience thermal transformations during fire. As a result, iron oxide maghemite is formed, which is a thermal transformation product of Fe hydroxides goethite/lepidocrocite/ferrihydrite in the presence of OM (Cornell and Schwertmann, 2003; Hanesch et al., 2006). Thus, well-established strong surface enhancement with ferrimagnetic mineral maghemite as a result of thermal transformations of weakly magnetic oxyhydroxides (goethite, ferryhidrate, and lepidocrocite) is observed in top layers of the burnt soils (Mullins and Tite, 1973; Ketterings et al., 2000; Chaparro et al., 2006; Oldfield and Crowther, 2007; Clement et al., 2011). Another source of magnetic minerals in burnt soils is the residual ash, which contains a significant amount of strongly magnetic iron oxides (Lu et al., 2000). As shown by Lu et al. (2000), the degree of enhancement strongly depends on the type of plants—C3 or C4. The magnetic susceptibility value for the ash residue of C4 plants is $532 \pm 61 \times 10^{-8}$ m^3/kg, while susceptibility of the C3 ash is only $120 \pm 65 \times 10^{-8}$ m^3/kg. Consequently, the ash residues from burning grassland (which consists mainly of C4 plants) alone may enhance soil magnetic susceptibility up to 30−40% (Lu et al., 2000). On the other hand, some authors report insignificant magnetic enhancement of burnt soils (Roman et al., 2013). The latter study is based on detailed mineral magnetic characterization of topsoil samples from prairie and oak savanna fields, subjected to a series of prescribed burns. The authors report the presence of significant magnetic enhancement with SP particles only for the severely burnt locations, while moderate burn did not produce magnetic enhancement, which could be distinguished from the natural pedogenic magnetic signal. These results may also be explained by the wide variability of the soil temperature, which is attained during the prescribed fires, ranging from less than 40 to 480°C (Penman and Towerton, 2008; Write and Clark, 2008).

The sensitivity of the soil magnetic mineralogy on fire influence is a promising tool for reconstructions of past fire regimes in combination with other methods (Conedera et al., 2009). To discriminate between natural magnetic enhancement of soils and fire-induced enhancement, Oldfield and Crowther (2007) demonstrate that magnetic properties, related to concentration and grain size of the fine SP particles, could be successfully applied. Biplot of magnetic ratios of χ_{arm}/χ_{fd} and χ_{arm}/χ well differentiate burnt soils from unburnt soils, as well as from sediments and extracellular magnetite (Oldfield and Crowther, 2007).

The published literature up to now on the magnetism of fire-affected soils considers depth variations in the magnetic properties usually at 1-cm interval. The next example from our study of fire-affected soil shows the magnetic signature along the depth of a soil profile under prescribed burn of a grassland area along the Danube River (Fig. 10.5.1). The soil is Haplic Chernozem developed on loess. Depth variations of the magnetic susceptibility along native unburnt soil profile in close proximity to the burnt plot exhibit a typical picture of pedogenic magnetic enhancement in the upper 100 cm with a maximum value of $\chi = 101 \times 10^{-8}$ m^3/kg at the topmost part of the profile (Fig. 10.5.1a).

Sampling of the burnt soil at 0.5-cm interval of the uppermost 2 cm reveals a higher-resolution picture of the effect of grass fire on the magnetic signature of the burnt Chernozem. Clearly, only the uppermost 0.5 cm exhibits significantly higher values of magnetic susceptibility and anhysteretic remanence. However, the relative enhancement with SP grains is higher, which is deduced from the decrease of the ratio χ_{arm}/χ in the upper 0.5 cm (Fig. 10.5.1b). Because of the finding that the maximum enhancement is observed at the topmost 0.5 cm, which is represented 80% by grass ashes, magnetic enhancement due to the high content of strongly magnetic minerals in the ash is more probable than thermal transformations of the soil-forming minerals from the mineral horizon. The presence of significant amounts of magnetic fraction in ash residues is reported by Lu et al. (2000), which is in line with the chemistry of ashes from biomass combustion (Biedermann and Obernberger, 2005).

10.6 ENVIRONMENTAL MAGNETISM APPLIED TO LANDMINE CLEARANCE AND FORENSIC STUDIES

Landmines are often used in armed conflicts, representing a severe hazard to civil society long after the end of the conflict. Subsequent clearance operations need proper detection of the buried objects. Electromagnetic induction metal detectors are among the most commonly used equipment in humanitarian operations because they are relatively inexpensive and easy to use. The response signal of the metal detector is

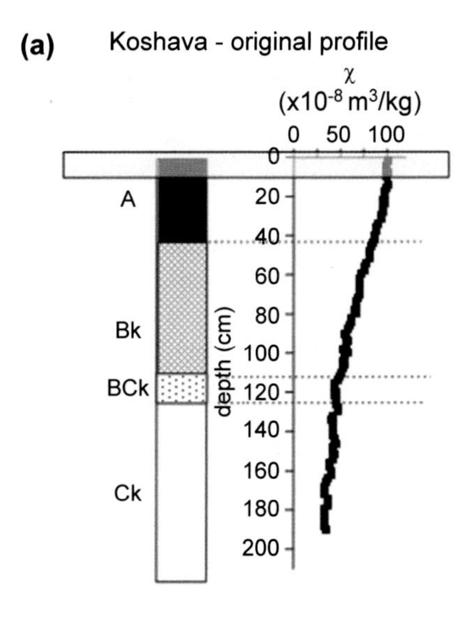

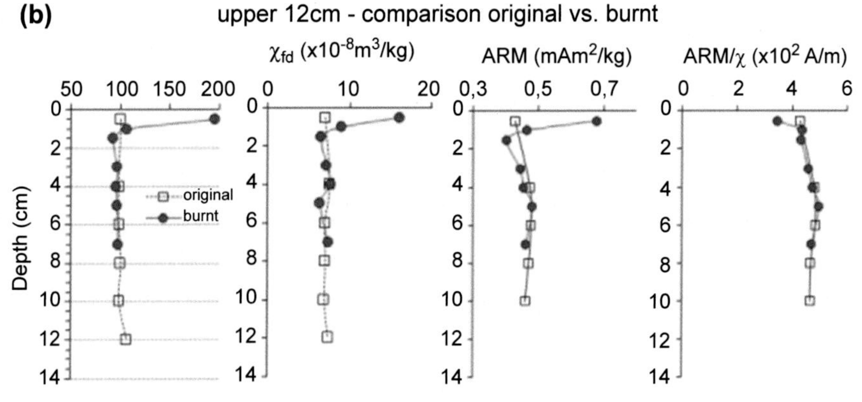

FIGURE 10.5.1

(a) Depth variations of mass-specific magnetic susceptibility (χ) along natural (nonaffected by burning) profile of Haplic Chernozem; (b) comparison of the depth variations of magnetic parameters (magnetic susceptibility, frequency-dependent magnetic susceptibility, ARM and the ratio ARM/χ for the upper 12 cm of the original and burnt soils.

influenced by a number of factors, such as detector technology, target type, orientation, and the properties of the surrounding soil (Butler, 2003). The most influential parameters are soil magnetic susceptibility and frequency-dependent magnetic susceptibility (Preetz et al., 2009a,b; Igel et al., 2012). Modern antipersonnel mines have a low metal content mostly within the ignition device, which causes a low contrast between the mine's signal and the background signal of the soil. Recent, more-sophisticated metal detectors are designed for a ground compensation mode of operation, which takes into account the signal of the surrounding soil, but in some cases this causes a reduction in sensitivity or false alarms or the identification of low-metal content mines is impossible. Another class of detectors work in a time domain (pulse induction) and are sensitive to the presence of viscous SP/SSD ferromagnetic grains. The response of this SP fraction to the detector is similar to an eddy current, which is induced by metallic targets (mines), thus causing false alarms. The latter strongly increases the time, cost, and uncertainty in landmine detection during clearance operations (Hannam and Dearing, 2008). A major task during planning of demining operations is to choose the most suitable detector type, according to the expected degree of influence of the soil magnetic signal on the sensitivity and effectiveness of the detector for landmine detection. This requires a prediction (estimation) of the spatial variability and intensity of the soil magnetic susceptibility. Due to the worldwide scale of the problem with landmine clearance, a special proposal for the development of a world soil database related to the soil electromagnetic characteristics (magnetic susceptibility and electrical conductivity) has been proposed (Das et al., 2003). In line with this initiative, extensive studies of soil magnetic properties from tropics have been carried out (van Dam et al., 2005; Preetz et al., 2009a,b; Igel et al., 2012). The emphasis on tropical soils is given because most of the affected countries by landmines and unexploded ordnance (UXO) are in the tropics. A total of 511 samples from lateritic soils developed on different parent materials were used for magnetic susceptibility and frequency-dependent magnetic susceptibility measurements by Preetz et al. (2009a,b). The authors consider the effect of parent rock lithology as a primary factor in controlling the magnetic susceptibility of the soils and, thus, they define six groups of soils, developed on different parent rock materials—ultrabasic, basic, acid, clays and clay slates, phyllites, and sandstones. The obtained values of the volume magnetic susceptibility vary in a wide range—from 1 to $11,800 \times 10^{-5}$ SI. The highest values with a median of about 1500×10^{-5} SI are observed for the soils developed on ultrabasic and basic rocks. The greatest scatter is also typical for these groups, which is explained by the various types of rocks from different locations and the natural variability in the Fe content of the mafic magma, as well as the presence of rock pieces at the surface, etc. Classification of sites for landmine clearance exclusively according to parent rock is developed and defines the median and the upper 90% quantile of the magnetic susceptibility for each group, corresponding to various severities of the effects on landmine detectors—very severe ($K > 2000 \times 10^{-5}$ SI); severe ($500 < K < 2000 \times 10^{-5}$ SI), moderate ($50 < K < 500 \times 10^{-5}$ SI), and neutral ($K < 50 \times 10^{-5}$ SI) (Preetz et al., 2009a,b). More precise assessment can be obtained taking into account the degree of weathering of the different soil types. The latter is related to the magnetic pedogenic enhancement, which further leads to increased values of magnetic susceptibility for certain climate conditions (see Section 10.1 of this chapter). Based on this classification system, Preetz et al. (2009b) established maps of predicted soil susceptibilities of Angola based on the 90% quantiles of susceptibility data and geological maps. Igel et al. (2012) consider the distribution and values of the frequency-dependent magnetic susceptibility of the same data set, reported by Preetz et al. (2009a,b), and define a classification system based on the values of volume frequency-dependent magnetic susceptibility (in 10^{-5} SI units). Two groups of soils are considered: (1) soils, developed on ultramafic and mafic rocks, and (2) soils developed on intermediate, felsic, and sedimentary rocks. The first group exhibits a very severe effect on the detector performance ($K_{fd\ median} = 26 \times 10^{-5}$ SI with 90% quantile of 390), while the soils in the second group have various effects (from severe to neutral), depending on the parent rock- and soil-forming factors. Hannam and Dearing (2008), in a study of soils from Bosnia-Herzegovina, present a classification of soils based on mass-specific magnetic susceptibility corresponding to the efficacy of Schiebel AN-19/2 metal detector with categories "very high" for $\chi > 1500 \times 10^{-8}$ m^3/kg, "high" $376 < \chi < 1500 \times 10^{-8}$ m^3/kg, "medium" $37.6 < \chi < 376 \times 10^{-8}$ m^3/kg, "low" $37.6 < \chi < 375 \times 10^{-8}$ m^3/kg, and "very low" $0 < \chi < 37.5 \times 10^{-8}$ m^3/kg.

Establishment of spatial maps of soil magnetic susceptibility and frequency-dependent susceptibility provides demining authorities with valuable information on the likelihood of having problems with detectors performance for landmine detection in areas, covered by different soil types.

Another recent application of soil magnetism is its utilization in the forensic research. Forensic searches for buried objects (clandestine graves; illegally buried weapons, drugs, illegal waste disposals) use various conventional and geoforensic methods (remote sensing, site study, dogs, soil probes, near surface geophysics, etc.) (Pringle et al., 2012; Pirrie et al., 2013). Pringle et al. (2012a, 2015) report the application of magnetic susceptibility measurements of soil, covering experimental clandestine graves, different forensic objects (bread knife, gun, ammunition box, mortar shell, etc.) and observed well-expressed anomalies in the magnetic susceptibility measured on the surface profiles crossing the buried items. The objects were buried at 15-cm depth and measurements were performed with Bartington MS2B sensor. Taking into account the penetration depth of the sensor and the metallic character of part of the objects, their direct strongly magnetic signal is measured at the surface. In case of the simulated clandestine grave, the observed anomaly is probably due to the contrast in the soil magnetic properties of the topsoil and the subsoil, which were disturbed (mixed) during digging of the grave (Pringle et al., 2015).

10.7 ENVIRONMENTAL MAGNETISM IN ARCHEOLOGY

Soils at archaeological sites are routinely studied in archaeological practice and increasing involvement of multidisciplinary approaches and more sophisticated and precise analytical methods give wide possibilities for gaining information about human activities and use of the landscape (Canti and Huisman, 2015). A comprehensive review of soil science application in archeology is presented by Walkington (2010), who emphasized that only a combined multidisciplinary approach could generate thorough reconstruction of the ancient environment and human activities. The author presents major steps for a correct application of soil properties in an archaeological context, pointing out that soil processes, related to pedogenesis, diagenetic change, or anthropogenic inputs/removals/transfers or transformations, are in the core of an accurate interpretation. Along with the well-established and emerging chemical methods (Pastor et al., 2016), magnetic methods are also widely applied in archeology (Fassbinder, 2015). The two reviews (Fassbinder, 2015; Pastor et al., 2016) provide an excellent and systematic view on the recent achievements of the major branches in the natural sciences, providing a significant advance in the interpretation of archaeological sites. Here, a brief outline of the key areas of possible application of soil magnetism in archaeological research will be given. These could be classified in the following groups:

- Aiding in interpretation of the magnetic prospection data
- Identification of archaeological features/levels affected by fire and recognition of anthropogenically influenced areas
- Identification of material sources (in conjunction with classic chemical methods). Sourcing obsidians.

- Determination of firing temperature and atmosphere (reducing–oxidizing) of burnt clay remains (pottery, kilns, ovens, etc.)

Apart from the soil magnetism applications mentioned earlier, archaeomagnetism—the reconstruction of the geomagnetic field elements (direction and intensity) since the Neolithic—is used for archaeomagnetic dating and synchronizations (e.g., Herve et al., 2013; Kovacheva et al., 2014; Pavón-Carrasco et al., 2014), making another important contribution to archeology.

10.7.1 SOIL MAGNETISM APPRAISAL IN MAGNETIC PROSPECTION

The role of soil mineral magnetic properties for precise and reliable interpretation of magnetic prospection results is discussed in a number of studies. A major contribution of mineral magnetism consists in its ability to distinguish between remanent and induced magnetization components into the total magnetization signal (Dalan, 2008; Fassbinder, 2015 and references therein). Since a magnetic survey does not distinguish between the induced and remanent magnetizations, application of rock-magnetic techniques becomes important. Magnetic susceptibility, as the simplest and easy to obtain (even in the field) parameter, is widely applied for resolving this issue. Interpretation of positive and negative magnetic anomalies and related soil/artifact response is discussed in detail in Fassbinder (2015). Identification of the kind of magnetic oxides (magnetite, maghemite, hematite), their magnetic grain sizes, and concentration yield detailed information, aiding in identifying more reliably the archaeological feature buried in the soil. Depending on the particular magnetic mineral detected, its grain-size distribution, and other magnetic characteristics, employed in soil magnetism (as in Chapters 2–7), it is possible to draw conclusions about the possible origin of the feature, which creates a given magnetic anomaly. Knowledge about the most characteristic magnetic properties of each specific soil type (Chapters 2–7) and the pedogenic minerals formed under different environmental conditions allow distinguishing the natural soil signature from human-induced alterations (Walkington, 2010).

10.7.2 IDENTIFICATION OF ARCHAEOLOGICAL FEATURES/LEVELS AFFECTED BY FIRE AND RECOGNITION OF ANTHROPOGENICALLY INFLUENCED AREAS

As discussed in Section 10.5, fire severely influences the soil's physical and chemical properties, generally leading to creation of strongly magnetic maghemite—a thermal alterations product of several iron oxyhydroxides (lepidocrocite, goethite, ferrihydrite) (Cornell and Schwertmann, 2003; Hanesch et al., 2006), as well as in Fe-bearing clay minerals (Murad, 1998; Riccardi et al., 1999; Zhang et al., 2012; Jiang et al., 2015). Since the concentration and grain size of the newly created magnetic phase reflect very sensitively the thermodynamic conditions during firing (atmosphere, heating duration, heating rate, maximum temperature achieved, etc.), magnetic characteristics such as saturation magnetization (M_s), magnetic susceptibility (χ), magnetic remanences (ARM, IRM), thermomagnetic analyses of magnetic susceptibility, etc. could be used to resolve the inverse problem. Most often, burnt clay remains contain a much larger amount of strongly magnetic minerals with prevailing SP fraction compared with natural (pedogenically enhanced) soil (Oldfield and Crowther, 2007). As established in a number of studies, the degree of magnetic enhancement during laboratory heating depends on the raw clay type and (free) iron available for thermal transformations (Tite, 1972; Jordanova et al., 2001; Van Klinken, 2001; Kostadinova-Avramova and Kovacheva, 2013), as well as on the heating atmosphere (Murad and Wagner, 1998). Therefore, mineral magnetic studies of depth profiles from archaeological sites could sensitively detect the presence of a cultural layer containing fired remains (Dalan, 2008; Rosendahl et al., 2014; Fassbinder, 2015). A field-based plow zone magnetic susceptibility mapping of an archaeological site in Georgia (Bigman, 2014) successfully delineated anthropogenically affected areas and allowed delineation of the site's extension to be made. Mixing of anthropogenically altered archaeological levels inside the plow zone caused higher soil magnetic susceptibility, positively correlated with the abundance of archaeological artifacts. In another study, Nolan (2014) uses integrated magnetic susceptibility/phosphorous content mapping in an area of supposed prehistoric non-intensive agricultural fields. A high correlation between magnetic susceptibility and phosphate was obtained, and the magnetic susceptibility mapping and laboratory measurements of soil samples are suggested as a useful and cost-effective tool for prospecting for non-intensive agricultural fields (Nolan, 2014).

10.7.3 IDENTIFICATION OF MATERIAL SOURCES (IN CONJUNCTION WITH CLASSIC CHEMICAL METHODS). SOURCING OBSIDIANS

Provenance studies are mostly based on comparison between the elemental composition of a "source" and different artifacts made of materials coming from it (Pastor et al., 2016). In addition, mineral magnetic techniques are successfully applied for sourcing different archaeological materials, like raw fuel (Church et al., 2007), obsidians (Frahm and Feinberg, 2013; references therein; Frachm et al., 2014), or clay for brick production (Scalenghe et al., 2015). Church et al. (2007) performed laboratory-controlled burning experiments with four main fuel types (well-humified peat, peaty turf, fibrous peat, and wood). Based on a set of six magnetic parameters, the authors carry out discriminant analysis and show good discrimination between the well-humified peat ash and wood ash, with some overlap between the fibrous-upper peat ash and peat turf ash. Further, Church et al. (2007) compare these laboratory results with the magnetic properties of materials gathered from different archaeological sites in a wider area and make conclusions regarding the most used fuel source and related implications of the sites' archaeological evolution. Magnetic characterization of other artifacts—obsidians—is proved to give valuable additional information (to the classic chemical elemental analyses and geochronology) related to matching artifacts to locations within a specific flow and different quarries used by ancient people (Frahm and Feinberg, 2013; references therein; Frahm et al., 2014).

10.7.4 DETERMINATION OF FIRING TEMPERATURE AND ATMOSPHERE (REDUCING–OXIDIZING) OF BURNT CLAY REMAINS (POTTERY, KILNS, OVENS, ETC.)

The estimation of firing temperature used in the production of pottery and the relationship between firing atmosphere and color are among the major subjects in studies on ancient production technologies (Tite, 2008). Appraisal of the maximum firing temperatures is important for establishing firing technology, used in a society and further to distinguish between such technologies either in space or in time. The analytical methods applied generally evaluate mineralogy and microstructure of the object as a function of firing temperature. Temperature estimation through rock magnetic parameters (Spassov and Hus, 2006; Linford and Platzman, 2004) is based on the assumption that a sample that had been heated in the past to a certain temperature T1 will not change its rock magnetic properties if again heated to T1 in the laboratory under similar conditions. This is valid only if the magnetic minerals already reached chemical equilibrium at T1. When the sample is heated above T1, magnetic minerals are no longer in chemical equilibrium, resulting in rock magnetic property changes. Alteration index (Hrouda et al., 2003) obtained from partial thermomagnetic curves of magnetic susceptibility during heating—cooling cycles is an additional tool for evaluation of paleotemperatures. Investigations of thermomagnetic behavior of magnetic susceptibility applying different heating rates (Jordanova and Jordanova, 2016) could give additional information about the stability and magnetic phase composition of fired archaeological remains. Rock magnetic techniques could be successfully applied for resolving firing temperatures, as evidenced by a number of studies (Linford and Platzman, 2004; Spassov and Hus, 2006; Oldfield and Crowther, 2007; Church et al., 2007; Carrancho and Villalain, 2011; Rasmussen et al., 2012). The core of the methodology is based on the thermal behavior of different forms of iron oxides (Cornell and Schwertmann, 2003) and related transformations in their magnetic properties (Dunlop and Özdemir, 1997; Zboril et al., 2002). As noted by Rassmussen et al. (2012), however, any method is vulnerable to the uncertainty introduced by the unknown firing history. Except for maximum firing temperature, many other factors influence the process of firing—heating rate, cooling rate, duration of firing, possible re-firing after the initial fire, fuel type. A very important initial condition is the raw clay type—its mineralogy and texture.

The described various aspects and possibilities for application of soil and mineral magnetism in archeology open many new opportunities for reinforcement of multidisciplinary approach in archaeological investigations and for obtaining new and probably underestimated information that can give clues for unexpected valuable knowledge.

REFERENCES

Achat, D., Pousse, N., Nicolas, M., Brédoire, F., Augusto, L., 2016. Soil properties controlling inorganic phosphorus availability: general results from a national forest network and a global compilation of the literature. Biogeochemistry 127, 255—272.

Adhikari, D., Yang, Y., 2015. Selective stabilization of aliphatic organic carbon by iron oxide. Sci. Rep. 5, 11214. http://dx.doi.org/10.1038/srep11214.

Adhikari, D., Poulson, S., Sumaila, S., Dynes, J., McBeth, J., Yang, Y., 2016. Asynchronous reductive release of iron and organic carbon from hematite—humic acid complexes. Chem. Geol. 430, 13—20.

Alloway, B.J., 1995. Heavy Metals in Soils. Chapman and Hall, London, UK.

Ameen, N.N., Kluegein, N., Appel, E., Petrovský, E., Kapler, A., Leven, C., 2014. Effect of hydrocarbon-contaminated fluctuating groundwater on magnetic properties of shallow sediments. Stud. Geophys. Geod. 58, 442—460.

Armstrong, A., Quinton, J.N., Maher, B.A., 2012. Thermal enhancement of natural magnetism as a tool for tracing eroded soil. Earth Surf. Process. Landforms 37, 1567—1572.

Ayoubi, S., Ahmadi, M., Abdi, M.R., Afshar, F.A., 2012. Relationships of Cs-137 inventory with magnetic measures of calcareous soils of hilly region in Iran. J. Environ. Radioact. 112, 45—51.

Balsam, W.L., Ellwood, B.B., Ji, J., Williams, E.R., Long, X., El Hassani, A., 2011. Magnetic susceptibility as a proxy for rainfall: worldwide data from tropical and temperate climate. Quat. Sci. Rev. 30, 2732—2744.

Barrón, V., Torrent, J., 2002. Evidence for a simple pathway to maghemite in Earth and Mars soils. Geochim. Cosmochim. Acta 66, 2801—2806.

Barrón, V., Torrent, J., De Grave, E., 2003. Hydromaghemite, an intermediate in the hydrothermal transformation of 2-line ferrihydrite into hematite. Am. Mineral. 88 (11), 1679—1688.

Biedermann, F., Obernberger, I., 2005. Ash-related problems during biomass combustion and possibilities for a sustainable ash utilization. In: Proceedings of International Conference of World Renewable Energy Congress (WREC), Aberdeen, UK.

Bigman, D., 2014. Mapping plow zone soil magnetism to delineate disturbed archaeological site boundaries. J. Archaeol. Sci. 42, 367–372.

Bloemendal, J.C., Liu, X., 2005. Rock magnetism and geochemistry of two plio-pleistocene Chinese loess-paleosol sequences-implications for quantitative palaeoprecipitation reconstruction. Palaeogeogr. Palaeoclimatol. Palaeoecol. 226, 149–166.

Blundell, A., Hannam, J.A., Dearing, J.A., Boyle, J.F., 2009. Detecting atmospheric pollution in surface soils using magnetic measurements: a reappraisal using an England and Wales database. Environ. Pollut. 157, 2878–2890.

Bodi, M., Martin, D.A., Santin, C., Balfour, V., Doerr, S.H., Pereira, P., Cerda, A., Mataix-Solera, J., 2014. Wildland fire ash: production, composition and eco-hydro- geomorphic effects. Earth-Sci. Rev. 130, 103–127.

Bond, W.J., 2015. Fires in the Cenozoic: a late flowering of flammable ecosystems. Front. Plant Sci. 5, Article 749.

Borch, T., Fendorf, S., 2008. Phosphate interactions with iron (hydr)oxides: mineralization pathways and phosphorus retention upon bioreduction. In: Barnett, M.O., Kent, D.B. (Eds.), Developments in Earth & Environmental Sciences, vol. 7. Elsevier B.V., pp. 321–348. http://dx.doi.org/10.1016/S1571-9197(07)07012-7

Braadbaart, F., Poole, I., 2008. Morphological, chemical and physical changes during charcoalification of wood and its relevance to archaeological contexts. J. Archaeol. Sci. 35, 2434–2445.

Butler, D.K., 2003. Implications of magnetic backgrounds for unexploded ordnance detection. J. Appl. Geophys. 54 (1–2), 111–125.

Canti, M., Huisman, D., 2015. Scientific advances in geoarchaeology during the last twenty years. J. Archaeol. Sci. 56, 96–108.

Carrancho, Á., Villalaín, J.J., 2011. Different mechanisms of magnetisation recorded in experimental fires: archaeomagnetic implications. Earth Planet. Sci. Lett. 312, 176–187.

Certini, G., 2005. Effects of fire on properties of forest soils: a review. Oecologia 143, 1–10.

Chadwick, O.A., Chorover, J., 2001. The chemistry of pedogenic thresholds. Geoderma 100, 321–353.

Chaparro, M.A.E., Gogorza, C.S.G., Chaparro, M.A.E., Irurzun, M.A., Sinito, A.M., 2006. Review of magnetism and heavy metal pollution studies of various environments in Argentina. Earth Planet. Space 58, 1411–1422.

Chen, L.M., Zhang, G.L., Effland, W., 2011. Soil characteristic response times and pedogenic thresholds during the 1000-Year evolution of a paddy soil chronosequence. Soil Sci. Soc. Am. J. 75, 1807–1820.

Chen, L.M., Zhang, G.L., Rossiter, D.G., Cao, Z.H., 2015. Magnetic depletion and enhancement in the evolution of paddy and non-paddy soil chronosequences. Eur. J. Soil Sci. 66 (5), 886–897.

Chlupáčová, M., Hanák, J., Müller, P., 2010. Magnetic susceptibility of cambisol profiles in the vicinity of the Vír Dam, Czech Republic. Stud. Geophys. Geod. 54, 153–184.

Church, M.J., Peters, C., Batt, C.M., 2007. Sourcing fire ash on archaeological sites in the Western and Northern Isles of Scotland, using mineral magnetism. Geoarchaeol. Int. J. 22 (7), 747–774.

Clement, B.M., Javier, J., Sah, J.P., Ross, M.S., 2011. The effects of wildfires on the magnetic properties of soils in the Everglades. Earth Surf. Process. Landforms 36, 460–466.

Conedera, M., Tinner, W., Neff, C., Meurer, M., Dickens, A.F., Krebs, P., 2009. Reconstructing past fire regimes: methods, applications, and relevance to fire management and conservation. Quat. Sci. Rev. 28, 555–576.

Conrad, R., Frenzel, P., 2002. Flooded soils. In: Britton, G. (Ed.), Encyclopedia of Environmental Microbiology. John Wiley & Sons, New York, pp. 1316–1333.

Cornell, R., Schwertmann, U., 2003. The Iron Oxides. Structure, Properties, Reactions, Occurrence and Uses (Weinheim, New York).

Dalan, R.A., 2008. A review of the role of magnetic susceptibility in archaeogeophysical studies in the USA: recent developments and prospects. Archaeol. Prospect 15, 1–31.

Das, Y., McFee, J.E., Russell, K.L., Cross, G., Katsube, T.J., September 15, 2003. Soil information requirements for humanitarian demining: the case for a soil properties database. In: Proc. SPIE 5089, Detection and Remediation Technologies for Mines and Minelike Targets VIII, vol. 1146. http://dx.doi.org/10.1117/12.486306.

De Alba, S., Lindstrom, M.J., Schumacher, T.E., Malo, D.D., 2004. Soil landscape evolution due to soil redistribution by tillage: a new conceptual model of soil catena evolution in agricultural landscapes. CATENA 58, 77–100.

de Jong, E., Nestor, P.A., Pennock, D.J., 1998. The use of magnetic susceptibility to measure long-term soil redistribution. CATENA 32, 23–35.

De Jong, E., Heck, R., Ponomarenko, E., 2005. Magnetic susceptibility of soil separates of Gleysolic and Chernozemic soils. Can. J. Soil Sci. 85, 233–244.

de Lima, J.L.M.P., Dinis, P.A., Souza, C.S., de Lima, M.I.P., Cunha, P.P., Azevedo, J.M., Singh, V.P., Abreu, J.M., 2011. Patterns of grain-size temporal variation of sediment transported by overland flow associated with moving storms: interpreting soil flume experiments. Nat. Hazards Earth Syst. Sci. 11, 2605–2615.

Dearing, J.A., Morton, R.I., Price, T.W., Foster, I.D.L., 1986. Tracing movements of topsoil by magnetic measurements: two case studies. Phys. Earth Planet. Inter. 42, 93–104.

Dearing, J., Lees, J., White, C., 1995. Mineral magnetic properties of acid gleyed soils under oak and Corsican Pine. Geoderma 68, 309–319.

DIN ISO 10694, 1995. Soil Quality – Determination of Organic and Total Carbon after Dry Combustion (Elementary Analysis). Beuth Verlag GmbH, Berlin.

DIN ISO 13878, 1998. Soil Quality – Determination of Total Nitrogen Content by Dry Combustion (Elemental Analysis). Beuth Verlag GmbH, Berlin.

Dunlop, D., Özdemir, Ö., 1997. Rock Magnetism. Fundamentals and frontiers. In: Edwards, D. (Ed.), Cambridge Studies in Magnetism. Cambridge University Press.

Fassbinder, J., 2015. Seeing beneath the farmland, steppe and desert soil: magnetic prospecting and soil magnetism. J. Archaeol. Sci. 56, 85–95.

Fialova, H., Maier, G., Petrovský, E., Kapička, A., Boyko, T., Scholger, R., MAGPROX Team, 2006. Magnetic properties of soils from sites with different geological and environmental settings. J. Appl. Geophys. 59, 273–283.

Frahm, E., Feinberg, J., 2013. From flow to quarry: magnetic properties of obsidian and changing the scale of archaeological sourcing. J. Archaeol. Sci. 40, 3706–3721.

Frahm, E., Feinberg, J., Schmidt-Magee, B., Wilkinson, K., Gasparyan, B., Yeritsyan, B., Karapetian, S., Meliksetian, K., Muth, M., Adler, D., 2014. Sourcing geochemically identical obsidian: multiscalar magnetic variations in the Gutansar volcanic complex and implications for Palaeolithic research in Armenia. J. Archaeol. Sci. 47, 164–178.

Geiss, C.E., Zanner, C.W., 2007. Sediment magnetic signature of climate in modern loessic soils from the Great Plains. Quat. Int. 162–163, 97–110.

Geiss, C.E., Egli, R., Zanner, C.W., 2008. Direct estimates of pedogenic magnetite as a tool to reconstruct past climates from buried soils. J. Geophys. Res. 113, B11102.

Gimeno-García, E., Andreu, V., Rubio, J.L., 2004. Spatial patterns of soil temperatures during experimental fires. Geoderma 118, 17–38.

Go1uchowska, B.J., 2001. Some factors affecting an increase in magnetic susceptibility of cement dusts. J. Appl. Geophys. 48 (2), 103–112.

Gong, Z.T., 1986. Origin, evolution and classification of paddy soils in China. Adv. Soil Sci. 5, 174–200.

Govers, G., Lobb, D., Quine, T.A., 1999. Tillage erosion and translocation: emergence of a new paradigm in soil erosion research. Soil Till. Res. 51, 167–174.

Grimley, D., Arruda, N., 2007. Observations of magnetite dissolution in poorly drained soils. Soil Sci. 172 (12), 968–982.

Guzman, G., Barron, V., Gomez, J.A., 2010. Evaluation of magnetic iron oxides as sediment tracers in water erosion experiments. CATENA 82, 126–133.

Han, G.-Z., Zhang, G.-L., 2013. Changes in magnetic properties and their pedogenetic implications for paddy soil chronosequences from different parent materials in south China. Eur. J. Soil Sci. 64 (4), 435–444.

Hanesch, M., Stanjek, H., Petersen, N., 2006. Thermomagnetic measurements of soil iron minerals: the role of organic carbon. Geophys. J. Int. 165, 53–61.

Hanesch, M., Rantitsch, G., Hemetsberger, S., Scholger, R., 2007. Lithological and pedological influences on the magnetic susceptibility of soil: their consideration in magnetic pollution mapping. Sci.Tot. Environ. 382, 351–363.

Hannam, J.A., Dearing, J.A., 2008. Mapping soil magnetic properties in Bosnia and Herzegovina for landmine clearance operations. Earth Planet. Sci. Lett. 274, 285–294.

Hansel, C., Benner, S., Netss, J., Dohnalkova, A., Kukkadapu, R., Fendorf, S., 2003. Secondary mineralization pathways induced by dissimilatory iron reduction of ferrihydrite under advective flow. Geochim. Cosmochim. Acta 67 (16), 2977–2992.

Hansel, C., Benner, S., Fendorf, S., 2005. Competing Fe(II)-induced mineralization pathways of ferrihydrite. Environ. Sci. Technol. 39, 7147–7153.

Hao, X., Liu, S., Wu, J., Hu, R., Tong, C., Su, Y., 2008. Effect of long-term application of inorganic fertilizer and organic amendments on soil organic matter and microbial biomass in three subtropical paddy soils. Nutr. Cycl. Agroecosyst. 81 (1), 17–24.

Heller, F., Liu, T., 1986. Palaeoclimate and sedimentary history from magnetic susceptibility of loess in China. Geophys. Res. Lett. 13, 1169–1172.

Heller, F., Strzyszcz, Z., Magiera, T., 1998. Magnetic record of industrial pollution in forest soils of Upper Silesia, Poland. J. Geophys. Res. 103 (B8), 17767–17774.

Hengl, T., 2006. Finding the right pixel size. Comput. Geosci. 32, 1283–1298.

Hervé, G., Chauvin, A., Lanos, P., 2013. Geomagnetic field variations in Western Europe from 1500 BC to AD 200. Part II: new intensity secular variation curve. Phys. Earth Planet. Inter. 218, 51–65.

Heslop, D., Roberts, A.P., 2013. Calculating uncertainties on predictions of palaeoprecipitation from the magnetic properties of soils. Glob. Planet. Change 110, 379–385.

Howard, J.L., Orlicki, K.M., 2016. Composition, micromorphology and distribution of microartifacts in anthropogenic soils, Detroit, Michigan, USA. CATENA 138, 103–116.

Hrouda, F., Müller, P., Hanák, J., 2003. Repeated progressive heating in susceptibility versus temperature investigation: a new palaeotemperature indicator? Phys. Chem. Earth 28, 653–657.

Hu, X.-F., Jiang, Y., Shu, Y., Hu, X., Liu, L., Luo, F., 2014. Effects of mining wastewater discharges on heavy metal pollution and soil enzyme activity of the paddy fields. J. Geochem. Explor. 147, 139–150.

Huang, L.M., Thompson, A., Zhang, G.L., Chen, L.M., Han, G.Z., Gong, Z.T., 2015. The use of chronosequences in studies of paddy soil evolution: a review. Geoderma 237–238, 199–210.

Hyland, E., Sheldon, N.D., Van der Voo, R., Badgley, C., Abrajevitch, A., 2015. A new paleoprecipitation proxy based on soil magnetic properties: implications for expanding paleoclimate reconstructions. Geol. Soc. Am. Bull. http://dx.doi.org/10.1130/B31207.1.

Igel, J., Preetz, H., Altfelder, S., 2012. Magnetic viscosity of tropical soils: classification and prediction as an aid for landmine detection. Geophys. J. Int. 190, 843–855.

IUSS Working Group WRB, 2006. World Reference Base for Soil Resources 2006. World Soil Resources Reports 103. FAO, Rome.

Jakšík, O., Kodešová, R., Kapička, A., Klement, A., Fér, M., Nikodem, A., 2016. Using magnetic susceptibility mapping for assessing soil degradation due to water erosion. Soil Water Res. 11 (2), 105–113.

Jelenska, M., Hasso-Agopsowicz, A., Kopcewicz, B., Sukhorada, A., Tyamina, K., Kacdzialko-Hofmokl, M., Matviishina, Z., 2004. Magnetic properties of the profiles of polluted and non-polluted soils. A case study from Ukraine. Geophys. J. Int. 159, 104–116.
Jiang, Z., Liu, Q., Zhao, X., Jin, C., Liu, C., Li, S., 2015. Thermal magnetic behaviour of Al-substituted haematite mixed with clay minerals and its geological significance. Geophys. J. Int. 200, 130–143.
Jordanova, D., Jordanova, N., 2016. Thermomagnetic behaviour of magnetic susceptibility—heating rate and sample size effects. Front. Earth Sci. 3, 90. http://dx.doi.org/10.3389/feart.2015.00090.
Jordanova, D., Petrovský, E., Jordanova, N., Evlogiev, J., Butchvarova, V., 1997. Rock magnetic properties of recent soils from Northeastern Bulgaria. Geophys. J. Int. 128, 474–488.
Jordanova, N., Petrovský, E., Kovacheva, M., Jordanova, D., 2001. Factors determining magnetic enhancement of burnt clay from archaeological sites. J. Archaeol. Sci. 28 (11), 1137–1148.
Jordanova, D., Veneva, L., Hoffmann, V., 2003. Magnetic susceptibility screening of anthropogenic impact on the Danube river sediments in Northwestern Bulgaria – preliminary results. Studi. Geophys. Geod. 47, 403–418.
Jordanova, D., Jordanova, N., Hoffmann, V., 2006. Magnetic mineralogy and grain-size dependence of hysteresis parameters of single spherules from industrial waste products. Phys. Earth Planet. Inter. 154, 255–265.
Jordanova, N., Jordanova, D., Tsacheva, T., 2008. Application of magnetometry for delineation of anthropogenic pollution in areas covered by various soil types. Geoderma 144, 557–571.
Jordanova, D., Jordanova, N., Petrov, P., Tsacheva, T., 2010. Soil development of three Chernozem-like profiles from North Bulgaria revealed by magnetic studies. CATENA 83 (2–3), 158–169.
Jordanova, D., Jordanova, N., Werban, U., 2013a. Environmental significance of magnetic properties of Gley soils near Rosslau (Germany). Environ. Earth Sci. 69 (5), 1719–1732.
Jordanova, D., Goddu, S.R., Kotsev, T., Jordanova, N., 2013b. Industrial contamination of alluvial soils near Fe–Pb mining site revealed by magnetic and geochemical studies. Geoderma 192 (1), 237–248.
Jordanova, D., Jordanova, N., Petrov, P., 2014. Pattern of cumulative soil erosion and redistribution pinpointed through magnetic signature of Chernozem soils. CATENA 120, 46–56.
Kapička, A., Jordanova, N., Petrovský, E., Ustjak, S., 2000. Magnetic stability of power-plant fly ash in different soil solutions. Phys. Chem. Earth 25 (5), 431–436.
Kapička, A., Petrovský, E., Fialova, H., Podrazsky, V., Dvorak, I., 2008. Mapping of anthropogenic pollution in the Giant mountains National Park using soil magnetometry. Stud. Geophys. Geod. 52, 271–284.
Kapička, A., Grison, H., Petrovský, E., Jakšík, O., Kodešová, R., 2015. Use of magnetic susceptibility for evaluation of soil erosion at two locations with different soil types. In: International Multidisciplinary Scientific GeoConference Surveying Geology and Mining Ecology Management, vol. 2. SGEM, pp. 417–424 issue (3).
Ketterings, Q., Bigham, J., Laperche, V., 2000. Changes in soil mineralogy and texture caused by slash-and-burn fires in Sumatra, Indonesia. Soil Sci. Soc. Am. J. 64, 1108–1117.
Kirk, G., 2004. The Biogeochemistry of Submerged Soils. Wiley, Chichester.
Klose, S., Tölle, R., Bäucker, E., Makeschin, F., 2003. Stratigraphic distribution of lignite-derived atmospheric deposits in forest soils of the Upper Lusatian Region, East Germany. Water Air Soil Pollut. 142, 3–25.
Kögel-Knabner, I., Amelung, W., Cao, Zh, Fiedler, S., Frenzel, P., Jahn, R., Kalbitz, K., Kölbl, A., Schloter, M., 2010. Biogeochemistry of paddy soils. Geoderma 157, 1–14.
Kölbl, A., Schad, P., Jahn, R., Amelung, W., Bannert, A., Cao, Z.H., Fiedler, S., Kalbitz, K., Lehndorff, E., Müller-Niggemann, C., Schloter, M., Schwark, L., Vogelsang, V., Wissing, L., Kögel-Knabner, I., 2014. Accelerated soil formation due to paddy management on marshlands (Zhejiang Province, China). Geoderma 228–229, 67–89.
Kostadinova-Avramova, M., Kovacheva, M., 2013. The magnetic properties of baked clays and their implications for past geomagnetic field intensity determinations. Geophys. J. Int. 195 (3), 1534–1550.

Kovacheva, M., Kostadinova-Avramova, M., Jordanova, N., Lanos, P., Boyadzhiev, Y., 2014. Extended and revised archaeomagnetic database and secular variation curves from Bulgaria for the last eight millennia. Phys. Earth Planet. Inter. 236, 79–94.

Kukkadapu, R., Zachara, J., Fredrickson, J., Kennedy, D., 2004. Biotransformation of two-line silica-ferrihydrite by a dissimilatory Fe(III)-reducing bacterium: formation of carbonate green rust in the presence of phosphate. Geochim. Cosmochim. Acta 68, 2799–2814.

Lecoanet, H., Leveque, F., Segura, S., 1999. Magnetic susceptibility in environmental applications: comparison of field probes. Phys. Earth Planet. Inter. 115, 191–204.

Li, J., Okin, G.S., Alvarez, L.J., Epstein, H.E., 2008. Effects of wind erosion on the spatial heterogeneity of soil nutrients in a desert grassland of southern New Mexico. Biogeochemistry 88, 73–88.

Li, Y., Guo, J., Velde, B., Zhang, G., Hu, F., Zhao, M., 2011. Evolution and significance of soil magnetism of basalt-derived chronosequence soils in tropical southern China. Agric. Sci. 2, 536–543.

Linford, N., Platzman, E., 2004. Estimating the approximate firing temperature of burnt archaeological sediments through an unmixing algorithm applied to hysteresis data. Phys. Earth Planet. Inter. 147, 197–207.

Liu, Q., Deng, C., Torrent, J., Zhu, R., 2007. Review of recent development in mineral magnetism of the Chinese loess. Quat. Sci. Rev. 26 (3–4), 368–385.

Liu, Q., Barrón, V., Torrent, J., Eeckhout, S., Deng, C., 2008. Magnetism of intermediate hydromaghemite in the transformation of 2-line ferrihydrite into hematite and its paleoenvironmental implications. J. Geophys. Res. 113, B01103. http://dx.doi.org/10.1029/2007JB005207.

Liu, Q., Hu, P., Torrent, J., Barrón, V., Zhao, X., Jiang, Z., Su, Y., 2010. Environmental magnetic study of a Xeralf chronosequence in northwestern Spain: indications for pedogenesis. Palaeogeogr. Palaeoclimatol. Palaeoecol. 293, 144–156.

Liu, Z., Liu, Q., Torrent, J., Barrón, V., Hu, P., 2013. Testing themagnetic proxy χfd/HIRM for quantifying paleoprecipitation in modern soil profiles from Shaanxi Province, China. Glob. Planet. Change 110, 368–378.

Liu, L., Zhang, K., Zhang, Z., Qiu, Q., 2015. Identifying soil redistribution patterns by magnetic susceptibility on the black soil farmland in Northeast China. CATENA 129, 103–111.

Logsdon, S.D., 2013. Depth dependence of chisel plow tillage erosion. Soil Till. Res. 128, 119–124.

Long, X., Ji, J., Balsam, W., 2011. Rainfall-dependent transformations of iron oxides in a tropical saprolite transect of Hainan Island, South China: spectral and magnetic measurements. J. Geophys. Res. 116, F03015.

Long, X., Ji, J., Balsam, W., Barrón, V., Torrent, J., 2015. Grain growth and transformation of pedogenic magnetic particles in red Ferralsols. Geophys. Res. Lett. 42, 5762–5770. http://dx.doi.org/10.1002/2015GL064678.

Lovley, D., Stolz, J., Nord, G., Phillips, E., 1987. Anaerobic production of magnetite by a dissimilatory iron-reducing microorganism. Nature 330, 252–254.

Lu, H., Liu, T., Gu, Z., Liu, B., Zhou, L., Han, J., Wu, N., 2000. Effect of burning C3 and C4 plants on the magnetic susceptibility signal in soils. Geophys. Res. Lett. 27 (13), 2013–2016.

Lu, S.G., Zhu, L., Yu, J.Y., 2012. Mineral magnetic properties of Chinese paddy soils and its pedogenic implications. CATENA 93, 9–17.

Lu, S., Yu, X., Chen, Y., 2016. Magnetic properties, microstructure and mineralogical phases of technogenic magnetic particles (TMPs) in urban soils: their source identification and environmental implications. Sci. Total Environ. 543, 239–247.

Lu, S.G., 2000. Characterization of subtropical soils by mineral magnetic measurements. Commun. Soil Sci. Plant Anal. 31 (1–2), 1–11.

Lützow, M., Kögel-Knabner, I., Ekschmitt, K., Matzner, E., Guggenberger, G., Marschner, D., Flessa, H., 2006. Stabilization of organic matter in temperate soils: mechanisms and their relevance under different soil conditions - a review. Eur. J. Soil Sci. 57, 426–445.

Mabit, L., Benmansour, M., Walling, D.E., 2008. Comparative advantages and limitations of the fallout radionuclides 137Cs, 210Pbex and 7Be for assessing soil erosion and sedimentation. J. Environ. Radioact. 99, 1799–1807.

Magiera, T., Zawadski, J., 2007. Using of high-resolution topsoil magnetic screening for assessment of dust deposition: comparison of forest and arable soil datasets. Environ. Monit. Assess. 125 (1–3), 19–28.

Magiera, T., Strzyszcz, Z., Kapička, A., Petrovský, E., 2006. Discrimination of lithogenic and anthropogenic influences on topsoil magnetic susceptibility in Central Europe. Geoderma 130, 299–311.

Magiera, T., Strzyszcz, Z., Rachwal, M., 2007. Mapping particulate pollution loads using soil magnetometry in urban forests in the Upper Silesia Industrial Region, Poland. For. Ecol. Manage. 248, 36–42.

Magiera, T., Kapička, A., Petrovský, E., Strzyszcz, Z., Fialova, H., Rachwal, M., 2008. Magnetic anomalies of forest soils in the upper Silesia–Northern Moravia region. Environ. Pollut. 156, 618–627.

Magiera, T., Jablonska, M., Strzyszcz, Z., Rachwal, M., 2011. Morphological and mineralogical forms of technogenic magnetic particles in industrial dusts. Atmos. Environ. 45, 4281–4290.

Magiera, T., Parzentny, H., Róg, L., Chybiorz, R., Wawer, M., 2015. Spatial variation of soil magnetic susceptibility in relation to different emission sources in southern Poland. Geoderma 255–256, 94–103.

Maher, B.A., Possolo, A., 2013. Statistical models for use of palaeosol magnetic properties as proxies of palaeorainfall. Glob. Planet. Change 111, 280–287.

Maher, B.A., Thompson, R., 1995. Paleorainfall reconstructions from pedogenic magnetic susceptibility variations in the Chinese loess and paleosols. Quat. Int. 44, 383–391.

Maher, B.A., Thompson, R., Zhou, L.P., 1994. Spatial and temporal reconstructions of changes in the Asian palaeomonsoon: a new mineral magnetic approach. Earth Planet. Sci. Lett. 125, 461–471.

Maher, B.A., 1998. Magnetic properties of modern soils and Quaternary loessic paleosols: paleoclimatic implications. Palaeogeogr. Palaeoclimatol. Palaeoecol. 137, 25–54.

Maher, B.A., 2011. The magnetic properties of Quaternary aeolian dusts and sediments, and their palaeoclimatic significance. Aeolian Res. 3, 87–144.

Maxbauer, D.P., Feinberg, J.M., Fox, D.L., 2016. Magnetic mineral assemblages in soils and paleosols as the basis for paleoprecipitation proxies: a review of magnetic methods and challenges. Earth-Science Rev. 155, 28–48.

Mokhtari Karchegani, P., Ayoubi, S., Lu, S.G., Honarju, N., 2011. Use of magnetic measures to assess soil redistribution following deforestation in hilly region. J. Appl. Geophys. 75, 227–236.

Molina, N., 2009. Verification of conceptual models of phosphorus, clay, sand and organic carbon distribution in ABt sola. Geoderma 150, 396–403.

Mueller-Niggemann, C., Bannert, A., Schloter, M., Lehndorff, E., Schwark, L., 2012. Intra- versus inter-site macroscale variation in biogeochemical properties along a paddy soil chronosequence. Biogeosciences 9, 1237–1251.

Mullins, C., Tite, M., 1973. Magnetic viscosity, quadrature susceptibility and frequency dependence of susceptibility in single-domain assemblies of magnetite and maghemite. J. Geophys. Res. 78 (5), 804–809.

Murad, E., Wagner, U., 1998. Clays and clay minerals: the firing process. Hyperfine Interact. 117, 337–356.

Murad, E., 1998. The characterization of soils, clays, and clay firing products. Hyperfine Interact. 117 (1–4), 251–259.

Nie, J., Stevens, T., Song, Y., King, J., Zhang, R., Ji, S., Gong, L., Cares, D., 2014. Pacific freshening drives Pliocene cooling and Asian monsoon intensification. Sci. Rep. 4, 5474. http://dx.doi.org/10.1038/srep05474.

Nolan, K., 2014. Prospecting for prehistoric gardens: results of a pilot study. Archaeol. Prospect. 21 (2), 147–154.

Oldfield, F., Crowther, J., 2007. Establishing fire incidence in temperate soils using magnetic measurements. Palaeogeogr. Palaeoclimatol. Palaeoecol. 249, 362–369.

Olson, K.R., Gennadiyev, A.N., Zhidkin, A.P., Markelov, M.V., Golosov, V.N., Lang, J.M., 2013. Use of magnetic tracer and radio-cesium methods to determine past cropland soil erosion amounts and rates. CATENA 104, 103–110.

Orgeira, M.J., Egli, R., Compagnucci, R.H., 2011. A quantitativemodel of magnetic enhancement in loessic soils. In: Petrovský, E., Ivers, D., Harinarayana, T., Herrero-Bervera, E. (Eds.), The Earth's Magnetic Interior. Springer, Netherlands, Dordrecht, pp. 361–397.

Pacyna, J.M., 1987. Atmospheric emissions of arsenic, cadmium, lead and mercury from high temperature processes in power generation and industry. In: Hutchinson, T.C., Meema, K.M. (Eds.), Lead, Mercury, Cadmium and Arsenic in the Environment. John Wiley & Sons Ltd, Chichester.

Paradelo, R., Barral, M.T., 2014. Magnetic susceptibility and trace element distribution in composts' size fractions. Span. J. Soil Sci. 4 (2), 204–210.

Pastor, A., Gallello, G., Cervera, M.L., de la Guardia, M., 2016. Mineral soil composition interfacing archaeology and chemistry. Trends Anal. Chem. 78, 48–59.

Pavón-Carrasco, F., Osete, M.L., Torta, J., de Santis, A., 2014. A geomagnetic field model for the Holocene based on archaeomagnetic and lava flow data. Earth Planet. Sci. Lett. 388, 98–109.

Penman, T.D., Towerton, A.L., 2008. Soil temperatures during autumn prescribed burning: implications for the germination of fire responsive species? Int. J. Wildland Fire 17, 572–578.

Piatak, N.M., Parsons, M.B., Seal II, R.R., 2015. Characteristics and environmental aspects of slag: a review. Appl. Geochem. 57, 236–266.

Pimentel, D., 2006. Soil erosion – a food and environmental threat. Environ. Dev. Sustain. 8, 119–137.

Pirrie, D., Ruffel, A., Dawson, L.A., 2013. Environmental and criminal geoforensics: an introduction. In: Pirrie, D., Ruffell, A., Dawson, L.A. (Eds.), Environmental and Criminal Geoforensics, vol. 384. Geological Society of London, Special Publications, pp. 1–7.

Porsch, K., Rijal, M.L., Borch, T., Troyer, L.D., Behrens, S., Wehland, F., Appel, E., Kappler, A., 2014. Impact of organic carbon and iron bioavailability on the magnetic susceptibility of soils. Geochim. Cosmochim. Acta 128, 44–57.

Preetz, H., Altfelder, S., Igel, J., 2009a. Tropical soils and landmine detection—an approach for a classification system. Soil Sci. Soc. Am. J. 72, 151–159.

Preetz, H., Altfelder, S., Hennings, V., Igel, J., 2009b. Classification of soil magnetic susceptibility and prediction of metal detector performance – case study of Angola. In: Proc. SPIE. 7303, Detection and Sensing of Mines, Explosive Objects, and Obscured Targets XIV, vol. 730313. http://dx.doi.org/10.1117/12.819394.

Pringle, J.K., Ruffell, A., Jervis, J.R., Donnelly, L., Hansen, J., Morgan, R., Pirrie, D., Harrison, M., 2012. The use of geoscience methods for terrestrial forensic searches. Earth Sci. Rev. 114, 108–123.

Pringle, J.K., Holland, C., Szkornik, K., Harrison, M., 2012a. Establishing forensic search methodologies and geophysical surveying for the detection of clandestine graves in coastal beach environments. Forensic Sci. Int. 219, 29–36.

Pringle, J.K., Giubertoni, M., Cassidy, N.J., Wisniewski, K.D., Hansen, J.D., Linford, N.T., Daniels, R.M., 2015. The use of magnetic susceptibility as a forensic search tool. Forensic Sci. Int. 246, 31–42.

Quijano, L., Chaparro, M.A.E., Marié, D.C., Gaspar, L., Navas, A., 2014. Relevant magnetic and soil parameters as potential indicators of soil conservation status of Mediterranean agroecosystems. Geophys. J. Int. 198, 1805–1817.

Quine, T.A., Desmet, P.J.J., Govers, G., Vandaele, K., Walling, D.E., 1994. A comparison of the roles of tillage and water erosion in landform development and sediment export on agricultural land near Leuven, Belgium. In: Olive, L.J., Loughran, R.J., Kesby, J.A. (Eds.), Variability in Stream Erosion and Sediment Transport, Proceedings of the Canberra Symposium. No. 224. IAHS Publication, pp. 77–86.

Rasmussen, K.L., De La Fuente, G., Bond, A., Mathiesen, K., Vera, S., 2012. Pottery firing temperatures: a new method for determining the firing temperature of ceramics and burnt clay. J. Archaeol. Sci. 39, 1705–1716.

Riccardi, M., Messiga, B., Duminuco, P., 1999. An approach to the dynamics of clay firing. Appl. Clay Sci. 15 (3–4), 393–409.

Rijal, M., Porsch, K., Appel, E., Kappler, A., 2012. Magnetic signature of hydrocarbon- contaminated soils and sediments at the former oil field Hänigsen, Germany. Stud. Geophys. Geod. 56, 889–908.

Roman, S.A., Johnson, W.C., Geiss, C.E., 2013. Grass fires—an unlikely process to explain the magnetic properties of prairie soils. Geophys. J. Int. 195 (3), 1566–1575.

Rosendahl, D., Lowe, K., Wallis, L., Ulm, S., 2014. Integrating geoarchaeology and magnetic susceptibility at three shell mounds: a pilot study from Mornington Island, Gulf of Carpentaria, Australia. J. Archaeol. Sci. 49, 21–32.

Royall, D., 2001. Use of mineral magnetic measurements to investigate soil erosion and sediment delivery in a small agricultural catchment in limestone terrain. CATENA 46, 15–34.

Royall, D., 2004. Particle-size and analytical considerations in the mineral magnetic interpretation of soil loss from cultivated landscapes. CATENA 57, 189–207.

Royall, D., 2007. A comparison of mineral-magnetic and distributed RUSLE modeling in the assessment of soil loss on a southeastern U.S. cropland. CATENA 69, 170–180.

Santín, C., Doerr, S.H., 2016. Fire effects on soils: the human dimension. Phil. Trans. R. Soc. B 371, 20150171.

Scalenghe, R., Barello, F., Saiano, F., Ferrara, E., Fontaine, C., Caner, L., Olivetti, E., Boni, I., Petit, S., 2015. Material sources of the Roman brick-making industry in the I and II century A.D. from Regio IX, Regio XI and Alpes Cottiae. Quat. Int. 357, 189–206.

Schibler, L., Boyko, T., Ferdyn, M., Gajda, B., Holl, S., Jordanova, N., Roesler, W., MAGPROX ream, 2001. Topsoil magnetic susceptibility mapping: data reproducibility and compatibility, measurement strategy. Stud. Geophys. Geod. 46, 43–57.

Schlesinger, W., Bernhardt, E., 2013. Biogeochemistry. An Analysis of Global Change, third ed. Academic Press, Elsevier Inc., ISBN 978-0-12-385874-0

Schwertmann, U., 1988. Occurrence and formation of iron oxides in various pedoenvironments. In: Stucki, J., Goodman, B., Schwertmann, U. (Eds.), Iron in Soils and Clay Minerals, NATO ASI Series, Series C: Mathematical and Physical Sciences, vol. 217. Reidel Publishing Company, pp. 267–308.

Shi, Z.H., Fang, N.F., Wub, F.Z., Wanga, L., Yue, B.J., Wua, G.L., 2012. Soil erosion processes and sediment sorting associated with transport mechanisms on steep slopes. J. Hydrol. 454–455, 123–130.

Singh, R.P., Singh, P., Ibrahim, H., Hashim, R., 2011. Land application of sewage sludge: physicochemical and microbial response. In: Whitacre, D.M. (Ed.), Reviews of Environmental Contaminination and Toxicology, vol. 214. Springer, NY, pp. 41–62.

Song, Y., Hao, Q., Ge, J., Zhao, D., Zhang, Y., Li, Q., Zuo, X., Lü, Y., Wang, P., 2014. Quantitative relationships between magnetic enhancement of modern soils and climatic variables over the Chinese Loess Plateau. Quat. Intern. 334–335, 119–131.

Spassov, S., Hus, J., 2006. Estimating baking temperatures in a Roman pottery kiln by rock magnetic properties: implications of thermochemical alteration on archaeointensity determinations. Geophys. J. Int. 167, 592–604.

Stoof, C.R., Moore, D., Fernandes, P.M., Stoorvogel, J.J., Fernandes, R.E.S., Ferreira, A.J.D., Ritsema, C.J., 2013. Hot fire, cool soil. Geophys. Res. Lett. 40, 1534–1539.

Szuszkiewicz, M., Magiera, T., Kapička, A., Petrovský, E., Grison, H., Gołuchowska, B., 2015. Magnetic characteristics of industrial dust from different sources of emission: a case study of Poland. J. Appl. Geophys. 116, 84–92.

Szuszkiewicz, M., Łukasik, A., Magiera, T., Mendakiewicz, M., 2016. Combination of geo- pedo- and technogenic magnetic and geochemical signals in soil profiles – diversification and its interpretation: a new approach. Environ. Pollut. 214, 464–477.

Taylor, R.M., Maher, B.A., Self, P.G., 1987. Magnetite in soils: I. The synthesis of singledomain and superparamagnetic magnetite. Clay Miner. 22, 411–422.

Thornton, P., Doney, S., Lindsay, K., Moore, J., Mahowald, N., Randerson, J., Fung, I., Lamarque, J.-F., Feddema, J., Lee, Y.-H., 2009. Carbon-nitrogen interactions regulate climate-carbon cycle feedbacks: results from an atmosphere-ocean general circulation model. Biogeosciences 6, 2099–2120.

Tite, M., 1972. The influence of geology on the magnetic susceptibility of soils on archaeological sites. Archaeometry 14 (2), 229–236.

Tite, M.S., 2008. Ceramic production, provenance and use e a review. Archaeometry 50 (2), 216–231.

Torrent, J., Barron, V., Liu, Q.S., 2006. Magnetic enhancement is linked to and precedes hematite formation in aerobic soil. Geophys. Res. Lett. 33, L02401.

Van Breemen, N., 1988. Long-term chemical, mineralogical and morphological effects of iron-redox processes in periodically flooded soils. In: Stucki, J., Goodman, B., Schwertmann, U. (Eds.), Iron in Soils and Clay Minerals, NATO ASI Series. D. Reidel Publishing Company.

van Dam, R.L., Hendrickx, J.M.H., Harrison, J.B.J., Borchers, B., 2005. Conceptual model for prediction of magnetic properties in tropical soils. Proc. SPIE 5794, 177–187.

Van Klinken, J., 2001. Magnetization of ancient ceramics. Archaeometry 43 (1), 49–57.

Van Oost, K., Govers, G., Desmet, P.J.J., 2000. Evaluating the effects of changes in landscape structure on soil erosion by water and tillage. Landsc. Ecol. 15 (6), 579–591.

Van Oost, K., Govers, G., de Alba, S., Quine, T.A., 2006. Tillage erosion: a review of controlling factors and implications for soil quality. Progr. Phys. Geogr. 4, 443–466.

Veneva, L., Hoffmann, V., Jordanova, D., Jordanova, N., Fehr, T., 2004. Rock magnetic, mineralogical and microstructural characterization of fly ashes from Bulgarian power plants and the nearby anthropogenic soils. Phys. Chem. Earth 29, 1011–1023.

Vieira, D.A.N., Dabney, S.M., 2009. Modeling landscape evolution due to tillage: model development. Trans. ASABE 52, 1505–1522.

Vieira, D.A.N., Dabney, S.M., 2011. Modeling edge effects of tillage erosion. Soil Till. Res. 111, 197–207.

Walkington, H., 2010. Soil science applications in archaeological contexts: a review of key challenges. Earth-Sci. Rev. 103, 122–134.

Wawer, M., Magiera, T., Ojha, G., Appel, E., Kusza, G., Hu, S., Basavaiah, N., 2015. Traffic-related pollutants in roadside soils of different countries in Europe and Asia. Water Air Soil Pollut. 226, 216. http://dx.doi.org/10.1007/s11270-015-2483-6.

Weber, K., Achenbach, L., Coates, J., 2006. Microorganisms pumping iron: anaerobic microbial iron oxidation and reduction. Nat. Rev. Microbiol. 4, 752–764.

Winkler, P., Kaiser, K., Kölbl, A., Kühn, T., Schad, P., Urbanski, L., Fiedler, S., Lehndorff, E., Kalbitz, K., Utami, S.R., Cao, Z., Zhang, G., Jahn, R., Kögel-Knabner, I., 2016. Response of Vertisols, Andosols, and Alisols to paddy management. Geoderma 261, 23–35.

Wolf, M., Lehndorff, E., Wiesenberg, G.L.B., Stockhausen, M., Schwark, L., Amelung, W., 2013. Towards reconstruction of past fire regimes from geochemical analysis of charcoal. Org. Geochem. 55, 11–21.

Wright, B.R., Clarke, P.J., 2008. Relationships between soil temperatures and properties of fire in feathertop spinifex (*Triodia schinzii* (Henrard) Lazarides) sandridge desert in central Australia. Rangel. J. 30, 317–325.

Xu, X., Thornton, P., Post, W., 2013. A global analysis of soil microbial biomass carbon, nitrogen and phosphorus in terrestrial ecosystems. Glob. Ecol. Biogeogr. 22, 737–749.

Yan, H.T., Hu, S.Y., Blaha, U., Rösler, W., Duan, X.M., Appel, E., 2011. Paddy soil—a suitable target for monitoring heavy metal pollution by magnetic proxies. J. Appl. Geophys. 75, 211–219.

Yang, J., Zhao, Y., Zyryanov, V., Zhang, J., Zheng, C., 2014. Physical–chemical characteristics and elements enrichment of magnetospheres from coal fly ashes. Fuel 135, 15–26.

Yu, J.Y., Zhao, W.S., Zhan, S.R., 1981. The magnetic susceptibility distinction of paddy soils in Taihu lake region. In: Proceedings of Symposium on Paddy Soils. Part II, vol. 1. Springer, Berlin, Heidelberg, ISBN 978-3-642-68143-1, pp. 493–497.

Zachara, J., Fredrickson, J., Li, S., Kennedy, D., Smith, S., Gassman, P., 1998. Bacterial reduction of crystalline Fe^{3+} oxides in single phase suspensions and subsurface materials. Am. Miner. 83, 1426–1443.
Zachara, J., Kukkadapu, R., Fredrickson, J., Gorby, Y., Smith, S., 2002. Biomineralization of poorly crystalline Fe(III) oxides by dissimilatory metal reducing bacteria (DMRB). Geomicrobiol. J. 19, 179–207.
Zboril, R., Mashlan, M., Petridis, D, 2002. Iron (III) oxides from thermal processes-synthesis, structural and magnetic properties, Mossbauer spectroscopy characterization, and applications. Chem. Mater. 14 (3), 969–982.
Zhang, C., Paterson, G., Liu, Q., 2012. A new mechanism for the magnetic enhancement of hematite during heating: the role of clay minerals. Stud. Geophys. Geod. 56 (3), 845–860.
Zhang, C., Appel, E., Qiao, Q., 2013. Heavy metal pollution in farmland irrigated with river water near a steel plant—magnetic and geochemical signature. Geophys. J. Int. 192, 963–974.
Zhang, G.-L., Gong, Z.-T., 2003. Pedogenic evolution of paddy soils in different soil landscapes. Geoderma 115 (1–2), 15–29.

FUTURE CHALLENGES IN SOIL MAGNETISM STUDIES

One major outcome from the conducted systematic magnetic studies of different soil profiles is that, in most cases, the room temperature magnetic measurements, including magnetic susceptibility and laboratory remanences, give adequate information about the magnetic mineralogy and grain-size variations, especially in case of dense (continuous) sampling along depth. The use of more sophisticated low-temperature techniques is necessary, however, when the presence of goethite and hematite of superparamagnetic (SP) grain size is suspected. On the other hand, an unfortunate drawback of the preferred use by many authors of only low-temperature analyses of various laboratory-synthesized analogs of pedogenic ferri/antiferromagnets is, for example, missing literature data on a simple parameter such as room temperature frequency-dependent susceptibility (χ_{fd}, $\chi_{fd}\%$) of SP hematite and goethite (with and/or without substitutions).

Recapitulating the information accumulated on different aspects of soil magnetism and its applications, a two-fold impression appears: on the one hand, a lot of knowledge is gained on the soil magnetic enhancement and its mechanisms in well-aerated soils from the temperate climate (Chernozems, Phaeozems, Luvisols) and to some degree of tropical soils, links with certain climate parameters, and elucidation of the role of each factor (precipitation, moisture, temperature) for pedogenic formation of different iron oxides in soils. Thus, the major efforts in environmental magnetic studies of modern soils are focused only on the link with climate parameters. However, investigations of other soil types seem underestimated and limited to a few single studies (e.g., studies of Podzols, Planosols, Vertisols, etc.). Therefore, an overall bias in the soil magnetism exists, which could limit its power and possible applications.

The following major challenges for the future work in soil magnetism could be envisaged:

1. More detailed investigations on the link between the pedogenic magnetic grain-size variations (SP, single-domain, pseudo-single-domain) along soil profiles, and the degree of pedogenic textural differentiation (caused by eluviation, illuviation, podzolization, etc.) could give more insight into (palaeo) environmental reconstructions, evaluation of the influence of local/regional factors on the magnetic signature of soils, etc.
2. Focused magnetic mineral studies on strongly eluviated and podzolized profiles could reveal the (geo)chemical driving forces leading to the observed split in the concentration of the SP fraction along the profile, as well as mechanisms and underlying processes responsible for the detected stable magnetic minerals in the eluvial horizons.
3. An emerging research question from the studies of high mountain soils (see Chapter 6) is whether the observed regularity in the abundance of the different magnetic grain-size fractions with the altitude is a common feature. Linking this magnetic fingerprint to a wealth of geochemical research on high Alpine soils could reveal possible interesting implications.

4. Because a non-negligible part of soils experience reductive conditions for prolonged time, more extensive mineral magnetic studies on well-controlled laboratory column experiments of reductive dissolution of different iron (hydr)oxides in combination with numerous already available geochemical studies could give important information on the significance and expression of these processes in the soil magnetic properties.
5. Widely applied magnetic methods for estimation of soil contamination would benefit from more attention to be paid to the specific soil type present in the study area. Hopefully, the results presented in Chapters 2–7 demonstrate that soil type matters!
6. Further research on the link between soil magnetism and the content of soil nutrients would give wide prospects for incorporation of the methodology in interdisciplinary studies on global carbon and nitrogen cycling, which is a great challenge for combating global warming.

Index

'*Note*: Page numbers followed by "f" indicate figures and "t" indicate tables.'

D

E

F

G

R

S

T

U

V

W

Z

Printed in the United States
By Bookmasters